Fourier Analysis

Fourier Analysis

JAMES S. WALKER

University of Wisconsin – Eau Claire

New York   Oxford
OXFORD UNIVERSITY PRESS
1988

Oxford University Press

Oxford   New York   Toronto
Delhi   Bombay   Calcutta   Madras   Karachi
Petaling Jaya   Singapore   Hong Kong   Tokyo
Nairobi   Dar es Salaam   Cape Town
Melbourne   Auckland

and associated companies in
Berlin   Ibadan

Library of Congress Cataloging-in-Publication Data
Walker, James S.
Fourier analysis.
Bibliography: p. 0   Includes index.   1. Fourier analysis.   I. Title.
QA403.5.W35 1988   515′.2433   87-10081
ISBN 0-19-504300-6

1 3 5 7 9 8 6 4 2

Printed in the United States of America
on acid-free paper

To My Wife
Dawn Manire

Preface

The deep study of nature is the most fruitful source of knowledge.
                                                    Joseph Fourier

Fourier's Theorie analytique de la chaleur is the bible of the mathematical physicist.
                                                    Arnold Sommerfeld

My purpose in writing this book is to explain the basic mathematical theory as well as some of the principal applications of Fourier analysis. Vibrations and sound, heat conduction, optics, and CAT scanning are just some of the many areas to which Fourier analysis contributes deep insights. Although there are many fine books that cover any one of these applications in detail, each such book must necessarily be rather brief in its coverage of mathematical techniques. Therefore, I feel that there is a real need for a text that covers most of the basic mathematics of Fourier analysis and gives concise discussions of how that mathematics is applied.

This book arose from the lecture notes that I have given to my students in courses and seminars at the University of Wisconsin at Eau Claire. These classes consisted mostly of seniors majoring in mathematics and physics. My intention has been to provide them with a firm foundation for the modern ideas and applications of Fourier analysis and not just stick to the simplest concepts. Hence the book should also be suitable for first-year graduate students. A previous course in advanced calculus would make the book most easily comprehensible; all the necessary background can be found in any of the following books: Bartle (1964), Buck (1978), Kaplan (1984), or Rudin (1964). I have found, however, that talented physics majors have been able to read the text with a fair degree of comprehension.

To make this book accessible to as wide an audience as possible I have not utilized the Lebesgue theory of integration. Those readers who are familiar with this sophisticated theory, however, should have no trouble making minor, obvious modifications in theorems and proofs to obtain more general results. Also, for reasons of space, I have not included a complete discussion of computer methods. I hope that the section on Fast Fourier Transforms in Chapter 7 will serve as a brief introduction to this aspect of Fourier analysis. Perhaps in a second edition I might expand this discussion. In any case, I have provided ample references to the literature of computerized Fourier analysis in §13 of Chapter 7.

When I first planned to write this book I wanted to include more material on the history of Fourier analysis, since it has united much of mathematical analysis and physics for nearly two hundred years. But such material would have made the book even longer than it already is. Readers who are interested in the rich history of Fourier analysis might begin by consulting the References for Chapter 1.

The exercises in the text range from straightforward applications of formulas to collections of problems aiming toward extensions of results in the text or describing additional concepts. It is very important that the reader try as many exercises as possible; the text cannot be comprehended without them. *Those exercises that are absolutely essential for understanding the text are marked by a star* (★). Some exercises that might be found difficult are marked by an asterisk (∗).

This book was written for a one-year course, in which case there should be time to cover the entire book. If the book is used for one semester, then one possible syllabus, which I have found effective, consists of Chapters 1, 2, §§1–4, 3, 2, §§6–8, 6, 7, and (optionally), 8, §§1–4. In classes with a high proportion of physics majors I have found it expedient to cover Chapter 3 after §§1–4 of Chapter 2, and then return to §§6–8 of Chapter 2. There is some flexibility for presentation of material, as indicated by the following diagram of the dependency of chapters.

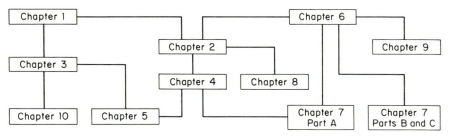

Dependency of Chapters.

Acknowledgments

I would like to thank Professors Gabriel Kojoian, Andrew Balas, and Kevin Gough for their participation in my classes in Fourier analysis and for their encouragement when I was writing this book. Thanks also go to Professor Walter "Doc" Reid for many enjoyable conversations about the book.

Professor S. G. Lipson was kind enough to send me the photographs in Chapter 7 and I thank him for that. Thanks also to Roland T. Katz for the moral support he has given me.

My students also deserve acknowledgment. I especially want to thank Sue Kelly, Doug Pearson, and Ken Dykema for their many helpful questions and suggestions. Ken Dykema must be thanked again for his help with the computer graphs.

I owe a debt of gratitude to my wife for all of her patience and kind support during the long hours, days, and months that it took to write this book.

May 1987   J.S.W.
Eau Claire, Wisconsin

Contents

Notation

The notation in the text is fairly standard—for example, a, b, c, d are usually real constants, j, k, m, n are usually integers, and t, u, v, x, y, z are usually real variables. We will sometimes write t_0, u_0, v_0, x_0, y_0, z_0 to denote fixed values of those variables. The letters f, g, h are reserved for functions; f', g', h' will denote derivatives of those functions when they depend on a single variable. The open interval $a < x < b$ will be denoted by (a, b) while the closed interval $a \le x \le b$ will be denoted by $[a, b]$.

The following is a list of some special notations that are employed.

$\hat{f}$ = Fourier transform of the function f

f_x, f_y, (f_z) = partial derivatives $\partial f/\partial x$, $\partial f/\partial y$, $(\partial f/\partial z)$ of a function f that depends on x, y, (z)

$f(x_0+)$ = one-sided limit from the right $\lim\limits_{\substack{u \to 0 \\ (u>0)}} f(x_0 + u) = \lim\limits_{u \to 0+} f(x_0 + u)$

$f(x_0-)$ = one-sided limit from the left $\lim\limits_{\substack{u \to 0 \\ (u<0)}} f(x_0 + u) = \lim\limits_{u \to 0-} f(x_0 + u)$

$f^>$ = Laplace transform of a function f

$f^\#$ = Radon transform of a function f (*This notation for Radon transforms will be used only in Chapter 9; it will occasionally be used in earlier chapters to denote auxiliary functions with no fixed meaning.*)

Π = function $\Pi(x) = \begin{cases} 1 & \text{if} & |x| < \frac{1}{2} \\ 0 & \text{if} & |x| > \frac{1}{2} \end{cases}$

Λ = function $\Lambda(x) = \begin{cases} 1 - |x| & \text{if} & |x| \le 1 \\ 0 & \text{if} & |x| > 1 \end{cases}$

sinc = function $\operatorname{sinc} x = \sin \pi x / \pi x$

Ψ = harmonic function

$\mathbb{R}$ = set of real numbers

$\mathbb{R}^2$, $(\mathbb{R}^3, \mathbb{R}^n)$ = set of ordered pairs (triples, n-tuples) of real numbers

Brief Review

We will now briefly discuss, through exercises, some concepts and techniques that will ease the burden of some of our later work.

1. Integrate the following integrals by parts.

   (a) $\displaystyle\int_0^1 x \sin 2\pi x \, dx$

   (b) $\displaystyle\int_{-\pi}^{\pi} e^x \sin 3x \, dx$

2. *Even and Odd Functions.* The definition of an *even function f* over an interval $(-a, a)$ symmetric about the origin is that

$$f(-x) = f(x) \qquad \text{for each } x \text{ in } (-a, a)$$

An *odd function f* over $(-a, a)$ is a function that satisfies

$$f(-x) = -f(x) \qquad \text{for each } x \text{ in } (-a, a)$$

Given these definitions, do the following exercises.

(a) State which of the following functions are odd, even, or neither, on the given interval.

$f(x) = x^2$ on $(-1, 1)$ $\qquad\qquad$ $g(x) = x^3 \cos 4x$ on $(-\pi, \pi)$

$h(x) = x^2 \cos 4x$ on $(-\pi, \pi)$ $\qquad$ $k(x) = x + \frac{1}{4}x^2$ on $(-2, 2)$

(b) Show that for all (continuous) even functions

$$\int_{-a}^{a} f(x) \, dx = 2\int_{0}^{a} f(x) \, dx,$$

and for all (continuous) odd functions

$$\int_{-a}^{a} f(x) \, dx = 0.$$

(c) Fill in the following table by writing odd, even, or neither, in the appropriate blanks.

f	g	fg	f/g	$f+g$	$f-g$
odd	odd				
even	even				
even	odd				
odd	even				

3. Using the results of Exercise 2, evaluate the following integrals.

(a) $\displaystyle\int_{-\pi}^{\pi} 3x \cos x \, dx$

(b) $\displaystyle\int_{-\frac{1}{2}\pi}^{\frac{1}{2}\pi} (x^3 + x + 3) \cos x \, dx$

[*Note*: For (b), split the integral into even and odd terms.]

4. *Kronecker's Rule.* With the aid of *Kronecker's rule*, many of the integrals needed in Fourier series are easier to evaluate. The following exercises are intended to explain Kronecker's rule.

(a) Let $p(x)$ be a polynomial in x of degree m, and $f(x)$ a continuous function. Let $F_1(x) = \int f(x)\, dx$, $F_2(x) = \int F_1(x)\, dx$, ..., $F_{m+1}(x) = \int F_m(x)\, dx$. And, let $p^{(j)}(x)$ be the jth derivative of $p(x)$. Prove that

$$\int p(x)f(x)\, dx = p(x)F_1(x) - p^{(1)}(x)F_2(x) + p^{(2)}(x)F_3(x)$$

$$- \cdots + (-1)^m p^{(m)}(x)F_{m+1}(x) + C$$

$$= \sum_{j=0}^{m} (-1)^j p^{(j)}(x)F_{j+1}(x) + C$$

[*Hint*: Integrate repeatedly by parts until $p^{(m+1)}(x) = 0$.] It follows immediately from this result that

$$\int_a^b p(x)f(x)\, dx = \left[\sum_{j=0}^{m} (-1)^j p^{(j)}(x)F_{j+1}(x) \right]\Bigg|_a^b$$

For example,

$$\int_{-\pi}^{\pi} x^3 \sin x \, dx = 2\int_0^{\pi} x^3 \sin x \, dx \qquad (x^3 \sin x \text{ is even})$$

$$= 2[x^3(-\cos x) - (3x^2)(-\sin x)$$

$$+ (6x)(\cos x) - (6)(\sin x)]|_0^{\pi}$$

$$= 2\pi^3 - 12\pi$$

(b) Using Kronecker's rule, evaluate the following integrals.

$$\int_{-\pi}^{\pi} x^2 \cos nx \, dx \qquad \int_{-1}^{1} (x^3 + 2x^2 + 1)\sin n\pi x \, dx$$

[*Note:* For the second integral, do not forget about even and odd functions.]

5. *Complex Numbers.* A complex number z is a quantity $a + ib$ where a and b are real numbers and $i^2 = -1$. The arithmetic of complex numbers is the usual one subject to the relations

$$(a + ib) \pm (c + id) = (a \pm c) + i(b \pm d)$$
$$(a + ib)(c + id) = (ac - bd) + i(bc + ad)$$

It is often convenient to think of z as the point (a, b) in the Cartesian plane. Therefore, the distance from z to $0 = (0, 0)$ is $(a^2 + b^2)^{1/2}$. We define $|z| = |a + ib|$ to equal that distance $(a^2 + b^2)^{1/2}$ and call $|z|$ the *modulus* of z. We call a the *real part* and b the *imaginary part* of z and write $a = \text{Re } z$ and $b = \text{Im } z$. Finally, we define $\bar{z}$ to be the complex number $a - ib$ and call it the *complex conjugate* of z. Using these definitions do the following exercises.

(a) Compute $3 - 2i - (4 + 3i)$ $(3 + 2i)(6 - 5i)$ $3 + i - (4 + i)i$

(b) Prove that $\overline{zw} = \bar{z}\,\bar{w}$ and that $|zw| = |z| \cdot |w|$ for all complex numbers z and w. Conclude that $|z^n| = |z|^n$ for all positive integers n. [*Hint:* $|z| = (z\bar{z})^{1/2}$.]

(c) Prove that the following inequalities hold for all complex numbers z and w

$$|z \pm w| \le |z| + |w| \qquad |\text{Re } z| \le |z| \qquad |\text{Im } z| \le |z|$$

(d) Suppose that $a + ib \ne 0 + i0$. We then define division as follows

$$\frac{c + id}{a + ib} = \frac{(c + id)(a - ib)}{a^2 + b^2} = \left(\frac{ac + bd}{a^2 + b^2}\right) + i\left(\frac{ad - bc}{a^2 + b^2}\right)$$

Using that definition, compute

$$\frac{3 - 2i}{4 + i} \qquad \text{and} \qquad \frac{2 - 4i}{2 - i}$$

(e) Prove that for all complex numbers z and w we have

$$z + w = w + z \qquad zw = wz$$

Fourier Analysis

1
Introduction to Fourier Series

In this chapter we will discuss, on an informal basis, some of the essential concepts in the study of Fourier series. Fourier series are the principal tool in the analysis of periodic functions, which play a major role in the solution of some fundamental problems in mathematical physics. The solution of physical problems will be discussed in Chapters 3 and 4.

The first two sections of this chapter describe the basic ideas involved in the formal definition of Fourier series. In §3 we deal with the special cases of even and odd functions[1] and their associated cosine and sine series, which play an important role in the solution of physical problems. Some precise conditions for a Fourier series to equal the function it represents are introduced in §4. Most of the theorems stated in §4 will not be proved until Chapter 2 because we want to take our time in discussing the subtle aspects of convergence of Fourier series. Uniform convergence, a basic tool in the application of Fourier series to differential equations, is introduced in §5. In §6, which is optional, we introduce Abel's Test, an interesting test for uniform convergence of Fourier series. The chapter concludes with a discussion of the complex form of Fourier series in §7.

§1. Periodic Functions, Fourier Series on $(-\pi, \pi)$

Fourier series are the main tool in the analysis of periodic functions. Therefore, we begin with the definition of such functions.

(1.1) Definition. A function f is *periodic* with period $p > 0$ if $f(x + p) = f(x)$ for each x value. ∎

For example, the functions $\sin x$ and $\cos x$ both have period 2π. While the

---

1. For the definition of even and odd functions, as well as some other preliminary material, see the Brief Review that precedes this chapter.

function $\sin(\pi x/3)$ has period 6 since

$$\sin[\pi(x+6)/3] = \sin[\pi x/3 + 2\pi] = \sin[\pi x/3]$$

A periodic function has more than one period. The following theorem illustrates this fact.

(1.2) Theorem.   If the function f has period p, then any integral multiple of p is also a period for f.

Proof.   If f has period p, then for each x value

$$f(x-p) = f(x-p+p) = f(x)$$
$$f(x+2p) = f(x+p+p) = f(x+p) = f(x)$$
$$f(x-2p) = f(x-2p+p) = f(x-p) = f(x)$$

Continuing in this way, we obtain $f(x-kp) = f(x)$ for any integer k.   ∎

(1.3) Corollary.   An important corollary of Theorem (1.2) is the following: The graph of f shifted to a new coordinate origin at $(x, y) = (kp, 0)$ equals the original graph of f. (See Figure 1.1.)

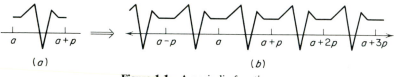

(a) (b)

Figure 1.1   A periodic function.

For the rest of this section, we will discuss the particular case of period 2π; more general periodic functions will be discussed in the next section.

The most important periodic functions are the sines and cosines; in particular, the following set of functions.

(1.4) Definition.   The *trigonometric system*, period 2π, is the following set of functions

$$1, \ \sin x, \ \cos x, \ \sin 2x, \ \cos 2x, \ \ldots, \ \sin nx, \ \cos nx, \ldots \quad ■$$

The functions $\sin nx$ and $\cos nx$ both have periods of $2\pi/n$. Therefore, by Theorem (1.2), these functions both have periods of 2π as well. Besides being the simplest periodic functions, the trigonometric system also possesses the following property, called orthogonality.[2] This property makes the trigonometric system useful for analyzing more general periodic functions.

(1.5) Theorem: Orthogonality of the Trigonometric System.   The trigonometric system, period 2π, satisfies the following *orthogonality relations* (in

---

2. The term *orthogonality* (in ancient Greek, *ortho* means "right" and *gonos* means "angle") originates from a formal correspondence between the integrals in Theorem (1.5) and the dot product of perpendicular (or right-angled) vectors in Euclidean space.

each formula m and n are positive integers):

(a) $\displaystyle\int_{-\pi}^{\pi} \cos nx \, dx = 0$ $\displaystyle\int_{-\pi}^{\pi} \sin nx \, dx = 0$

(b) $\displaystyle\int_{-\pi}^{\pi} \sin mx \cos nx \, dx = 0$

(c) $\displaystyle\int_{-\pi}^{\pi} \cos mx \cos nx \, dx = \begin{cases} 0 & \text{if } m \neq n \\ \pi & \text{if } m = n \end{cases}$

(d) $\displaystyle\int_{-\pi}^{\pi} \sin mx \sin nx \, dx = \begin{cases} 0 & \text{if } m \neq n \\ \pi & \text{if } m = n \end{cases}$

Proof. Relation (a) is easily verified. Relation (b) holds because all the integrands are odd functions on the interval $(-\pi, \pi)$. [See Exercise 2 in Brief Review.] Relations (c) and (d) follow from these trigonometric identities

(c′) $\begin{cases} \cos\theta \cos\phi = \frac{1}{2}\cos(\theta - \phi) + \frac{1}{2}\cos(\theta + \phi) \\ \cos^2\theta = \frac{1}{2} + \frac{1}{2}\cos 2\theta \end{cases}$

(d′) $\begin{cases} \sin\theta \sin\phi = \frac{1}{2}\cos(\theta - \phi) - \frac{1}{2}\cos(\theta + \phi) \\ \sin^2\theta = \frac{1}{2} - \frac{1}{2}\cos 2\theta \end{cases}$

For example, (c′) implies (c). If $m \neq n$, then let's suppose $m > n$. We then have

$$\int_{-\pi}^{\pi} \cos mx \cos nx \, dx = \frac{1}{2}\int_{-\pi}^{\pi} \cos(m-n)x \, dx + \frac{1}{2}\int_{-\pi}^{\pi} \cos(m+n)x \, dx$$

Since $m - n$ and $m + n$ are both positive integers, relation (a) implies that the right side of the equality above is zero. If $m = n$, then using (c′) and relation (a) yields

$$\int_{-\pi}^{\pi} \cos nx \cos nx \, dx = \frac{1}{2}\int_{-\pi}^{\pi} 1 \, dx + \frac{1}{2}\int_{-\pi}^{\pi} \cos 2nx \, dx = \pi + \frac{1}{2}0 = \pi$$

Thus, (c) is verified. The trigonometric relation (d′) implies the orthogonality relation (d) in a similar way.   ■

The basic idea underlying Fourier series consists in using the orthogonality relations for the trigonometric system to express periodic functions as infinite series of sines and cosines. Suppose the function f has period 2π, then our goal is to express $f(x)$ in the following way (where the symbols A_0, A_n, and B_n are numbers; the factor $\frac{1}{2}$ on A_0 is just a useful convention):

(1.6) $f(x) = \frac{1}{2}A_0 + \displaystyle\sum_{n=1}^{\infty} [A_n \cos nx + B_n \sin nx]$

Assuming that (1.6) is valid we now show how to determine what the numbers A_0, A_n, and B_n are. Integrating both sides of Eq. (1.6) we obtain (if we integrate the right side term by term[3])

$$\int_{-\pi}^{\pi} f(x)\, dx = \int_{-\pi}^{\pi} \tfrac{1}{2}A_0\, dx + \sum_{n=1}^{\infty} \left(\int_{-\pi}^{\pi} A_n \cos nx\, dx + \int_{-\pi}^{\pi} B_n \sin nx\, dx \right)$$

$$= A_0 \pi + \sum_{n=1}^{\infty} \left[A_n \int_{-\pi}^{\pi} \cos nx\, dx + B_n \int_{-\pi}^{\pi} \sin nx\, dx \right]$$

$$= A_0 \pi$$

Thus, we have the following formula for A_0

(1.7)
$$A_0 = \frac{1}{\pi} \int_{-\pi}^{\pi} f(x)\, dx$$

Multiplying (1.6) by $\cos nx$ yields (upon a change of summation index)

$$f(x) \cos nx = \tfrac{1}{2}A_0 \cos nx + \sum_{m=1}^{\infty} [A_m \cos mx \cos nx + B_m \sin mx \cos nx].$$

Integrating this equality term by term, then using the orthogonality relations (c) and (b) of Theorem (1.5), we obtain

$$\int_{-\pi}^{\pi} f(x) \cos nx\, dx = \tfrac{1}{2}A_0 \int_{-\pi}^{\pi} \cos nx\, dx + \sum_{m=1}^{\infty} \left[A_m \int_{-\pi}^{\pi} \cos mx \cos nx\, dx \right.$$

$$\left. + B_m \int_{-\pi}^{\pi} \sin mx \cos nx\, dx \right]$$

$$= A_n \pi$$

Thus, we have the following formula for A_n

(1.8)
$$A_n = \frac{1}{\pi} \int_{-\pi}^{\pi} f(x) \cos nx\, dx$$

A similar calculation, using orthogonality relations (b) and (d) from Theorem (1.5), yields the following formula for B_n

(1.9)
$$B_n = \frac{1}{\pi} \int_{-\pi}^{\pi} f(x) \sin nx\, dx$$

Since $f(x) \cos 0x = f(x)$ we can combine formulas (1.7) and (1.8); along with (1.9), this gives us the following definition.

---

3. The phrase *integrating term by term* amounts to the equality

$$\int_{-\pi}^{\pi} \left[\tfrac{1}{2}A_0 + \sum_{n=1}^{\infty} \left(A_n \cos nx + B_n \sin nx \right) \right] dx$$

$$= \int_{-\pi}^{\pi} \tfrac{1}{2}A_0\, dx + \sum_{n=1}^{\infty} \left(\int_{-\pi}^{\pi} A_n \cos nx\, dx + \int_{-\pi}^{\pi} B_n \sin nx\, dx \right)$$

Later we will show that integration term by term is valid for Fourier series.

(1.10) Definition: Fourier Coefficients. For a function f, period 2π, the *Fourier coefficients* of f are

$$A_n = \frac{1}{\pi} \int_{-\pi}^{\pi} f(x) \cos nx \, dx \qquad \text{for } n = 0, 1, 2, 3, \ldots$$

$$B_n = \frac{1}{\pi} \int_{-\pi}^{\pi} f(x) \sin nx \, dx \qquad \text{for } n = 1, 2, 3, \ldots$$

With these coefficients, it is conventional to write

$$f \sim \tfrac{1}{2} A_0 + \sum_{n=1}^{\infty} (A_n \cos nx + B_n \sin nx)$$

The formal series on the right is called the *Fourier series* for the function f. ∎

In (1.10), the symbol $\sim$, which denotes correspondence, has replaced the equality sign in (1.6), since we did not justify any of the steps leading to the definition above. A Fourier series is defined for a function f as long as the integrals in Definition (1.10) are defined (even if they are difficult, or impossible, to evaluate precisely). This is the case, for instance, if f is bounded on $[-\pi, \pi]$ and has only a finite number of discontinuities.

Let's consider a few examples.

(1.11) Example. Suppose that f has period 2π and $f(x) = 1$ for $x > 0$ and -1 for $x < 0$ on the interval $(-\pi, \pi)$. (See Figure 1.2.) Find the Fourier series for f.

Solution. Note that f is an odd function. Therefore, we have

$$A_0 = \frac{1}{\pi} \int_{-\pi}^{\pi} f(x) \, dx = 0$$

$$A_n = \frac{1}{\pi} \int_{-\pi}^{\pi} f(x) \cos nx \, dx = 0 \qquad \text{since } f(x) \cos nx \text{ is odd}$$

$$B_n = \frac{1}{\pi} \int_{-\pi}^{\pi} f(x) \sin nx \, dx = \frac{2}{\pi} \int_{0}^{\pi} f(x) \sin nx \, dx = \frac{2}{\pi} \int_{0}^{\pi} \sin nx \, dx$$

$$= \frac{-2}{n\pi} [(-1)^n - 1] = \begin{cases} 0 & \text{if } n \text{ is even} \\ 4/(n\pi) & \text{if } n \text{ is odd} \end{cases}$$

Thus, writing $2k + 1$ for an odd integer n,

$$f \sim \frac{4}{\pi} \sum_{k=0}^{\infty} \frac{\sin(2k+1)x}{2k+1} = \frac{4}{\pi} \left[\sin x + \frac{\sin 3x}{3} + \frac{\sin 5x}{5} + \cdots \right]$$

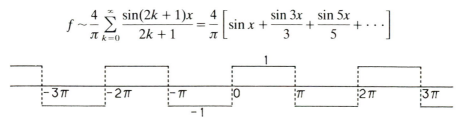

Figure 1.2 Graph of the function in Example (1.11).

(1.12) Example. Suppose that f has period 2π and $f(x) = x^4$ on the interval $(-\pi, \pi)$. Find the Fourier series for f.

Solution. Since f is even, the sine coefficient B_n vanishes due to $x^4 \sin nx$ being odd. But, $x^4 \cos nx$ is even, hence

$$A_0 = \frac{1}{\pi}\int_{-\pi}^{\pi} x^4\, dx = \frac{2}{\pi}\int_0^{\pi} x^4\, dx = \frac{2\pi^4}{5}$$

and, applying Kronecker's Rule (see Exercise 4, Review section)

$$A_n = \frac{2}{\pi}\left[(x^4)\left(\frac{\sin nx}{n}\right) - (4x^3)\left(\frac{-\cos nx}{n^2}\right) + (12x^2)\left(\frac{-\sin nx}{n^3}\right)\right.$$

$$\left. - (24x)\left(\frac{\cos nx}{n^4}\right) + (24)\left(\frac{\sin nx}{n^5}\right)\right]\Big|_0^{\pi}$$

$$= \frac{2}{\pi}\left[\frac{4\pi^3(-1)^n}{n^2} - \frac{24\pi(-1)^n}{n^4}\right] = 8(-1)^n(n^2\pi^2 - 6)/n^4$$

Thus,

$$f \sim \frac{\pi^4}{5} + \sum_{n=1}^{\infty} 8(-1)^n(n^2\pi^2 - 6)\frac{\cos nx}{n^4}$$

(1.13) Example. Suppose f is defined on the interval $(-\pi, \pi)$ by

$$f(x) = \begin{cases} 1 & \text{for } 0 < x < \pi/2 \\ 0 & \text{elsewhere} \end{cases}$$

and f has period 2π. Find the Fourier series for f.

Solution.

$$A_0 = \frac{1}{\pi}\int_0^{\pi/2} 1\, dx = \tfrac{1}{2}$$

$$A_n = \frac{1}{\pi}\int_0^{\pi/2} \cos nx\, dx = \frac{\sin(n\pi/2)}{n\pi}$$

$$B_n = \frac{1}{\pi}\int_0^{\pi/2} \sin nx\, dx = \frac{1 - \cos(n\pi/2)}{n\pi}$$

Thus

$$f \sim \tfrac{1}{4} + \frac{1}{\pi}\left[\cos x - \frac{\cos 3x}{3} + \frac{\cos 5x}{5} - \cdots\right]$$

$$+ \frac{1}{\pi}\left[\sin x + \sin 2x + \frac{\sin 3x}{3} + \frac{\sin 5x}{5} + \frac{\sin 6x}{3} + \cdots\right]$$

Exercises

(1.14) Sketch the graphs of the following periodic functions.

(a) $f(x) = x$ for $-1 < x < 1$ and $f(x + 2) = f(x)$

(b) $f(x) = \begin{cases} 0 & \text{for } -\pi < x < 0 \\ \cos x & \text{for } 0 < x < \pi \end{cases}$ and $f(x + 2\pi) = f(x)$

(c) $f(x) = \begin{cases} 0 & \text{for } -\pi < x < 0 \\ 2 & \text{for } 0 < x < 2\pi \end{cases}$ and $f(x + 3\pi) = f(x)$

(1.15) Verify orthogonality relation (d) in (1.5).

(1.16) Derive (1.9).

***(1.17)** Suppose $f(x)$ has period p. Prove that for every real number d

$$\int_d^{d+p} f(x)\,dx = \int_0^p f(x)\,dx$$

[*Hint:* You may use the fact that

$$\int_a^b f(x)\,dx = \int_a^c f(x)\,dx + \int_c^b f(x)\,dx$$

for *any a, b,* and *c* (*c* does not have to lie between *a* and *b*).] This result shows that *if we integrate a periodic function, period p, over any two intervals of length p, we always obtain the same result.*

(1.18) Find the Fourier coefficients of the functions given below. All are supposed to have period 2π. Sketch the graph of each function.

(a) $f(x) = x$ for $-\pi < x < \pi$
(b) $f(x) = x^2$ for $-\pi < x < \pi$

(c) $f(x) = \begin{cases} 0 & \text{for } -\pi < x < 0 \\ 1 & \text{for } 0 < x < \pi \end{cases}$

(d) $f(x) = x$ for $0 < x < 2\pi$

§2. Arbitrary Periods, Periodic Extensions

We now turn to the discussion of periodic functions with an arbitrary period $p = 2a$. The basis for Fourier series in this case, as in the previous section, is the orthogonality of the trigonometric system of functions.

(2.1) Theorem. The *trigonometric system*, period $2a$, is the set of functions

$$1, \sin\frac{\pi x}{a}, \cos\frac{\pi x}{a}, \sin\frac{2\pi x}{a}, \cos\frac{2\pi x}{a}, \ldots, \sin\frac{n\pi x}{a}, \cos\frac{n\pi x}{a}, \ldots$$

These functions all have period $2a$ and satisfy the following orthogonality relations (in each formula m and n are positive integers):

(a) $\quad \displaystyle\int_{-a}^{a} \cos\frac{n\pi x}{a}\, dx = 0 \qquad \int_{-a}^{a} \sin\frac{n\pi x}{a}\, dx = 0$

(b) $\quad \displaystyle\int_{-a}^{a} \sin\frac{m\pi x}{a}\cos\frac{n\pi x}{a}\, dx = 0$

(c) $\quad \displaystyle\int_{-a}^{a} \cos\frac{m\pi x}{a}\cos\frac{n\pi x}{a}\, dx = \begin{cases} 0 & \text{if } m \neq n \\ a & \text{if } m = n \end{cases}$

(d) $\quad \displaystyle\int_{-a}^{a} \sin\frac{m\pi x}{a}\sin\frac{n\pi x}{a}\, dx = \begin{cases} 0 & \text{if } m \neq n \\ a & \text{if } m = n \end{cases}$

Proof.   We have

$$\cos[n\pi(x + 2a)/a] = \cos[n\pi x/a + 2n\pi] = \cos[n\pi x/a]$$
$$\sin[n\pi(x + 2a)/a] = \sin[n\pi x/a + 2n\pi] = \sin[n\pi x/a]$$

so the functions in the trigonometric system all have period $2a$.

The orthogonality relations follow from those for the trigonometric system, period 2π, by a change of variable. Letting $y = \pi x/a$, we have

$$\int_{-a}^{a} \cos\frac{n\pi x}{a}\, dx = \frac{a}{\pi}\int_{-\pi}^{\pi} \cos ny\, dy = 0$$

$$\int_{-a}^{a} \sin\frac{n\pi x}{a}\, dx = \frac{a}{\pi}\int_{-\pi}^{\pi} \sin ny\, dy = 0$$

$$\int_{-a}^{a} \cos\frac{m\pi x}{a}\cos\frac{n\pi x}{a}\, dx = \frac{a}{\pi}\int_{-\pi}^{\pi} \cos my \cos ny\, dy = \begin{cases} 0 & \text{if } m \neq n \\ a & \text{if } m = n \end{cases}$$

Thus, (a) and (c) hold. Orthogonality relations (b) and (d) follow from the same change of variable.   ∎

Using Theorem (2.1) we can derive Fourier series for functions having period $2a$. The steps of the derivation are identical to those shown in §1. Thus, we obtain the following definition of Fourier series.

(2.2) Definition.   For a function f with period $2a$ the *Fourier coefficients* of f are

$$A_n = \frac{1}{a}\int_{-a}^{a} f(x)\cos\frac{n\pi x}{a}\, dx \qquad \text{for } n = 0, 1, 2, 3, \ldots$$

$$B_n = \frac{1}{a}\int_{-a}^{a} f(x)\sin\frac{n\pi x}{a}\, dx \qquad \text{for } n = 1, 2, 3, \ldots$$

With these coefficients, it is conventional to write

$$f \sim \tfrac{1}{2}A_0 + \sum_{n=1}^{\infty}\left[A_n \cos\frac{n\pi x}{a} + B_n \sin\frac{n\pi x}{a}\right]$$

The formal series on the right is called the *Fourier series* for the function *f*. ∎

(2.3) Example. Let *f* have period 6 and $f(x) = x^2$ on $[-3, 3]$. Find the Fourier series for *f*.

Solution. The period of *f* is $6 = 2 \cdot 3$ so $a = 3$. Thus, $A_0 = \frac{1}{3}\int_{-3}^{3} x^2 \, dx = 6$ and

$$A_n = \frac{1}{3}\int_{-3}^{3} x^2 \cos\frac{n\pi x}{3} \, dx = \frac{2}{3}\int_{0}^{3} x^2 \cos\frac{n\pi x}{3} \, dx$$

$$= \left[(x^2)\left(\frac{2\sin n\pi x/3}{n\pi}\right) - (2x)\left(\frac{-6\cos n\pi x/3}{n^2\pi^2}\right) + (2)\left(\frac{-18\sin n\pi x/3}{n^3\pi^3}\right) \right]\Big|_{0}^{3}$$

$$= (-1)^n 36/(n^2\pi^2)$$

$$B_n = \frac{1}{3}\int_{-3}^{3} x^2 \sin\frac{n\pi x}{3} \, dx = 0$$

Thus, the Fourier series for *f* is

$$f \sim 3 + \sum_{n=1}^{\infty} \frac{(-1)^n 36 \cos(n\pi x/3)}{n^2\pi^2}$$

If we look closely at the formulas in (2.2), we observe that the integrals do not require the periodicity of *f*; they only require that *f* be defined on $(-a, a)$. For example, suppose

$$f(x) = \begin{cases} 1 & \text{for } 0 < x < 7 \\ 0 & \text{for } -7 < x < 0 \end{cases}$$

[See Figure 1.3(a).] Then, using $a = 7$, we have

$$A_0 = \frac{1}{7}\int_{-7}^{7} f(x) \, dx = \frac{1}{7}\int_{0}^{7} 1 \, dx = 1$$

$$A_n = \frac{1}{7}\int_{-7}^{7} f(x) \cos\frac{n\pi x}{7} \, dx = \frac{1}{7}\int_{0}^{7} \cos\frac{n\pi x}{7} \, dx = 0$$

$$B_n = \frac{1}{7}\int_{0}^{7} \sin\frac{n\pi x}{7} \, dx = \frac{1 - \cos n\pi}{n\pi} = \begin{cases} 0 & \text{if } n \text{ is even} \\ 2/(n\pi) & \text{if } n \text{ is odd} \end{cases}$$

Thus,

$$f \sim \frac{1}{2} + \frac{2}{\pi}\left[\sin(\pi x/7) + \frac{\sin(3\pi x/7)}{3} + \frac{\sin(5\pi x/7)}{5} + \cdots \right]$$

What does this correspondence represent? The right side must have period $14 = 2 \cdot 7$, since each term has period 14. Therefore, the correspondence above should represent the Fourier series for the *periodic extension* of *f*. [See Figure 1.3(b).] We now define this concept of a periodic extension.

Figure 1.3   A function and its periodic extension. (See Figure 1.1 also.) (*a*) Graph of *f*. (*b*) Periodic extension of *f*.

(2.4) Definition.   Let *f* be a function defined on an interval of length 2*a* [such as $(-a, a)$]. The *periodic extension* of *f* is that function, denoted by f_P, which satisfies for $p = 2a$

$$f_P(x + kp) = f(x) \qquad \text{for each integer } k \qquad \blacksquare$$

It is geometrically obvious that the periodic extension of *f* is a periodic function (just draw its graph), and since the two functions equal each other on some interval of length 2*a* [such as $(-a, a)$] they must have *identical Fourier series*. For, if *f* is defined on $(-a, a)$ then

$$A_n = \frac{1}{a} \int_{-a}^{a} f(x) \cos \frac{n\pi x}{a} \, dx = \frac{1}{a} \int_{-a}^{a} f_P(x) \cos \frac{n\pi x}{a} \, dx \qquad (n = 0, 1, 2, \ldots)$$

$$B_n = \frac{1}{a} \int_{-a}^{a} f(x) \sin \frac{n\pi x}{a} \, dx = \frac{1}{a} \int_{-a}^{a} f_P(x) \sin \frac{n\pi x}{a} \, dx \qquad (n = 1, 2, 3, \ldots)$$

so the Fourier coefficients of *f* and its periodic extension f_P are the same.

(2.5) Example.   Let $f(x) = \sin x$ for $-\frac{1}{2}\pi < x < \frac{1}{2}\pi$. Find the Fourier series for *f* and graph the periodic extension of *f*.

Solution.   We have $a = \frac{1}{2}\pi$. Observing that *f* is odd on $(-\frac{1}{2}\pi, \frac{1}{2}\pi)$ we obtain $A_0 = 0$ and $A_n = 0$. Also

$$B_n = \frac{2}{\pi} \int_{-\frac{1}{2}\pi}^{\frac{1}{2}\pi} \sin x \sin 2nx \, dx = \frac{4}{\pi} \int_{0}^{\frac{1}{2}\pi} \sin x \sin 2nx \, dx$$

$$= \frac{2}{\pi} \int_{0}^{\frac{1}{2}\pi} \cos(2n - 1)x - \cos(2n + 1)x \, dx$$

$$= \frac{2}{\pi} \left[\frac{\sin(2n - 1)\frac{1}{2}\pi}{2n - 1} - \frac{\sin(2n + 1)\frac{1}{2}\pi}{2n + 1} \right]$$

Thus, the Fourier series for *f* is

$$f \sim \frac{2}{\pi} \left(\frac{4 \sin 2x}{3} - \frac{8 \sin 4x}{15} + \frac{12 \sin 6x}{35} - \cdots \right)$$

The graph of the periodic extension of *f* is shown in Figure 1.4.

Some minor problems do arise in the definition of a periodic extension of a function if closed intervals are considered [see Exercise (2.11)].

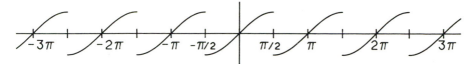

Figure 1.4   Graph of the periodic extension of the function in Example (2.5).

Exercises

(2.6)   Find the Fourier series for each of the following functions. Graph the periodic extension of each function.

(a) $f(x) = |x|$ for $-1 < x < 1$

(b) $f(x) = \begin{cases} -3 & \text{for } -2 < x < 0 \\ 3 & \text{for } 0 < x < 2 \end{cases}$

***(2.7)**   Suppose a Fourier series is needed for a function defined on an interval of length $p = 2a$, say $(c, c + 2a)$. Find formulas for the Fourier coefficients of such a function which involve integrals over $(c, c + 2a)$. Justify these formulas by orthogonality relations for trigonometric functions [again over $(c, c + 2a)$]. Prove that the Fourier series for such a function is the same as the Fourier series for its periodic extension where the integrals which define the Fourier coefficients for the periodic extension may be taken over *any* interval of length $2a$ [such as $(-a, a)$].
[*Hint*: See Exercise (1.17).]

(2.8)   Use the results of the previous exercise to obtain Fourier series for the following functions.

(a) $f(x) = x$ for $0 < x < 1$
(b) $f(x) = ax^2 + bx + c$ where a, b, and c are constants and $0 < x < 2\pi$
(c) $f(x) = ax^2 + bx + c$ for $-\pi < x < \pi$

(2.9)   Verify the following orthogonality relations.

(a) $\displaystyle \int_0^a \cos \frac{n\pi x}{a}\, dx = 0$

(b) $\displaystyle \int_0^a \cos \frac{m\pi x}{a} \cos \frac{n\pi x}{a}\, dx = \begin{cases} 0 & \text{if } m \neq n \\ \frac{1}{2}a & \text{if } m = n \end{cases}$

(c) $\displaystyle \int_0^a \sin \frac{m\pi x}{a} \sin \frac{n\pi x}{a}\, dx = \begin{cases} 0 & \text{if } m \neq n \\ \frac{1}{2}a & \text{if } m = n \end{cases}$

(2.10)   Using relation (c) from (2.9), show how a function f defined on $(0, a)$ can be related to a series of sines. Then show how the same function f can be related to a series of cosines plus a constant term.

(2.11)   For which of the following functions is it possible to define a periodic extension. For those which it is not, state how to restrict their domains so periodic extension is possible (and Fourier series are unaffected).

(a) $f(x) = x$ for $-1 \leq x \leq 1$
(b) $f(x) = x$ for $-1 \leq x < 1$
(c) $f(x) = x^2$ for $-1 \leq x \leq 1$
(d) $f(x) = x^3$ for $-2 \leq x \leq 2$

§3. Cosine and Sine Series

Expansions of a function into cosine and sine series will play an essential role in the physical problems discussed in Chapters 3 and 4. Sine series will be especially useful. The concepts of even and odd extensions, especially odd extension, will also be relied on greatly in those forthcoming chapters. In this section, therefore, we will discuss the basic ideas involved in cosine and sine series, and even and odd extensions.

Suppose that f is an even function on the open interval $(-a, a)$. Since $f(x)\cos(n\pi x/a)$ is even and $f(x)\sin(n\pi x/a)$ is odd, the Fourier coefficients of f assume the following form

$$A_n = \frac{2}{a} \int_0^a f(x) \cos \frac{n\pi x}{a} \, dx \qquad B_n = 0$$

Thus, *for an even function f over $(-a, a)$*, the Fourier series for f has the form

(3.1)
$$\begin{cases} f \sim \frac{1}{2}A_0 + \sum_{n=1}^{\infty} A_n \cos \frac{n\pi x}{a} \\[2mm] A_n = \frac{2}{a} \int_0^a f(x) \cos \frac{n\pi x}{a} \, dx & \text{for } n = 0, 1, 2, 3, \ldots \end{cases}$$

Formula (3.1) is called the *cosine series* for f over $(0, a)$. The integrals defining the cosine coefficients A_n in (3.1) only require that f be defined on $(0, a)$. Therefore, we can define a cosine series for f *even if f is defined only on $(0, a)$*.

(3.2) Example.   Let

$$f(x) = \begin{cases} 0 & \text{for } 0 < x < \frac{1}{2} \quad \text{and} \quad 1 < x < 3/2 \\ 1 & \text{for } \frac{1}{2} < x < 1 \end{cases}$$

Find the cosine series for f.

Solution.   The function f is defined only on $(0, 3/2)$. But, we can still use (3.1) to obtain

$$A_0 = \frac{2}{3/2} \int_0^{3/2} f(x) \, dx = \frac{4}{3} \int_{1/2}^1 1 \, dx = \frac{2}{3}$$

$$A_n = \frac{4}{3} \int_{1/2}^1 \cos \frac{2n\pi x}{3} \, dx = \frac{2}{n\pi} \left[\sin \frac{2n\pi}{3} - \sin \frac{n\pi}{3} \right]$$

Thus, the cosine series for f is

$$f \sim \frac{1}{3} - \frac{2\sqrt{3}}{\pi}\left[\frac{\cos(4\pi x/3)}{2} - \frac{\cos(8\pi x/3)}{4} + \frac{\cos(16\pi x/3)}{8} - \cdots\right]$$

Suppose, on the other hand, that f is an odd function over $(-a, a)$. Since $f(x)\cos(n\pi x/a)$ is odd and $f(x)\sin(n\pi x/a)$ is even, the Fourier coefficients for f assume the following form:

$$A_n = 0 \qquad \text{and} \qquad B_n = \frac{2}{a}\int_0^a f(x)\sin\frac{n\pi x}{a}\,dx$$

Thus, *for an odd function f on $(-a, a)$ the Fourier series for f has the form*

(3.3) $$f \sim \sum_{n=1}^{\infty} B_n \sin\frac{n\pi x}{a} \qquad B_n = \frac{2}{a}\int_0^a f(x)\sin\frac{n\pi x}{a}\,dx$$

Formula (3.3) is called the *sine series* for f on $(0, a)$. The integrals defining the sine coefficients B_n in (3.3) only require that f be defined on $(0, a)$. Therefore, we can define a sine series for f even if f is defined only on $(0, a)$.

(3.4) Example.   Let f be the same function as in Example (3.2). Find the sine series for f.

Solution.

$$B_n = \frac{4}{3}\int_0^{3/2} f(x)\sin\frac{2n\pi x}{3}\,dx = \frac{4}{3}\int_{1/2}^{1}\sin\frac{2n\pi x}{3}\,dx$$

$$= \frac{2}{n\pi}\left[\cos\frac{n\pi}{3} - \cos\frac{2n\pi}{3}\right]$$

Thus, the sine series for f is

$$f \sim \frac{2}{\pi}\left[\sin(2\pi x/3) - 2\frac{\sin(2\pi x)}{3} + \frac{\sin(10\pi x/3)}{5} - \cdots\right]$$

Examples (3.2) and (3.4) show that it is possible to find both a cosine and a sine series for a given function. But, what do these two different series represent? For the cosine series all the terms are even functions; therefore, the cosine series must represent an even function. On the other hand, for the sine series all the terms are odd functions; therefore, the sine series must represent an odd function. The following definition states precisely what these functions are.

(3.5) Definition.   Let the function f be defined on $(0, a)$. The *even extension* f_e and the *odd extension* f_o of f are the following functions

$$f_e(x) = \begin{cases} f(x) & \text{for } 0 < x < a \\ f(-x) & \text{for } -a < x < 0 \end{cases}$$

$$f_o(x) = \begin{cases} f(x) & \text{for } 0 < x < a \\ -f(-x) & \text{for } -a < x < 0 \end{cases} \qquad \blacksquare$$

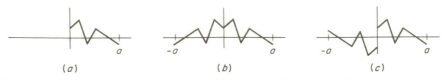

Figure 1.5 Even and odd extensions of a function. (*a*) Graph of *f*. (*b*) Even extension f_e. (*c*) Odd extension f_o.

That the even extension is an even function, and the odd extension is an odd function, is made clear by drawing their graphs. (See Figure 1.5.)

(3.6) Theorem. Let the function *f* be defined on $(0, a)$. The Fourier series for f_e is identical to the cosine series for *f*. The Fourier series for f_o is identical to the sine series for *f*.

Proof. Since f_e is even, the Fourier series for f_e has the form

$$(3.7) \qquad f_e \sim \tfrac{1}{2}A_0 + \sum_{n=1}^{\infty} A_n \cos \frac{n\pi x}{a} \qquad A_n = \frac{2}{a} \int_0^a f_e(x) \cos \frac{n\pi x}{a} \, dx$$

But, $f_e(x) = f(x)$ for $0 < x < a$. Therefore,

$$A_n = \frac{2}{a} \int_0^a f_e(x) \cos \frac{n\pi x}{a} \, dx = \frac{2}{a} \int_0^a f(x) \cos \frac{n\pi x}{a} \, dx$$

Hence, the Fourier series for f_e shown in (3.7) is identical, term for term, with the cosine series for *f*. A similar argument, using the oddness of f_o and that $f_o(x) = f(x)$ for $0 < x < a$, proves that the Fourier series for f_o and the sine series for *f* are identical. ∎

(3.8) Example. Let $f(x) = \cos x$ on $(0, \pi)$. Find the sine series for *f*; graph the odd extension f_o and the periodic extension of f_o (called the *odd periodic extension* of *f*).

Solution.

$$B_n = \frac{2}{\pi} \int_0^\pi \cos x \sin nx \, dx = \frac{1}{\pi} \int_0^\pi [\sin(n+1)x + \sin(n-1)x] \, dx$$

$$= \begin{cases} 4n[(n^2 - 1)\pi]^{-1} & \text{if } n \text{ is even} \\ 0 & \text{if } n \text{ is odd} \end{cases}$$

Thus, the sine series for $\cos x$ is

$$\cos x \sim \frac{8}{\pi} \left[\frac{\sin 2x}{3} + \frac{2 \sin 4x}{15} + \frac{3 \sin 6x}{35} + \frac{4 \sin 8x}{63} + \cdots \right]$$

The graphs of the odd extension of *f* and the odd periodic extension are shown in Figure 1.6.

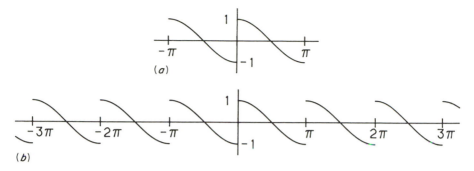

Figure 1.6 Odd and odd periodic extensions of the function in Example (3.8).

Exercises

Remark. The formulas for the cosine and sine series for f could have been derived by using orthogonality relations as in §§1 and 2. This was the point of Exercise (2.10). The reader who did not complete that exericse should try it again.

(3.9) Find cosine and sine series for the following functions. Sketch the graphs of the even and odd extensions and their periodic extensions.

(a) $f(x) = 1$ on $(0, a)$
(b) $f(x) = x$ on $(0, a)$
(c) $f(x) = \sin x$ on $(0, 1)$
(d) $f(x) = \sin x$ on $(0, \pi)$

(3.10) Find Fourier series for the following functions:

(a) $f(x) = x$ on $(-2, 2)$
(b) $f(x) = \begin{cases} x & \text{for } -\frac{1}{2} < x < \frac{1}{2} \\ 1 - x & \text{for } \frac{1}{2} < x < 3/2 \end{cases}$

§4. Convergence of Fourier Series

Up to now we have treated Fourier series on a somewhat informal and nonrigorous basis. The question of when a function f actually equals its Fourier series, for instance, has been left open. This question is a very difficult one to answer in general; we will state a partial theorem in this section which is of wide applicability.

Because a Fourier series usually consists of an infinite number of terms, we must discuss the question of convergence of series having an infinite number of terms. We will consider one type of convergence, known as pointwise convergence. A second type of convergence, mean square convergence, will be discussed in Chapter 2.

The definition of pointwise convergence of a series of functions is as follows.

(4.1) Definition: Pointwise Convergence. An infinite series of functions $\sum_{n=1}^{\infty} f_n$ *converges* at the point x_0 if the *numerical* series $\sum_{n=1}^{\infty} f_n(x_0)$ converges. The *sum* S of $\sum_{n=1}^{\infty} f_n$ is that function which has as its domain all those points for which $\sum_{n=1}^{\infty} f_n$ converges, and takes the value $S(x_0) = \sum_{n=1}^{\infty} f_n(x_0)$ at x_0. If the numerical series $\sum_{n=1}^{\infty} f_n(x_0)$ diverges, then $\sum_{n=1}^{\infty} f_n$ is said to *diverge* at x_0. ∎

(4.2) Example.

(a) The series $\sum_{n=0}^{\infty} x^n = 1 + x + x^2 + \cdots + x^n + \cdots$ converges for every real (or complex) number x such that $|x| < 1$, and for each such x

$$\sum_{n=0}^{\infty} x^n = \frac{1}{1-x}$$

The series diverges for all x such that $|x| \geq 1$.

(b) The series $\sum_{n=0}^{\infty} x^n/n! = 1 + x + x^2/2 + x^3/6 + \cdots$ converges for every real (or complex) number x, and for each such x, $\sum_{n=0}^{\infty} x^n/n! = e^x$.

(c) The series $\pi/2 + \sum_{n=1}^{\infty} (\sin nx)/n^2$ converges for each real number x. This is because for each fixed $x = x_0$ we have $|(\sin nx_0)/n^2| \leq 1/n^2$ and $\sum_{n=1}^{\infty} 1/n^2$ converges; hence, we may apply the Comparison Test.

(d) The series $\sum_{n=1}^{\infty} (\cos nx)/n$ diverges at each integral multiple of 2π, since for each such x, the series equals $\sum_{n=1}^{\infty} 1/n$ which diverges. Although we will not take the time now, it can be shown that the series converges for all other values of x. We will show this in §6.

Unfortunately, Fourier series do not always converge. In Appendix A it is shown that some continuous functions have Fourier series which diverge at certain values of x. For a rather large class of functions, however, the Fourier series will converge. Those functions are the *piecewise smooth* functions. We will now discuss the definition of this class of functions and then state a theorem on the convergence of their Fourier series.

We begin by defining a piecewise continuous function; a piecewise smooth function f is then just a function whose derivative f' is piecewise continuous. A piecewise continuous function is essentially just a function that has only a finite number of noninfinite discontinuities in any given interval. (See Figure 1.7). These discontinuities, however, are of a special type that we now define.

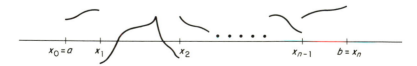

Figure 1.7 A piecewise continuous function.

(4.3) Definition.   A function f has a *jump discontinuity* at the point x_0 if the one sided limits

$$f(x_0 -) = \lim_{x \to x_0 -} f(x) \qquad f(x_0 +) = \lim_{x \to x_0 +} f(x)$$

both exist but are unequal. ■

It can happen that a function has one-sided limits that are equal but the function is undefined at the point in question. Such a point is called a *removable discontinuity*. For example, the function $f(x) = (\sin x)/x$ has a removable discontinuity at $x = 0$. *We will always assume that removable discontinuities are removed by redefining $f(x_0)$ so that $f(x_0) = f(x_0 +) = f(x_0 -)$.* For example, we assume that for $f(x) = (\sin x)/x$ we have $f(0) = 1$.

We can now state the definitions of a piecewise continuous function and a piecewise smooth function.

(4.4) Definition.   A function is *piecewise continuous on* (a, b) if it has only a finite number of jump discontinuities in (a, b) and the left-hand limit exists at b and the right hand limit exists at a. A function is *piecewise smooth on* (a, b) if it and its derivative are piecewise continuous on (a, b). A function is *piecewise continuous (smooth)* if it is piecewise continuous (smooth) on every finite open interval (a, b). ■

For example, the function f defined by

$$f(x) = \begin{cases} 1 & \text{for } -1 < x < 2 \\ x & \text{for } 2 < x < 3 \end{cases}$$

is piecewise continuous on $(-1, 3)$. Its derivative

$$f'(x) = \begin{cases} 0 & \text{for } -1 < x < 2 \\ 1 & \text{for } 2 < x < 3 \end{cases}$$

is piecewise continuous on $(-1, 3)$. Therefore, f is piecewise smooth on $(-1, 3)$.

A continuous function is always piecewise continuous. Furthermore, a function which is piecewise continuous (smooth) on an open interval of length $2a$ has a periodic extension of period $2a$ which is piecewise continuous (smooth).

We can now state our main theorem on convergence of Fourier series.

(4.5) Theorem: Convergence of Fourier Series.   If f is a piecewise smooth function with period $2a$, then the Fourier series for f converges for all x values. If $x = x_0$ is a point of continuity for f, then the Fourier series converges to $f(x_0)$ at x_0, that is

$$\tfrac{1}{2}A_0 + \sum_{n=1}^{\infty} [A_n \cos(n\pi x_0/a) + B_n \sin(n\pi x_0/a)] = f(x_0)$$

If $x = x_0$ is a point of discontinuity for f, then the Fourier series for f

converges to the average of the right-hand and left-hand limits at x_0, that is

$$\tfrac{1}{2}A_0 + \sum_{n=1}^{\infty} [A_n \cos(n\pi x_0/a) + B_n \sin(n\pi x_0/a)] = \tfrac{1}{2}[f(x_0+) + f(x_0-)]$$

We will prove this theorem in Chapter 2 (see §3 of that chapter).

If f is piecewise smooth on $(-a, a)$, or any open interval of length $2a$, then its periodic extension f_P satisfies the hypotheses of our convergence theorem. Since the Fourier series for f is identical to that for f_P, it will converge to the periodic extension f_P at all points of continuity of f_P. While at points of discontinuity of f_P the Fourier series for f will converge to the average of the right-hand and left-hand limits of f_P. The best way to determine the convergence of the Fourier series for a piecewise smooth function is *by graphing its periodic extension and applying Theorem* (4.5). We will use this dictum in the following examples.

(4.6) Example.   Let the function f be defined by

$$f(x) = \begin{cases} \tfrac{1}{2}\pi & \text{for } 0 < x < \pi \\ -\tfrac{1}{2}\pi & \text{for } -\pi < x < 0 \end{cases}$$

This function is piecewise smooth on $(-\pi, \pi)$ as can be seen from graphing its periodic extension, shown in Figure 1.8. The Fourier series for f

$$2\sum_{k=0}^{\infty} \frac{\sin(2k+1)x}{2k+1} = 2\sin x + \frac{2\sin 3x}{3} + \frac{2\sin 5x}{5} + \cdots$$

converges to the periodic extension f_P at all points of continuity of f_P. At each point of discontinuity of f_P the Fourier series converges to the average of the right-hand and left-hand limits of f_P. In this case, these averages are all zero. Figure 1.9 shows some of the graphs of the *partial sums* of the Fourier series for f.

(4.7) Example.   Let f be the function defined by $f(x) = \tfrac{1}{2}x$ for $0 < x < 2\pi$. The Fourier series for f is

$$\tfrac{1}{2}\pi - \sum_{n=1}^{\infty} \frac{\sin nx}{n}$$

which converges to f at all points in $(0, 2\pi)$. The graph of the function to which the Fourier series converges for all x values is shown in Figure 1.10.

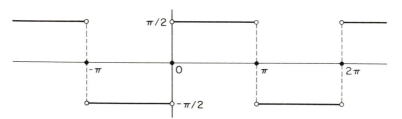

Figure 1.8   Graph of the function in Example (4.6).

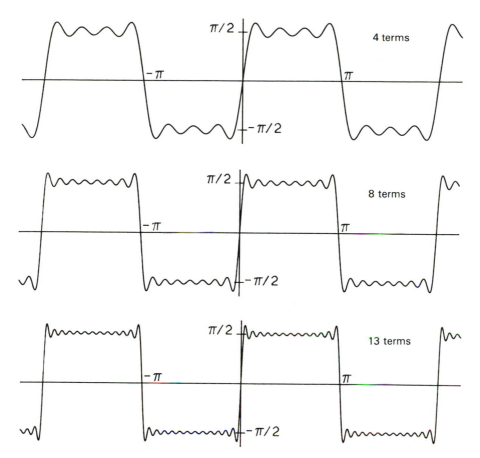

Figure 1.9 Graphs of some of the partial sums of $2\sum_{k=0}^{\infty}[\sin(2k+1)x]/(2k+1)$. (Compare with Figure 1.8.) Notice the sharp peaks (and valleys) in the partial sums of the Fourier series near the jump discontinuities of the function. This is an example of *Gibb's phenomenon*, according to which the maximum (minimum) values of the partial sums of the Fourier series tend to *overshoot* (*undershoot*) the one-sided limits of the function at its jump discontinuities. We will discuss Gibb's phenomenon in Chapter 2, §5.

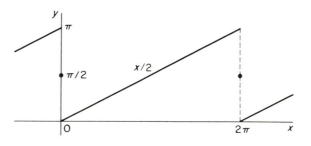

Figure 1.10 Graph of the function in Example (4.7).

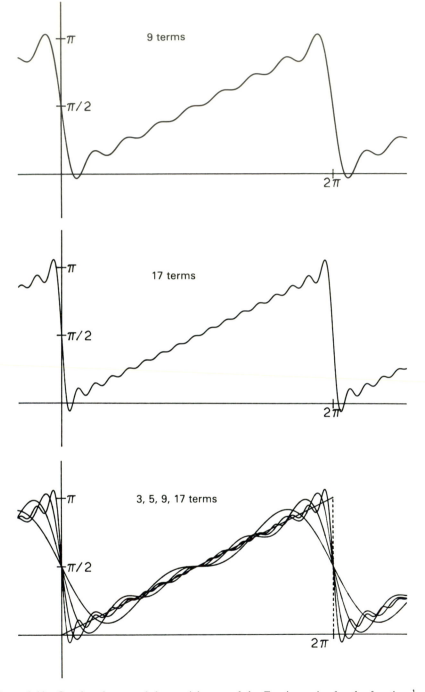

Figure 1.11   Graphs of some of the partial sums of the Fourier series for the function $\frac{1}{2}x$ on $(0, 2\pi)$. (See Figure 1.10.) Note that in the third graph above Gibb's phenomenon occurs again.

In Figure 1.11 the graphs of some of the partial sums of the Fourier series for f are shown.

We conclude this section with a few remarks concerning piecewise continuous functions. *For the rest of this book, unless stated otherwise, we will assume that all functions are piecewise continuous.*

An integral of a piecewise continuous function is performed piecewise. To see this, let

$$x_1 < x_2 < \cdots < x_{n-1}$$

denote the jump discontinuities (if any) of f on (a, b) and let $x_0 = a$ and $x_n = b$. Then, by defining $f(x_{j-1}) = f(x_{j-1}+)$ and $f(x_j) = f(x_j-)$ the function f becomes continuous on each interval $[x_{j-1}, x_j]$ for $j = 1, 2, \ldots, n$. We then have

(4.8) $$\int_a^b f(x)\, dx = \int_a^{x_1} f(x)\, dx + \int_{x_1}^{x_2} f(x)\, dx + \cdots + \int_{x_{n-1}}^b f(x)\, dx$$

The integrals on the right side of (4.8) all exist since f is continuous on each of the separate intervals. Formula (4.8) can be thought of as obvious if we interpret integration in terms of area (where areas above the x axis are regarded as positive and areas below the x axis are regarded as negative). (See Figure 1.12.)

Alternatively, we can use the right side of (4.8) as the *definition* of the left side of (4.8). In this case, all of the familiar properties of integrals from calculus remain true (such as, the integral of a sum being the sum of the integrals, integration by parts, etc.).

Another important property of piecewise continuous functions is that sums and products of piecewise continuous functions are piecewise continuous. To be more precise, we have the following theorem.

(4.9) Theorem.   If f and g are piecewise continuous functions, and c and d are constants, then $cf + dg$ is piecewise continuous and so is $f \cdot g$.

Proof.   Let $a = x_0 < x_1 < \cdots < x_{n-1} < x_n = b$ be the discontinuities (if any) of either f or g in some given interval $[a, b]$. Then f and g are both continuous on $[x_{j-1}, x_j]$ if defined by their one-sided limits at the end points. Hence, we know from calculus that $cf + dg$ and $f \cdot g$ are continuous on each interval $[x_{j-1}, x_j]$. Therefore, $cf + dg$ and $f \cdot g$ are piecewise continuous on the given interval $[a, b]$. Thus, $cf + dg$ and $f \cdot g$ are piecewise continuous. ∎

It follows from this theorem, and the previous discussion, that $f \cdot g$ and f^2 are both integrable over $[a, b]$ if f and g are piecewise continuous on (a, b).

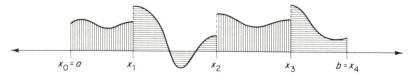

Figure 1.12   An integral of a piecewise continuous function is done piecewise.

Exercises

(4.10)   Check each function below to see if it is piecewise smooth. If it is, state the value to which its Fourier series converges at each x value in the given interval and at the end points. Sketch the graph of the Fourier series.

(a) $f(x) = 2|x| + x$ $-1 < x < 1$
(b) $f(x) = x \cos x$ $-\pi < x < \pi$

(c) $f(x) = \begin{cases} 0 & \text{for } 1 < x < 3 \\ 1 & \text{for } -1 < x < 1 \\ x & \text{for } -3 < x < -1 \end{cases}$

(4.11)   State convergence theorems for sine and cosine series of piecewise smooth functions on the interval $(0, a)$. Apply these theorems to the sine and cosine series for $f(x) = x^2 + x$ on the interval $(0, 2)$; sketch graphs of these sine and cosine series on the interval $[-4, 4]$.

(4.12)   The convergence theorem for Fourier series can be used to evaluate the sums of some numerical series. Given the Fourier series below, find the values of the given numerical series by evaluating the Fourier series at an appropriate value of x.

(a) $|x| = \frac{1}{2} - \frac{4}{\pi^2} \sum_{k=0}^{\infty} \frac{\cos(2k+1)\pi x}{(2k+1)^2}$ for $-1 \le x \le 1$

$$1 + \frac{1}{9} + \frac{1}{25} + \cdots + \frac{1}{(2k+1)^2} + \cdots = ?$$

(b) $|\sin x| = \frac{2}{\pi} - \frac{4}{\pi} \sum_{n=1}^{\infty} \frac{\cos 2nx}{4n^2 - 1}$ for $-\pi \le x \le \pi$

$$\frac{1}{3} - \frac{1}{15} + \cdots + \frac{(-1)^{n+1}}{4n^2 - 1} + \cdots = ?$$

(c)   Using the Fourier series for x^2 over $(-\pi, \pi)$ and Theorem (4.5), show that $\sum_{n=1}^{\infty} 1/n^2 = \pi^2/6$.

(4.13)   Give an example of a piecewise continuous function which is not piecewise smooth.

(4.14) *Integration by Parts.*   Prove that if f and g are continuous on the closed interval $[a, b]$ with piecewise continuous derivatives f' and g', then

$$\int_a^b f(x)g'(x)\, dx = f(x)g(x)\Big|_a^b - \int_a^b f'(x)g(x)\, dx$$

[*Hint*: Split the integral on the left into a sum of integrals over intervals where f' and g' are continuous.]

§5. Uniform Convergence

The concept of the uniform convergence of an infinite series of continuous functions is of fundamental importance in Fourier analysis and the theory of differential equations. We will introduce this concept and discuss a few applications in this section; many other applications will be discussed in the sequel, especially in Chapters 3 and 4.

(5.1) Definition. The series of continuous functions $\sum_{n=1}^{\infty} f_n$ *converges uniformly* on the closed interval $[a, b]$ if the following two conditions are satisfied: *First*, the series converges to the sum $S(x)$ for each x value in the interval; *second*, given any number $\delta > 0$, no matter how small, there exists some positive constant M such that for all $m \geq M$ the partial sums $S_m = \sum_{n=1}^{m} f_n$ satisfy $|S(x) - S_m(x)| \leq \delta$ for every x value in the interval. ∎

Uniform convergence means that for all sufficiently large integers m the graphs of S and S_m are less than δ apart, where δ is any preassigned small positive number. The two graphs are *uniformly close* on the interval. (See Figure 1.13.) An essential aspect of Definition (5.1) is that once δ is chosen, the choice of M is independent of all x values in the interval. *The constant M depends only on δ.* Furthermore, the reader might observe that the choice of a *closed* interval in Definition (5.1) is not really essential; the definition applies as well to any interval (including the whole real line) and sometimes we will speak of uniform convergence on intervals that are not closed.

The following theorem gives an easily applied condition for determining uniform convergence. It is especially useful in its application to Fourier series.

(5.2) Theorem: Weierstrass' M-Test. If for all points x in $[a, b]$ we have $|f_n(x)| \leq M_n$ from a certain $n = k$ on, and the series of positive numbers $\sum_{n=k}^{\infty} M_n$ converges, then the series $\sum_{n=1}^{\infty} f_n$ converges uniformly on $[a, b]$.

Proof. For each x value in the interval $[a, b]$, since $|f_n(x)| \leq M_n$ and $\sum_{n=k}^{\infty} M_n$ converges, the series $\sum_{n=k}^{\infty} f_n(x)$ converges absolutely by the Comparison Test. Therefore, the series $\sum_{n=k}^{\infty} f_n$ converges for each x in $[a, b]$.

Given some $\delta > 0$, no matter how small, we have [due to possible cancellation of positive and negative values of $f_n(x)$] for each x in $[a, b]$

$$\left| \sum_{n=m+1}^{\infty} f_n(x) \right| \leq \sum_{n=m+1}^{\infty} |f_n(x)| \leq \sum_{n=m+1}^{\infty} M_n$$

If $m \geq M$ is sufficiently large, then the last member of the inequalities above

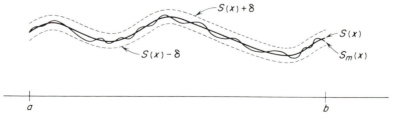

Figure 1.13 The graph of S_m is uniformly close to the graph of S.

will be less than δ in magnitude. Thus, for all $m \geq M$

$$|S(x) - S_m(x)| \leq \delta \qquad \text{for every } x \text{ in } [a, b] \qquad \blacksquare$$

Here is an example of the application of Weierstrass' M-test.

(5.3) Example.   Consider the infinite series

$$\tfrac{1}{2} + \sum_{n=1}^{\infty} \frac{-4 \cos(2n - 1)x}{\pi^2(2n - 1)^2}$$

The reader may wish to verify that this series is the Fourier series for the function $f(x) = 1/\pi \, |x|$ on the interval $(-\pi, \pi)$. Since

$$\left| \frac{-4 \cos(2n - 1)x}{\pi^2(2n - 1)^2} \right| \leq \frac{4}{\pi^2} \frac{1}{(2n - 1)^2}$$

and $4/\pi^2 \sum_{n=1}^{\infty} [1/(2n - 1)^2]$ converges, it follows from Weierstrass' M-test that the Fourier series converges uniformly on $[-\pi, \pi]$. We know from our convergence theorem for Fourier series that this particular Fourier series converges for all x. Figure 1.14 illustrates the uniformity of this convergence. The periodicity of f implies that the uniformity of this convergence actually holds on the whole real line, and not just on $[-\pi, \pi]$.

We will now discuss two other very important theorems concerning uniform convergence. The next two theorems are quite standard. Their proofs can be found in any of the advanced calculus texts mentioned in the introduction.

(5.4) Theorem.   If the series of continuous functions $\sum_{n=1}^{\infty} f_n$ converges uniformly on the interval $[a, b]$, then

(a)   The sum $S = \sum_{n=1}^{\infty} f_n$ of the series is continuous on $[a, b]$
(b)   The sum S can be integrated term by term, that is

$$\int_a^b \sum_{n=1}^{\infty} f_n(x) \, dx = \int_a^b S(x) \, dx = \sum_{n=1}^{\infty} \left[\int_a^b f_n(x) \, dx \right]$$

(5.5) Theorem.   If the series of continuous functions $\sum_{n=1}^{\infty} f_n$ converges, if each term f_n of the series is differentiable with f_n' continuous on $[a, b]$, and if the series of continuous derivatives $\sum_{n=1}^{\infty} f_n'$ converges uniformly on $[a, b]$, then $\sum_{n=1}^{\infty} f_n$ is continuously differentiable term by term, that is

$$\left[\sum_{n=1}^{\infty} f_n \right]' = S' = \sum_{n=1}^{\infty} f_n'$$

Let's now consider a few examples of these theorems.

(5.6) Example.   The series

(5.7)
$$\sum_{n=2}^{\infty} \frac{\sin nx}{n^3 - 3}$$

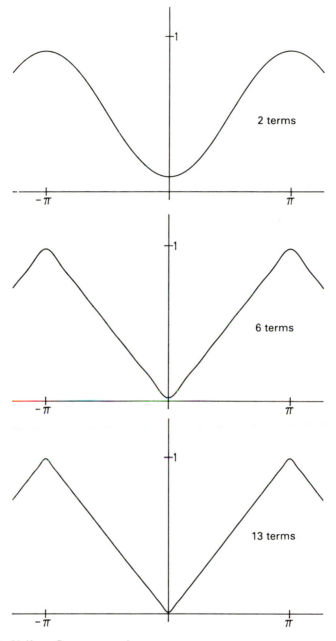

2 terms

6 terms

13 terms

Figure 1.14   Uniform Convergence of

$$\tfrac{1}{2} + \sum_{n=1}^{\infty} \frac{-4 \cos(2n-1)x}{\pi^2(2n-1)^2}$$

(The reader might observe that in the graphs above, in contrast to those in Figures 1.9 and 1.11, Gibb's phenomenon does not occur. This is due to the *uniformity* of the convergence of the Fourier series for $\frac{1}{\pi}|x|$.)

converges uniformly on $[-\pi, \pi]$ to a continuous function S since

$$\left|\frac{\sin nx}{n^3 - 3}\right| \leq \frac{1}{n^3 - 3}$$

for $n \geq 2$ and $\sum_{n=2}^{\infty}[1/(n^3 - 3)]$ converges. Moreover,

$$S'(x) = \sum_{n=2}^{\infty} \frac{n \cos nx}{n^3 - 3}$$

We deduce this last result by applying Weierstrass' M-test and Theorem (5.5).

Although we do not have an explicit expression for the function S, other than the series in (5.7), it is possible to show that the series in (5.7) is the Fourier series for S.[4] Since

$$S(x) \cos nx = \sum_{m=2}^{\infty} \frac{\sin mx \cos nx}{m^3 - 3}$$

converges uniformly by the Weierstrass' M-test, we can apply Theorem (5.4b) and integrate term by term. Hence

$$A_n = \frac{1}{\pi} \int_{-\pi}^{\pi} S(x) \cos nx \, dx = \frac{1}{\pi} \sum_{m=2}^{\infty} \left[\frac{1}{m^3 - 3} \int_{-\pi}^{\pi} \sin mx \cos nx \, dx\right]$$

$$= \frac{1}{\pi} \sum_{m=2}^{\infty} 0 = 0$$

Similarly, we have

$$B_n = \frac{1}{\pi} \int_{-\pi}^{\pi} S(x) \sin nx \, dx = \frac{1}{\pi} \sum_{m=2}^{\infty} \left[\frac{1}{m^3 - 3} \int_{-\pi}^{\pi} \sin mx \sin nx \, dx\right]$$

$$= \frac{1}{\pi} \frac{1}{n^3 - 3} \pi = \frac{1}{n^3 - 3}$$

Thus, the series in (5.7) is the Fourier series for S.

(5.8) Example.   Using the results of Example (4.6) we have for $-\pi \leq x \leq \pi$

$$(5.9) \qquad 2\sum_{k=0}^{\infty} \frac{\sin(2k + 1)x}{2k + 1} = \begin{cases} \frac{1}{2}\pi & \text{for } 0 < x < \pi \\ 0 & \text{for } x = 0, \pm\pi \\ -\frac{1}{2}\pi & \text{for } -\pi < x < 0 \end{cases}$$

Therefore, the series

$$2\sum_{k=0}^{\infty} \frac{\sin(2k + 1)x}{2k + 1}$$

---

4. Just because the series in (5.7) is a sum of sines does not, a priori, make it a Fourier series. The coefficients $1/(n^3 - 3)$ must be Fourier coefficients for some function. We aim to show that this function is S.

does *not* converge uniformly on $[-\pi, \pi]$ since it converges to a discontinuous function. For, by Theorem (5.4a), a uniformly convergent series of continuous functions must converge to a continuous function.[5]

(5.10) Example.   The following theorem can be proved using the theorems discussed above and our convergence theorem for Fourier series.

Theorem.   The Fourier series for a continuous, piecewise smooth, periodic function converges uniformly to that function on the whole real line.

We will prove this theorem in Chapter 2 (see §4). As an instance of this theorem, take $f(x) = |x|^3$ on the interval $(-1, 1)$, then the Fourier series for f converges uniformly to f_P, the periodic extension of f, on the whole real line.

Exercises

(5.11)   Determine whether or not the Fourier series for the following functions converge uniformly. Sketch the graph of each function, and the graph of the function to which the Fourier series converges. (You are not required to compute these series.)

(a) $f(x) = e^x$ on $(-1, 1)$
(b) $f(x) = \sin x + |x|$ on $(-\pi, \pi)$
(c) $f(x) = x(x^2 - 1)$ on $(-1, 1)$

(5.12)   Suppose $\{a_n\}$ and $\{b_n\}$ are bounded for all n, that is, some positive constant M exists such that $|a_n| \leq M$ and $|b_n| \leq M$ for all n. Then, given $c > 0$ prove that the series

$$a_0 + \sum_{n=1}^{\infty} [a_n \cos nx + b_n \sin nx]e^{-cn}$$

converges uniformly on every closed interval.

⋆(5.13)

(a)   Prove, under the same hypotheses as in (5.12), that

$$y = a_0 + \sum_{n=1}^{\infty} [a_n \cos nx + b_n \sin nx] \cdot e^{-cn}$$

is twice continuously differentiable, term by term, for every x.

---

5. While it is true that

$$\left| \frac{2 \sin(2k+1)x}{2k+1} \right| \leq \frac{2}{2k+1} \qquad \text{but} \qquad \sum_{k=0}^{\infty} \frac{2}{2k+1}$$

diverge, this is not the reason for the nonuniform convergence of the series in (5.9). It can be shown that the series in (5.9) converges uniformly on any closed interval which does not include an integral multiple of π. [See Exercise (6.10).]

(b)   Suppose x is held fixed and $c > 0$ is allowed to vary. Prove that the series in part (a) is twice continuously differentiable, term by term, with respect to c, for all $c \geq \delta > 0$. Therefore, letting δ tend to zero, the result obtains for all $c > 0$.

(c)   Show that the series in (a) satisfies

$$\frac{\partial^2 y}{\partial x^2} = -\frac{\partial^2 y}{\partial c^2}$$

for all real x and all $c > 0$.

$\star$**(5.14)**   Prove the following theorem.

Theorem.   If the series $C_0 + \sum_{n=1}^{\infty} [C_n \cos nx + D_n \sin nx]$ converges uniformly on the interval $[-\pi, \pi]$, then the series is the Fourier series for its sum, the function S.

*§6. Abel's Test

In this section we discuss a very useful test for uniform convergence known as Abel's test. We will use Abel's test in our treatment of Gibb's phenomenon in Chapter 2.

Abel's test is sometimes useful when Weierstrass' M-test does not work. For instance, the series

(6.1)
$$\sum_{n=1}^{\infty} \frac{\cos nx}{n}$$

is a typical such case. Weierstrass' M-test does not work for this series, since $\sum_{n=1}^{\infty} 1/n$ diverges. For, the condition $|f_n(x)| \leq M_n$ where $\sum M_n$ converges, implies the *absolute* convergence of $\sum f_n(x)$ for each x; and yet if we let $x = \pi$, for example, then

$$\sum_{n=1}^{\infty} \frac{\cos n\pi}{n} = \sum_{n=1}^{\infty} \frac{(-1)^n}{n}$$

converges but not absolutely. We will soon see, however, that Abel's test can be used to show that the series in (6.1) converges uniformly on any closed interval which does not contain an integral multiple of 2π.

Before we can state Abel's test we need a few preliminary definitions and a lemma. We say that a series of functions $\sum_{n=1}^{\infty} f_n$ is *uniformly bounded* on the interval $[a, b]$ if there exists a positive constant M such that for all positive integers N we have

$$\left| \sum_{n=1}^{N} f_n(x) \right| \leq M$$

for every x in $[a, b]$. The reader might note that it is not necessary for a series to converge in order for it to be uniformly bounded. We say that a sequence $\{a_n\}$ of numbers *decreases monotonically* if

$$a_1 \geq a_2 \geq a_3 \geq \cdots \geq a_n \geq \cdots$$

such a sequence is said to *decrease monotonically to zero* if

$$\lim_{n \to \infty} a_n = 0$$

as well. Abel's Test depends on the following lemma, known as Abel's lemma.

(6.2) Lemma: Abel's Lemma. Suppose that the sequence of numbers $\{a_n\}_{n=1}^N$ decreases monotonically and that there are two constants m and M such that the sequence $\{b_n\}_{n=1}^N$ satisfies

$$m \leq b_1 + b_2 + \cdots + b_N \leq M$$

then

$$a_1 m \leq a_1 b_1 + a_2 b_2 + \cdots + a_N b_N \leq a_1 M$$

Proof. Let $s_k = b_1 + \cdots + b_k$. Then

$$a_1 b_1 + a_2 b_2 + \cdots + a_N b_N = a_1 s_1 + a_2(s_2 - s_1) + \cdots + a_N(s_N - s_{N-1})$$
$$= (a_1 - a_2)s_1 + (a_2 - a_3)s_2 + \cdots$$
$$+ (a_{N-1} - a_N)s_{N-1} + a_N s_N$$

Since each number in parentheses in the last line above is positive or zero, the sum is not decreased if each s_k is replaced by the larger number M, and this yields

$$(a_1 - a_2)M + (a_2 - a_3)M + \cdots + (a_{N-1} - a_N)M + a_N M = a_1 M$$

Hence,

$$a_1 b_1 + a_2 b_2 + \cdots + a_N b_N \leq a_1 M$$

Similarly, we obtain

$$a_1 m \leq a_1 b_1 + a_2 b_2 + \cdots + a_N b_N \quad \blacksquare$$

(6.3) Corollary. If the series $\sum_{n=1}^{\infty} a_n b_n$ converges (a_n and b_n as in Abel's lemma), then

$$a_1 m \leq \sum_{n=1}^{\infty} a_n b_n \leq a_1 M$$

$$\left| \sum_{n=1}^{\infty} a_n b_n - \sum_{n=1}^{N} a_n b_n \right| \leq a_{N+1}(M - m)$$

For instance, if the sequence $\{a_n\}$ decreases monotonically to zero, then the series $\sum_{n=1}^{\infty} a_n b_n$ does converge and the inequalities above do hold.

Proof. The first set of inequalities follows from Abel's lemma by letting N tend to infinity. As for the second inequality, we note that

$$\sum_{n=1}^{\infty} a_n b_n - \sum_{n=1}^{N} a_n b_n = \sum_{n=N+1}^{\infty} a_n b_n$$

Since we have for each nonnegative integer p

$$m \le \sum_{n=1}^{N+p} b_n \le M$$

we also have

$$m - M \le \sum_{n=N+1}^{n=N+p} b_n \le M - m$$

Thus, the first result of this corollary applied to the series $\sum_{n=N+1}^{\infty} a_n b_n$ yields

$$\left| \sum_{n=N+1}^{\infty} a_n b_n \right| \le a_{N+1}(M - m)$$

Finally, suppose that the sequence $\{a_n\}$ tends to zero. Then, using a result from the proof of Abel's lemma

(6.4) $$\sum_{n=1}^{N} a_n b_n - a_N s_N = (a_1 - a_2)s_1 + (a_2 - a_3)s_2 + \cdots + (a_{N-1} - a_N)s_{N-1}$$

If we let N tend to infinity, then the sums on the right side of (6.4) converge absolutely. For

$$|(a_1 - a_2)s_1| + |(a_2 - a_3)s_2| + \cdots + |(a_{N-1} - a_N)s_{N-1}|$$
$$\le (a_1 - a_2)M + (a_2 - a_3)M + \cdots + (a_{N-1} - a_N)M = a_1 M - a_N M$$

Hence, when N goes to infinity we obtain in the limit

$$\sum_{n=1}^{\infty} |(a_n - a_{n+1})s_n| \le \sum_{n=1}^{\infty} (a_n - a_{n+1})M = a_1 M$$

(The last series has *telescoped* to the sum $a_1 M$.) Since

$$|a_N s_N| = a_N |s_N| \le a_N M$$

and $a_N M$ tends to zero as N tends to infinity, we have $\lim_{N \to \infty} a_N s_N = 0$. Therefore, letting N tend to infinity in (6.4) yields

$$\sum_{n=1}^{\infty} a_n b_n = \sum_{n=1}^{\infty} (a_n - a_{n+1})s_n$$

Thus, $\sum_{n=1}^{\infty} a_n b_n$ converges. ∎

For example, consider the following alternating series

$$-1 + \frac{1}{3^3} - \frac{1}{5^3} + \frac{1}{7^3} - \cdots = \sum_{k=0}^{\infty} \frac{(-1)^{k+1}}{(2k+1)^3}$$

We have

$$-1 \leq \sum_{k=0}^{N} (-1)^{k+1} \leq 0$$

for each positive integer N, and $\{1/(2k+1)^3\}$ decreases monotonically to zero. Hence, by Corollary (6.3)

$$\left| \sum_{k=0}^{\infty} \frac{(-1)^{k+1}}{(2k+1)^3} - \sum_{k=0}^{N} \frac{(-1)^{k+1}}{(2k+1)^3} \right| \leq \frac{1}{(2N+3)^3}$$

For instance, by setting $1/(2N+3)^3 < 0.000125$ and solving for N we determine that

$$\sum_{k=0}^{9} \frac{(-1)^{k+1}}{(2k+1)^3} = -0.968884556 \cdots$$

approximates

$$\sum_{k=0}^{\infty} \frac{(-1)^{k+1}}{(2k+1)^3}$$

to within an error of ± 0.000125. See Exercise (6.12) for a generalization of this result; see also Exercise (6.8) where the exact sum of the alternating series is found.

We now come to Abel's test.

(6.5) Theorem: Abel's Test. Suppose that the partial sums of the series of functions $\sum_{n=1}^{\infty} f_n$ are uniformly bounded on the interval $[a, b]$ and the sequence of positive numbers $\{a_n\}$ decreases monotonically to zero, then $\sum_{n=1}^{\infty} a_n f_n(x)$ converges uniformly on $[a, b]$.

Proof. Let x be some number in the interval $[a, b]$ and let b_n equal the number $f_n(x)$. Then the series $\sum_{n=1}^{\infty} f_n(x) = \sum_{n=1}^{\infty} b_n$ and the sequence $\{a_n\}$ satisfy the hypotheses of Abel's lemma. Hence, Corollary (6.3) implies that $\sum_{n=1}^{\infty} a_n f_n(x)$ converges and

(6.6)
$$\left| \sum_{n=1}^{\infty} a_n f_n(x) - \sum_{n=1}^{N} a_n f_n(x) \right| \leq 2 a_{N+1} M$$

(where $|\sum_{n=1}^{N} f_n(x)| \leq M$ for each x in $[a, b]$). Since a_n tends to zero as n tends to infinity, the quantity $2a_{N+1} M$ tends to zero as N tends to infinity. Therefore, given $\delta > 0$, there exists a positive integer K so large that for all $N \geq K$ we have $2a_{N+1} M \leq \delta$. This fact, along with (6.6), implies that for all $N \geq K$

$$\left| \sum_{n=1}^{\infty} a_n f_n(x) - \sum_{n=1}^{N} a_n f_n(x) \right| \leq \delta \qquad \text{for each } x \text{ in } [a, b]$$

This proves that $\sum_{n=1}^{\infty} a_n f_n(x)$ converges uniformly on $[a, b]$ since the choice of K does not depend on x. ∎

Let's now apply Abel's test to the series, shown in (6.1), with which we began this discussion. We will show that for any closed interval $[a, b]$ contained in the open interval $(0, 2\pi)$ the series in (6.1) converges uniformly on $[a, b]$.

For x in the interval $[a, b]$, we first prove that

(6.7) $$\cos x + \cos 2x + \cdots + \cos Nx = \frac{\sin(N + \frac{1}{2})x - \sin \frac{1}{2}x}{2 \sin \frac{1}{2}x}$$

Multiplying the left side of (6.7) by $2 \sin \frac{1}{2}x$, and then applying the trigonometric identity $2 \sin \theta \cos \phi = \sin (\phi + \theta) - \sin (\phi - \theta)$, we have

$$(2 \sin \tfrac{1}{2}x)(\cos x + \cos 2x + \cdots + \cos Nx)$$

$$= \left[\sin \frac{3x}{2} - \sin \tfrac{1}{2}x \right] + \left[\sin \frac{5x}{2} - \sin \frac{3x}{2} \right] + \cdots + [\sin(N + \tfrac{1}{2})x - \sin(N - \tfrac{1}{2})x]$$

$$= \sin(N + \tfrac{1}{2})x - \sin \tfrac{1}{2}x$$

Dividing by $2 \sin \frac{1}{2}x$, which is greater than zero on the interval $(0, 2\pi)$, we obtain (6.7). From (6.7), we have

$$\left| \sum_{n=1}^{N} \cos nx \right| = \frac{|\sin(N + \frac{1}{2})x - \sin \frac{1}{2}x|}{2 \sin \frac{1}{2}x} \leq \frac{|\sin(N + \frac{1}{2})x| + |\sin \frac{1}{2}x|}{2 \sin \frac{1}{2}x}$$

$$\leq \frac{1}{\sin \frac{1}{2}x}$$

Since $[a, b]$ is a closed interval contained in the open interval $(0, 2\pi)$, it is easily seen (either by calculus or graphically) that the maximum M of $f(x) = 1/\sin \frac{1}{2}x$ on $[a, b]$ is either $1/\sin \frac{1}{2}a$ or $1/\sin \frac{1}{2}b$. Therefore, $|\sum_{n=1}^{N} \cos nx| \leq M$ for every x in the interval $[a, b]$. Thus, the partial sums of $\sum_{n=1}^{\infty} \cos nx$ are uniformly bounded on the interval $[a, b]$. Letting $a_n = 1/n$, $f_n(x) = \cos nx$, we apply Abel's test to conclude that $\sum_{n=1}^{\infty} n^{-1} \cos nx$ converges uniformly on the closed interval $[a, b]$ contained in the open interval $(0, 2\pi)$. By periodicity, the series converges uniformly on any closed interval which does not include an integral multiple of 2π.

It can be shown that for x not equal to an integral multiple of 2π

$$\sum_{n=1}^{\infty} \frac{\cos nx}{n} = -\ln |2 \sin \tfrac{1}{2}x|$$

See Exercise (9.14), Chapter 4.

Exercises

(6.8)   Show that

$$\sum_{k=0}^{\infty} \frac{(-1)^{k+1}}{(2k + 1)^3} = -\frac{\pi^3}{32}$$

Compare this result with the approximate result obtained in the text. [*Hint:* Consider the sine series for $f(x) = [x^3 - 3\pi x^2 + 2\pi^2 x]/12$ over $(0, \pi)$.]

⋆(6.9)   Prove that the series $\sum_{n=1}^{\infty} \sin nx/n$ converges uniformly on any closed interval $[a, b]$ contained in the open interval $(0, 2\pi)$. Then prove that $\pi/2 - \sum_{n=1}^{\infty} n^{-1} \sin nx$ converges uniformly to $x/2$ on any closed interval $[a, b]$ contained in $(0, 2\pi)$.

(6.10)   Prove that the series

$$2 \sum_{k=0}^{\infty} \frac{\sin(2k + 1)x}{2k + 1}$$

converges uniformly on any closed interval which does not contain an integral multiple of π.

(6.11)   Interpret Figures 1.9 and 1.11 in the light of Exercises (6.10) and (6.9). Draw diagrams to illustrate your discussion.

(6.12)   Prove the following theorem.

Theorem: Alternating Series Test.   If the sequence of positive numbers $\{a_n\}$ decreases monotonically to zero, then the numerical series $\sum_{n=1}^{\infty} (-1)^n a_n$ converges and

$$\left| \sum_{n=1}^{\infty} (-1)^n a_n - \sum_{n=1}^{N} (-1)^n a_n \right| \le a_{N+1}$$

(6.13)   Show that the series in (6.1) does *not* converge *absolutely* at any rational multiple of π.

(6.14)   Prove that the series $\sum_{n=2}^{\infty} \sin nx/\log n$ converges uniformly on any closed interval that does not include an integral multiple of 2π.

⋆(6.15)   Use the Maclaurin expansion of $\sin y$ to show that

$$\frac{2}{\pi} \int_0^{\pi} \frac{\sin y}{y} \, dy = 1.1789 \cdots$$

to four decimal places.

(6.16)   Assume that the partial sums of the trigonometric series $\sum_{n=1}^{\infty} \sin nx/n$ are uniformly bounded on $[0, 2\pi]$. Prove that

$$\sum_{n=1}^{\infty} \frac{\sin nx}{n \log(n + 1)}$$

converges uniformly on $[0, 2\pi]$; hence, uniformly on the whole real line. [In Exercise (7.26), Chapter 2, we will prove the assumption about $\sum \sin nx/n$.]

(6.17)   Prove the following theorem.

Theorem: Abel's Theorem.   Suppose that the numerical series $\sum_{n=0}^{\infty} a_n$ converges, then $\sum_{n=0}^{\infty} a_n x^n$ converges uniformly for all x values in the closed interval $[0, 1]$. Hence,

$$\lim_{x \to 1} \sum_{n=0}^{\infty} a_n x^n = \sum_{n=0}^{\infty} a_n$$

(6.18)   Prove that the series $\sum_{k=1}^{\infty} (-1)^{k+1} (\sin kx)/k$ converges uniformly to $\frac{1}{2}x$ over any closed interval $[a, b]$ contained in the open interval $(-\pi, \pi)$.

§7. Complex Form of Fourier Series

Fourier series are often expressed in terms of *complex exponentials* instead of sines and cosines. The basis for this approach is *Euler's formula*

(7.1) $e^{i\phi} = \cos \phi + i \sin \phi$ (ϕ real)

[see Exercise (7.13)]. From (7.1) we obtain by simple algebra

(7.2) $\cos \phi = \frac{1}{2}e^{i\phi} + \frac{1}{2}e^{-i\phi}, \qquad \sin \phi = \frac{i}{2}e^{-i\phi} - \frac{i}{2}e^{i\phi}$

Suppose that f is a (piecewise continuous) function with period 2π. Using (7.2), with ϕ replaced by nx, we rewrite the Nth partial sum S_N of the Fourier series for f as follows

(7.3)
$$S_N(x) = \frac{1}{2}A_0 + \sum_{n=1}^{N} (A_n \cos nx + B_n \sin nx)$$

$$= \frac{1}{2}A_0 + \sum_{n=1}^{N} \left(\frac{A_n - iB_n}{2} e^{inx} + \frac{A_n + iB_N}{2} e^{-inx} \right)$$

Therefore, if we define the constants c_n by

(7.4) $c_0 = \frac{1}{2}A_0 \qquad c_n = \frac{A_n - iB_n}{2} \qquad c_{-n} = \frac{A_n + iB_n}{2}$

then (7.3) becomes

(7.5) $S_N(x) = c_0 + \sum_{n=1}^{N} [c_n e^{inx} + c_{-n} e^{-inx}] = \sum_{n=-N}^{N} c_n e^{inx}$

Formula (7.5) expresses the partial sums of the Fourier series for f as *symmetric* (from $n = -N$ to $n = +N$) sums of complex exponentials.

Since (7.5) holds for each N we can write

(7.6) $f \sim \frac{1}{2}A_0 + \sum_{n=1}^{\infty} [A_n \cos nx + B_n \sin nx] = \sum_{n=-\infty}^{+\infty} c_n e^{inx}$

Using the definitions of A_n and B_n, formula (7.4) tells us that for each positive integer n

$$c_n = \frac{1}{2\pi} \int_{-\pi}^{\pi} f(x) \cos nx \, dx - \frac{i}{2\pi} \int_{-\pi}^{\pi} f(x) \sin nx \, dx$$

$$= \frac{1}{2\pi} \int_{-\pi}^{\pi} f(x)[\cos nx - i \sin nx] \, dx$$

$$= \frac{1}{2\pi} \int_{-\pi}^{\pi} f(x) e^{-inx} \, dx$$

A similar argument for negative n allows us to write

$$c_n = \frac{1}{2\pi} \int_{-\pi}^{\pi} f(x) e^{-inx} \, dx \qquad (n = 0, \pm 1, \pm 2, \cdots)$$

Thus we have established the following *complex form* of the Fourier series for f

(7.7) $$f \sim \sum_{n=-\infty}^{+\infty} c_n e^{inx} \qquad c_n = \frac{1}{2\pi} \int_{-\pi}^{\pi} f(x) e^{-inx} \, dx$$

Also we have shown that the partial sums of the complex form of the Fourier series for f should be viewed as symmetric sums, as shown in (7.5).

(7.8) Remark. For the real valued functions that we have been using up to now, it is useful to keep in mind that c_n and c_{-n} are *complex conjugates*. This is clear from formula (7.4).

All of our discussion above generalizes to functions of period $2a$. In particular, the complex Fourier series for such a function f has the form

(7.9) $$f \sim \sum_{n=-\infty}^{+\infty} c_n e^{i(n\pi x/a)} \qquad c_n = \frac{1}{2a} \int_{-a}^{a} f(x) e^{-i(n\pi x/a)} \, dx$$

We could also have derived (7.9) from the following theorem, in a manner similar to the one used in §1 to derive a real form of Fourier series.

(7.10) Theorem: Orthogonality of Complex Exponentials. The complex exponentials $\{e^{i(n\pi x/a)}\}$ satisfy the following *orthogonality relations*

$$\int_{-a}^{a} e^{i(m\pi x/a)} e^{-i(n\pi x/a)} \, dx = \begin{cases} 0 & \text{for } m \neq n \\ 2a & \text{for } m = n \end{cases}$$

where m and n are integers.

Proof. From (7.1) we obtain

$$e^{i(m\pi x/a)} e^{-i(n\pi x/a)} = \cos \frac{(m-n)\pi x}{a} + i \sin \frac{(m-n)\pi x}{a}$$

by using the sine and cosine addition formulas. We then have

$$\int_{-a}^{a} e^{i(m\pi x/a)} e^{-i(n\pi x/a)} \, dx = \int_{-a}^{a} \cos \frac{(m-n)\pi x}{a} \, dx + i \int_{-a}^{a} \sin \frac{(m-n)\pi x}{a} \, dx$$

$$= \begin{cases} 0 & \text{for } m \neq n \\ 2a & \text{for } m = n \end{cases} \qquad \blacksquare$$

Complex exponentials are our first examples of *complex valued functions*. In general, we say a function f is *complex valued* if $f(x) = f_R(x) + if_I(x)$ where f_R and f_I are real valued functions, called the real and imaginary parts of f, respectively. A complex valued function is called *piecewise continuous* (*continuous, piecewise smooth*) if its real and imaginary parts are both piecewise continuous (continuous, piecewise smooth). If f is a complex valued function on $[a, b]$, then we write

$$\int_{a}^{b} f(x) \, dx = \int_{a}^{b} f_R(x) \, dx + i \int_{a}^{b} f_I(x) \, dx$$

The *complex conjugate* of f is denoted by $\bar{f}$ and equals $f_R - if_I$. We also write $|f|$ to denote the function $(f\bar{f})^{1/2} = [f_R^2 + f_I^2]^{1/2}$.

Exercises

(7.11) Find the Fourier series in complex form for each of the following functions.

(a) $f(x) = |x|$ on $(-\pi, \pi)$
(b) $f(x) = x$ on $(-a, a)$
(c) $f(x) = x - ix^2$ on $(-\pi, \pi)$
(d) $f(x) = 3x^2 + ix^3$ on $(-\frac{1}{2}, \frac{1}{2})$.

$\star$(7.12) Prove that for a complex valued function f on $(-a, a)$, the Fourier series for f is the sum of the Fourier series for the real part of f plus i times the Fourier series for the imaginary part of f.

(7.13) Derive (7.1) by substituting $i\phi$ in place of x in the Maclaurin expansion of e^x [see (4.2b)] and separating into real and imaginary parts.

$\star$(7.14) Suppose that f is a piecewise continuous complex valued function on (a, b). Prove that $|\int_a^b f(x) \, dx| \leq \int_a^b |f(x)| \, dx$.
Hint: Consider the real part of $\int_a^b \bar{c} f(x) \, dx$ where c is the complex constant $\int_a^b f(s) \, ds / |\int_a^b f(s) \, ds|$.

(7.15) Show that Weierstrass' M-test and Theorems (5.4) and (5.5) remain valid if series of continuous *complex valued* functions are used.

References

Some general references on the theory of Fourier series include Bary (1964), Churchill and Brown (1978), Davis (1963), Tolstov (1976), and Zygmund (1968).

The first and last references are very complete and advanced treatises. For more on uniform convergence, see Titchmarsh (1939), Chapter I. For some history of Fourier series, see the excellent article on Fourier analysis in Davis and Hersh (1982). Other historical material can be found in the historical introduction to Carslaw (1950) and in Eves (1983), Kline (1972), Kramer (1981), Langer (1947), Mackey (1980), and Struik (1967).

2
Convergence of Fourier Series

In this chapter we shall consider a few approaches to the problem of convergence of Fourier series. In the first two sections we treat some important preliminaries: Bessel's inequality and Dirichlet's integral form for the partial sums of a Fourier series. Pointwise convergence will be discussed in §3, where we will our prove our main theorem on convergence from Chapter 1. In §4, we discuss the inequalities of Schwarz and Cauchy, and prove a theorem on uniform convergence of Fourier series. A fascinating defect of pointwise convergence, Gibb's phenomenon, is treated in §5. Two methods of overcoming the defects of pointwise convergence are examined in §§6 and 7, *mean square convergence* and summation of Fourier series by *arithmetic means*. Summation by methods other than arithmetic means will be treated in §8; *that section is very important for the remainder of the book.*

As we showed in the last section of Chapter 1, Fourier series can be expressed in either real or complex form. In this chapter we will mainly use the real form and treat the complex forms of our results in occasional exercises.

§1. Bessel's Inequality

The inequality called Bessel's inequality can be deduced by considering the trigonometric systems as orthogonal sets of functions.

(1.1) Definition. An *orthogonal system* (*or set*) *of functions* over the closed interval [a, b] is a set of real valued functions $\{g_n\}_{n=0}^{\infty}$ such that

$$\int_a^b g_m(x)g_n(x)\, dx = 0 \quad \text{if } m \neq n \qquad \int_a^b g_n^2(x)\, dx > 0 \qquad \text{for each } n$$

If $\int_a^b g_n^2(x)\, dx = 1$ for each n, then $\{g_n\}_{n=0}^{\infty}$ is called an *orthonormal* system of functions over [a, b]. ∎

The trigonometric system, period 2π, is an orthogonal system over $[-\pi, \pi]$ because of Theorem (1.5), Chapter 1. It is also clear that

$$\frac{1}{\sqrt{2\pi}}, \frac{1}{\sqrt{\pi}}\sin x, \frac{1}{\sqrt{\pi}}\cos x, \ldots, \frac{1}{\sqrt{\pi}}\sin nx, \frac{1}{\sqrt{\pi}}\cos nx, \ldots$$

is an orthonormal set over $[-\pi, \pi]$; we shall refer to it as the *orthonormal trigonometric system, period* 2π. Another orthonormal set of functions is the set $\{\sqrt{\frac{2}{\pi}}\sin nx\}_{n=1}^{\infty}$ considered over the interval $[0, \pi]$. Several other examples of orthogonal and orthonormal sets of functions will be given throughout the text, as well as in the exercises for this section.

(1.2) Definition.   If $\{g_n\}$ is an orthogonal set of functions over $[a, b]$, then the nth (generalized) *Fourier coefficient* of a piecewise continuous function f relative to this set is

$$c_n = \int_a^b f(x)g_n(x)\,dx \Big/ \int_a^b g_n^2(x)\,dx$$

The (generalized) *Fourier series* for f is defined as the formal series on the right side of the following correspondence

$$f \sim \sum_{n=0}^{\infty} c_n g_n(x) \qquad \blacksquare$$

The reader might note that for the trigonometric system, period 2π, Definition (1.2) yields the same Fourier series that we defined in §1 of Chapter 1.

If $\{g_n\}$ is an orthonormal set, then the Fourier coefficients for a function f take the simple form

$$c_n = \int_a^b f(x)g_n(x)\,dx$$

For example, for the orthonormal trigonometric system, period 2π, we have

$$c_0 = \frac{1}{\sqrt{2\pi}}\int_{-\pi}^{\pi} f(x)\,dx$$

$$c_1 = \frac{1}{\sqrt{\pi}}\int_{-\pi}^{\pi} f(x)\sin x\,dx \qquad c_2 = \frac{1}{\sqrt{\pi}}\int_{-\pi}^{\pi} f(x)\cos x\,dx$$

$$c_{2k-1} = \frac{1}{\sqrt{\pi}}\int_{-\pi}^{\pi} f(x)\sin kx\,dx \qquad c_{2k} = \frac{1}{\sqrt{\pi}}\int_{-\pi}^{\pi} f(x)\cos kx\,dx \qquad (k=2, 3, \ldots)$$

which upon formation of the Fourier series for f yields *the same Fourier series that we previously defined*

$$f \sim \left[\frac{1}{\sqrt{2\pi}} \int_{-\pi}^{\pi} f(x)\,dx \right] \frac{1}{\sqrt{2\pi}} + \sum_{n=1}^{\infty} \left\{ \left[\frac{1}{\sqrt{\pi}} \int_{-\pi}^{\pi} f(x) \sin nx\,dx \right] \frac{1}{\sqrt{\pi}} \sin nx \right.$$

$$\left. + \left[\frac{1}{\sqrt{\pi}} \int_{-\pi}^{\pi} f(x) \cos nx\,dx \right] \frac{1}{\sqrt{\pi}} \cos nx \right\}$$

$$= \frac{1}{2} \left[\frac{1}{\pi} \int_{-\pi}^{\pi} f(x)\,dx \right] + \sum_{n=1}^{\infty} \left\{ \left[\frac{1}{\pi} \int_{-\pi}^{\pi} f(x) \sin nx\,dx \right] \sin nx \right.$$

$$\left. + \left[\frac{1}{\pi} \int_{-\pi}^{\pi} f(x) \cos nx\,dx \right] \cos nx \right\}$$

$$= \tfrac{1}{2} A_0 + \sum_{n=1}^{\infty} (A_n \cos nx + B_n \sin nx)$$

Bessel's inequality is a consequence of the following lemma which we shall also apply in §6.

(1.3) Lemma. Let $\{g_n\}$ be an orthonormal set of functions over $[a, b]$. For each set of numbers $a_0, a_1, \ldots, a_N$ the following identity holds:

$$\int_a^b \left[f(x) - \sum_{n=0}^{N} a_n g_n(x) \right]^2 dx = \int_a^b f^2(x)\,dx - \sum_{n=0}^{N} c_n^2 + \sum_{n=0}^{N} (c_n - a_n)^2$$

where $c_0, c_1, \ldots, c_N$ are the first $N+1$ Fourier coefficients of f relative to $\{g_n\}$.

Proof. Let the function s be defined by $s(x) = \sum_{n=0}^{N} a_n g_n(x)$. Then

(1.4)
$$\int_a^b \left[f(x) - \sum_{n=0}^{N} a_n g_n(x) \right]^2 dx = \int_a^b [f(x) - s(x)]^2 dx$$

$$= \int_a^b f^2(x)\,dx - 2 \int_a^b f(x) s(x)\,dx + \int_a^b s^2(x)\,dx$$

Using the definitions of $s(x)$ and c_n, we have

$$\int_a^b f(x) s(x)\,dx = \int_a^b f(x) \sum_{n=0}^{N} a_n g_n(x)\,dx = \sum_{n=0}^{N} a_n \int_a^b f(x) g_n(x)\,dx$$

$$= \sum_{n=0}^{N} a_n c_n$$

Expanding the last term of (1.4) yields

$$\int_a^b s^2(x)\,dx = \int_a^b \left[\sum_{n=0}^{N} a_n g_n(x) \right] \left[\sum_{k=0}^{N} a_k g_k(x) \right] dx$$

$$= \sum_{n=0}^{N} \sum_{k=0}^{N} \left[a_n a_k \int_a^b g_n(x) g_k(x)\,dx \right]$$

$$= \sum_{n=0}^{N} a_n^2$$

where we have used the orthonormality of $\{g_n\}$ to obtain the last equality above. Substituting the results of the last two calculations into the right side of (1.4), plus some algebra, yields

$$\int_a^b \left[f(x) - \sum_{n=0}^N a_n g_n(x) \right]^2 dx = \int_a^b f^2(x)\, dx - 2 \sum_{n=0}^N a_n c_n + \sum_{n=0}^N a_n^2$$

$$= \int_a^b f^2(x)\, dx - \sum_{n=0}^N c_n^2 + \sum_{n=0}^N (c_n - a_n)^2 \quad \blacksquare$$

(1.5) Corollary: Bessel's Inequality. Given an orthonormal set $\{g_n\}$ over $[a, b]$, the following inequality holds:

$$\sum_{n=0}^\infty c_n^2 \le \int_a^b f^2(x)\, dx$$

Proof. Setting each a_n equal to c_n in Lemma (1.3) yields

$$\int_a^b \left[f(x) - \sum_{n=0}^N c_n g_n(x) \right]^2 dx = \int_a^b f^2(x)\, dx - \sum_{n=0}^N c_n^2$$

Since the integrand on the left side of the equality above is nonnegative, we obtain

$$0 \le \int_a^b f^2(x)\, dx - \sum_{n=0}^N c_n^2$$

Thus, for each positive integer N

$$\sum_{n=0}^N c_n^2 \le \int_a^b f^2(x)\, dx$$

Letting N tend to infinity on the left side of the inequality above we obtain Bessel's inequality. $\blacksquare$

In §6 we will see that Bessel's inequality is actually an equality for many orthonormal systems. If we apply Bessel's inequality to the orthonormal trigonometric system, period 2π, we get

$$\sum_{n=0}^\infty c_n^2 \le \int_{-\pi}^\pi f^2(x)\, dx$$

where

$$c_0 = \frac{1}{\sqrt{2\pi}} \int_{-\pi}^\pi f(x)\, dx$$

and

$$c_{2k} = \frac{1}{\sqrt{\pi}} \int_{-\pi}^\pi f(x) \cos kx\, dx \qquad c_{2k-1} = \frac{1}{\sqrt{\pi}} \int_{-\pi}^\pi f(x) \sin kx\, dx$$

for $k = 1, 2, 3, \ldots$. This last inequality implies that

(1.6)
$$\tfrac{1}{2}A_0^2 + \sum_{n=1}^{\infty} (A_n^2 + B_n^2) \le \frac{1}{\pi} \int_{-\pi}^{\pi} f^2(x) \, dx$$

where A_n and B_n are the usual Fourier coefficients for f. Since the left side of (1.6) is a convergent infinite series, its general term $A_n^2 + B_n^2$ must tend to zero as n tends to infinity. Therefore, because

$$0 \le A_n^2 \le A_n^2 + B_n^2 \qquad 0 \le B_n^2 \le A_n^2 + B_n^2$$

for each n, we know that both A_n^2 and B_n^2 tend to zero as n tends to infinity. This last result implies that both A_n and B_n tend to zero as n tends to infinity. Thus, we have shown that for each piecewise continuous function f on $(-\pi, \pi)$

(1.7)
$$\lim_{n\to\infty} \frac{1}{\pi} \int_{-\pi}^{\pi} f(x) \sin nx \, dx = 0 \qquad \lim_{n\to\infty} \frac{1}{\pi} \int_{-\pi}^{\pi} f(x) \cos nx \, dx = 0$$

We can apply the first limit in (1.7) to the function $\tfrac{1}{2}f(x) \cos \tfrac{1}{2}x$ and the second limit to the function $\tfrac{1}{2}f(x) \sin \tfrac{1}{2}x$. Then upon adding the results, and applying the sine addition formula, we obtain for each piecewise continuous function on $(-\pi, \pi)$

(1.8)
$$\lim_{n\to\infty} \frac{1}{2\pi} \int_{-\pi}^{\pi} f(x) \sin(n + \tfrac{1}{2})x \, dx = 0$$

Formula (1.8) will be needed for our discussion of pointwise convergence in §3.

Exercises

(1.9)   Verify (1.6).

(1.10)   Prove that for an orthonormal system $\{g_n\}$ over $[a, b]$, and f piecewise continuous on (a, b)

$$\lim_{n\to\infty} \int_a^b f(x)g_n(x) \, dx = 0$$

(1.11)   Prove that $\{\sqrt{(1/\pi)} \sin(n + \tfrac{1}{2})x\}_{n=0}^{\infty}$ is an orthonormal set over $[-\pi, \pi]$. Use this result to give an alternative proof of (1.8).

(1.12)   Show that Bessel's inequality applied to $\{\sqrt{(2/\pi)} \sin nx\}_{n=1}^{\infty}$ yields

$$\sum_{n=1}^{\infty} B_n^2 \le \frac{2}{\pi} \int_0^{\pi} f^2(x) \, dx$$

where

$$B_n = \frac{2}{\pi} \int_0^{\pi} f(x) \sin nx \, dx$$

(1.13)   By substituting $x = at/\pi$, prove that

$$\tfrac{1}{2}A_0^2 + \sum_{n=1}^{\infty} [A_n^2 + B_n^2] \leq \frac{1}{a} \int_{-a}^{a} f^2(x)\, dx$$

where A_n and B_n are the Fourier coefficients for a piecewise continuous function f of period $2a$.

(1.14)   Prove that $\{\sin mx \sin ny\}_{m,n=1}^{\infty}$ is orthogonal over the rectangular region $R = \{(x, y): 0 \leq x \leq \pi, 0 \leq y \leq \pi\}$ in the sense that

$$\iint_R [\sin mx \sin ny][\sin jx \sin ky]\, dx\, dy = \begin{cases} 0 & \text{if } j \neq m \text{ or } k \neq n \\ \tfrac{1}{4}\pi^2 & \text{if } j = m \text{ and } k = n \end{cases}$$

State formulas for Fourier coefficients and generalized Fourier series.

(1.15)   Verify that the functions $P_0(x) = 1$, $P_1(x) = x$, $P_2(x) = \tfrac{1}{2}(3x^2 - 1)$, $P_3(x) = \tfrac{1}{2}(5x^3 - 3x)$ form a finite orthogonal system over $[-1, 1]$. (These functions are the first four *Legendre polynomials*, see Chapter 8.)

(1.16)   Prove that if the piecewise continuous function f has the complex Fourier series $\sum c_n e^{inx}$ then

$$\sum_{n=-\infty}^{+\infty} |c_n|^2 \leq \frac{1}{2\pi} \int_{-\pi}^{\pi} |f(x)|^2\, dx$$

[*Hint*: Use real and imaginary parts.]

§2. Dirichlet's Integral

A very useful tool in Fourier analysis is *Dirichlet's integral form* for the nth partial sum

(2.1) $$S_n(x_0) = \tfrac{1}{2}A_0 + \sum_{m=1}^{n} (A_m \cos mx_0 + B_m \sin mx_0)$$

of the Fourier series for f at $x = x_0$. We will show in this section how this integral form is derived.

If we use t rather than x as the variable of integration, then the Fourier coefficients of the function f are

$$A_m = \frac{1}{\pi} \int_{-\pi}^{\pi} f(t) \cos mt\, dt \qquad B_m = \frac{1}{\pi} \int_{-\pi}^{\pi} f(t) \sin mt\, dt$$

Substituting these expressions into each term in (2.1), and then bringing the constants $\tfrac{1}{2}$, $\cos mx_0$, and $\sin mx_0$ inside the integrals, we obtain

(2.2)
$$\begin{aligned} S_n(x_0) = \frac{1}{\pi} \int_{-\pi}^{\pi} f(t)\tfrac{1}{2}\, dt + \sum_{m=1}^{n} \left[\frac{1}{\pi} \int_{-\pi}^{\pi} f(t) \cos mt \cos mx_0\, dt \right. \\ \left. + \frac{1}{\pi} \int_{-\pi}^{\pi} f(t) \sin mt \sin mx_0\, dt \right] \end{aligned}$$

Writing the sum of integrals in (2.2) as an integral of a sum, and then applying the cosine subtraction formula, we have

$$S_n(x_0) = \frac{1}{\pi} \int_{-\pi}^{\pi} f(t) \left[\frac{1}{2} + \sum_{m=1}^{n} \cos m(t - x_0) \right] dt$$

If we substitute $u = t - x_0$ into the integral above we get

(2.3) $$S_n(x_0) = \frac{1}{\pi} \int_{-\pi - x_0}^{\pi - x_0} f(x_0 + u) \left[\frac{1}{2} + \sum_{m=1}^{n} \cos mu \right] du$$

The sum $\frac{1}{2} + \sum_{m=1}^{n} \cos mu$ has period 2π since each term has that period.

If we now assume that f has period 2π, then $f(x_0 + u)$ has period 2π as a function of u. Hence, the integrand in formula (2.3) has period 2π. Since the interval of integration $[-\pi - x_0, \pi - x_0]$ has length 2π, we conclude from Exercise (1.17), Chapter 1, that

$$\int_{-\pi - x_0}^{\pi - x_0} f(x_0 + u) \left[\frac{1}{2} + \sum_{m=1}^{n} \cos mu \right] du = \int_{-\pi}^{\pi} f(x_0 + u) \left[\frac{1}{2} + \sum_{m=1}^{n} \cos mu \right] du$$

Thus, formula (2.3) assumes the form

(2.4) $$S_n(x_0) = \frac{1}{\pi} \int_{-\pi}^{\pi} f(x_0 + u) \left[\frac{1}{2} + \sum_{m=1}^{n} \cos mu \right] du$$

for a function f having a period of 2π.

We now show that

(2.5) $$\frac{1}{2} + \sum_{m=1}^{n} \cos mu = \frac{\sin(n + \frac{1}{2})u}{2 \sin(\frac{1}{2}u)}$$

Multiply the left side of (2.5) by $2 \sin(\frac{1}{2}u)$. If we then use the identity $2 \sin(\frac{1}{2}u) \cos mu = \sin(m + \frac{1}{2})u - \sin(m - \frac{1}{2})u$, and regroup terms, we obtain through cancellation

$$2 \sin(\frac{1}{2}u) \left[\frac{1}{2} + \sum_{m=1}^{n} \cos mu \right] = \sin(\frac{1}{2}u) + \sum_{m=1}^{n} [\sin(m + \frac{1}{2})u - \sin(m - \frac{1}{2})u]$$
$$= \sin(n + \frac{1}{2})u$$

Dividing by $2 \sin(\frac{1}{2}u)$ we obtain (2.5) for $0 < |u| \le \pi$. Formula (2.5) holds for $u = 0$ if we interpret its right side as a limit for u tending to 0, the formula then holds for all values of u by periodicity.

Using the right side of (2.5) in (2.3), and factoring out $\frac{1}{2}$, we obtain *Dirichlet's integral form* for $S_n(x_0)$

(2.6) $$S_n(x_0) = \frac{1}{2\pi} \int_{-\pi}^{\pi} f(x_0 + u) \frac{\sin(n + \frac{1}{2})u}{\sin(\frac{1}{2}u)} du$$

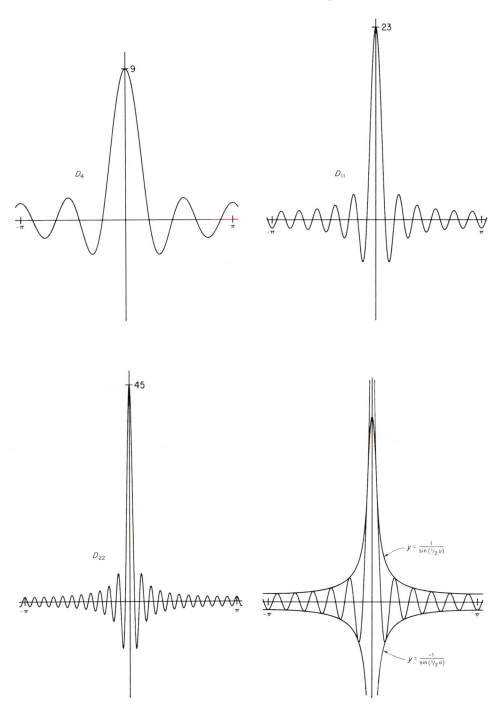

Figure 2.1 Graphs of the Dirichlet kernel. At the left are graphs of $D_n(u)$ for $n = 4$ and 22. On the right are graphs of D_{11} which illustrate how, except for u near zero, each D_n is just a sine function $\sin(n + \frac{1}{2})u$ that oscillates within an envelope given by $\pm 1/[\sin(\frac{1}{2}u)]$.

The function

(2.7) $$D_n(u) = \frac{\sin(n + \frac{1}{2})u}{\sin(\frac{1}{2}u)} \qquad (n = 0, 1, 2, \ldots)$$

is known as *Dirichlet's kernel*;[1] we will often use Dirichlet's kernel to express (2.6) in the form

(2.8) $$S_n(x_0) = \frac{1}{2\pi} \int_{-\pi}^{\pi} f(x_0 + u) D_n(u) \, du$$

In Figure 2.1 the reader will find some graphs of Dirichlet's kernel for a few values of n.

Exercises

*(2.9) Prove that for each n

$$\frac{1}{2\pi} \int_{-\pi}^{\pi} D_n(u) \, du = 1$$

[*Hint*: Multiply Eq. (2.5) by 2 and integrate both sides.]

*(2.10) Prove that

$$\frac{1}{2\pi} \int_{-\pi}^{0} D_n(u) \, du = \frac{1}{2} = \frac{1}{2\pi} \int_{0}^{\pi} D_n(u) \, du$$

[*Hint*: D_n is an even function.]

(2.11) Prove that $S_n(x_0)$ can also be expressed in the two alternative forms

$$S_n(x_0) = \frac{1}{2\pi} \int_{-\pi}^{\pi} f(x_0 - v) D_n(v) \, dv = \frac{1}{2\pi} \int_{-\pi}^{\pi} f(v) D_n(x_0 - v) \, dv$$

(2.12) Discuss why a typical function D_n must have a graph like the one shown in Figure 2.2. Prove that *all* the extreme values (maxima and minima) of D_n have absolute values of at least 1. Also, prove that there are at least $2n + 1$ extreme values for D_n in the interval $[-\pi, \pi]$.

§3. Pointwise Convergence

In this section, we will prove several theorems on convergence of Fourier series. Our purpose is not to obtain the most general, or most difficult, theorems. Rather, we will obtain a few simple yet widely applicable tests for convergence while illustrating the essential ideas involved in all convergence theorems.

---

1. The term kernel is used because D_n is the essential part of (2.8): for *all* functions we obtain $S_n(x_0)$ by computing an *integral average* over $[-\pi, \pi]$ of $f(x_0 + u)$ *times* $D_n(u)$.

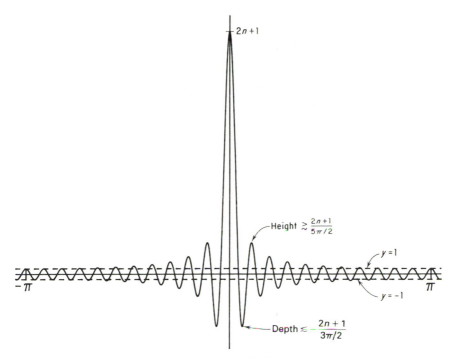

$2n+1$

Height $\gtrsim \dfrac{2n+1}{5\pi/2}$

$y=1$

$y=-1$

$-\pi$

π

Depth $\leq -\dfrac{2n+1}{3\pi/2}$

Figure 2.2   General form of Dirichlet's kernel.

Our first goal is to show that under certain conditions the nth partial sum of the Fourier series for a function f of period 2π evaluated at x_0

(3.1)
$$S_n(x_0) = \frac{1}{2\pi} \int_{-\pi}^{\pi} f(x_0 + u) D_n(u)\, du$$

tends to $f(x_0)$ as n tends to infinity. From Exercise (2.9) we have

$$1 = \frac{1}{2\pi} \int_{-\pi}^{\pi} D_n(u)\, du$$

Multiplying this last equation by $f(x_0)$, and then bringing this constant inside the integral sign, we obtain

(3.2)
$$f(x_0) = \frac{1}{2\pi} \int_{-\pi}^{\pi} f(x_0) D_n(u)\, du$$

Subtracting Eq. (3.2) from Eq. (3.1), and collecting terms under the integral sign, yields

(3.3)
$$S_n(x_0) - f(x_0) = \frac{1}{2\pi} \int_{-\pi}^{\pi} [f(x_0 + u) - f(x_0)] D_n(u)\, du$$

For notational convenience, we define the function ϕ by

(3.4)
$$\phi(u) = \frac{f(x_0 + u) - f(x_0)}{\sin(\tfrac{1}{2}u)}$$

Using this definition of ϕ along with the definition of $D_n(u)$ in Eq. (2.7), we obtain the following form for Eq. (3.3)

(3.5)
$$S_n(x_0) - f(x_0) = \frac{1}{2\pi} \int_{-\pi}^{\pi} \phi(u) \sin(n + \tfrac{1}{2})u \, du$$

If ϕ were piecewise continuous on the interval $(-\pi, \pi)$, then formula (1.8), applied to ϕ rather than f, implies that

$$\lim_{n \to \infty} [S_n(x_0) - f(x_0)] = 0$$

Hence, we would have $\lim_{n \to \infty} S_n(x_0) = f(x_0)$ which is our desired result.

Unfortunately, things are not that simple. For one thing $\phi(0)$ has the indeterminate form $0/0$. However, we do have the following partial result.

(3.6) Lemma. Suppose that f is a piecewise continuous function with period 2π. Then ϕ is piecewise continuous on the intervals $(-\pi, -\delta)$ and on (δ, π) for each number $0 < \delta < \pi$.

Proof. The function ϕ equals $h_1 h_2$ where

(3.7)
$$h_1(u) = \frac{1}{\sin(\tfrac{1}{2}u)} \qquad h_2(u) = f(x_0 + u) - f(x_0)$$

The function h_1 is continuous on the closed intervals $[-\pi, -\delta]$ and $[\delta, \pi]$. Therefore, h_1 satisfies the definition of a piecewise continuous function on $(-\pi, -\delta)$ and on (δ, π). As for h_2, the graph of h_2 is just the graph of f shifted to the left by the amount x_0 and shifted down by the amount $f(x_0)$. Therefore, h_2 is piecewise continuous on $(-\pi, -\delta)$ and on (δ, π). Since ϕ is a product of piecewise continuous functions, we conclude from Theorem (4.9), Chapter 1, that ϕ is piecewise continuous on $(-\pi, -\delta)$ and on (δ, π). ∎

(3.8) Corollary. If $0 < \delta < \pi$, then

$$\lim_{n \to \infty} \frac{1}{2\pi} \int_{-\pi}^{-\delta} \phi(u) \sin(n + \tfrac{1}{2})u \, du = 0 = \lim_{n \to \infty} \frac{1}{2\pi} \int_{\delta}^{\pi} \phi(u) \sin(n + \tfrac{1}{2})u \, du$$

Proof. We define g by

$$g(u) = \begin{cases} \phi(u) & \text{for } -\pi \le u \le -\delta \\ 0 & -\delta < u \le \pi \end{cases}$$

Then g is piecewise continuous on $(-\pi, \pi)$ because ϕ is piecewise continuous on $(-\pi, -\delta)$. Applying (1.8) to g instead of f we obtain the first limit above, since the integral over $(-\pi, \pi)$ reduces to one over $(-\pi, -\delta)$. A similar argument demonstrates the second limit above. ∎

The integral in (3.5) can be split into a sum of three integrals

(3.9)

$$S_n(x_0) - f(x_0) = \frac{1}{2\pi} \int_{-\pi}^{-\delta} \phi(u) \sin(n + \tfrac{1}{2})u \, du$$

$$+ \frac{1}{2\pi} \int_{-\delta}^{\delta} \phi(u) \sin(n + \tfrac{1}{2})u \, du$$

$$+ \frac{1}{2\pi} \int_{\delta}^{\pi} \phi(u) \sin(n + \tfrac{1}{2})u \, du$$

By considering Corollary (3.8) together with formula (3.9) we can see that

$$\lim_{n \to \infty} S_n(x_0) = f(x_0)$$

is reduced to examining the middle integral on the right side of (3.9). The essential aspect of all pointwise convergence proofs is the necessity of stating conditions on $f(x)$ for x at or near x_0 so that the middle integral in (3.9) can be ignored as negligibly small in magnitude. The simplest of these conditions is that f be differentiable at x_0.

(3.10) Theorem. Suppose that f is a piecewise continuous function with period 2π. If f is differentiable at x_0, then the Fourier series for f converges to $f(x_0)$ at x_0.

Proof. Let $\varepsilon > 0$ be some small number. For $0 < |u| \leq \pi$, we can write

$$\phi(u) = \frac{u}{\sin(\tfrac{1}{2}u)} \cdot \frac{f(x_0 + u) - f(x_0)}{u}$$

We then have

$$\lim_{u \to 0} \phi(u) = \lim_{u \to 0} \frac{u}{\sin(\tfrac{1}{2}u)} \lim_{u \to 0} \frac{f(x_0 + u) - f(x_0)}{u}$$

$$= 2f'(x_0)$$

Hence, for $\delta > 0$ sufficiently small, we certainly have for all $0 < |u| < \delta$

$$|\phi(u) - 2f'(x_0)| < 1$$

Therefore, using the triangle inequality for absolute values

$$|\phi(u)| = |[\phi(u) - 2f'(x_0)] + 2f'(x_0)|$$

$$\leq |\phi(u) - 2f'(x_0)| + |2f'(x_0)|$$

$$< 1 + 2|f'(x_0)|$$

Letting M stand for the quantity $1 + 2|f'(x_0)|$ we have for each $n \geq 0$

$$\left| \frac{1}{2\pi} \int_{-\delta}^{\delta} \phi(u) \sin(n + \tfrac{1}{2})u \, du \right| \leq \frac{1}{2\pi} \int_{-\delta}^{\delta} |\phi(u) \sin(n + \tfrac{1}{2})u| \, du$$

$$\leq \frac{1}{2\pi} \int_{-\delta}^{\delta} |\phi(u)| \, du$$

$$< \frac{\delta}{\pi} M$$

If we have chosen δ so small that $(\delta/\pi)M < \epsilon$, then for all $n \geq 0$

$$\left| \frac{1}{2\pi} \int_{-\delta}^{\delta} \phi(u) \sin(n + \tfrac{1}{2})u \, du \right| < \epsilon$$

Recalling the rigorous definition of limit, Corollary (3.8) tells us that for some N sufficiently large, if $n \geq N$

$$\left| \frac{1}{2\pi} \int_{-\pi}^{-\delta} \phi(u) \sin(n + \tfrac{1}{2})u \, du \right| < \epsilon \qquad \left| \frac{1}{2\pi} \int_{\delta}^{\pi} \phi(u) \sin(n + \tfrac{1}{2})u \, du \right| < \epsilon$$

Using the triangle inequality and these last three inequalities, we obtain from formula (3.9) that for all $n \geq N$

$$|S_n(x_0) - f(x_0)| < \epsilon + \epsilon + \epsilon = 3\epsilon$$

Since ϵ is arbitrarily small in magnitude, we have proved that

$$\lim_{n \to \infty} S_n(x_0) = f(x_0) \qquad \blacksquare$$

(3.11) Remark.   The reader may have observed that in the proof above the crucial point was that for $\delta > 0$ sufficiently small

$$|\phi(u)| < M \qquad (0 < |u| < \delta)$$

In words, $\phi(u)$ *is bounded in magnitude for all u near enough to zero.* We used the assumption that f is differentiable at x_0 only in order to show that $\phi(u)$ was bounded in magnitude for all u near enough to zero.

A much less restrictive assumption than differentiability is that there exists a positive constant C such that for all u near enough to zero (i.e., $0 < |u| < \delta$ for some $\delta > 0$)

(3.12) $$|f(x_0 + u) - f(x_0)| < C |u|$$

Condition (3.12) is called a *Lipschitz condition, order one,* for f at x_0. The proof of the following Theorem is left as an exercise for the reader.

(3.13) Theorem.   If f satisfies a Lipschitz condition, order one, at x_0, then the Fourier series for f converges to $f(x_0)$ at x_0.

Theorem (3.10), while not as general as Theorem (3.13), is nevertheless a very useful test for convergence of Fourier series. A couple of examples should suffice as illustrations of the usefulness of Theorem (3.10).

(3.14) Examples

(a) Let $f(x) = x$ on the interval $(-\pi, \pi)$. The periodic extension of f is differentiable for all x values which are not odd integral multiples of π. Therefore, the Fourier series for f converges to the periodic extension of f for all such x-values. In particular, for $-\pi < x < \pi$.

$$x = \sum_{n=1}^{\infty} \frac{2(-1)^{n+1}}{n} \sin nx$$

(b) Suppose that f is a piecewise smooth function with period 2π. For each point x_0 where f is differentiable, the Fourier series for f converges to $f(x_0)$ at x_0. Thus, we can write

$$f(x) = \tfrac{1}{2}A_0 + \sum_{n=1}^{\infty} [A_n \cos nx + B_n \sin nx]$$

and the equality will hold for all but a finite number of x values in every finite interval.

The equality displayed in (b) above will actually hold for all x values where f is continuous. For, even if $f'(x_0)$ does not exist, f' is piecewise continuous so the one-sided limits

$$f'(x_0 -) = \lim_{x \to x_0-} f'(x) \qquad f'(x_0 +) = \lim_{x \to x_0+} f'(x)$$

both exist. Therefore, by l'Hôpital's rule

$$\lim_{u \to 0+} \phi(u) = \lim_{u \to 0+} \frac{f(x_0 + u) - f(x_0)}{\sin(\tfrac{1}{2}u)} = \lim_{u \to 0+} \frac{f'(x_0 + u)}{\tfrac{1}{2}\cos(\tfrac{1}{2}u)} = 2f'(x_0 +)$$

$$\lim_{u \to 0-} \phi(u) = 2f'(x_0 -)$$

Since $\phi(u)$ tends to finite limits as u approaches zero from either side, it follows that $\phi(u)$ is bounded in magnitude for all u near enough to zero. Therefore, as we noted in Remark (3.11), the Fourier series for f converges to $f(x_0)$ at x_0.

For example, if $f(x) = |x|$ on $(-\pi, \pi)$, then the Fourier series for f will converge to the periodic extension of f at all x values, even though at integral multiples of π the periodic extension of f is not differentiable.

We have now proved Theorem (4.5), Chapter 1, for the case of all points of continuity of a piecewise smooth function of period 2π. If the piecewise smooth function f has period $2a$, then we make the substitution $x = (a/\pi)t$. Thus, we define the function g by

(3.15)
$$g(t) = f\left(\frac{a}{\pi}t\right)$$

It may be observed that g is piecewise smooth with period 2π. Moreover, since $x = (a/\pi)t$ is just a change of scale, the points of continuity for $f(x)$ become points of continuity for $g(t) = f[(a/\pi) \cdot t]$ and vice versa. Also, the values of one-sided limits are unchanged: $g(t-) = f(x-)$ and $g(t+) = f(x+)$. Therefore, we know that for each point of continuity t of g

$$g(t) = \tfrac{1}{2}A_0 + \sum_{n=1}^{\infty} [A_n \cos nt + B_n \sin nt]$$

where

$$A_n = \frac{1}{\pi} \int_{-\pi}^{\pi} g(t) \cos nt \, dt \qquad B_n = \frac{1}{\pi} \int_{-\pi}^{\pi} g(t) \sin nt \, dt$$

Substituting $t = (\pi/a)x$ into the series and the integrals above yields

$$f(x) = \tfrac{1}{2}A_0 + \sum_{n=1}^{\infty} \left[A_n \cos \frac{n\pi x}{a} + B_n \sin \frac{n\pi x}{a} \right]$$

where

$$A_n = \frac{1}{a} \int_{-a}^{a} f(x) \cos \frac{n\pi x}{a} \, dx \qquad B_n = \frac{1}{a} \int_{-a}^{a} f(x) \sin \frac{n\pi x}{a} \, dx$$

This establishes the desired result for all points of continuity of f. Thus, Theorem (4.5), Chapter 1, is completely proved for points of continuity of piecewise smooth periodic functions.

(3.16) Remark. In the sequel we shall always prove theorems about Fourier series only for the case of functions with period 2π. This is because a substitution argument, such as the one we used above, can always be used to prove the theorems for functions with period $2a$.

The only remaining point, as far as piecewise smooth periodic functions are concerned, is the convergence of Fourier series at points of discontinuity. This case is very similar to that of points of continuity, so we will just outline the arguments leaving some of the details to the reader as an exercise.

Suppose that f is piecewise smooth, period 2π, and x_0 is a point of discontinuity for f. In this case, we will prove that

(3.17) $$\lim_{n \to \infty} S_n(x_0) = \tfrac{1}{2}[f(x_0+) + f(x_0-)]$$

Consider the expression for $S_n(x_0)$ given in (3.1). If we split the integral over $[-\pi, \pi]$ into a sum of two integrals we get

$$S_n(x_0) = S_n^-(x_0) + S_n^+(x_0)$$

(3.18) $$S_n^-(x_0) = \frac{1}{2\pi} \int_{-\pi}^{0} f(x_0 + u)D_n(u) \, du$$

$$S_n^+(x_0) = \frac{1}{2\pi} \int_{0}^{\pi} f(x_0 + u)D_n(u) \, du$$

First, we will show that the quantity $S_n^-(x_0)$ tends to $\frac{1}{2}f(x_0-)$ as n tends to infinity. Then a similar argument can be used to show that $S_n^+(x_0)$ tends to $\frac{1}{2}f(x_0+)$ as n tends to infinity.

From Exercise (2.10), we have that

$$\tfrac{1}{2} = \frac{1}{2\pi} \int_{-\pi}^{0} D_n(u)\,du$$

Multiplying this equality by the constant $f(x_0-)$ we obtain

$$\tfrac{1}{2}f(x_0-) = \frac{1}{2\pi} \int_{-\pi}^{0} f(x_0-)D_n(u)\,du$$

Subtracting the equation above from the equation defining $S_n^-(x_0)$ in formula (3.18) yields, after some algebra

(3.19) $$S_n^-(x_0) - \tfrac{1}{2}f(x_0-) = \frac{1}{2\pi} \int_{-\pi}^{0} \phi_1(u) \sin(n+\tfrac{1}{2})u\,du$$

where

$$\phi_1(u) = \frac{f(x_0+u)-f(x_0-)}{\sin(\tfrac{1}{2}u)}$$

l'Hôpital's rule implies that

$$\lim_{u \to 0-} \phi_1(u) = \lim_{u \to 0-} \frac{f'(x_0+u)}{\tfrac{1}{2}\cos(\tfrac{1}{2}u)} = 2f'(x_0-)$$

It follows that $\phi_1(u)$ is bounded in magnitude for all u near enough to zero. Hence, using the same type of arguments as in the proof of Theorem (3.10) and Remark (3.11), we conclude that

(3.20) $$\lim_{n \to \infty} S_n^-(x_0) = \tfrac{1}{2}f(x_0-)$$

Similarly, it can be shown that

(3.21) $$\lim_{n \to \infty} S_n^+(x_0) = \tfrac{1}{2}f(x_0+)$$

Combining (3.20) and (3.21) with (3.18) yields

$$\lim_{n \to \infty} S_n(x_0) = \lim_{n \to \infty} [S_n^-(x_0) + S_n^+(x_0)] = \tfrac{1}{2}f(x_0-) + \tfrac{1}{2}f(x_0+)$$

Thus, Theorem (4.5), Chapter 1, is completely proved.

Exercises

*(3.22) Suppose that f is piecewise smooth on the interval $(0, a)$. Prove that both the cosine and sine series for f converge to $f(x_0)$ if x_0 is a point of continuity for f, or if x_0 is a point of discontinuity for f the two series converge to $\frac{1}{2}[f(x_0+)+f(x_0-)]$ at x_0.

(3.23) Prove (3.13) and complete the proofs of (3.20) and (3.21).

(3.24) We say that f has a *right-hand derivative* at x_0 if

$$\lim_{x \to x_0+} \frac{f(x)-f(x_0+)}{x-x_0}$$

exists; we say that f has a *left-hand derivative* if

$$\lim_{x \to x_0-} \frac{f(x)-f(x_0-)}{x-x_0}$$

exists. *Prove the following theorem.*

Theorem. Suppose that f is piecewise continuous and has period 2π. If f has left- and right-hand derivatives at x_0, then the Fourier series for f converges to $\frac{1}{2}[f(x_0+)+f(x_0-)]$ at x_0.

(3.25) Explain why the Fourier series for $\sqrt[3]{x}$, over the interval $(-\pi, \pi)$, converges to $\sqrt[3]{x}$ for all x in the interval $(-\pi, \pi)$.
[*Note*: $f(x) = \sqrt[3]{x}$ is *not* piecewise smooth on the interval $(-\pi, \pi)$.]

Our next two exercises treat a fairly broad generalization of Theorem (3.10).

$\star$**(3.26)** *Lipschitz functions.* A function f is *Lipschitz*, order $\alpha > 0$, at x_0 if in some open interval which includes x_0 we have

$$|f(x)-f(x_0)| \le C\,|x-x_0|^\alpha$$

where C is some positive constant.[2]

  (a) Show that $f(x) = x^{1/3}$ is Lipschitz, order $1/3$, at 0.
  (b) Show that $f(x) = x\cos(1/x)$ for $x \ne 0$ and 0 for $x = 0$ is Lipschitz, order 1, at 0.
  (c) Prove that if f is differentiable at x_0, then f is Lipschitz, order 1, at x_0.
  (d) Show that the functions in (a) and (b) are Lipschitz at all x-values (the order may vary for different x values).

$\star$**(3.27)** In this exercise, we will outline the proof of the following theorem.

Theorem. Suppose that f is piecewise continuous on $(-\pi, \pi)$, and has period 2π. If f is Lipschitz, order α, at x_0 then the Fourier series for f converges to $f(x_0)$ at x_0.

  (a) Show that if $0 \le |v| \le \frac{1}{2}\pi$, then $\frac{2}{\pi}\,|v| \le |\sin v|$.

---

2. Some authors say that f is *Hölder with exponent* α.

(b)   Using (a) and the Lipschitz condition, order α, at x_0, prove that for $\delta > 0$ sufficiently small the function ϕ defined in (3.4) satisfies

$$|\phi(u)| \leq C\pi \, |u|^{-1+\alpha}$$

for some positive constant C and $0 \leq |u| \leq \delta$.

(c)   Using (b), prove that given $\epsilon > 0$ we can find a $\delta > 0$ so small that for all $n \geq 0$

$$\left| \frac{1}{2\pi} \int_{-\delta}^{\delta} \phi(u) \sin(n + \tfrac{1}{2})u \, du \right| < \epsilon$$

(d)   Using (c) and Corollary (3.8), prove the theorem stated above.

⋆(3.28)   Prove that if f is a piecewise smooth *complex valued* function of period 2π, then the complex Fourier series, $\sum c_n e^{inx}$, for f converges to $\frac{1}{2}[f(x +) + f(x -)]$ for all x values. Explain why all the theorems in this section remain true for complex Fourier series of complex valued functions.

§4. The Cauchy–Schwarz Inequalities, Uniform Convergence

Before we discuss some deeper properties of convergence, we need to discuss Schwarz's inequality and Cauchy's inequality. These inequalities are very useful tools in Fourier analysis; in this section we shall use Cauchy's inequality to prove a theorem on uniform convergence of Fourier series.

(4.1) Theorem: Schwarz's Inequality.   If f and g are piecewise continuous on the interval (a, b), then

$$\left[\int_a^b f(x)g(x) \, dx \right]^2 \leq \left[\int_a^b f^2(x) \, dx \right]\left[\int_a^b g^2(x) \, dx \right]$$

Or, expressed another way

$$\left| \int_a^b f(x)g(x) \, dx \right| \leq \left[\int_a^b f^2(x) \, dx \right]^{1/2}\left[\int_a^b g^2(x) \, dx \right]^{1/2}$$

Proof.[3]   Let t be a real variable. Due to the nonnegativity of the integrand $[f(x) + tg(x)]^2$ we have

$$0 \leq \int_a^b [f(x) + tg(x)]^2 \, dx$$

Expanding the integral in the last inequality yields (since t is a constant,

---

3. Those readers who have had a course in linear algebra might recognize Schwarz's inequality and Cauchy's inequality [Theorem (4.3)] as particular cases of the Cauchy–Schwarz inequality for inner products. In which case, they might wish to skip the proofs of these inequalities and proceed directly to Theorem (4.4).

relative to x, we may bring it outside the integral signs)

(4.2) $$0 \le \int_a^b f^2(x)\, dx + 2t \int_a^b f(x)g(x)\, dx + t^2 \int_a^b g^2(x)\, dx$$

Letting

$$A = \int_a^b g^2(x)\, dx \qquad B = \int_a^b f(x)g(x)\, dx \qquad C = \int_a^b f^2(x)\, dx$$

formula (4.2) becomes $At^2 + 2Bt + C \ge 0$ for all real numbers t. That is, the quadratic $q(t) = At^2 + 2Bt + C$ never takes on negative values. In particular $q(t)$ *cannot have two real roots*, because if it did it would take on negative values between those roots.

Therefore, the discriminant of $q(t)$ cannot be positive; if it were the quadratic formula would yield two distinct real roots for $q(t)$. Hence, calculating the discriminant of $q(t)$ yields

$$(2B)^2 - 4AC \le 0$$

Dividing this last inequality by 4, bringing AC to the other side, and substituting in the definitions of A, B, and C given above, yields

$$\left[\int_a^b f(x)g(x)\, dx \right]^2 \le \left[\int_a^b f^2(x)\, dx \right] \left[\int_a^b g^2(x)\, dx \right]$$

Taking positive square roots of both sides of the inequality above gives the other form of Schwarz's inequality. ∎

We now turn to Cauchy's inequality, which is closely related to Schwarz's inequality.

(4.3) Theorem: Cauchy's Inequality.   Let $\{a_n\}_{n=1}^N$ and $\{b_n\}_{n=1}^N$ be two finite sets of real numbers. Then

$$\left[\sum_{n=1}^N a_n b_n \right]^2 \le \left[\sum_{n=1}^N a_n^2 \right] \left[\sum_{n=1}^N b_n^2 \right]$$

or, expressed another way

$$\left| \sum_{n=1}^N a_n b_n \right| \le \left[\sum_{n=1}^N a_n^2 \right]^{1/2} \left[\sum_{n=1}^N b_n^2 \right]^{1/2}$$

Proof.   Follow the line of argument of the proof of Schwarz's inequality; using

$$q(t) = \left[\sum_{n=1}^N a_n^2 \right] + 2 \left[\sum_{n=1}^N a_n b_n \right] t + \left[\sum_{n=1}^N b_n^2 \right] t^2 \ge 0$$

which is obtained from $0 \le \sum_{n=1}^N [a_n + tb_n]^2$. The rest of the details are left to the reader. ∎

We will now prove a theorem on uniform convergence of Fourier series by applying Cauchy's inequality and Bessel's inequality.

(4.4) Theorem. Let f be a continuous, piecewise smooth, periodic function. The Fourier series for f converges uniformly to f over the whole real line.

Proof. We will discuss the case of period 2π; the general case of period $2a$ follows by a substitution argument [see Remark (3.16)].

We have for all $n \geq 1$

$$|A_n \cos nx + B_n \sin nx| \leq |A_n| + |B_n|$$

It will now be shown that $\sum_{n=1}^{\infty} [|A_n| + |B_n|]$ converges; this will allow us to apply Weierstrass' M-test to the Fourier series for f.

The derivative of f is piecewise continuous. If we let A_n' and B_n' stand for the Fourier coefficients of f', then

$$A_0' = \frac{1}{\pi} \int_{-\pi}^{\pi} f'(x) \, dx = \frac{1}{\pi} f(x) \bigg|_{-\pi}^{\pi} = 0$$

due to the periodicity of f. Integrating by parts, we have

$$A_n' = \frac{1}{\pi} \int_{-\pi}^{\pi} f'(x) \cos nx \, dx = \frac{1}{\pi} f(x) \cos nx \, |_{-\pi}^{\pi}$$

$$+ \frac{n}{\pi} \int_{-\pi}^{\pi} f(x) \sin nx \, dx$$

$$= \frac{n}{\pi} \int_{-\pi}^{\pi} f(x) \sin nx \, dx = nB_n$$

Thus, $A_n' = nB_n$ and a similar calculation shows that $B_n' = -nA_n$. From Cauchy's inequality and Bessel's inequality *applied to f'*, we deduce that for each m

$$\sum_{n=1}^{m} \left[|A_n| + |B_n| \right] = \sum_{n=1}^{m} \frac{1}{n} |A_n'| + \sum_{n=1}^{m} \frac{1}{n} |B_n'|$$

$$\leq \left[\sum_{n=1}^{m} \frac{1}{n^2} \right]^{1/2} \left[\sum_{n=1}^{m} |A_n'|^2 \right]^{1/2} + \left[\sum_{n=1}^{m} \frac{1}{n^2} \right]^{1/2} \left[\sum_{n=1}^{m} |B_n'|^2 \right]^{1/2}$$

$$\leq \left[\sum_{n=1}^{\infty} \frac{1}{n^2} \right]^{1/2} \left[\left(\sum_{n=1}^{\infty} |A_n'|^2 \right)^{1/2} + \left(\sum_{n=1}^{\infty} |B_n'|^2 \right)^{1/2} \right]$$

$$\leq \left[\sum_{n=1}^{\infty} \frac{1}{n^2} \right]^{1/2} \left[2 \left(\frac{1}{\pi} \int_{-\pi}^{\pi} f'^2(x) \, dx \right)^{1/2} \right]$$

Letting m tend to infinity, we see that $\sum_{n=1}^{\infty} [|A_n| + |B_n|]$ converges to a sum no greater than the last quantity in the inequalities above. Therefore, by Weierstrass' M-test, the Fourier series for f

$$\tfrac{1}{2} A_0 + \sum_{n=1}^{\infty} [A_n \cos nx + B_n \sin nx]$$

converges uniformly on the whole real line. We know that the Fourier series converges to f, since f is continuous and piecewise smooth. ∎

We close this section by introducing a useful notation. The *square integral norm* of the piecewise continuous function f over the interval (a, b) is denoted by $\|f\|_2$ and is defined by (f is *real valued*)

$$\|f\|_2 = \left[\int_a^b f^2(x) \, dx \right]^{1/2}$$

The following theorem summarizes some of this norm's basic properties.

(4.5) Theorem.   The square integral norm enjoys the following properties:

(a) $\|f\|_2 \geq 0$ for each piecewise continuous function f.
(b)   If $f(x) = 0$ on $[a, b]$, then $\|f\|_2 = 0$.
(c)   For each constant c, $\|cf\|_2 = |c| \, \|f\|_2$.
(d)   *Triangle inequality.* $\|f + g\|_2 \leq \|f\|_2 + \|g\|_2$ for all functions f, g.

Proof.   We will only prove the triangle inequality; properties (a) through (c) are left as exercises for the reader.

If we square $\|f + g\|_2$ then we obtain

$$\|f + g\|_2^2 = \int_a^b [f(x) + g(x)]^2 \, dx$$

$$= \int_a^b f^2(x) \, dx + 2 \int_a^b f(x)g(x) \, dx + \int_a^b g^2(x) \, dx$$

$$= \|f\|_2^2 + 2 \int_a^b f(x)g(x) \, dx + \|g\|_2^2$$

Since a real number is never greater than its absolute value, Schwarz's inequality gives us (using the norm notation)

$$2 \int_a^b f(x)g(x) \, dx \leq 2 \left| \int_a^b f(x)g(x) \, dx \right| \leq 2 \|f\|_2 \|g\|_2$$

Substituting this last inequality, dropping the middle quantity, into the last line of the equalities above yields

$$\|f + g\|_2^2 \leq \|f\|_2^2 + 2 \|f\|_2 \|g\|_2 + \|g\|_2^2$$
$$= [\|f\|_2 + \|g\|_2]^2$$

Taking positive square roots of the first and last quantities of this last inequality gives us (d), the triangle inequality. ∎

Exercises

(4.6)   Complete the proof of Cauchy's inequality and Theorem (4.5).

(4.7)   Prove that if f is piecewise continuous on (a, b) and $\|f\|_2 = 0$, then $f(x) = 0$ throughout (a, b) except possibly for a finite number of points.

(4.8) Give another proof of Cauchy's inequality using the inequality

$$\sum_{n=1}^{N} \sum_{m=1}^{N} (a_n b_m - a_m b_n)^2 \ge 0$$

(4.9) The *integral norm* $\|f\|_1$ is defined by $|f\|_1 = \int_a^b |f(x)|\, dx$. Show that Theorem (4.5) holds with $\|f\|_1$ in place of $\|f\|_2$.

*(4.10) Prove the following theorem.

Theorem. Let f be a periodic function. If f has k continuous derivatives, then for some positive constant C the Fourier coefficients of f satisfy for every positive integer n

$$|A_n| \le \frac{C}{n^k} \qquad |B_n| \le \frac{C}{n^k}$$

*(4.11) If f is *complex valued*, then we *define* $\|f\|_2 = (\int_a^b |f(x)|^2)^{1/2}$. Prove that when f and g are complex valued and piecewise continuous on (a, b) *then* $|\int_a^b f(x)\overline{g(x)}\, dx| \le \|f\|_2 \|g\|_2$ *and* Theorem (4.5) remains true. Show also that Theorem (4.4) remains true for complex Fourier series.

*§5. Gibb's Phenomenon

When we first discussed pointwise convergence in Chapter 1 we noted the occurrence of Gibb's phenomenon (see Figures 1.9 and 1.11). The fact that Fourier series overshoot or undershoot the right- and left-hand limits of functions at their points of discontinuity is a surprising observation that is usually credited to the physicist Gibb (1899), although it was observed by Wilbraham in 1848.

Besides its intrinsic interest, Gibb's phenomenon is important for two reasons. First, from a mathematical standpoint, it shows precisely how Fourier series fail to converge uniformly for discontinuous piecewise smooth functions. Second, from a physical standpoint, it is associated with the phenomenon called "ringing" in optics (see §10, Chapter 7) and electric circuits (a high voltage circuit that is suddenly switched on will throw a spark because of this phenomenon). We will first discuss Gibb's phenomenon for a specific function where the mathematics is as simple as possible, then we will prove that it occurs for all discontinuous piecewise smooth functions.

Let's consider the function g that satisfies $g(x) = \frac{1}{2}(\pi - x)$ for $0 < x < 2\pi$ and has period 2π. This function g consists of $\frac{1}{2}\pi$ added to the negative of the function discussed in Example (4.7), Chapter 1. The Fourier expansion of g is

(5.1) $$g(x) = \sum_{k=1}^{\infty} \frac{\sin kx}{k} \qquad (x \ne 0, \pm 2\pi, \pm 4\pi, \dots)$$

(5.2) Lemma. For the function g described by (5.1), Gibb's phenomenon occurs at $x = 0$. More precisely,

$$\frac{1}{\pi} \lim_{n\to\infty} \left(\operatorname*{maximum}_{0\le x\le \pi/n} S_n(x) - \tfrac{1}{2}\pi \right)$$

is approximately 0.089. Thus, for large enough n the first peak for S_n on the right side of 0 overshoots $g(0+)$ by about 9% relative to π, the size of the discontinuity of g at 0 (i.e., the overshooting *does not go away* as $n \to \infty$).

Proof. Because $S_n'(x) = \sum_{k=1}^{n} \cos kx = \tfrac{1}{2}D_n(x) - \tfrac{1}{2}$ we have

$$(5.3) \qquad\qquad S_n(x) = \int_0^x \tfrac{1}{2}D_n(u)\,du - \tfrac{1}{2}x$$

Using some algebra (5.3) turns into

$$(5.4) \qquad \begin{aligned} S_n(x) &= \int_0^x \frac{\sin(n+\tfrac{1}{2})u}{u}\,du + w_n(x) \\ &= \int_0^{(n+\tfrac{1}{2})x} \frac{\sin u}{u}\,du + w_n(x) \end{aligned}$$

where

$$w_n(x) = \int_0^x \left(\frac{1}{2\sin\tfrac{1}{2}u} - \frac{1}{u} \right) \sin(n+\tfrac{1}{2})u\,du - \tfrac{1}{2}x$$

Applying l'Hôpital's rule, we obtain

$$\lim_{u\to 0+} \left(\frac{1}{2\sin\tfrac{1}{2}u} - \frac{1}{u} \right) = \lim_{u\to 0+} \frac{u - 2\sin\tfrac{1}{2}u}{2u\sin\tfrac{1}{2}u} = 0$$

Therefore, given $\epsilon > 0$ we have for $0 \le x \le \delta$

$$(5.5) \qquad |w_n(x)| \le \int_0^x \left| \frac{1}{2\sin\tfrac{1}{2}u} - \frac{1}{u} \right| du + \tfrac{1}{2}x < \epsilon$$

provided δ is sufficiently small. From (5.4) and (5.5) we have

$$(5.6) \qquad S_n(x) - \tfrac{1}{2}\pi \doteq \int_0^{(n+\tfrac{1}{2})x} \frac{\sin u}{u}\,du - \tfrac{1}{2}\pi \qquad (0 \le x \le \delta)$$

where $\doteq$ stands for "within $\pm\epsilon$." Due to the alternation in sign of $\sin u/u$ we conclude that the largest value of the function $I(v) = \int_0^v (\sin u)/u\,du$ is attained at $v = \pi$. Moreover, using the result of Exercise (6.15), Chapter 1, we have

$$I(\pi) = \int_0^\pi \frac{\sin u}{u}\,du = \tfrac{1}{2}\pi(1.1789\cdots) = 1.8518\cdots$$

Setting v equal to $(n+\tfrac{1}{2})x$ for $x = \pi/(n+\tfrac{1}{2})$ we get $v = \pi$ and (5.6) yields, in the light of the considerations above, for all n so large that $\pi/n < \delta$

$$\frac{1}{\pi} \operatorname*{maximum}_{0 \le x \le \pi/n} \left[S_n(x) - \tfrac{1}{2}\pi \right] \doteq \frac{1}{\pi} \operatorname*{maximum}_{0 \le x \le \pi/n} \left[\int_0^{(n+\frac{1}{2})x} \frac{\sin u}{u} \, du - \tfrac{1}{2}\pi \right]$$

$$= \frac{1}{\pi} [1.8158 - \tfrac{1}{2}\pi] = 0.089$$

which proves our lemma. ∎

(5.7) Remark. Due to the oddness of g, the Fourier series for g undershoots $g(0-) = -\tfrac{1}{2}\pi$ by about 9%. Moreover, it was established in Exercise (6.9), Chapter 1, as a simple consequence of Abel's lemma, that the series in (5.1) converges uniformly to g on every closed interval $[a, b]$ contained in the open interval $(0, 2\pi)$. This last result shows that Gibb's phenomenon does not occur for g on $[a, b]$. By periodicity, Gibb's phenomenon occurs for g precisely at its discontinuities $(0, \pm 2\pi, \pm 4\pi, \ldots)$.

Using Lemma (5.2) and Theorem (4.4) we can show that Gibb's phenomenon always occurs at the discontinuities of piecewise smooth functions. First, however, we prove a preliminary result of such importance that we prefer to state it separately.

(5.8) Theorem. Let f be piecewise smooth with period 2π. If the closed interval $[a, b]$ contains no discontinuities of f, then the Fourier series for f converges uniformly to f on $[a, b]$.

Proof. Let $\{a_j\}_{j=1}^m$ be the discontinuities for f on the interval $[-\pi, \pi)$ and let c_j stand for the jump $f(a_j +) - f(a_j -)$ for $j = 1, \ldots, m$. We define the function G by the following combination of shifts of the function g described by (5.1)

$$G(x) = \sum_{j=1}^m (c_j/\pi) g(x - a_j)$$

Then $f - G$ is continuous and piecewise smooth with period 2π. Theorem (4.4) tells us that the Fourier series for $f - G$ converges uniformly to $f - G$ over $\mathbb{R}$. Moreover, based on Remark (5.7), we conclude that the Fourier series for G converges uniformly to G on every closed interval $[a, b]$ that contains no discontinuities of f. Thus, it follows that the Fourier series for $f = (f - G) + G$ converges uniformly to f on each such interval $[a, b]$. ∎

(5.9) Theorem: Gibb's Phenomenon. Let f be a piecewise smooth function with period 2π. At each discontinuity x_0 of f the Fourier series for f either overshoots or undershoots $f(x_0 +)$ by about 9% in comparison to the magnitude $|f(x_0 +) - f(x_0 -)|$ of the jump discontinuity. Overshooting (undershooting) occurs when $f(x_0 +) - f(x_0 -)$ is positive (negative).

Proof. Define G as in the proof of Theorem (5.8). Since the 9% overshoot occurs for g at $0+$, it occurs for $g(x - a_j)$ at $a_j +$ $(j = 1, \ldots, m)$. Therefore, the same *relative* overshoot (undershoot) occurs for $(c_j/\pi) g(x - a_j)$ at each $a_j +$ if c_j is positive (negative). Since $f - G$ has a uniformly convergent Fourier series, we conclude that there is a 9% relative overshoot or undershoot for $f = G + (f - G)$ at each $a_j +$ according to whether $c_j = f(a_j +) - f(a_j -)$ is positive or negative. By periodicity we conclude that the theorem is proved. ∎

(5.10) Remark. In Remark (5.7) we noted that the Fourier series for g undershoots $g(0 -)$ by about 9%, that is

$$\frac{1}{\pi} \lim_{n \to \infty} \text{maximum}_{-\pi/n \leq x \leq 0} [-\tfrac{1}{2}\pi - S_n(x)]$$

is about 9%. It follows that the Fourier series for the function f in Theorem (5.9) either undershoots or overshoots $f(x_0 -)$ by about 9%, undershooting (overshooting) occurring when $f(x_0 +) - f(x_0 -)$ is positive (negative).

Because of Gibb's phenomenon, as well as the divergence of Fourier series for some continuous functions, we will take up some alternative approaches to convergence of Fourier series in the next two sections. The method of *mean square convergence* will show that, although the difference between S_n and f at individual points may be significant, the integral of $|S_n - f|^2$ over $[-\pi, \pi]$ will diminish to zero as n tends to $+\infty$. The method of *arithmetic means,* on the other hand, rejects the sequence of partial sums $\{S_n\}$ in order to examine the sequence of their arithmetic means. By taking arithmetic means, the problems involving pointwise convergence (divergence or Gibb's phenomenon) will disappear.

Exercises

(5.11) Show that for the function g in (5.1), $\lim_{n \to \infty} S_n(\pi/n) - \tfrac{1}{2}\pi$ is about 0.28. This is a less precise way of stating Gibb's phenomenon.

(5.12) Prove that $\int_0^\infty (\sin u)/u \, du = \tfrac{1}{2}\pi$.

(5.13) Let f be as in Theorem (5.9). Prove that as *curves in the plane* the curves $y = S_n(x)$ converge uniformly to the curve formed from the graph of f with vertical line segments centered at the abscissas of each discontinuity of f and having lengths of about 18% larger than the magnitude of the jumps of f at those discontinuities.

$\star$**(5.14)** Explain why Gibb's phenomenon might be expected as a necessary consequence of the formula $S_n(x) = 1/2\pi \int_{-\pi}^{\pi} f(u)D_n(u - x) \, du$. [*Hint*: Interpret the integral geometrically.]

(5.15) Show that Theorem (5.9) implies that Gibb's phenomenon (in particular, the 9% overshoot or undershoot) occurs for all piecewise smooth, discontinuous functions of period $2a$.

(5.16) Let the integral norms $\| \ \|_2$ and $\| \ \|_1$ be defined over some interval $[a, b]$ having length 2π. Prove that for f as in Theorem (5.9)

$$\lim_{n \to \infty} \|f - S_n\|_2 = 0 \qquad \text{and} \qquad \lim_{n \to \infty} \|f - S_n\|_1 = 0$$

(5.17)   Prove that the partial sums of the sine series in (5.1) are uniformly bounded.

§6. Mean Square Convergence

A second approach to convergence of Fourier series is provided by considering *mean square convergence*. This approach is necessary because pointwise convergence, while straightforward in definition, is in reality a very complicated process. For example, there are continuous functions whose Fourier series diverge at all rational multiples of π. See Appendix A for a discussion of these pathological functions. We should remember, of course, that none of these pathological functions is piecewise smooth (or even Lipschitz).

On the other hand, as we shall see below, mean square convergence applies to all piecewise continuous functions. Mean square convergence is very widely used in applications. For example, it is an essential part of the mathematics used in quantum mechanics. We will apply mean square convergence in Chapter 7 when we discuss the *sampling theorem,* an essential tool in modern communication theory.

The definition of mean square convergence is based on the *Least Squares Theorem.*

(6.1) Theorem: Least Squares.   If $\{g_n\}$ is an orthogonal system of functions on $[a, b]$, then for each set of numbers $a_0, a_1, \ldots, a_N, \ldots$

$$\left\| f - \sum_{n=0}^{N} c_n g_n \right\|_2 \leq \left\| f - \sum_{n=0}^{N} a_n g_n \right\|_2$$

where $c_0, c_1, \ldots, c_N, \ldots$ are the Fourier coefficients of f relative to $\{g_n\}$.

Proof.   First, let's suppose that $\{g_n\}$ is orthonormal. Then, expressing Lemma (1.3) in norm notation yields

$$\left\| f - \sum_{n=0}^{N} a_n g_n \right\|_2^2 = \left[\|f\|_2^2 - \sum_{n=0}^{N} c_n^2 \right] + \sum_{n=0}^{N} (c_n - a_n)^2$$

Observing that the last term above is nonnegative and equals 0 only when $a_n = c_n$ for each n, we conclude that

$$\left\| f - \sum_{n=0}^{N} c_n g_n \right\|_2^2 = \|f\|_2^2 - \sum_{n=0}^{N} c_n^2 \leq \left\| f - \sum_{n=0}^{N} a_n g_n \right\|_2^2$$

Taking positive square roots of the inequality above we obtain the theorem for orthonormal sets of functions.

Second, suppose that $\{g_n\}$ is merely orthogonal. If we define the function f_n by $f_n(x) = g_n(x)/\|g_n\|_2$, then $\{f_n\}$ is an orthonormal set over $[a, b]$. Letting d_n denote the nth generalized Fourier coefficient for f relative to

$\{f_n\}$ we obtain by a straightforward calculation

(6.2) $$d_n = \|g_n\|_2 c_n$$

where c_n is the nth generalized Fourier coefficient for f relative to $\{g_n\}$. Applying the least squares theorem, just proved, to $\{f_n\}$ and the constants $a_0 \|g_0\|_2, \ldots, a_N \|g_N\|_2$ yields

$$\left\| f - \sum_{n=0}^{N} d_n f_n \right\|_2 \le \left\| f - \sum_{n=0}^{N} (a_n \|g_n\|_2) f_n \right\|_2$$

which, using (6.2) and the definition of f_n, yields the desired result. ∎

(6.3) Remark.   In the course of the proof above we derived the important equality

$$\left\| f - \sum_{n=0}^{N} c_n g_n \right\|_2 = \left[\|f\|_2^2 - \sum_{n=0}^{N} c_n^2 \right]^{1/2}$$

for an orthonormal set $\{g_n\}$ over $[a, b]$. We leave it as an exercise for the reader to verify that if $\{g_n\}$ is merely orthogonal, then

$$\left\| f - \sum_{n=0}^{N} c_n g_n \right\|_2 = \left[\|f\|_2^2 - \sum_{n=0}^{N} c_n^2 \|g_n\|_2^2 \right]^{1/2}$$

When applied to trigonometric Fourier series the least squares theorem says that

$$\left\| f - \left[\tfrac{1}{2} A_0 + \sum_{n=0}^{N} (A_n \cos nx + B_n \sin nx) \right] \right\|_2$$
$$\le \left\| f - \left[C_0 + \sum_{n=0}^{N} (C_n \cos nx + D_n \sin nx) \right] \right\|_2$$

where A_n and B_n are the usual Fourier coefficients for some function f over $(-\pi, \pi)$ while C_n and D_n are any set of constants.

The least squares theorem gives an intrinsic characterization of Fourier coefficients. They are the numbers which minimize

$$\left\| f - \sum_{n=0}^{N} a_n g_n \right\|_2$$

for each N. From the least squares theorem we deduce that the sequence of square integral norms $\|f - \sum_{n=0}^{N} c_n g_n\|_2$ decreases as N increases. [See Exercise (6.15).] Since the sequence is bounded below by zero, we know from calculus that its limit exists as N tends to infinity; moreover

$$\lim_{N \to \infty} \left\| f - \sum_{n=0}^{N} c_n g_n \right\|_2 \ge 0$$

This last result is the basis for the following definition.

(6.4) Definition. The generalized Fourier series for f is said to *mean square converge* to f if

$$\lim_{N \to \infty} \left\| f - \sum_{n=0}^{N} c_n g_n \right\|_2 = 0$$

If the limit above holds for all continuous functions f on $[a, b]$, then $\{g_n\}$ is called *complete*. ∎

Using the notation $S_N(x) = \sum_{n=0}^{N} c_n g_n(x)$ the property of completeness can also be expressed by

(6.5) $$\lim_{N \to \infty} \| f - S_N \|_2 = 0$$

for each continuous function f on $[a, b]$.

(6.6) Examples

(a) The trigonometric system of functions, period 2π, is complete. We will prove it in §7. A substitution argument shows that the trigonometric system, period $2a$, is complete. [See Exercise (6.16).]

(b) The sine and cosine systems on $(0, a)$ are complete. [See Exercise (6.16).]

(c) The Legendre polynomials, which we shall discuss in Chapter 8, are complete as an orthogonal system over $[-1, 1]$.

Completeness of an orthogonal system imparts to that system a number of useful properties. Two of the most important are uniqueness of Fourier coefficients, and the validity of term by term integration of Fourier series (whether or not the series converges uniformly).

(6.7) Theorem: Uniqueness of Fourier Coefficients. If two continuous functions have the same Fourier coefficients relative to a complete orthogonal system on $[a, b]$, then the two functions are identical on $[a, b]$.

Proof. Consider the continuous function h defined by $h(x) = f(x) - g(x)$ where f and g are the two continuous functions referred to in the statement of the theorem. The Fourier coefficients of h relative to $\{g_n\}$ are all zero. Thus, the Fourier series for h reduces to zero, that is, for each n

$$h - \sum_{m=0}^{n} c_m g_m = h$$

Because the completeness of $\{g_n\}$ implies that the Fourier series for h converges to h in mean square, we have from the equality above

$$\lim_{n \to \infty} \| h \|_2 = 0$$

Therefore, $\| h \|_2 = 0$, so $\| h \|_2^2 = \int_a^b h^2(x)\, dx = 0$. Since $h^2(x)$ is continuous and nonnegative on $[a, b]$, it follows from the last equality that $h^2(x) = 0$ for all x values in $[a, b]$. Consequently, $h(x) = 0$ on $[a, b]$. Thus, using the definition of h, we conclude that $f(x) = g(x)$ on $[a, b]$. ∎

Another way of expressing Theorem (6.7) is that a complete orthogonal system of continuous functions on $[a, b]$ is *maximally orthogonal*. That means that no nonzero continuous function can be included into the system that is orthogonal to all the other members of the system.

A second important property of complete orthogonal systems is that their Fourier series may be integrated term by term. Before we prove this theorem, however, we need to extend the definition of completeness to include all piecewise continuous functions.

(6.8) Theorem. An orthogonal system $\{g_n\}$ on $[a, b]$ is complete if and only if

$$\lim_{n \to \infty} \|f - S_n\|_2 = 0$$

holds for each piecewise continuous function f on (a, b).

Proof. Since a continuous function on $[a, b]$ is piecewise continuous on (a, b), one part of the theorem follows immediately. We now proceed to the other part of the theorem.

Suppose that f is piecewise continuous on (a, b). Let $\epsilon > 0$ be given. There exists a continuous function g_ϵ on $[a, b]$ such that

(6.9) $$\|f - g_\epsilon\|_2 < \epsilon$$

We define g_ϵ as follows. Let

$$a = x_0 < x_1 < x_2 < \cdots < x_{K-1} < x_K = b$$

be defined so that f is continuous on each interval $[x_{j-1}, x_j]$ if defined by its one-sided limits at the end points. On each subinterval $[x_{j-1}, x_j]$ we let $g_\epsilon(x)$ equal $f(x)$ until x is quite close (within $\delta > 0$) to either of the end points of the subinterval, and then extend the graph of $g_\epsilon(x)$ linearly to zero at the end points (see Figure 2.3).

Now, letting M stand for the maximum value of $|f(x)|$ on the closed interval $[a, b]$, we have

$$\|f - g_\epsilon\|_2^2 = \sum_{j=1}^{K} \int_{x_{j-1}}^{x_j} [f(x) - g_\epsilon(x)]^2 \, dx < \sum_{j=1}^{K} 8\delta M^2 = 8\delta M^2 K$$

We then take $\delta < \epsilon^2/(8M^2K)$ and (6.9) must hold.

We now let S_N^g stand for the Nth partial sum of the Fourier series for g_ϵ, as distinguished from S_N, the Nth partial sum of the Fourier series for f. Since $\{g_n\}$ is complete, we have $\lim_{N \to \infty} \|g_\epsilon - S_N^g\|_2 = 0$. That is, if A is

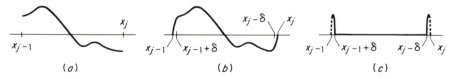

Figure 2.3 Definition of g_ϵ. (a) Graph of f on $[x_{j-1}, x_j]$. (b) Graph of g_ϵ on $[x_{j-1}, x_j]$. (c) Graph of $(f - g_\epsilon)^2$ on $[x_{j-1}, x_j]$.

sufficiently large, then for all $N \geq A$

(6.10) $$\|g_\epsilon - S_N^g\|_2 < \epsilon$$

From (6.9) and (6.10) we conclude, using the triangle inequality for the square integral norm, that for all $N \geq A$

$$\|f - S_N^g\|_2 = \|[f - g_\epsilon] + [g_\epsilon - S_N^g]\|_2$$
$$\leq \|f - g_\epsilon\|_2 + \|g_\epsilon - S_N^g\|_2$$
$$< 2\epsilon$$

By the theorem of least squares, we must have for all $N \geq A$

$$\|f - S_N\|_2 \leq \|f - S_N^g\|_2 < 2\epsilon$$

Since ϵ can be taken to be arbitrarily small in magnitude, we have proved that $\lim_{N \to \infty} \|f - S_N\|_2 = 0$. ∎

(6.11) Corollary. If $\{g_n\}$ is a complete orthogonal system of functions on $[a, b]$, then the Fourier series for a piecewise continuous function f can be integrated term by term, whether or not the series converges pointwise.

Proof. Suppose f has the Fourier series $\sum_{n=0}^{\infty} c_n g_n(x)$. Then we shall prove that

$$\int_c^d f(x) \, dx = \sum_{n=0}^{\infty} c_n \int_c^d g_n(x) \, dx$$

where $[c, d]$ is any interval contained in $[a, b]$.

For each finite N, we have

$$\left| \int_c^d f(x) \, dx - \sum_{n=0}^{N} c_n \int_c^d g_n(x) \, dx \right| = \left| \int_c^d \left[f(x) - \sum_{n=0}^{N} c_n g_n(x) \right] dx \right|$$
$$\leq \int_c^d \left| f(x) - \sum_{n=0}^{N} c_n g_n(x) \right| dx$$

Then Schwarz's inequality tells us that (*Note*: $[c, d]$ is contained in $[a, b]$.)

$$\left| \int_c^d f(x) \, dx - \sum_{n=0}^{N} c_n \int_c^d g_n(x) \, dx \right| \leq \int_a^b 1 \left| f(x) - \sum_{n=0}^{N} c_n g_n(x) \right| dx$$
$$\leq \|1\|_2 \left\| f - \sum_{n=0}^{N} c_n g_n \right\|_2$$

Letting N tend to infinity the last part of the inequality above tends to zero; proving that

$$\lim_{N \to \infty} \sum_{n=0}^{N} c_n \int_c^d g_n(x) \, dx = \int_c^d f(x) \, dx \qquad ∎$$

For example, over the interval $[0, 2\pi]$

$$\tfrac{1}{2}x \sim \tfrac{1}{2}\pi - \sum_{n=1}^{\infty} \frac{\sin nx}{n}$$

Therefore, since the trigonometric system is complete, we can integrate the expression above term by term on the right side, obtaining

$$\int_0^\pi \tfrac{1}{2}x \, dx = \frac{\pi^2}{4} = \tfrac{1}{2}\pi^2 + \sum_{n=1}^\infty \frac{(-1)^n - 1}{n^2} = \tfrac{1}{2}\pi^2 - 2 \sum_{k=0}^\infty \frac{1}{(2k+1)^2}$$

Or, if we let x be a variable upper limit of integration

$$\int_0^x \tfrac{1}{2}t \, dt = \frac{x^2}{4} = \tfrac{1}{2}\pi x + \sum_{n=1}^\infty \frac{\cos nx - 1}{n^2} \qquad (0 < x < 2\pi)$$

Using $\sum_{n=1}^\infty 1/n^2 = \pi^2/6$ [see (4.12c), Chapter 1] we obtain

$$\frac{x^2}{4} - \tfrac{1}{2}\pi x + \frac{\pi^2}{6} = \sum_{n=1}^\infty \frac{\cos nx}{n^2} \qquad (0 < x < 2\pi)$$

which the reader may verify by computing a Fourier series over $[0, 2\pi]$.

Our final topic for this section is the result known as *Parseval's equality*.

(6.12) Theorem: Parseval's Equality.   Let $\{g_n\}$ be an orthonormal system of functions over $[a, b]$. The system $\{g_n\}$ is complete if and only if Bessel's inequality is an equality, that is

$$\sum_{n=0}^\infty c_n^2 = \int_a^b f^2(x) \, dx$$

holds for each piecewise continuous (continuous) function f.

Proof.   Since $\|f\|_2^2 = \int_a^b f^2(x) \, dx$, Remark (6.3) tells us that

$$\left\| f - \sum_{n=0}^N c_n g_n \right\|_2 = \left(\int_a^b f^2(x) \, dx - \sum_{n=0}^N c_n^2 \right)^{1/2}$$

Letting N tend to infinity the convergence of either side of the equality above to zero implies the convergence of the other side to zero.   ■

If $\{g_n\}$ is merely orthogonal, then dividing each function g_n by its square integral norm $\|g_n\|_2$, along with some algebra, proves that *Parseval's equality*

(6.13) $$\sum_{n=0}^\infty c_n^2 \|g_n\|_2^2 = \int_a^b f^2(x) \, dx$$

holds if and only if $\{g_n\}$ is a complete orthogonal system on $[a, b]$.

Since the trigonometric system is complete, Bessel's inequality in the form given in (1.6) is an equality. Thus

(6.14) $$\tfrac{1}{2}A_0^2 + \sum_{n=1}^\infty [A_n^2 + B_n^2] = \frac{1}{\pi} \int_{-\pi}^\pi f^2(x) \, dx$$

where A_n and B_n are the usual Fourier coefficients for f.

Exercises

(6.15) Prove that the sequence of square integral norms $\|f - \sum_{n=0}^{N} c_n g_n\|_2$ decreases as N increases. Also, verify (6.13).

(6.16) Given that (6.14) holds, prove that the trigonometric system, period $2a$, is a complete orthogonal system over $[-a, a]$ and that the sine and cosine systems are complete orthogonal systems over $[0, a]$.

(6.17) Parseval's equality can be used to evaluate sums of numerical series. For example, use (6.14) with the given function below to find the sum of the given numerical series.

$$f(x) = \frac{\pi^2 - 3x^2}{12} \qquad \sum_{n=1}^{\infty} \frac{1}{n^4} \qquad \left[Answer \sum_{n=1}^{\infty} \frac{1}{n^4} = \frac{\pi^4}{90} \right]$$

(6.18) Show that $\{\sin nx\}_{n=2}^{\infty}$ is an *in*complete orthogonal system on $[0, \pi]$.

(6.19) *Parseval.* Prove that an orthonormal system of functions $\{g_n\}$ over $[a, b]$ is complete if and only if for every two piecewise continuous functions f and g on (a, b) we have $\sum c_n d_n = \int_a^b f(x)g(x)\, dx$ (where c_n and d_n are the Fourier coefficients of f and g relative to $\{g_n\}$). Generalize this result to orthogonal systems of functions.

*(6.20) Using (6.16), and (7.12) from Chapter 1, prove the following results for f and g piecewise continuous complex valued functions on $(-a, a)$

(a) $f \sim \sum c_n e^{in\pi x/a}$ implies $\sum |c_n|^2 = (2a)^{-1} \int_{-a}^{a} |f(x)|^2\, dx$.
(b) $f \sim \sum c_n e^{in\pi x/a}$ implies $\lim_{N \to +\infty} \|f - \sum_{n=-N}^{+N} c_n e^{in\pi x/a}\|_2 = 0$

[using the notation of Exercise (4.11)].

§7. Summation of Fourier Series by Arithmetic Means

The method of summation by arithmetic means provides a third approach to the problem of convergence of Fourier series. This method also neatly connects the previous approaches of pointwise and mean square convergence.

(7.1) Definition. The *arithmetic means* of an infinite series $\sum_{m=0}^{\infty} a_m$ are the sequence of numbers σ_n defined by

$$\sigma_n = \frac{s_0 + s_1 + \cdots + s_{n-1}}{n} = \frac{1}{n} \sum_{k=0}^{n-1} s_k \qquad (n = 1, 2, 3, \ldots)$$

where each $s_k = \sum_{m=0}^{k} a_m$ is a partial sum of the infinite series. If $\lim_{n \to \infty} \sigma_n$ exists, then we say that the series is *summable by arithmetic means* to this limit, which is often denoted by σ. ∎

For example, consider the series

$$\sum_{m=0}^{\infty} (-1)^m = 1 - 1 + 1 - 1 + 1 \cdots$$

This series diverges; however, it is summable by arithmetic means. Its arithmetic means are

$$\sigma_1 = 1, \quad \sigma_2 = \tfrac{1}{2}, \quad \sigma_3 = \tfrac{2}{3}, \ldots, \quad \sigma_{2k} = \tfrac{1}{2}, \sigma_{2k+1} = \frac{k+1}{2k+1}, \cdots$$

Hence, $\lim_{n \to \infty} \sigma_n = 1/2$. The series is summable by arithmetic means to the value $\sigma = 1/2$.

Fortunately, if an infinite series converges, then it is also summable by arithmetic means and to the same value as its sum as a convergent series. To be precise, we have the following theorem which proves that the method of summation by arithmetic means provides a generalization of the usual process of summing an infinite series.

(7.2) Theorem.   If an infinite series converges to the sum s, then the series is summable by arithmetic means to this same value s.

Proof.   Due to the convergence of the series $\sum_{m=0}^{\infty} a_m$, given any $\epsilon > 0$, we can find a positive integer M so large that for all $k \geq M$

(7.3) $|s - s_k| < \epsilon$

But, if $n > M$ then

$$s - \sigma_n = \frac{ns}{n} - \frac{1}{n}\sum_{k=0}^{n-1} s_k$$

$$= \frac{1}{n}\sum_{k=0}^{M-1}(s - s_k) + \frac{1}{n}\sum_{k=M}^{n-1}(s - s_k)$$

Using the triangle inequality for absolute values several times yields

$$|s - \sigma_n| \leq \frac{1}{n}\sum_{k=0}^{M-1}|s - s_k| + \frac{1}{n}\sum_{k=M}^{n-1}|s - s_k|$$

$$< \frac{1}{n}\sum_{k=0}^{M-1}|s - s_k| + \frac{\epsilon + \epsilon + \cdots + \epsilon}{n}$$

where the last term involves epsilons because for $k \geq M$ we can apply (7.3). Since the number of epsilons in that last term is less than n, we obtain for $n > M$

(7.4) $|s - \sigma_n| < \frac{1}{n}\sum_{k=0}^{M-1}|s - s_k| + \epsilon$

Since M is fixed, if we let n grow even larger, then for all $n \geq N \ (>M)$ sufficiently large we will have

$$\frac{1}{n}\sum_{k=0}^{M-1}|s - s_k| < \epsilon$$

Combining this last inequality with (7.4) yields for all $n \geq N$

$$|s - \sigma_n| < 2\epsilon$$

Since ϵ can be made arbitrarily small in magnitude, we have proved that $\lim_{n \to \infty} \sigma_n = s$. ∎

Unfortunately, not all series are summable by arithmetic means. For example, the series $\sum_{m=1}^{\infty} 1/m$ is not summable by this method. [See Exercise (7.22).]

We now apply Definition (7.1) to the Fourier series of a function f with period 2π. In this case, each arithmetic mean σ_n is a function of x

$$\sigma_n(x) = \frac{1}{n} \sum_{k=0}^{n-1} S_k(x)$$

where for each k

$$S_k(x) = \tfrac{1}{2}A_0 + \sum_{m=1}^{k} \left[A_m \cos mx + B_m \sin mx \right]$$

is a partial sum of the Fourier series for f. We leave it as an exercise for the reader [see Exercise (7.23)] to verify that

(7.5) $$\sigma_n(x) = \tfrac{1}{2}A_0 + \sum_{k=1}^{n} \left(1 - \frac{k}{n}\right)\left[A_k \cos kx + B_k \sin kx \right]$$

It is interesting to compare the form of $\sigma_n(x)$ in (7.5) with $S_n(x)$; σ_n is identical in form to S_n except for the factor $[1 - (k/n)]$ on each term. These factors are often called *convergence factors*, since they will sometimes cause the sequence $\{\sigma_n(x)\}$ to converge when the sequence $\{S_n(x)\}$ does not.

We will now discuss a convenient integral form for σ_n. If we fix an x value, say x_0, then Dirichlet's integral form for $S_k(x_0)$ given in (2.8) yields

$$\sigma_n(x_0) = \frac{1}{n} \sum_{k=0}^{n-1} \left[\frac{1}{2\pi} \int_{-\pi}^{\pi} f(x_0 + u)D_k(u)\, du \right]$$

$$= \frac{1}{2\pi} \int_{-\pi}^{\pi} f(x_0 + u)\left[\frac{1}{n} \sum_{k=0}^{n-1} D_k(u) \right] du$$

We now define the function F_n, for each n, by

(7.6) $$F_n(u) = \frac{1}{n} \sum_{k=0}^{n-1} D_k(u)$$

The last expression for $\sigma_n(x_0)$ then assumes the form

(7.7) $$\sigma_n(x_0) = \frac{1}{2\pi} \int_{-\pi}^{\pi} f(x_0 + u)F_n(u)\, du$$

We shall now prove that

(7.8) $$F_n(u) = \frac{1}{n}\left[\frac{\sin(nu/2)}{\sin(u/2)} \right]^2$$

If we multiply the equality in (7.6) by $2 \sin^2(u/2)$ and then use the definition of $D_k(u)$ [see (2.7)] we obtain

$$F_n(u) \cdot 2 \sin^2(u/2) = \frac{1}{n} \sum_{k=0}^{n-1} D_k(u) \cdot 2 \sin^2(u/2)$$

$$= \frac{1}{n} \sum_{k=0}^{n-1} 2 \sin(k + \tfrac{1}{2})u \, \sin(\tfrac{1}{2}u)$$

Using the trigonometric identity

(7.9) $2 \sin \theta \sin \phi = \cos(\theta - \phi) - \cos(\theta + \phi)$

we obtain from the result above

$$F_n(u) \cdot 2 \sin^2(u/2) = \frac{1}{n} \Big\{ (\cos 0 - \cos u) + (\cos u - \cos 2u) + \cdots$$

$$+ [\cos(n-1)u - \cos nu] \Big\}$$

$$= \frac{1}{n} \Big[1 - \cos nu \Big]$$

$$= \frac{2}{n} \sin^2(nu/2)$$

where we have applied the half angle identity for cosines [set $\phi = \theta$ in (7.9)] in the last step. Dividing through by $2 \sin^2(u/2)$ in the equalities above yields (7.8).

Combining (7.7) and (7.8), we have the following result, due to Fejér

(7.10)
$$\begin{cases} \sigma_n(x_0) = \dfrac{1}{2\pi} \displaystyle\int_{-\pi}^{\pi} f(x_0 + u) F_n(u) \, du \\[2mm] F_n(u) = \dfrac{1}{n} \left[\dfrac{\sin(nu/2)}{\sin(u/2)} \right]^2 \qquad (n = 1, 2, 3, \ldots) \end{cases}$$

Figure 2.4 Graphs of the Fejér kernel. We have sketched the graphs of $F_n(u)$ for $n = 5, 9$, and 12. (The reader might find it interesting to compare the graphs above with those of Dirichlet's kernel in Figure 2.1.)

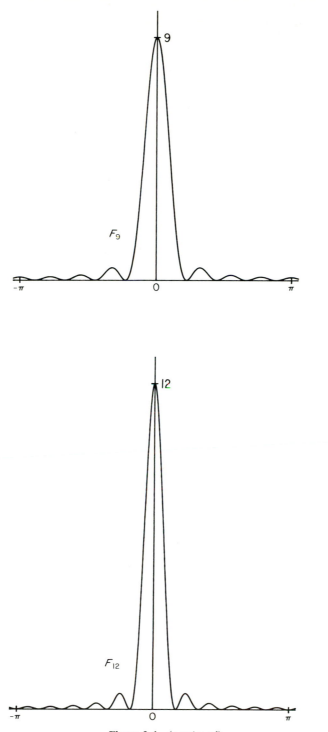

Figure 2.4   (*continued*)

The function F_n is called *Fejér's kernel*; it plays the essential part in the integral formula (7.10) that describes summation by arithmetic means of Fourier series. The reader may recall that a similar terminology was used in §2 when we derived Dirichlet's kernel D_n, which describes the process of *term by term* summation of Fourier series.

Some graphs of Fejér's kernel for various values of n are shown in Figure 2.4.

The following lemma describes the main properties of Fejér's kernel.

(7.11) Lemma.   Fejér's kernel satisfies the following three properties

(A_1) $F_n(u) \geq 0$ for each n.

(A_2) $\dfrac{1}{2\pi} \displaystyle\int_{-\pi}^{\pi} F_n(u)\, du = 1$ for each n.

(A_3)   Given $\epsilon > 0$ and $\delta > 0$, we can have

$$0 \leq F_n(u) < \epsilon$$

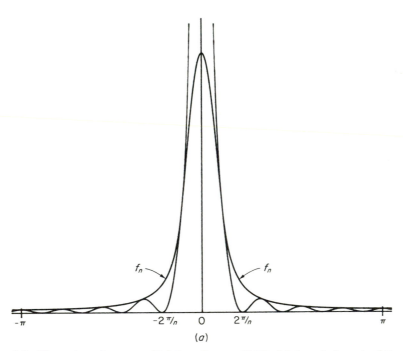

(a)

Figure 2.5   Illustration of property (A_3) from Lemma (7.11). (a) A typical graph of F_n which illustrates how, except for u near zero, each F_n is just $\sin^2(nu/2)$ oscillating within an envelope described by the graph of

$$f_n(u) = [n \sin^2(u/2)]^{-1}$$

If $\pi \geq |u| \geq \delta$ where $\delta > 0$, then (b) shows how successive graphs of f_n (for $n = 3, 12$, and 91) are tending toward zero. The points marked on the vertical axis are the maxima for $f_n(u)$ where $\pi \geq |u| \geq \delta$ (for $n = 3, 12$, and 91) which decrease to zero as n increases to infinity. Thus, the graph of $F_n(u)$ is pushed down toward zero by the graph of $f_n(u)$ when $\pi \geq |u| \geq \delta > 0$.

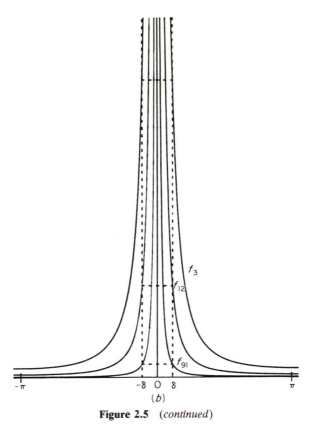

f_3

f_{12}

f_{91}

$-\pi$ $-\delta$ 0 δ π

(b)

Figure 2.5   (*continued*)

provided $\pi \geq |u| \geq \delta$ and $n \geq N$ for some N of sufficiently large magnitude. (See Figure 2.5.)

Proof. Property (A_1) is obvious from formula (7.8). To prove (A_2) we integrate in formula (7.6) and apply the result of Exercise (2.9); thus

$$\frac{1}{2\pi} \int_{-\pi}^{\pi} F_n(u)\, du = \frac{1}{n} \sum_{k=0}^{n-1} \frac{1}{2\pi} \int_{-\pi}^{\pi} D_k(u)\, du$$

$$= \frac{1}{n} \sum_{k=0}^{n-1} 1 = 1$$

As for (A_3), suppose that $\delta > 0$ and $\epsilon > 0$ are given. If $\pi \geq |u| \geq \delta$, then by (A_1) and because the sine function increases on $[\delta, \frac{1}{2}\pi]$

$$0 \leq F_n(u) \leq \frac{1}{n} \frac{\sin^2(nu/2)}{\sin^2(\delta/2)} \leq \frac{1}{n} \frac{1}{\sin^2(\delta/2)}$$

Since δ is fixed, if we let n tend toward infinity, then the right side of the set of inequalities above will tend to zero. In other words, we can find a constant N so large that for all $n \geq N$, we have

$$\frac{1}{n} \frac{1}{\sin^2(\delta/2)} < \epsilon$$

Therefore, for all $n \geq N$

$$0 \leq F_n(u) < \epsilon \qquad \blacksquare$$

We are now in a position to prove the main theorem for this section.

(7.12) Theorem. Suppose that f is a continuous function with period 2π. Then for each x value the Fourier series of f is summable by arithmetic means to $f(x)$.

Proof. Let $\epsilon > 0$ be given. Let M be the maximum value of $|f(x)|$ for x in the closed interval $[-\pi, \pi]$. Then for all x in $[-\pi, \pi]$

(7.13) $0 \leq |f(x)| \leq M$

By periodicity, (7.13) holds for all x values. Using property (A_2) from Lemma (7.11) we have, since x is a constant relative to the variable of integration

$$f(x) = \frac{1}{2\pi} \int_{-\pi}^{\pi} f(x) F_n(u)\, du$$

Therefore, using (7.10) and the equality above, we have

$$\sigma_n(x) - f(x) = \frac{1}{2\pi} \int_{-\pi}^{\pi} [f(x+u) - f(x)] F_n(u)\, du$$

Hence, applying absolute values and (A_1)

$$|\sigma_n(x) - f(x)| \leq \frac{1}{2\pi} \int_{-\pi}^{\pi} |f(x+u) - f(x)|\, F_n(u)\, du$$

Splitting the integral on the right side into a sum of three integrals over $[-\pi, -\delta]$, $[-\delta, \delta]$, and $[\delta, \pi]$ we obtain

(7.14)
$$|\sigma_n(x) - f(x)| \leq \frac{1}{2\pi} \int_{|u| \geq \delta} |f(x+u) - f(x)|\, F_n(u)\, du$$
$$+ \frac{1}{2\pi} \int_{-\delta}^{\delta} |f(x+u) - f(x)|\, F_n(u)\, du$$

where $\int_{|u| \geq \delta}$ is used as shorthand for $\int_{-\pi}^{-\delta} + \int_{\delta}^{\pi}$.

Since f is continuous at x we can choose δ so small that

(7.15) $|(x+u) - x| = |u| < \delta$ implies that $|f(x+u) - f(x)| < \varepsilon$

Moreover, formula (7.13) implies that

(7.16) $|f(x+u) - f(x)| \leq |f(x+u)| + |f(x)| \leq 2M$

Once we have chosen δ so that (7.15) holds, we can then apply (7.16), followed by property (A_3) from Lemma (7.11), to the first integrand in inequality (7.14). Hence, for all $n \geq N$

$$\frac{1}{2\pi} \int_{|u| \geq \delta} |f(x+u) - f(x)| \, F_n(u) \, du \leq \frac{1}{2\pi} \int_{|u| \geq \delta} 2MF_n(u) \, du$$

$$< \frac{1}{2\pi} \int_{|u| \geq \delta} 2M\epsilon \, du$$

$$< 2M\epsilon$$

Moreover, if we apply (7.15) along with (A_1) and (A_2), then the second integral in (7.14) satisfies

$$\frac{1}{2\pi} \int_{-\delta}^{\delta} |f(x+u) - f(x)| \, F_n(u) \, du < \frac{\epsilon}{2\pi} \int_{-\delta}^{\delta} F_n(u) \, du$$

$$< \frac{\epsilon}{2\pi} \int_{-\pi}^{\pi} F_n(u) \, du$$

$$= \epsilon$$

These last two calculations, when their results are applied to (7.14), imply that for all $n \geq N$

$$|\sigma_n(x) - f(x)| < 2M\epsilon + \epsilon = [2M + 1]\epsilon$$

Since ϵ can be taken to be arbitrarily small in magnitude, we have proved that

$$\lim_{n \to \infty} \sigma_n(x) = f(x) \quad \blacksquare$$

In proving Theorem (7.12) we actually demonstrated considerably more than the statement of the theorem. This is because the choice of δ in the proof above can be made independently of all x values. To be more precise, it is possible to choose a number $\delta > 0$ so small that (7.15) holds for all x values. We can do this because f is *uniformly continuous* on the interval $[-\pi, \pi]$, and so (7.15) then holds for all x values because f has period 2π. Once δ is chosen, the choice of N in the proof above was made using property (A_3) from Lemma (7.11); consequently, the choice of N depended only on δ and not on any x values. Thus, we say that $\lim_{n \to \infty} \sigma_n(x) = f(x)$ holds *uniformly* for all x values, meaning that given $\epsilon > 0$ we have for all $n \geq N$

(7.17) $$|f(x) - \sigma_n(x)| < \epsilon$$

and *the choice of N can be made independently of all x values.* Thus, we have proved the following theorem.

(7.18) Theorem: Fejér's Theorem. If f is continuous and has period 2π, then

$$\lim_{n \to \infty} \sigma_n(x) = f(x)$$

holds uniformly for all x values.

An important consequence of Fejér's theorem is the completeness of the trigonometric system, period 2π.

(7.19) Theorem.   The trigonometric system, period 2π, is a complete orthogonal system of functions on $[-\pi, \pi]$.

Proof.   Suppose that f is a continuous function on $[-\pi, \pi]$ and $f(-\pi) = f(\pi)$. If we apply Fejér's theorem to f_P, the continuous periodic extension of f, we obtain for all $n \geq N$ and all x values

$$|f_P(x) - \sigma_n(x)| < \epsilon$$

where ϵ is any preassigned positive number. By restricting x to the interval $[-\pi, \pi]$ we have $|f(x) - \sigma_n(x)| < \epsilon$ for all $n \geq N$. Therefore, for all $n \geq N$

$$\|f - \sigma_n\|_2^2 = \int_{-\pi}^{\pi} |f(x) - \sigma_n(x)|^2 \, dx$$

$$< \int_{-\pi}^{\pi} \epsilon^2 \, dx = 2\pi\epsilon^2$$

Hence, for all $n \geq N$, we have $\|f - \sigma_n\|_2 < \sqrt{2\pi}\epsilon$. Since formula (7.5) shows that σ_n is a finite sum of sines and cosines, it follows from the least squares theorem (6.1) that for all $n \geq N$

$$\|f - S_n\|_2 \leq \|f - \sigma_n\|_2 < \sqrt{2\pi}\epsilon$$

where, as usual, S_n denotes the nth partial sum of the Fourier series for f. Thus, if f is continuous and $f(-\pi) = f(\pi)$ we have proved that

(7.20) $$\lim_{n \to +\infty} \|f - S_n\|_2 = 0$$

All that remains is to show that (7.20) holds for all functions which are continuous on $[-\pi, \pi]$, in particular, those functions for which $f(-\pi) \neq f(\pi)$.

Suppose that f is a continuous function on $[-\pi, \pi]$ and that $f(-\pi) \neq f(\pi)$. We then define a continuous function g_ϵ which equals f until quite close (within $\delta > 0$) to the end points of the interval $[-\pi, \pi]$, and then extend the graph of g_ϵ linearly to zero at $\pm\pi$. (See Figure 2.3, from §6, where a similar function g_ϵ was defined.) Then, by an argument identical to the one which concluded the proof of Theorem (6.8) we obtain (7.20) for every continuous function f on $[-\pi, \pi]$. ■

Because of Theorem (6.8) we also know that (7.20) holds for every piecewise continuous function f on $(-\pi, \pi)$. Moreover, the following result follows easily from (7.19). [See Exercise (6.20(b).]

(7.21) Theorem.   The complex exponential system $\{e^{in\pi x/a}\}_{n=-\infty}^{+\infty}$ is complete over $[-a, a]$. That is, for every complex valued piecewise continuous function f on $(-a, a)$ we have

$$\lim_{N \to +\infty} \|f - S_N\|_2 = \lim_{N \to +\infty} \left[\int_{-a}^{a} |f(x) - S_N(x)|^2 \, dx \right]^{1/2} = 0$$

where $S_N(x) = \sum_{n=-N}^{+N} c_n e^{in\pi x/a}$ is the Nth partial sum of the complex Fourier series for f.

Exercises

(7.22) Show that $\sum_{m=1}^{\infty} 1/m$ is not summable by arithmetic means, that is, $\lim_{n\to\infty} \sigma_n$ fails to exist for this series.

(7.23) Verify (7.5).

(7.24) Suppose that f is piecewise continuous with period 2π. Prove that if $m \le f(x) \le M$, then $m \le \sigma_n(x) \le M$ for all x values.

(7.25) Prove that Gibb's phenomenon does not occur for discontinuous piecewise continuous functions when arithmetic means are used.

(7.26) Prove that for each m

$$\left| \sum_{n=1}^{m} \frac{\sin nx}{n} \right| \le \tfrac{1}{2}\pi + 1$$

[*Hint*: Compare the arithmetic means of the Fourier series $\sum_{n=1}^{\infty} n^{-1} \sin nx$ with its partial sums, using (7.5).]

(7.27) *Weierstrass Approximation Theorem.* Suppose that f is a continuous function on $[-\pi, \pi]$ and that $f(-\pi) = f(\pi)$. Prove that, given $\epsilon > 0$, there exists a finite trigonometric sum

$$T_\epsilon(x) = C_0 + \sum_{n=1}^{m} [C_n \cos nx + D_n \sin nx]$$

such that $|f(x) - T_\epsilon(x)| < \epsilon$ for all x values in $[-\pi, \pi]$.

(7.28) Suppose that f is a piecewise continuous function with period 2π. Prove that if x is a point of continuity for f, then the Fourier series for f evaluated at x is summable by arithmetic means to $f(x)$.

$\star$(7.29) Suppose that f is a piecewise continuous function with period 2π. Prove that if x is a point of discontinuity for f, then the Fourier series for f evaluated at x is summable by arithmetic means to the value $\tfrac{1}{2}[f(x+) + f(x-)]$.

(7.30) Suppose that x is a point of continuity for a piecewise continuous function f with period 2π. Prove that if the Fourier series for f converges at x, then it converges to the value $f(x)$.

$\star$(7.31) Prove that $\sigma_n(x)$ can also be expressed in the two alternative forms

$$\sigma_n(x) = \frac{1}{2\pi} \int_{-\pi}^{\pi} f(x-v) F_n(v) \, dv = \frac{1}{2\pi} \int_{-\pi}^{\pi} f(v) F_n(x-v) \, dv$$

What do these two integrals describe geometrically?

(7.32)   Show that σ_n has the complex form

$$\sigma_n(x) = \sum_{k=-n}^{n} \left(1 - \frac{|k|}{n}\right) c_k e^{ikx}$$

when $f \sim \sum c_k e^{ikx}$. Generalize Fejér's theorem to complex valued functions.

§8. Summation Kernels

Fejér's kernel, in §7, was our first example of a *summation kernel*. *Summation kernels are a fundamental tool of mathematical analysis; they will be used frequently in this book.* In this section we shall discuss their principal properties.

The reader should compare the following definition with Lemma (7.11).

(8.1) Definition.   A sequence (family) of piecewise continuous functions K_τ on the interval $(-a, a)$, parameterized by the positive real variable τ, is called a *summation kernel* if it has the following properties

(A₁) $K_\tau(u) \geq 0$ for each $\tau > 0$

(A₂) $\dfrac{1}{2a} \displaystyle\int_{-a}^{a} K_\tau(u)\, du = 1$ for each $\tau > 0$

(A₃)   For every $\epsilon > 0$ and $\delta > 0$ we have

$$0 \leq K_\tau(u) < \epsilon$$

provided $a \geq |u| \geq \delta$ and τ is taken close enough to 0. ■

Property (A₃) can also be expressed by saying that $\lim_{\tau \to 0+} K_\tau(u) = 0$ holds *uniformly* on $[-a, -\delta]$ and $[\delta, a]$ for each $\delta > 0$.

(8.2) Examples.   Here are a few examples of summation kernels. The verifications of (8.1) are left to the reader as an exercise.

(a)   *Point Impulse Kernel.* Define K_τ for $0 < \tau < \frac{1}{2}$ by

$$K_\tau(u) = \begin{cases} \dfrac{1}{2\tau} & \text{for } |u| \leq \tau \\ 0 & \text{for } |u| > \tau \end{cases}$$

Then K_τ is a summation kernel on $(-\frac{1}{2}, \frac{1}{2})$.

(b)   Define K_τ for $0 < \tau < \frac{1}{2}$ by

$$K_\tau(u) = \begin{cases} \tau^{-1}(1 - |x|/\tau) & \text{for } |x| \leq \tau \\ 0 & \text{for } |x| \geq \tau \end{cases}$$

Then K_τ is a summation kernel on $(-\frac{1}{2}, \frac{1}{2})$.

(c) Let g be a nonnegative continuous function satisfying $\frac{1}{2a} \int_{-a}^{a} g(x)\, dx = 1$. Define K_τ for $0 < \tau < 1$ by

$$K_\tau(u) = \begin{cases} \tau^{-1} g(u/\tau) & \text{for } |u| \leq a\tau \\ 0 & \text{for } |u| \geq a\tau \end{cases}$$

Then K_τ is a summation kernel on the interval $(-a, a)$.

(d) *Fejér kernel.* Define K_τ by

$$K_\tau(u) = \tau \left[\frac{\sin(\frac{1}{2} u \tau^{-1})}{\sin(\frac{1}{2} u)} \right]^2$$

If we assign τ the values n^{-1} for each positive integer n, then K_τ equals the Fejér kernel F_n. Lemma (7.11) shows that K_τ is a summation kernel over $[-\pi, \pi]$.

By modifying slightly the proofs of Theorems (7.12) and (7.18), the reader may prove the following theorem.

(8.3) Theorem. Let K_τ be a summation kernel on the interval $(-a, a)$. Suppose that f is a continuous function with period $2a$. Then

$$\lim_{\tau \to 0+} \frac{1}{2a} \int_{-a}^{a} f(x - u) K_\tau(u)\, du = f(x)$$

holds uniformly for all x values.

The previous theorem has an extensive generalization. Let the function f be called *absolutely integrable* (over $[-a, a]$) if $\int_{-a}^{a} |f(u)|\, du$ is finite (converges). The reader may take this definition to as high a level of generality as possible; f may be viewed as either piecewise continuous on $(-a, a)$, or $|f|$ as properly or improperly Riemann integrable over $(-a, a)$, or f as Lebesgue integrable on $(-a, a)$.

(8.4) Theorem. Let K_τ be a summation kernel on $(-a, a)$. Suppose f has period $2a$ and f is absolutely integrable over $[-a, a]$, then

$$\lim_{(x,\tau) \to (x_0, 0+)} \frac{1}{2a} \int_{-a}^{a} f(x - u) K_\tau(u)\, du = f(x_0)$$

provided that x_0 is a point of continuity of f.

Proof. For simplicity of notation, we will assume that the period of f is 1 $(a = \frac{1}{2})$. The reader should have no trouble modifying the argument to cover the general case of period $2a$. Using properties (A$_1$) and (A$_2$), and proceeding as we did when we proved Theorem (7.12), we obtain for

$0 < \delta < \tfrac{1}{2}$

$$\left| \int_{-\frac{1}{2}}^{\frac{1}{2}} f(x-u) K_\tau(u)\, du - f(x_0) \right| \le \int_{-\frac{1}{2}}^{\frac{1}{2}} |f(x-u) - f(x_0)|\, K_\tau(u)\, du$$

$$= \int_{|u| \ge \delta} f(x-u) - f(x_0)|\, K_\tau(u)\, du + \int_{|u| < \delta} |f(x-u) - f(x_0)|\, K_\tau(u)\, du$$

(8.5)

$$\le \int_{|u| \ge \delta} |f(x-u)|\, K_\tau(u)\, du + \int_{|u| \ge \delta} |f(x_0)|\, K_\tau(u)\, du$$

$$+ \int_{|u| < \delta} |f(x-u) - f(x_0)|\, K_\tau(u)\, du$$

To get the last inequality above we used the triangle inequality in the form $|f(x-u) - f(x_0)| \le |f(x-u)| + |f(x_0)|$.

Now, let $\epsilon > 0$ be given. Suppose $|x - x_0| < \delta$ and $|u| < \delta$, then $|(x-u) - x_0| < 2\delta$. Hence, choosing δ small enough, the continuity of f at x_0 forces $|f(x-u) - f(x_0)| < \epsilon$. This last inequality implies

(8.6)
$$\int_{|u| < \delta} |f(x-u) - f(x_0)|\, K_\tau(u)\, du < \epsilon \int_{|u| < \delta} K_\tau(u)\, du < \epsilon$$

where (A_1) and (A_2) were used to get the last inequality.

Using (8.6), and applying (A_3) to the first two integrals in the final inequality in (8.5), we have for τ close enough to 0

$$\left| \int_{-\frac{1}{2}}^{\frac{1}{2}} f(x-u) K_\tau(u)\, du - f(x_0) \right|$$

(8.7)
$$< \epsilon \left[\int_{|u| \ge \delta} |f(x-u)|\, du + \int_{|u| \ge \delta} |f(x_0)|\, du + 1 \right]$$

$$\le \epsilon \left[\int_{-\frac{1}{2}}^{\frac{1}{2}} |f(x-u)|\, du + |f(x_0)| + 1 \right]$$

By a change of variable and the periodicity of f we have

$$\int_{-\frac{1}{2}}^{\frac{1}{2}} |f(x-u)|\, du = \int_{-\frac{1}{2}}^{\frac{1}{2}} |f(u)|\, du$$

Therefore, for $|x - x_0| < \delta$ and τ close enough to 0

$$\left| \int_{-\frac{1}{2}}^{\frac{1}{2}} f(x-u) K_\tau(u)\, du - f(x_0) \right| < \epsilon \left[\int_{-\frac{1}{2}}^{\frac{1}{2}} |f(u)|\, du + |f(x_0)| + 1 \right]$$

which, since ϵ may be taken arbitrarily small, yields the desired limit. ■

We now turn to the problem of describing summation kernels over the infinite interval $\mathbb{R} = (-\infty, +\infty)$. Our definition is almost identical to Definition (8.1).

(8.8) Definition. A sequence (family) of piecewise continuous functions K_τ, parameterized by the positive real variable τ, is called a *summation kernel* (over $\mathbb{R}$) if it has the following properties

(A₁) $K_\tau(u) \geq 0$ for each $\tau > 0$
(A₂) $\int_{-\infty}^{+\infty} K_\tau(u)\, du = 1$ for each $\tau > 0$
(A₃) For every $\epsilon > 0$ and $\delta > 0$ we have

$$0 \leq K_\tau(u) < \epsilon \qquad 0 \leq \int_{|u| \geq \delta} K_\tau(u)\, du < \epsilon$$

provided $|u| \geq \delta$ and τ is taken close enough to 0. ∎

(8.9) Examples. Here are some examples, the verifications are left to the reader.

(a) The point impulse kernel, and the kernel in Example (8.2b), are also summation kernels over $\mathbb{R}$ for all $\tau > 0$. The general kernel described in (8.2c) also works for $\mathbb{R}$ if the value $a = \frac{1}{2}$ is used.

(b) *Poisson kernel.* Define K_τ by

$$K_\tau(u) = \frac{1}{\pi} \frac{\tau}{\tau^2 + u^2}$$

for $\tau > 0$. Then K_τ is a summation kernel over $\mathbb{R}$.

For summation kernels over $\mathbb{R}$ we have the following two results. By an absolutely integrable function f over $\mathbb{R}$ we mean that $\int_{-\infty}^{+\infty} |f(u)|\, du$ is finite (converges).

(8.10) Theorem. Let K_τ be a summation kernel over $\mathbb{R}$. If f is absolutely integrable over $\mathbb{R}$, then

$$\lim_{(x,\tau) \to (x_0, 0+)} \int_{-\infty}^{+\infty} f(x - u) K_\tau(u)\, du = f(x_0)$$

provided x_0 is a point of continuity for f.

Proof. First, notice that the first inequality in (A₃) shows that K_τ is a bounded function for each *fixed* τ. Therefore, the integral in the limit above is defined (convergent) for each τ. Now, let $\epsilon > 0$ be given. Noting the similarities between Definitions (8.1) and (8.8), we see that (8.5) and (8.6) continue to hold (with $\int_{-\infty}^{+\infty}$ in place of $\int_{-\frac{1}{2}}^{\frac{1}{2}}$). Thus, for $|x - x_0| < \delta$ we have

$$\left| \int_{-\infty}^{+\infty} f(x - u) K_\tau(u)\, du - f(x_0) \right|$$

$$< \int_{|u| \geq \delta} |f(x - u)|\, K_\tau(u)\, du + \int_{|u| \geq \delta} |f(x_0)|\, K_\tau(u)\, du + \epsilon$$

Applying the first inequality in (A₃) to the first integral on the right above, and the second inequality in (A₃) to the second integral above, we have for

τ close enough to 0 (and $|x - x_0| < \delta$)

$$\left| \int_{-\infty}^{+\infty} f(x - u)K_\tau(u)\, du - f(x_0) \right| < \epsilon \left[\int_{|u| \geq \delta} |f(x - u)|\, du + |f(x_0)| + 1 \right]$$

$$\leq \epsilon \left[\int_{-\infty}^{+\infty} |f(u)|\, du + |f(x_0)| + 1 \right]$$

Since ϵ can be taken arbitrarily small, this yields the desired limit. ∎

(8.11) Theorem.   Let K_τ be a summation kernel over $\mathbb{R}$. If f is a bounded continuous function, then for every point x_0 in $\mathbb{R}$

$$\lim_{(x, \tau) \to (x_0, 0+)} \int_{-\infty}^{+\infty} f(x - u)K_\tau(u)\, du = f(x_0)$$

Moreover, if $\lim_{|x| \to +\infty} f(x) = 0$ then the limit holds uniformly for all x_0 in $\mathbb{R}$.

Proof.   Suppose $|f(v)| \leq M$ for all v in $\mathbb{R}$, where M is a positive constant. Let $\epsilon > 0$ be given. Combining (8.5) and (8.6), we have for $|x - x_0| < \delta$

$$\left| \int_{-\infty}^{+\infty} f(x - u)K_\tau(u)\, du - f(x_0) \right| < 2M \int_{|u| \geq \delta} K_\tau(u)\, du + \epsilon$$

Taking τ close enough to 0 we have

(8.12) $$\left| \int_{-\infty}^{+\infty} f(x - u)K_\tau(u)\, du - f(x_0) \right| < \epsilon(2M + 1)$$

which proves the limit. If $\lim_{|x| \to +\infty} f(x) = 0$ then f is *uniformly continuous* over all of $\mathbb{R}$ (not just over any finite closed interval). Then the choice of δ that makes (8.6) hold can be made uniformly for all x_0 in $\mathbb{R}$, and (8.12) will then hold uniformly as well. ∎

(8.13) Notation.   If g_1 is a function on $(-a, a)$ and g_2 is a function of period $2a$, then we write $g_1 * g_2$ for the function defined by

$$g_1 * g_2(x) = \frac{1}{2a} \int_{-a}^{a} g_1(u)g_2(x - u)\, du$$

The integral above is called the *convolution (integral)* of g_1 with g_2. Using this notation, we can express the result of Theorem (8.4) as

$$\lim_{(x, \tau) \to (x_0, 0+)} K_\tau * f(x) = f(x_0)$$

This notation carries over to $\mathbb{R}$ if we write

$$g_1 * g_2(x) = \int_{-\infty}^{+\infty} g_1(u)g_2(x - u)\, du$$

for two functions g_1 and g_2 over $\mathbb{R}$. We will discuss convolution over $\mathbb{R}$ at some length in Chapter 7.

Remark. Frequently it is said that a summation kernel *converges to the Dirac delta function*, that is $\lim_{\tau \to 0+} K_\tau(x) = \delta(x)$. The *Dirac delta function* $\delta(x)$ is supposed to satisfy

(a) $\delta * f(x) = f(x)$ (convolution identity)

for each absolutely integrable, continuous function f. Formula (a) must, however, be interpreted carefully. As the reader can easily check, for each of the kernels in (8.2) and (9.9), $\lim_{\tau \to 0+} K_\tau(x) = 0$ for $x \neq 0$. [This also follows from (A_3) in (8.1) or (8.8).] Thus $\delta(x) = 0$ for $x \neq 0$ and (a) makes no sense. One way of meaningfully interpreting (a) is to view it as a convenient shorthand for

(b) $$\lim_{\tau \to 0+} K_\tau * f(x) = f(x)$$

which we know to be correct. This is often done in physics and engineering, and it usually saves a lot of time. A firm mathematical foundation for interpreting (a) has been discovered only fairly recently and is beyond the scope of this book. [See Gel'fand and Shilov (1964) or Vladimirov (1971).] Formulas (a) and (b) reveal why summation kernels are also called *approximations to the identity*.

Exercises

(8.14) Prove Theorem (8.3) and check that all the kernels in Examples (8.2) and (8.9) are summation kernels.

*(8.15)** Let f, g, and h be piecewise continuous functions of period $2a$. Prove the following results, concerning convolution:

(a) *Commutativity.* $f * g = g * f$.
(b) *Associativity.* $f * (g * h) = (f * g) * h$.
(c) *Multiplier property.* If f has the complex Fourier series $\sum c_n e^{i(n\pi x/a)}$ and g has the complex Fourier series $\sum d_n e^{i(n\pi x/a)}$, then $f * g$ has the complex Fourier series $\sum c_n d_n e^{i(n\pi x/a)}$.

(8.16) *Landau's kernel.* Let $a_n = [\int_{-1}^{1} (1 - x^2)^n \, dx]^{-1}$ for each positive integer n. Landau's kernel L_n is defined by $L_n(u) = 2a_n(1 - u^2)^n$ for $|u| \leq 1$. Prove that L_n is a summation kernel over $[-1, 1]$ (where $L_n = K_\tau$ for $\tau = 1/n$). [*Hint*: For (A_3) replace $1 - u^2$ by $1 - |u|$ to estimate a_n.]

(8.17) Using Landau's kernel prove the following theorem.

Theorem: Weierstrass' Polynomial Approximation. Let f be a continuous function on a closed interval $[a, b]$. For each $\epsilon > 0$ there is a polynomial p_ϵ such that $|f(x) - p_\epsilon(x)| < \epsilon$ holds for all x values in $[a, b]$.

(*Hint*: First, let $[a, b] = [-\frac{1}{2}, \frac{1}{2}]$, make L_n and f have period 2, use (8.3) and (8.15a). Change variables to treat the general interval $[a, b]$.)

(8.18) *Gauss–Weierstrass kernel.* Let $K_\tau(u) = \tau^{-1} e^{-\pi u^2/\tau^2}$ for all $\tau > 0$. Prove that K_τ is a summation kernel over $\mathbb{R}$. [*Hint*: $\int_{-\infty}^{+\infty} e^{-\pi x^2} \, dx = 1$.]

$\star$**(8.19)** *Hann's kernel.* Let $f \sim \frac{1}{2} A_0 + \sum [A_k \cos kx + B_k \sin kx]$ where f is a piecewise continuous function with period 2π. Define R_n by

$$(*) \qquad R_n(x) = \frac{1}{2} A_0 + \sum_{k=1}^{n} \left[\frac{1}{2} + \frac{1}{2} \cos\left(\frac{k\pi}{n}\right) \right] \{A_k \cos kx + B_k \sin kx\}$$

for each positive integer n. Prove that

$$(**) \qquad \begin{cases} R_n(x) = \dfrac{1}{2\pi} \displaystyle\int_{-\pi}^{\pi} f(x - u) H_n(u) \, du \\[2mm] H_n(u) = \frac{1}{2} D_n(u) + \frac{1}{4} D_n\left(u - \dfrac{\pi}{n}\right) + \frac{1}{4} D_n\left(u + \dfrac{\pi}{n}\right) \end{cases}$$

where D_n is the Dirichlet kernel. [*Remark. Hann's kernel H_n is used in the processing of telephone signals.*]

$\star$**(8.20)** Using a computer, plot the Fourier series partial sum S_n, the arithmetic mean σ_n, and R_n [see $(8.19*)$] for a sequence of values of n for each of the functions in Examples (4.6), (4.7), and (5.3) in Chapter 1.

(8.21) Let K_τ be a summation kernel on $(-a, a)$ and assume that K_τ is an even function for each $\tau > 0$. Prove that for all x values

$$\lim_{\tau \to 0+} \frac{1}{2a} \int_{-a}^{a} f(x - u) K_\tau(u) \, du = \frac{1}{2}[f(x+) + f(x-)]$$

provided f is a piecewise continuous function with period $2a$. Generalize this result to summation kernels over $\mathbb{R}$.

(8.22) Give examples which show that $\lim_{(x,\tau) \to (x_0, 0+)} K_\tau * f(x)$ does *not* exist when x_0 is a point of discontinuity of a piecewise continuous function f. [Compare this with Exercise (8.21).]

$\star$**(8.23)** Generalize the definition of K_τ in (8.1) as follows

(A_1') $\displaystyle\int_{-a}^{a} |K_\tau(u)| \, du \leq M$ a constant, for each $\tau > 0$

(A_2') $\dfrac{1}{2a} \displaystyle\int_{-a}^{a} K_\tau(u) \, du = 1$ for each $\tau > 0$

(A_3') For every $\epsilon > 0$ and $\delta > 0$ we can have $0 \leq |K_\tau(u)| < \epsilon$ provided $a \geq |u| \geq \delta$ and τ is close enough to 0.

Prove that Theorems (8.3) and (8.4) remain true, using (A_1') through (A_3'). Generalize Definition (8.8) and Theorems (8.10) and (8.11) in a similar way.

(8.24) *de la Vallée Poussin's kernel.* Let V_{2m} be defined by $V_{2m} = 2F_{2m+1} - F_m$ where F_n is Fejér's kernel. Show that V_{2m} is a summation kernel over $(-\pi, \pi)$ in the generalized sense of (8.23). (Here $V_{2m} = K_\tau$ for $\tau = 1/2m$.)

(8.25) Let f be a piecewise continuous function with period 2π. Show that

$$\frac{1}{2\pi} \int_{-\pi}^{\pi} f(x+u)V_{2m}(u)\, du = S_{m-1} + 2 \sum_{k=m}^{2m} \left(1 - \frac{k}{2m}\right)\left[A_k \cos kx + B_k \sin kx\right]$$

where V_{2m} is de la Vallée Poussin's kernel [see (8.24)].

(8.26) Suppose that g is a continuous function over $\mathbb{R}$ that satisfies

(a) $\displaystyle\int_{-\infty}^{+\infty} g(v)\, dv = 1$

(b) $\displaystyle |g(v)| \leq \frac{B}{1+v^2}$ for all v in $\mathbb{R}$.

Show that $K_\tau(u) = \tau^{-1}g(u/\tau)$ defines a summation kernel over $\mathbb{R}$ in the generalized sense of Exercise (8.23).

*(8.27) Is Hann's kernel H_n [see (8.19)] a summation kernel over $(-\pi, \pi)$, either in the sense of (8.1) or (8.23)?

*(8.28)

(a) Let $f \sim \frac{1}{2}A_0 + \sum_{k=1}^{\infty} [A_k \cos kx + B_k \sin kx]$, where f is piecewise continuous and has period 2π. Let g be an even, piecewise continuous, function satisfying $g(0) = 1$ and $g(x) = 0$ for $|x| > \pi$. Prove that

$$(*) \quad \frac{1}{2}A_0 + \sum_{k=1}^{n} g\left(\frac{k\pi}{n}\right)\left[A_k \cos kx + B_k \sin kx\right]$$
$$= \frac{1}{2\pi} \int_{-\pi}^{\pi} f(x-u)G_n(u)\, du$$

for some function G_n.

(b) Show that $(*)$ has the complex form

$$(**) \quad \sum_{k=-n}^{+n} g\left(\frac{k\pi}{n}\right)c_k e^{ikx} = \frac{1}{2\pi} \int_{-\pi}^{\pi} f(x-u)G_n(u)\, du$$

*(c) Is G_n a summation kernel (for $\tau = 1/n$), either in the sense of (8.1) or (8.23)? *Remark.* In communications, and in time series theory, the function g is sometimes called a *window*, the function G_n is called a *spectral* (or *frequency*) *window*.

(d) Graph the windows and spectral windows that describe pointwise convergence, summation by arithmetic means, summation by Hann's method [see (8.19)], and summation by de la Vallee Poussin's method [see (8.25)].

(8.29) Generalize Definition (8.8) to $\mathbb{R}^2$ and show that the *point impulse kernel*

$$K_\tau(u, v) = \begin{cases} (\pi\tau^2)^{-1} & \text{for } [u^2 + v^2]^{1/2} \leq \tau \\ 0 & \text{for } [u^2 + v^2]^{1/2} > \tau \end{cases}$$

and the *Poisson kernel*

$$K_\tau(u, v) = \frac{1}{2\pi} \frac{\tau}{[u^2 + v^2 + \tau^2]^{3/2}}$$

are both summation kernels over $\mathbb{R}^2$.

(8.30)   Show that Dirichlet's kernel is *not* a summation kernel, either in the sense of (8.1) or (8.23). In particular, $\int_{-\pi}^{\pi} |D_n(u)| \, du \to +\infty$ as $n \to +\infty$.

(8.31)   Some authors replace (A₃) by

$$(A_3') \qquad \lim_{\tau \to 0+} \int_{|u| \geq \delta} K_\tau(u) \, du = 0 \qquad \text{for each } \delta > 0$$

Show that Theorems (8.3) and (8.11) remain true after this replacement, but (8.4) and (8.10) may be false.

References

For further information the reader might consult Natanson (1964/1965), Titchmarsh (1939), Davis (1975), Wheeden and Zygmund (1977), Zygmund (1968), or Carslaw (1950).

3

Applications of Fourier Series

In this chapter we shall describe the application of Fourier series to some elementary problems in mathematical physics. We will focus on the vibrations of elastic strings and heat conduction through a finite rod. In §1 we derive the equations and problems that we shall consider throughout the chapter. In §2 we look at the solution by Fourier series of the problem of the force-free vibration of elastic strings. The mathematical verification of that solution is discussed in §3 where we describe D'Alembert's solution of the vibrating string problem. In §4 we discuss the problem that gave rise to much of our subject: Fourier's solution of the problem of heat conduction through a finite rod. We return to vibration problems in §5 where we discuss the applied force problem for vibrating strings. That problem is important physically because the phenomenon of resonance occurs.

§1. Derivation of Some Problems

We now derive the mathematical forms for several physical problems that will be considered in this chapter. We begin with the one-dimensional wave equation.

Wave Equation

Consider an elastic string of length L, such as a guitar string or a piano string, which is fastened at its endpoints. Let $y(x, t)$ stand for the height of the string above the horizontal x axis at the point x and time $t \geq 0$. (See Figure 3.1.) The condition of fastened ends can be expressed as $y(0, t) = 0$ and $y(L, t) = 0$ for each time t. Let's assume as well that the string has a given *initial position* at the time $t = 0$ for which we write $y(x, 0) = f(x)$ for $0 \leq x \leq L$. And we assume that the string has an *initial velocity*; this velocity

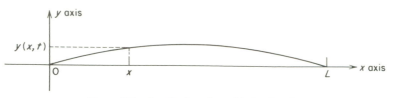

Figure 3.1   An elastic string with fixed ends.

can be expressed through a *partial derivative in time* y_t, that is, $y_t(x, 0) = g(x)$. Thus, we have

(1.1) $y(0, t) = 0$ $y(L, t) = 0$ $y(x, 0) = f(x)$ $y_t(x, 0) = g(x)$

The first two conditions in (1.1) are called *boundary conditions*, whereas the second two conditions are called *initial conditions*.

We now derive the wave equation. The following assumptions will be needed

(1.2)

 (a) The string has a constant linear density ρ.
 (b) The length L of the string is essentially constant.
 (c) The magnitude of tension T at all points of the string is essentially constant.
 (d) No force is acting on the string (force-free case).
 (e) The string only moves up or down, never horizontally.

Although these assumptions are never truly realized; they are a good approximation to the truth when the string is short, the tension is high, and the string is only deflected slightly. This is typically the case when guitar or violin strings are played (but is not the case with most piano strings).

A diagram of the forces acting on a small element of the string at a fixed time is shown below in Figure 3.2. Since we have assumed in (1.2e) that there is only vertical motion, we need to consider only the vertical components of the forces acting on the string element. Let ϕ be the angle between the tension along the tangent line to the string and the horizontal direction as shown in Figure 3.2. The mass of the string element is $\rho \Delta x$ and its acceleration is approximately $y_{tt}(x, t)$ provided Δx is small enough in magnitude. Thus, Newton's second law (force equals mass times acceleration) yields

(1.3) $$T \sin(\phi + \Delta\phi) - T \sin \phi \doteq \rho \Delta x y_{tt}(x, t)$$

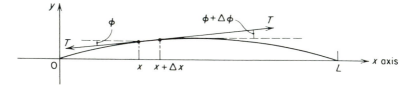

Figure 3.2   Diagram of the forces acting on a small element of a string.

Here $\doteq$ stands for *closely approximates*. To get (1.3) we have assumed that Δx is very small in magnitude and that $y_{tt}(x, t)$ is a continuous function of x [so that $y_{tt}(x, t) \doteq y_{tt}(x + \Delta x, t)$]. The assumed constant length of the string allows only a small amount of bending, hence both ϕ and $\phi + \Delta\phi$ are very small in magnitude. Since $\tan \phi / \sin \phi$ tends to 1 as ϕ tends to 0, (1.3) becomes

(1.4) $$T[\tan(\phi + \Delta\phi) - \tan \phi] \doteq \rho \Delta x y_{tt}$$

The slope of a tangent line to a functional curve can be expressed as $\tan \phi$, therefore Eq. (1.4) becomes

(1.5) $$T[y_x(x + \Delta x, t) - y_x(x, t)] \doteq \rho \Delta x y_{tt}$$

Dividing (1.5) by $\rho \Delta x$ and remembering that Δx is meant to be negligibly small in magnitude we obtain

(1.6) $$c^2 y_{xx} = y_{tt} \qquad (c^2 = T/\rho)$$

where we have dropped the approximate equality and introduced the notation c^2 for the positive constant T/ρ.

Equation (1.6) is called the one-dimensional *wave equation*. In the next section we will solve it subject to the boundary and initial conditions in (1.1).

The wave equation should be interpreted as a *mathematical model* for the vibrations of a string. It, or its generalizations to higher dimensions, also model other forms of vibratory motion such as the motion of sound waves in air and other media, or the motion of light waves through the electromagnetic field. Under many circumstances, solutions to Eq. (1.6) will compare well with experience. When they do not, then one or more of the assumptions in (1.2) might be relaxed or dropped altogether. For example, assumption (1.2d), that there are no forces acting on the string, is clearly unreasonable under many circumstances. Let $F(x, t)$ stand for a force *per unit length* in the y direction applied to the string over the course of time. The balance of forces for a string element leads to

(1.7) $$T[\sin(\phi + \Delta\phi) - \sin \phi] + F(x, t)\Delta x \doteq \rho \Delta x y_{tt}$$

Proceeding as we did above, we obtain the following equation from (1.7):

(1.8) $$c^2 y_{xx} + F(x, t)/\rho = y_{tt} \qquad (c^2 = T/\rho)$$

Equation (1.8) is called the *inhomogeneous wave equation*.[1] The inhomogeneous wave equation, along with the boundary and initial conditions in (1.1), is called the *applied force problem for the vibrating string*. We will solve it in §5.

---

1. Equation (1.8) is called *inhomogeneous* in order to distinguish it from (1.6) from which it differs by only the term $F(x, t)/\rho$ which is independent of y.

Heat Equation

Conduction of heat through matter can often be described by an equation known as the heat equation. We shall now derive that equation.

Suppose that we have a cylindrical rod, homogeneous in its physical properties (density, material, cross sections, etc.), whose lateral surface is insulated. Choose the x axis to lie along the central axis of the rod and let $u(x, t)$ stand for the temperature at time t of the cross section of the rod which passes through the point x. Consider the element C of the rod which lies between x and $x + \Delta x$ (See Figure 3.3).

For physical reasons we assume if Δt is a small enough quantity of time, then at each point x the temperature u will not vary significantly from time t to time $t + \Delta t$. Many experiments with different materials and rods of various densities have shown that the amount of heat flow q through a rod whose ends are at constant temperatures is proportional to the cross-sectional area A of the rod, to the difference in temperatures between the ends of the rod, and to the length of the time interval Δt. Moreover, q is inversely proportional to the length of the rod. Applying all these facts to the rod element C, and noting that heat flow is *opposite* to the direction of the temperature increase, we obtain

(1.9)
$$q = - KA\Delta t \frac{u(x + \Delta x, t) - u(x, t)}{\Delta x}$$

The constant of proportionality K is called the *thermal conductivity* of the material of the rod. See Table 3.1a for some experimentally determined values of K.

If we let Δx tend to zero, then we obtain from (1.9) that the amount of heat $Q(x)$ flowing through the cross section at x in the positive x direction during the time interval is

(1.10)
$$Q(x) = - KA\Delta t u_x(x, t)$$

At the point $x + \Delta x$, we have by the same reasoning that

(1.11)
$$Q(x + \Delta x) = - KA\Delta t u_x(x + \Delta x, t).$$

From (1.10) and (1.11) we conclude that the net amount of heat flowing *into* the element C in the time interval Δt is given by (see Figure 3.4).

(1.12)
$$\Delta Q = KA\Delta t[u_x(x + \Delta x, t) - u_x(x, t)]$$

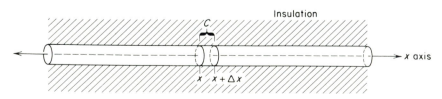

Figure 3.3   A homogeneous cylindrical rod, lateral surface insulated.

Table 3.1a Thermal Conductivity for some Common Materials

Material	K (Btu/hr ft °F)
Cast iron	27.6
Granite	1.7
Brick	0.4
Water	0.36

Table 3.1b Thermal Diffusivity for some Common Materials

Material	a^2 (cm²/sec)
Cast iron	0.12
Granite	0.011
Brick	0.0038
Water	0.00144

It has also been experimentally verified that the heat flow ΔQ into the element C results in a temperature change Δu which is directly proportional to ΔQ and inversely proportional to the mass m of the element. Thus, we have

(1.13)
$$\Delta u = \frac{1}{c}\frac{\Delta Q}{m} = \frac{\Delta Q}{c\rho A \Delta x}$$

where ρ is the density of the rod and the constant of proportionality $1/c$ is called the *specific heat* for the material of the rod. If we take Δx to be small enough in magnitude, then we may assume that the temperature u is essentially constant along the cross sections of the element C. Therefore, we can measure Δu by the temperature change for the cross section at x, obtaining $\Delta u = u(x, t + \Delta t) - u(x, t)$. Hence, (1.13) becomes

$$u(x, t + \Delta t) - u(x, t) = \frac{\Delta Q}{c\rho A \Delta x}$$

Solving for ΔQ in the equation above and then substituting into (1.12)

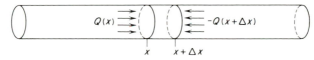

Figure 3.4 Illustration of formula (1.12).

yields upon dividing by $c\rho A \Delta t \Delta x$

$$\frac{u(x, t + \Delta t) - u(x, t)}{\Delta t} = \frac{K}{c\rho} \frac{u_x(x + \Delta x, t) - u_x(x, t)}{\Delta x}$$

Letting Δx and Δt tend to zero in the equation above we obtain the one-dimensional *heat equation*

(1.14) $u_t = a^2 u_{xx}$ $(a^2 = K/c\rho)$

The constant a^2 is called the *thermal diffusivity* for the material of the rod. See Table 3.1b for some experimentally determined values of a^2.

The reader may observe that the considerations above clearly apply to any homogeneous bar with congruent cross sections; hence the heat equation is valid for those objects as well.

We will now derive a few boundary and initial conditions for (1.14). Assume that the rod has a length L with the left end of the rod located at $x = 0$ and the right end of the rod located at $x = L$. If the rod is held at zero temperature, say in degrees Celcius, at both ends[2] with an initial temperature distribution $f(x)$ at time $t = 0$, then we have the following conditions

(1.15) $u(0, t) = 0$ $u(L, t) = 0$ $u(x, 0) = f(x)$

The first two conditions in (1.15) are called boundary conditions, whereas the third is called an initial condition.

If the ends of the rod are free to exchange heat with another medium, then we apply *Newton's Law of Cooling*. That law states that *the temperature change across the boundary between two mediums is proportional to their difference in temperatures.* Thus, letting h be the positive constant of proportionality we obtain

$$-u_x(0, t) = h[u(0, t) - M]   u_x(L, t) = h[u(L, t) - M]$$

where M is the temperature of the medium. If we set the temperature scale so that M is the zero point on that scale, then the equations above simplify to

$$-u_x(0, t) = hu(0, t)   u_x(L, t) = hu(L, t)$$

Assuming an initial temperature distribution as above we have the following boundary and initial conditions

(1.16) $u_x(0, t) + hu(0, t) = 0$ $u_x(L, t) - hu(L, t) = 0$ $u(x, 0) = f(x)$

where $h > 0$ is a constant.

Exercises

(1.17)   Let $T(x)$ denote the tension at the point x, where $0 \le x \le L$, on an elastic string. Let $\rho(x)$ denote the linear density of the string at the point x.

---

2. This might be done by placing large *heat reservoirs* at both ends of the bar.

Assuming that $T(x)$, $T'(x)$, and $\rho(x)$ are all continuous for $0 \le x \le L$ derive the following *generalized wave equation* for $y(x, t)$ the height of the string at the point x and the time t

$$[T(x)y_x]_x = \rho(x)y_{tt}$$

(1.18) Prove that if y_1 and y_2 solve the wave equation (1.6), then $b_1 y_1 + b_2 y_2$ solves (1.6) where b_1 and b_2 are real constants. Prove that a similar result holds for (1.14), the heat equation, but does not hold for (1.8).

(1.19) Show that if a force $F(x, t)$ per unit length is applied to an elastic string, under the same conditions as in (1.17), then

$$[T(x)y_x]_x + F(x, t) = \rho(x)y_{tt}$$

(1.20) Suppose that instead of an insulated cylindrical rod, as considered in the text, the lateral surface of the rod exchanges heat with a surrounding medium of temperature T. Assuming Newton's Law of Cooling, derive the following equation for $u(x, t)$ the temperature at point x and time t

$$u_t = a^2 u_{xx} - h(u - T)$$

where $h > 0$ is a constant.

(1.21) Consider the insulated rod discussed in the text. Suppose that the ends of the rod exchange heat with surrounding gaseous mediums according to Newton's Law of Cooling. Show that if the temperature of the medium at the left end of the rod is A, while the temperature at the right end of the rod is B, then the rod's temperature $u(x, t)$ satisfies

$$u_t = a^2 u_{xx} \qquad u_x(0, t) + hu(0, t) = hA$$
$$u_x(L, t) - hu(L, t) = -hB \qquad u(x, 0) = f(x)$$

(1.22) Suppose that an insulated rod, as considered in the text, with a temperature at its left end having a constant value A whereas its right end has a constant temperature of B has reached a state of equilibrium in terms of its temperature. Show that its temperature $u = u(x)$ satisfies

$$u''(x) = 0 \qquad u(0) = A \qquad u(L) = B$$

Determine $u(x)$.

(1.23) Suppose that instead of an insulated cylindrical rod, as considered in the text, the lateral surface of the rod is being heated (cooled) by some heat source. Assuming that the heat is absorbed by the material of the rod and over a time period Δt there is a temperature change δu given by

$$\delta u = \frac{1}{c} \frac{Q(x, t)}{\rho A} \Delta t$$

show that the temperature function $u(x, t)$ satisfies

$$u_t = a^2 u_{xx} + \frac{1}{c\rho A} Q(x, t)$$

§2. Solution of the Force-Free Vibrating String Problem

Our goal is to solve the following problem for y

(2.1)
$$\begin{cases} y_{tt} = c^2 y_{xx} & \text{(wave equation, } c^2 = T/\rho) \\ y(0, t) = 0 \qquad y(L, t) = 0 & \text{(boundary conditions)} \\ y(x, 0) = f(x) & \text{(initial position)} \\ y_t(x, 0) = g(x) & \text{(initial velocity)} \end{cases}$$

Our method for solving (2.1) is due to D. Bernoulli. The discussion will be nonrigorous; in the next section we shall present proofs that our results are valid under fairly general conditions.

Bernoulli's method for solving (2.1) is called *separation of variables*. It consists of two parts: (1) determining solutions with separated variables, and (2) superposition of those separated solutions. We will describe the method of separation of variables in great detail, our hope being that in the sequel the reader will be able to apply it as well.

We begin by assuming that $y(x, t)$ has the special form $y(x, t) = X(x)T(t)$. Then we substitute that form into the wave equation in (2.1) obtaining

$$X(x)T''(t) = c^2 X''(x)T(t)$$

where the primes denote differentiation with respect to the variable in parentheses. Dividing the equation above by $X(x)T(t)$ we obtain

(2.2)
$$\frac{T''(t)}{T(t)} = c^2 \frac{X''(x)}{X(x)}$$

The sides of (2.2) are functions of separate variables. By substituting particular values of x or t then letting t or x vary, we see that the two sides of (2.2) are equal to a constant. Denoting that constant by λ we have

(2.3)
$$\frac{T''(t)}{T(t)} = \lambda = c^2 \frac{X''(x)}{X(x)}$$

The constant λ is called a *separation constant*. Using a little algebra we obtain two separate equations from (2.3)

(2.4)
$$X''(x) = \frac{\lambda}{c^2} X(x) \qquad T''(t) = \lambda T(t)$$

Substituting $y = X(x)T(t)$ into the boundary conditions in (2.1) we get

$$X(0)T(t) = 0 \qquad X(L)T(t) = 0$$

Since these last two equations must hold for *all* $t \geq 0$ we divide $T(t)$ out,[3] hence $X(0) = X(L) = 0$.

Our analysis thus far has resulted in two separate problems

(2.5)   (a) $\begin{cases} X''(x) = \dfrac{\lambda}{c^2} X(x) \\ X(0) = 0 = X(L) \end{cases}$ (b) $T''(t) = \lambda T(t)$

We now solve these two problems. The ordinary differential equation in (2.5a) is a linear one with constant coefficients. There are three cases corresponding to $\lambda > 0$, $\lambda = 0$, $\lambda < 0$.

Case 1 ($\lambda > 0$).   The solution to (2.5a) is

$$X(x) = A e^{\sqrt{\lambda}x/c} + B e^{-\sqrt{\lambda}x/c}$$

where A and B are constants subject to

$$X(0) = 0 = A + B \qquad X(L) = 0 = A e^{\sqrt{\lambda}L/c} + B e^{-\sqrt{\lambda}L/c}$$

The determinant of the coefficients of A and B in the simultaneous linear equations above is

$$e^{-\sqrt{\lambda}L/c} - e^{\sqrt{\lambda}L/c} = -2\sinh(\sqrt{\lambda}L/c) \neq 0$$

since $\sqrt{\lambda}L/c \neq 0$. Therefore, we must have $A = B = 0$. Hence, $X(x) = 0$ so $y(x, t) = 0T(t) = 0$. *We have no need of such a trivial solution.*

Case 2 ($\lambda = 0$).   The differential equation in (2.5a) reduces to $X''(x) = 0$. Hence, $X(x) = Ax + B$. The reader may easily check that $X(0) = 0 = X(L)$ leads us to $A = B = 0$. Thus, $y(x, t) = 0$ again.

Case 3 ($\lambda < 0$).   The solution to (2.5a) is

$$X(x) = A\cos(\sqrt{-\lambda}x/c) + B\sin(\sqrt{-\lambda}x/c)$$

where A and B are constants subject to

$$0 = A \qquad 0 = A\cos(\sqrt{-\lambda}L/c) + B\sin(\sqrt{-\lambda}L/c)$$

Hence, either $B = 0$ or $\sin(\sqrt{-\lambda}L/c) = 0$. If $B = 0$ then $y(x, t) = 0$ again. Therefore, setting $\sin(\sqrt{-\lambda}L/c)$ equal to zero and solving for λ we obtain

$$\lambda = -\left(\frac{c\pi}{L}\right)^2, \quad -\left(\frac{2c\pi}{L}\right)^2, \quad \ldots, \quad -\left(\frac{nc\pi}{L}\right)^2, \ldots$$

Thus, we have found a sequence of special values for the separation constant, we shall call these values *eigenvalues* and denote them by λ_n where $\lambda_n = -(nc\pi/L)^2$. Associated with each eigenvalue is a particular

---

3. If $T(t) = 0$ for all t, then we would have $y(x, t) = X(x)0 = 0$ which is a trivial solution of no use for us.

solution X_n of (2.5a) which we call an *eigenfunction*.[4] Our sequence of eigenvalues and eigenfunctions is

(2.6) $\lambda_n = -\left(\dfrac{nc\pi}{L}\right)^2 \qquad X_n(x) = \sin\dfrac{n\pi x}{L} \qquad (n = 1, 2, 3, \ldots)$

Each eigenvalue λ_n yields a special form for (2.5b)

$$T_n''(t) = -\left(\dfrac{nc\pi}{L}\right)^2 T_n(t) \qquad (n = 1, 2, 3, \ldots)$$

The most general solution to that equation is

$$T_n(t) = A_n \cos\left(\dfrac{nc\pi}{L}t\right) + B_n \sin\left(\dfrac{nc\pi}{L}t\right) \qquad (n = 1, 2, 3, \ldots)$$

The constants A_n and B_n are arbitrary; hence the subscript n since their values might vary from one value of n to another.

We now define the function y_n to be the product of the functions X_n and T_n found above. Thus, for $n = 1, 2, 3, \ldots$

(2.7) $y_n(x, t) = \left[A_n \cos\left(\dfrac{nc\pi}{L}t\right) + B_n \sin\left(\dfrac{nc\pi}{L}t\right)\right] \sin\dfrac{n\pi x}{L}$

The reader might wish to check that each function y_n satisfies the first two conditions in problem (2.1), which are

(2.8) $y_{tt} = c^2 y_{xx} \qquad y(0, t) = 0 \qquad y(L, t) = 0$

It is also easily checked that any finite sum of the y_n's

$$\sum_{n=1}^{N} \left[A_n \cos\left(\dfrac{nc\pi}{L}t\right) + B_n \sin\left(\dfrac{nc\pi}{L}t\right)\right] \sin\dfrac{n\pi x}{L}$$

also satisfies (2.8).[5] We now let N tend to infinity and write down the following formal series as our most general solution to (2.8)

(2.9) $y(x, t) = \sum_{n=1}^{\infty} \left[A_n \cos\left(\dfrac{nc\pi}{L}t\right) + B_n \sin\left(\dfrac{nc\pi}{L}t\right)\right] \sin\dfrac{n\pi x}{L}$

Using the series in (2.9) we can now satisfy the two remaining conditions in (2.1). If we let t equal zero then, using the initial position condition in (2.1), Eq. (2.9) becomes

$$f(x) = \sum_{n=1}^{\infty} A_n \sin\dfrac{n\pi x}{L}$$

---

4. The terms *characteristic function* and *characteristic values* are also used.

5. Problem (2.8) is said to be *linear* since any *linear combination* of the fundamental solutions $\cos(nc\pi t/L)\sin(n\pi x/L)$ and $\sin(nc\pi t/L)\sin(n\pi x/L)$, as displayed in the line just below (2.8), is also a solution to (2.8). This is similar to the case of linear problems in ordinary differential equations.

Since we have a sine series for f displayed above, we define the coefficients A_n to be the coefficients in the sine series for f over $(0, L)$

(2.10) $$A_n = \frac{2}{L} \int_0^L f(x) \sin \frac{n\pi x}{L} \, dx \qquad (n = 1, 2, 3, \ldots)$$

If we differentiate (2.9) term by term with respect to t we obtain

$$y_t(x, t) = \sum_{n=1}^{\infty} \left[-\frac{nc\pi}{L} A_n \sin\left(\frac{nc\pi}{L} t\right) + \frac{nc\pi}{L} B_n \cos\left(\frac{nc\pi}{L} t\right) \right] \sin \frac{n\pi x}{L}$$

Letting t equal zero in the equation above and using the initial velocity condition in (2.1) we have

$$g(x) = \sum_{n=1}^{\infty} \frac{nc\pi}{L} B_n \sin \frac{n\pi x}{L}$$

Therefore, we define $(nc\pi/L)B_n$ to be the nth sine coefficient in the sine series for g over $(0, L)$. Upon solving for B_n we obtain

(2.11) $$B_n = \frac{2}{nc\pi} \int_0^L g(x) \sin \frac{n\pi x}{L} \, dx \qquad (n = 1, 2, 3, \ldots)$$

Combining equations (2.9) through (2.11) we have the following *solution to the force-free vibrating string problem* (2.1):

(2.12)
$$y(x, t) = \sum_{n=1}^{\infty} \left[A_n \cos\left(\frac{nc\pi}{L} t\right) + B_n \sin\left(\frac{nc\pi}{L} t\right) \right] \sin \frac{n\pi x}{L}$$
$$A_n = \frac{2}{L} \int_0^L f(x) \sin \frac{n\pi x}{L} \, dx \qquad B_n = \frac{2}{nc\pi} \int_0^L g(x) \sin \frac{n\pi x}{L} \, dx$$

Bernoulli's method as applied to problem (2.1) can be summarized as follows.

Summary

First, we determined solutions with variables separated [see Eq. (2.7)] to the *linear part* of problem (2.1). The linear part of problem (2.1) is shown in Eq. (2.8); see also footnote number 5 following that equation. Second, we formed a superposition of all those separated solutions resulting in the form for y shown in (2.9). Forming that superposition allowed us to satisfy the remaining conditions in (2.1).[6] [The reader also should take note of the *eigenvalues and eigenfunctions* that resulted from our solution of problem

---

6. Those remaining conditions $y(x, 0) = f(x)$ and $y_t(x, 0) = g(x)$ are called *nonhomogeneous* to distinguish them from the linear part of (2.1).

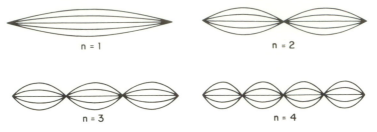

Figure 3.5   Harmonics of order n for $n = 1, 2, 3,$ and 4.

(2.5a); the occurrence of eigenvalues and eigenfunctions is typical for the case of such *boundary value* problems (data given on the boundary $x = 0$ and $x = L$ of $0 \leq x \leq L$).]

Before we apply our general solution to some sample problems, let's first interpret our result in terms of the sound that such a string would make vibrating in air. The nth term of the series for y shown in (2.12) consists of two terms of the form

$$A_n \cos\left(\frac{nc\pi}{L}t\right) \sin\frac{n\pi x}{L} \qquad B_n \sin\left(\frac{nc\pi}{L}t\right) \sin\frac{n\pi x}{L}$$

We call each of these terms *nth order harmonics*. Let's consider the nth order harmonic shown on the left above; similar considerations apply to the harmonic on the right. The factor $A_n \cos[(nc\pi/L)t]$ can be considered as a variable amplitude for $\sin(n\pi x/L)$. By plotting the various sine curves that result for various t values the reader will see that the nth order harmonic has graphs like the ones sketched above in Figure 3.5.

Since the variable amplitude $A_n \cos[(nc\pi/L)t]$ has a smallest period of

$$2\pi/(nc\pi/L) = 2L/nc$$

it follows that the nth order harmonic has a smallest period in the time variable t of $T = 2L/nc$. This period T is called the *period of vibration* for the harmonic. The quantity $2\pi/T = nc\pi/L$ is called the *frequency* of the harmonic.[7] The frequency equals the number of times, in a time interval of 2π sec, that the harmonic returns to its position at the beginning of the interval. The vibrations of the harmonic cause an alternating compression and expansion of the air near the string; the frequency of this compression and expansion matches precisely the frequency of the harmonic. Moreover, the expansion and contraction near the string induce further expansion and contraction of the air farther away; thus a *sound wave* is created of frequency equal to that of the harmonic and of speed (of transmission) equal to the speed of sound in air (1080 ft/sec).[8] The sound we hear from

---

7. Usually the quantity $1/T$ is called the frequency, but our definition matches nicely with the eigenvalues found above.

8. We are neglecting the effect of air resistance; see Exercise (2.24).

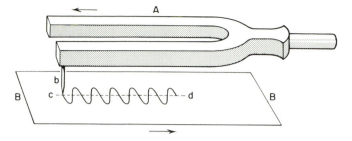

Figure 3.6 A tuning fork A is drawn at a uniform speed across the paper B. As the marker b attached to the tuning fork vibrates, a sinusoidal curve is drawn from d to c, proving that the fork vibrates like one of the harmonics described in the text. From Helmholtz (1954), p. 20. (Reproduced with permission of Dover Press.)

such sound waves are called *pure tones*; that is why the nth order harmonics are sometimes called pure tones of the nth order. *Since the series for y is a superposition of nth order harmonics, the sound it creates will be a superposition of pure tones.*

Other vibratory motion also can induce pure tones. Very well-designed tuning forks induce pure tones.[8] Helmholtz describes experiments with tuning forks in which a trace of their vibrations is obtained by attaching a marker to one of the tines of the fork and drawing the vibrating fork at a uniform speed across a paper. A uniform sinusoidal curve results, as can be seen in Figure 3.6.

We now apply method (2.12) to some sample problems.

(2.13) Example: Plucked string. Suppose an elastic string of length L is bent from the middle, held for a moment, and then released. Determine the subsequent motion of the string. (See Figure 3.7.)

Solution. The force free vibrating string problem is

$$y_{tt} = c^2 y_{xx} \qquad y(0, t) = 0 \qquad y(L, t) = 0$$

$$y(x, 0) = \begin{cases} 2hx/L & \text{for } 0 \leq x \leq \tfrac{1}{2}L \\ -2h(x - L)/L & \text{for } \tfrac{1}{2}L \leq x \leq L \end{cases} \qquad y_t(x, 0) = 0$$

Hence (2.12) tells us that each $B_n = 0$ and

$$A_n = \frac{4h}{L^2} \int_0^{\frac{1}{2}L} x \sin \frac{n\pi x}{L} \, dx - \frac{4h}{L^2} \int_{\frac{1}{2}L}^L (x - L) \sin \frac{n\pi x}{L} \, dx$$

$$= \frac{8h}{n^2 \pi^2} \sin \frac{n\pi}{2}$$

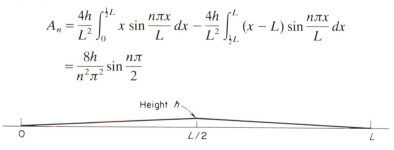

Height h

0	$L/2$	L

Figure 3.7 Initial position of the plucked string in Example (2.13).

Thus, the motion of the string is described for $t > 0$ by

$$y(x, t) = \frac{8h}{\pi^2} \left[\cos\left(\frac{c\pi}{L}t\right) \sin\frac{\pi x}{L} - \frac{1}{9}\cos\left(\frac{3c\pi}{L}t\right) \sin\frac{3\pi x}{L} \right.$$

$$\left. + \frac{1}{25}\cos\left(\frac{5c\pi}{L}t\right) \sin\frac{5\pi x}{L} - \cdots + \cdots \right]$$

From the series above we can see that only those harmonics of odd order will contribute to the sound the string creates. Furthermore, it is known that the *intensity of a pure tone is proportional to the square of the constant coefficient A_n (or B_n) of the harmonic that induces that tone.*[9] In this example, these squares are

$$\left(\frac{8h}{\pi^2}\right)^2 \quad 0 \quad \left(\frac{8h}{\pi^2}\right)^2 \frac{1}{9^2} \quad 0 \quad \left(\frac{8h}{\pi^2}\right)^2 \frac{1}{25^2} \quad 0 \quad \left(\frac{8h}{\pi^2}\right)^2 \frac{1}{49^2} \quad \cdots$$

Hence the tone from the first-order harmonic will be 81 times more intense than the tone from the third-order harmonic, and 625 times more intense that the tone from the fifth-order harmonic, and so on.

The first-order harmonic is called (when it is nonzero) the *fundamental harmonic*; its period of vibration determines the period of vibration for the string's motion [see Exercise (2.27)] and hence the overall tone of the sound created by the string's vibration. The other harmonics are said to contribute to the color (or timbre, or quality) of the sound.

(2.14) Example. Suppose an elastic string with fixed ends is at rest in the horizontal position. If the string is struck in such a manner that it receives an initial velocity

$$y_t(x, 0) = \begin{cases} 4ax/L & \text{for } 0 \le x \le \frac{1}{4}L \quad (a \text{ is a constant}) \\ 4a(\frac{1}{2}L - x)/L & \text{for } \frac{1}{4}L \le x \le \frac{1}{2}L \\ 0 & \text{for } \frac{1}{2}L \le x \le L \end{cases}$$

Determine the subsequent motion of the string.

Solution.   In this case $y(x, 0) = 0$ so $A_n = 0$. But

$$B_n = \frac{8a}{nc\pi L} \int_0^{\frac{1}{4}L} x \sin\frac{n\pi x}{L} dx + \frac{8a}{nc\pi L} \int_{\frac{1}{4}L}^{\frac{1}{2}L} (\tfrac{1}{2}L - x) \sin\frac{n\pi x}{L} dx$$

$$= \frac{16aL}{n^3 c\pi^3} \sin\frac{n\pi}{4} - \frac{8aL}{n^3 c\pi^3} \sin\frac{n\pi}{2}$$

Thus, the motion of the string is described for $t > 0$ by

$$y(x, t) = \frac{8aL}{c\pi^3} [(\sqrt{2} - 1) \sin\left(\frac{c\pi}{L}t\right) \sin\frac{\pi x}{L} + \frac{1}{4}\sin\left(\frac{2c\pi}{L}t\right) \sin\frac{2\pi x}{L}$$

$$+ \frac{(\sqrt{2} + 1)}{27} \sin\left(\frac{3c\pi}{L}t\right) \sin\frac{3\pi x}{L} + \cdots]$$

---

9. See Morse (1948), p. 88. The coefficient A_n (B_n) is called the *amplitude* of the harmonic.

*(2.15) Example: Struck String.   Suppose an impulse of magnitude P is applied to the midpoint of an elastic string of length L with fixed ends. Determine the subsequent motion of the string.

Solution.   In classical mechanics an *impulse* of magnitude P imparts to a body an instantaneous momentum of magnitude P; thus the impulse imparts to a body of mass m an instantaneous velocity of P/m. One way of interpreting the given problem is that an instantaneous momentum of magnitude P is imparted to the string over a very small element of the string, centered at $\frac{1}{2}L$.[10] Therefore, we will solve the following problem

$$y_{tt} = c^2 y_{xx} \qquad y(0, t) = 0 = y(L, t) \qquad y(x, 0) = 0$$

$$y_t(x, 0) = \begin{cases} 0 & \text{for } |x - \frac{1}{2}L| > \epsilon \\ P/(2\rho\epsilon) & \text{for } |x - \frac{1}{2}L| < \epsilon \end{cases} \qquad (\epsilon > 0)$$

where ρ is the density of the string and ϵ *tends to zero*. If we write $(P/\rho)\delta(x - \frac{1}{2}L)$ in place of $y_t(x, 0)$, then for each continuous function h on $[0, L]$ we have

(2.16)
$$\int_0^L h(x) \frac{P}{\rho} \delta(x - \tfrac{1}{2}L)\, dx = \frac{P}{\rho} h(\tfrac{1}{2}L)$$

Formula (2.16) is a shorthand version of the following limit process

(2.16′)
$$\begin{cases} \displaystyle \int_0^L h(x) \frac{P}{\rho} \delta(x - \tfrac{1}{2}L)\, dx = \lim_{\epsilon \to 0} \int_{\frac{1}{2}L - \epsilon}^{\frac{1}{2}L + \epsilon} h(x) \frac{P}{2\rho\epsilon}\, dx \\[3mm] \displaystyle \qquad\qquad = \frac{P}{\rho} \lim_{\epsilon \to 0} \left[\frac{1}{2\epsilon} \int_{\frac{1}{2}L - \epsilon}^{\frac{1}{2}L + \epsilon} h(x)\, dx \right] \\[3mm] \displaystyle \qquad\qquad = \frac{P}{\rho} h(\tfrac{1}{2}L) \end{cases}$$

The last equality in (2.16′) holds because $h(x) \doteq h(\frac{1}{2}L)$ for $\frac{1}{2}L - \epsilon \le x \le \frac{1}{2}L + \epsilon$ when ϵ is quite small in magnitude, due to the assumed continuity of h.

Using formula (2.16), the given problem is easily solved. Substituting $(P/\rho)\delta(x - \frac{1}{2}L)$ for $y_t(x, 0)$ we obtain from (2.12) that

$$B_n = \frac{2}{nc\pi} \int_0^L \sin \frac{n\pi x}{L} \frac{P}{\rho} \delta(x - \tfrac{1}{2}L)\, dx = \frac{2P}{\rho nc\pi} \sin \frac{n\pi}{2}$$

while, since $y(x) = 0$, we have $A_n = 0$. Thus, the motion of the string for $t > 0$ is described by

(2.17)
$$y(x, t) = \frac{2P}{\rho c\pi} \sum_{k=0}^{\infty} \frac{(-1)^k}{2k + 1} \sin\left[\frac{(2k + 1)\pi c}{L} t \right] \sin \frac{(2k + 1)\pi x}{L}$$

---

10. This might occur, for example, if a guitar string is struck with a guitar pick.

where $n = 2k + 1$ denotes an odd integer n. One way of interpreting this result is that (2.17) is proposed as an approximate solution to the given problem when ϵ is a very small positive number. The symbol $(P/\rho)\delta(x - \frac{1}{2}L)$ is called a *Dirac delta function* [of magnitude (P/ρ), centered at $\frac{1}{2}L$.]

Exercises

(2.18)   Discuss the motion of the nth order harmonics

$$A_n \cos\left(\frac{nc\pi}{L}t\right)\sin\frac{n\pi x}{L} \qquad B_n \sin\left(\frac{nc\pi}{L}t\right)\sin\frac{n\pi x}{L}$$

In particular verify (a) the statements made in the text between (2.12) and Example (2.13), (b) for an harmonic of order greater than one there are points of the string which *do not move* at $x = (k/n)L$ where $k = 1, 2, 3, \ldots$, $n - 1$; those points are called *nodes*.

(2.19)   An elastic string 2 ft long is fastened at its ends. The string is stretched to an initial position $y(x, 0) = 0.03x(x - 2)$ and then released with no initial velocity. Determine the subsequent motion of the string.

(2.20)   Find the displacement $y(x, t)$ of an elastic string, held fixed at its ends, which is set in motion with no initial velocity from the initial position

$$y(x, 0) = \begin{cases} bx & \text{for } 0 \leq x \leq \frac{1}{4}L \quad (b \text{ is a constant}) \\ \frac{1}{4}bL & \text{for } \frac{1}{4}L \leq x \leq 3L/4 \\ b(L - x) & \text{for } 3L/4 \leq x \leq L \end{cases}$$

(2.21)   Suppose an elastic string, with ends fixed, and initially in a horizontal position is struck in such a manner that the initial velocity of the string $y_t(x, 0)$ is the same function as in (2.20) above. Determine the subsequent motion of the string.

(2.22)   Consider an elastic string fastened at the points $x = 0$ and $x = L$. Suppose that at time $t = 0$, the point $x = a$ (where $0 < a < L$) is displaced by an amount h and then released (a plucked string). Determine the subsequent motion of the string.

(2.23)   Suppose that in problem (2.22) the value of a is equal to $(k/m)L$ where k is a positive integer less than m. Prove that no harmonics of order equal to an integral multiple of m contribute to the subsequent motion of the string. Explain physically why this happens.

(2.24)   *Air Resistance.*   In this exercise we take into account the effect of air resistance on the vibrating string. The exercise has four parts.

(a)   Given that air resistance, *as well as other frictional forces*, causes a *dampening* force proportional to the momentum of the string, that is, a force equal to $-k\rho\Delta xy_t$, on an element of the string (k is a positive constant); prove that $y(x, t)$ satisfies

$$c^2 y_{xx} = y_{tt} + k y_t \qquad (k > 0)$$

(b) Suppose that in addition to the equation in part (a) the function y also satisfies (1.1). Find a series form for y.

(c) Using the series form for y found in (b) give a clear, but not necessarily rigorous, discussion of why the vibration of an elastic string with fixed ends eventually *dies down* under the dampening influence of air resistance.

(d) Give a *qualitative* physical discussion of the sound which results from such damped vibration of strings. In particular, explain why the discussion of the sound waves from freely vibrating strings given in the text is valid for a short time interval after $t = 0$, that is, it was alright to neglect air resistance in that discussion.

(2.25) Using separation of variables find solutions to the following equations

(a) $\quad 2u_x + u_y = 0$

(b) $\quad 3u_{xy} = u$

(c) $\quad 2x^2 u_{yy} + 3y^2 u = 0$

***(2.26)** A string is fastened at the points $x = 0$ and $x = \pi$. If the string is initially straight with a velocity $y_t(x, 0) = b_n \sin nx$ where b_n is a constant, then determine the subsequent motion of the string. Generalize.

(2.27) Prove that the solution $y(x, t)$ displayed in (2.12) has a period of $2L/c$ in the time variable t [i.e., $y(x, t + 2L/c) = y(x, t)$ for $t > 0$].

(2.28) Explain the following physical phenomenon by appeal to the Fourier series solution in (2.12) of the vibrating string problem.

(a) If a string is stretched tighter, tension T increased, then the notes from the string are higher in pitch

(b) If a string is cut short, by pressure from a finger or bridge or some such device, then the notes from the string are higher in pitch

(c) Thicker strings make lower pitched notes than thinner strings of the same material. [*Note*: the quantity ρ is a *linear density*.]

(2.29) Suppose two people are singing a duet. If the two singers hit the same note, then a harmony results. Explain, using Fourier analysis, why the resultant note is *purer in tone* than either singer alone could achieve. Also, the intensity of the note goes up. What happens with a trio, or a choir?[11] Harmony also results if one singer is an octave higher than the other; what do you think occurs in that case?

---

11. If the choir has 30 members, then we do not *perceive* the sound to be 30 times as intense. We *perceive* it as $\ln(30)$ times as intense [see Leigh Silver (1971), p. 353].

*(2.30) Consider an elastic string of length L fastened at both ends. Suppose that at time $t = 0$ the string receives an impulse of magnitude P at the point $x = a$ where $0 < a < L$. Determine the subsequent motion of the string.

§3. D'Alembert's Solution of the Vibrating String Problem

In this section we shall investigate the mathematical validity of the general solution to the force-free, vibrating string problem found in the previous section. We will show that the previous solution (2.12) can be expressed in another form; originally found by D'Alembert.

Let's first consider the form of the series given for the proposed solution y in formula (2.12). That series for y can be formally split into two series as follows

$$
\begin{cases}
y(x, t) = v(x, t) + w(x, t) \\[2mm]
v(x, t) = \sum_{n=1}^{\infty} A_n \cos\left(\dfrac{nc\pi}{L} t\right) \sin \dfrac{n\pi x}{L} \\[2mm]
w(x, t) = \sum_{n=1}^{\infty} B_n \sin\left(\dfrac{nc\pi}{L} t\right) \sin \dfrac{n\pi x}{L}
\end{cases}
$$

(3.1)

First, we examine the series for v given above. Using the trigonometric identity $\cos\theta \sin\phi = \frac{1}{2}\sin(\phi + \theta) + \frac{1}{2}\sin(\phi - \theta)$ we have

$$
v(x, t) = \sum_{n=1}^{\infty} \left[\tfrac{1}{2}A_n \sin\frac{n\pi}{L}(x + ct) + \tfrac{1}{2}A_n \sin\frac{n\pi}{L}(x - ct) \right]
$$

The series above suggests a sum of half-multiples of the sine series for f, evaluated at $x + ct$ and $x - ct$, since

$$
A_n = \frac{2}{L} \int_0^L f(s) \sin\frac{n\pi s}{L}\, ds
$$

As we saw in Chapter 1, the sine series for f is identical to the Fourier series for F where F is *the odd periodic extension of f.* (See Figure 1.6, Chapter 1.) Moreover, *if we assume that F is continuous and piecewise smooth,* then Theorem (4.5) from Chapter 1 assures us that

$$
F(s) = \sum_{n=1}^{\infty} A_n \sin\frac{n\pi s}{L} \qquad \left[A_n = \frac{2}{L} \int_0^L F(s) \sin\frac{n\pi s}{L}\, ds \right]
$$

holds for all real s values. Substituting $x + ct$ for s, followed by $x - ct$ for s, into the equation above we obtain from the equation right after (3.1)

(3.2) $v(x, t) = \tfrac{1}{2}F(x + ct) + \tfrac{1}{2}F(x - ct)$

Second, we consider the series for w given in (3.1). Using the formula for B_n

from (2.12) we have

$$w(x, t) = \sum_{n=1}^{\infty} \left[\frac{L}{nc\pi} b_n \right] \sin\left(\frac{nc\pi}{L} t \right) \sin \frac{n\pi x}{L}$$

$$b_n = \frac{2}{L} \int_0^L g(s) \sin \frac{n\pi s}{L} ds$$

Using the identity $\sin \phi \sin \theta = \frac{1}{2} \cos(\theta - \phi) - \frac{1}{2} \cos(\theta + \phi)$ we have

$$(3.3) \quad w(x, t) = \sum_{n=1}^{\infty} \frac{1}{2c} \left[\frac{L}{n\pi} b_n \cos \frac{n\pi}{L} (x - ct) - \frac{L}{n\pi} b_n \cos \frac{n\pi}{L} (x + ct) \right]$$

Because of the form of the series for w above, we consider for all real s values the series

$$\sum_{n=1}^{\infty} \frac{L}{n\pi} b_n \cos \frac{n\pi s}{L}$$

If we could differentiate that series term by term, we would obtain $-\sum_{n=1}^{\infty} b_n \sin(n\pi s/L)$. From the formula for b_n given above, we deduce that this last series is the negative of the sine series for g over $[0, L]$. Therefore, as we did above for f, if we let G denote the odd periodic extension of g then *provided that G is continuous and piecewise smooth* we have for all real s values

$$(3.4) \qquad\qquad -G(s) = \sum_{n=1}^{\infty} -b_n \sin \frac{n\pi s}{L}$$

The right side of (3.4) is the Fourier series for $-G$. Although differentiation term by term may not be valid, Corollary (6.11), Chapter 2, says that we may integrate term by term. Moreover, since *trigonometric* Fourier series are periodic, we may do this over *all* intervals of finite length. Hence, from (3.4) we obtain for all real r values[12]

$$(3.5) \qquad -\int_0^r G(s) \, ds = \sum_{n=1}^{\infty} \left[\frac{L}{n\pi} b_n \cos \frac{n\pi r}{L} - \frac{L}{n\pi} b_n \right]$$

Substituting $x + ct$, and then $x - ct$, in place of r in Eq. (3.5) we obtain from (3.3)

$$(3.6) \qquad w(x, t) = \frac{1}{2c} \int_0^{x+ct} G(s) \, ds - \frac{1}{2c} \int_0^{x-ct} G(s) \, ds$$

due to the cancellation of each constant term $-\frac{L}{n\pi} b_n$.

Finally, we note that if f and g are both continuous and piecewise smooth on $(0, L)$ and

$$(3.7) \qquad\qquad f(0) = f(L) = 0 \qquad g(0) = g(L) = 0$$

then it follows that their odd periodic extensions F and G are both continuous and piecewise smooth. Combining this last observation with

---

12. See Exercise (3.15) for another way of obtaining (3.5).

Eqs. (3.2) and (3.6), and the assumptions we used to derive them, we have the following result.

> If f and g are continuous, piecewise smooth, and satisfy
>
> $$f(0) = f(L) = 0 \qquad g(0) = g(L) = 0$$
>
> then the series for y given in formula (2.12) satisfies
>
> $$y(x, t) = \tfrac{1}{2}[F(x + ct) + F(x - ct)]$$
>
> **(3.8)** $$+ \frac{1}{2c}\left[\int_0^{x+ct} G(s)\, ds - \int_0^{x-ct} G(s)\, ds\right]$$
>
> where F and G are the odd periodic extensions of f and g.

A useful alternative form for (3.8) is obtained by combining the integrals, thus

(3.8′) $$y(x, t) = \tfrac{1}{2}[F(x + ct) - F(x - ct)] + \frac{1}{2c}\int_{x-ct}^{x+ct} G(s)\, ds$$

Formula (3.8) will play a key role in the following discussion of the mathematical validity of the series proposed in §2 as solutions to various vibrating string problems. When we say that the series in formula (2.12) *rigorously solves* problem (2.1), we mean that the following physically motivated conditions are satisfied. (a) *Two continuous partial derivatives of y can be taken with respect to x and t for $0 \le x \le L$ and $t \ge 0$ and the wave equation is satisfied.* (b) *The boundary and initial conditions are satisfied.* Moreover, because (a) implies that both y and y_t are continuous in t we must have for $0 \le x \le L$

$$\lim_{t \to 0+} y(x, t) = f(x) \qquad \lim_{t \to 0+} y_t(x, t) = g(x)$$

The physical basis for all these requirements is that the string does not break or "jump" during its motion.

If the function y displayed in (3.8) is to be a rigorous solution of problem (2.1), then we must be able to take at least two continuous partial derivatives of y with respect to x and t. Because of the form of the right side of (3.8), we want F to be at least twice continuously differentiable and (because of the integration) G to be at least once continuously differentiable. Those considerations lead us to formulate the hypotheses for f and g which are given in the following theorem.

(3.9) Theorem. Suppose that f is twice continuously differentiable on $[0, L]$ with $f(0) = f(L) = 0$ and $f''(0) = f''(L) = 0$. Also, suppose that g is continuously differentiable on $[0, L]$ with $g(0) = g(L) = 0$. Then, the function y defined in (2.12) is a rigorous solution to the vibrating string problem.

Proof. Because of the hypotheses imposed on f and g the reader may check that F is twice continuously differentiable[13] and that G is continuously differentiable. Moreover, the functions f and g satisfy the hypotheses for (3.8); hence we shall examine that form for y. Using the Chain Rule, we have

(3.10) $y_t(x, t) = \frac{1}{2}cF'(x + ct) - \frac{1}{2}cF'(x - ct) + \frac{1}{2}G(x + ct) + \frac{1}{2}G(x - ct)$

and applying the Chain Rule again to (3.10) we get

(3.11) $y_{tt}(x, t) = \frac{1}{2}c^2F''(x + ct) + \frac{1}{2}c^2F''(x - ct) + \frac{1}{2}cG'(x + ct) - \frac{1}{2}cG'(x - ct)$

From the right sides of (3.8), (3.10), and (3.11) we infer that y, y_t, and y_{tt} are all continuous for $0 \le x \le L$ and $t \ge 0$. In particular, we have

$$\lim_{t \to 0+} y(x, t) = \frac{1}{2}F(x) + \frac{1}{2}F(x) + \frac{1}{2c}\left[\int_0^x G(s)\, ds - \int_0^x G(s)\, ds\right]$$

$$= F(x)$$

$$\lim_{t \to 0+} y_t(x, t) = \frac{1}{2}cF'(x) - \frac{1}{2}cF'(x) + \frac{1}{2}G(x) + \frac{1}{2}G(x)$$

$$= G(x)$$

Thus, if $0 \le x \le L$, we have

$$\lim_{t \to 0+} y(x, t) = f(x) \qquad \lim_{t \to 0+} y_t(x, t) = g(x)$$

By applying the Chain Rule twice, we obtain

$$y_x(x, t) = \frac{1}{2}F'(x + ct) + \frac{1}{2}F'(x + ct) + \frac{1}{2c}G(x + ct) - \frac{1}{2c}G(x - ct)$$

(3.12)

$$y_{xx}(x, t) = \frac{1}{2}F''(x + ct) + \frac{1}{2}F''(x - ct) + \frac{1}{2c}G'(x + ct) - \frac{1}{2c}G'(x - ct)$$

From (3.12) we see that y_x and y_{xx} are both continuous. Moreover, from the expressions for y_{xx} and y_{tt} we see immediately that the wave equation is satisfied. Finally substituting $x = 0$ or $x = L$ into the series form for y in formula (2.12) we see that the boundary conditions in (2.1) are satisfied. Thus, y as displayed in either (2.12) or (3.8) is a rigorous solution to the vibrating string problem (2.1). ∎

(3.13) Remark. The form for y displayed in (3.8), or (3.8') is known as *D'Alembert's solution* to the force-free vibrating string problem. D'Alembert's solution is of great importance for two reasons: (1) The solution y is expressed in *closed form*, its dependence on the initial data f and g is quite explicit [see Exercise (3.17) for an important consequence of

---

13. To show that F' is continuous the reader should first demonstrate that it is an even function.

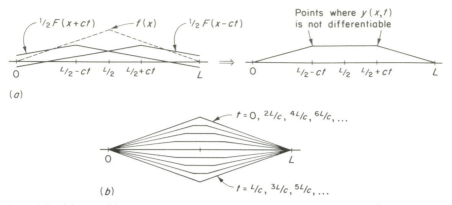

$\frac{1}{2}F(x+ct)$ $f(x)$ $\frac{1}{2}F(x-ct)$ Points where $y(x,t)$ is not differentiable

(a)

$t=0,\, 2L/c,\, 4L/c,\, 6L/c,\, \ldots$

$t=L/c,\, 3L/c,\, 5L/c,\, \ldots$

(b)

Figure 3.8 (a) and (b) illustrate how y is a sum of two sawtooth waves $\frac{1}{2}F(x+ct)$ plus $\frac{1}{2}F(x-ct)$ moving in opposite directions along the x axis at speed c. (a) Construction of the graph $y(x,t)$ at time t. (b) Graph of y for various times t. (The top curve is attained at $t=0$, $2L/c, 4L/c, \ldots$, while the bottom curve is attained at $t=L/c, 3L/c, 5L/c, \ldots$.)

this]. (2) The series solution (2.12) cannot be verified as a solution to the wave equation by term by term differentiation.[14]

There are some obvious limitations to Theorem (3.9). For instance, let's consider the solution found for Example (2.13). Since the initial position function f in that example has a continuous, piecewise smooth, odd periodic extension F and the initial velocity function g is zero, it follows that the series for y equals the form described by (3.8). Namely,

$$y(x,t) = \tfrac{1}{2}F(x+ct) + \tfrac{1}{2}F(x-ct)$$

But, in this case, the function f is not differentiable when $x = \frac{1}{2}L$. Hence, the solution $y(x,t)$ will not be differentiable when

(3.14) $\begin{cases} x + ct = \pm\frac{1}{2}L,\ \pm\frac{3}{2}L,\ \pm\frac{5}{2}L,\ \ldots \\ x - ct = \pm\frac{1}{2}L,\ \pm\frac{3}{2}L,\ \pm\frac{5}{2}L,\ \ldots \end{cases}$

(See Figure 3.8.) Hence, at points lying on those lines in the x–t plane the wave equation is *not* solved by y. Whereas if (x,t) is a point lying within one of the regions marked off by these lines, then the wave equation is solved by y. The lines given in (3.14) are called *characteristics*. (See Figure 3.9.) We have shown that the series found in Example (2.13) does not constitute a rigorous solution to the vibrating string problem. As we shall see in §5, by the Uniqueness Theorem (5.15), *if* a rigorous solution existed for the problem in Example (2.13) it would have to equal the series found for that problem. Therefore, no solution exists, in the rigorous sense, for the problem posed in Example (2.13). The question then arises: What good is the series found for that example? In Exercises (3.17) through (3.19) we

---

14. Because the terms of y would acquire factors of n^2 after two differentiations and then would not generally tend to zero let alone converge when summed in an infinite series.

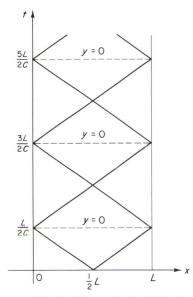

Figure 3.9 Regions of solvability for y marked off by the characteristics. (The dotted lines represent the time when the string is in a horizontal position along the x axis.)

present a partial answer to that question; our results will show that the series found in Example (2.13) can be interpreted as a *limit* of rigorous solutions to problem (2.1).

Exercises

(3.15) Prove that the left side of (3.5) is an even function of r, compute its cosine series, and show that the two sides of (3.5) are equal. [*Hint*: Integrating by parts based on $(d/dr) \int_0^r G(s) \, ds = G(r)$ is helpful.]

(3.16) Check the statement made at the beginning of the proof of Theorem (3.9) that F is twice continuously differentiable and G is continuously differentiable.

(3.17) Let f, $f^\#$ and g, $g^\#$ denote functions on $[0, L]$ which satisfy the requirements stated for (3.8). Let y and $y^\#$ be defined by (3.8) using f, g and $f^\#$, $g^\#$, respectively. Prove the following:

(a) Given $\epsilon > 0$, if $|f(x) - f^\#(x)| < \epsilon$ and $|g(x) - g^\#(x)| < \epsilon$ for $0 \le x \le L$, then

$$|y(x, t) - y^\#(x, t)| < \epsilon[1 + 2L/c]$$

for $0 \le x \le L$ and $0 \le t \le 2L/c$.

(b) The inequality in (a) actually holds for all $t \geq 0$. [*Hint*: Use periodicity in time.] We say that *D'Alembert's solution to* (2.1) *depends continuously on the initial data*: f and g.

(3.18) With the same hypotheses as in Exercise (3.17), prove that for all time $t \geq 0$

$$\|y - y^\#\|_2 = \left[\int_0^L |y(x, t) - y^\#(x, t)|^2 \, dx \right]^{1/2} < A\epsilon$$

for some positive constant A. [*Hint*: Use (3.17b).]

(3.19) Let S_n^f and S_n^g denote the nth partial sums of the sine series for f and g over $[0, L]$ where f and g satisfy the requirements of (3.8). Let y_n be defined by formula (3.8) using S_n^f and S_n^g in place of f and g respectively. Prove the following:

(a) Each y_n is a *rigorous* solution to the vibrating string problem (2.1) if S_n^f and S_n^g are used as initial conditions.
(b) For $0 \leq x \leq L$ and $t \geq 0$ we have

$$|y(x, t) - y_n(x, t)| < A\epsilon$$

for some positive constant A, if n is sufficiently large.

Remark. The result of Exercise (3.19b) shows that y is a limit of rigorous solutions to problem (2.1). For example, as we noted at the end of this section, the "solution" to Example (2.13) is a limit of rigorous solutions to (2.1). We call such a "solution" a *generalized solution* to (2.1).

(3.20) Prove that the series solution to Example (2.14) is not a rigorous solution, but that series is a generalized solution. Sketch a diagram of the characteristics and regions of solvability, as in Figure 3.9.

(3.21) Which problems from the exercises for §2 have a rigorous solution and which have a generalized solution?

(3.22) Prove that if f and g are piecewise continuous then for $t \geq 0$ we have [using the notations of Exercises (3.18) and (3.19)]

$$\|y - y_n\|_2 < A\epsilon$$

for all n sufficiently large. We say that y_n tends to y *in the mean*. In this case y is also called a generalized solution to (2.1).

(3.23) Prove that the function y in (3.8) can be written as $y = \Psi(x + ct) + \Phi(x - ct)$. Thus y can be interpreted as the sum of two waves moving in opposite directions with speed c; y is called a *standing wave*.

§4. Heat Conduction

We shall now discuss some problems of heat conduction. First, we derive the general solution to Eq. (1.14) with the boundary and initial conditions given in (1.15). The full problem is

(4.1)
$$\begin{cases} u_t = a^2 u_{xx} & \text{(heat equation)} \\ u(0, t) = 0 \qquad u(L, t) = 0 & \text{(boundary conditions)} \\ u(x, 0) = f(x) & \text{(initial temperature)} \end{cases}$$

We solve (4.1) by the method of separation of variables. If we let $u(x, t) = X(x)T(t)$, then the first two equations in (4.1) are transformed into two separate problems

(4.2)   (a)
$$\begin{cases} X''(x) = \dfrac{\lambda}{a^2} X(x) \\ X(0) = 0 \qquad X(L) = 0 \end{cases}$$

(b) $T'(t) = \lambda T(t)$

where λ is a separation constant. Problem (4.2a) was solved in §2. We obtained the following eigenvalues and eigenfunctions

$$\lambda_n = -\left(\frac{na\pi}{L}\right)^2 \qquad X_n(x) = \sin\frac{n\pi x}{L} \qquad (n = 1, 2, 3, \ldots)$$

Substituting λ_n into (4.2b) we obtain $T_n(t) = B_n \exp[-n^2(\pi a/L)^2 t]$ where exp stands for the exponential function, $\exp s = e^s$, and B_n is an arbitrary constant. If we define u_n as $X_n T_n$, that is,

$$u_n(x, t) = B_n \exp\left[-n^2\left(\frac{\pi a}{L}\right)^2 t\right] \sin\frac{n\pi x}{L}$$

then the reader may check that u_n satisfies the linear part of (4.1), that is, the heat equation and the boundary conditions. Moreover, any finite sum of these functions

$$\sum_{n=1}^{N} B_n \exp\left[-n^2\left(\frac{\pi a}{L}\right)^2 t\right] \sin\frac{n\pi x}{L}$$

also satisfies the linear part of (4.1). Letting N tend to infinity, we write as our most general solution

(4.3)
$$u(x, t) = \sum_{n=1}^{\infty} B_n \exp\left[-n^2\left(\frac{\pi a}{L}\right)^2 t\right] \sin\frac{n\pi x}{L}$$

The arbitrary constants B_n can be determined from the initial temperature condition in (4.1). Substituting $t = 0$ into both sides of (4.3) we have

$$f(x) = \sum_{n=1}^{\infty} B_n \sin\frac{n\pi x}{L}$$

Therefore, we define

(4.4)
$$B_n = \frac{2}{L} \int_0^L f(x) \sin\frac{n\pi x}{L} \, dx$$

Combining (4.3) and (4.4), our solution to problem (4.1) is

(4.5)
$$u(x, t) = \sum_{n=1}^{\infty} B_n \exp\left[-n^2\left(\frac{\pi a}{L}\right)^2 t \right] \sin \frac{n\pi x}{L}$$

$$B_n = \frac{2}{L} \int_0^L f(x) \sin \frac{n\pi x}{L} \, dx$$

We will now consider a couple of examples and then prove that the series given in (4.5) does provide a rigorous solution to (4.1).

(4.6) Example.   Suppose that we have a constant initial temperature T for the cylindrical rod. Find the temperature function u for all subsequent time.

Solution.   Method (4.5) yields

$$B_n = \frac{2T}{L} \int_0^L \sin \frac{n\pi x}{L} \, dx = \begin{cases} 4T/(n\pi) & \text{if } n \text{ is odd} \\ 0 & \text{if } n \text{ is even} \end{cases}$$

Hence, letting $2k + 1$ denote an odd integer n, our solution is

(4.7) $$u(x, t) = \frac{4T}{\pi} \sum_{k=0}^{\infty} \frac{\exp[-(2k+1)^2(\pi a/L)^2 t]}{2k+1} \sin\left[\frac{(2k+1)\pi x}{L}\right]$$

One nice result for the series above is that the exponential factors in each term tend very rapidly to zero as t tends to infinity.[15] Moreover, the larger the value of k, the larger $(2k+1)^2(\pi a/L)^2$ is; hence, it follows that $u(x, t)$ will tend to zero as t tends to infinity; we shall give a rigorous proof of this later. This last result is in accord with our physical experience which tells us that the rod will cool (if $T > 0$) or warm (if $T < 0$) until an equilibrium state of zero temperature is reached.

Furthermore the considerations of the preceding paragraph apply just as well to the series on the right below

$$u(x, t) - \frac{4T}{\pi} \exp\left[-\left(\frac{\pi a}{L}\right)^2 t \right] \sin \frac{\pi x}{L}$$

$$= \frac{4T}{\pi} \sum_{k=1}^{\infty} \frac{\exp[-(2k+1)^2(\pi a/L)^2 t]}{2k+1} \sin\left[\frac{(2k+1)\pi x}{L}\right]$$

Therefore it is possible to approximate $u(x, t)$, for t greater than some positive constant δ, by its first term (See Figure 3.10.)

$$u(x, t) \doteq \frac{4T}{\pi} \exp\left[-\left(\frac{\pi a}{L}\right)^2 t \right] \sin \frac{\pi x}{L} \qquad (t \geq \delta > 0)$$

It turns out that δ need not be too large in order for excellent approxima-

---

15. More rapidly than $1/t^m$ for any positive power m.

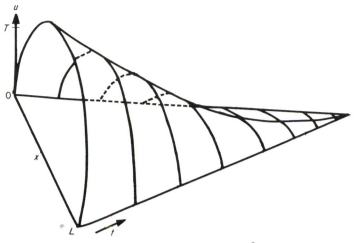

Figure 3.10 Graph of $(4T/\pi) \exp[-(\pi a/L)^2 t] \sin(\pi x/L)$.

tions to obtain. We will discuss these approximations further, after we discuss the proof that method (4.5) gives a rigorous solution to (4.1), for the two concepts are closely related.

(4.8) Example. Suppose that $u(0, t) = A$ and $u(L, t) = B$ where A and B are constants. Let $u(x, 0) = 0$ for $0 \leq x \leq L$. Find the temperature function u for $0 \leq x \leq L$ and $t > 0$.

Solution. We expect that as t increases toward infinity the temperature u will tend toward a steady-state (equilibrium) temperature. Let's denote this temperature distribution by $u_*(x)$. Then, since $(u_*)_t = 0$ we have

$$u_*''(x) = 0 \qquad u_*(0) = A \qquad u_*(L) = B$$

The solution of the problem above is easily seen to be $u_*(x) = [(B - A)/L]x + A$. Consider $v = u - u_*$. The reader should check that v satisfies

(4.9)
$$v_t = a^2 v_{xx} \qquad v(0, t) = 0 \qquad v(L, t) = 0$$

$$v(x, 0) = -\frac{B - A}{L} x - A$$

We can apply method (4.5) to problem (4.9) obtaining

$$v(x, t) = \sum_{n=1}^{\infty} \frac{2}{n\pi} [B(-1)^n - A] \exp\left[-n^2 \left(\frac{\pi a}{L}\right)^2 t \right] \sin \frac{n\pi x}{L}$$

Since $v = u - u_*$ we obtain for our solution to the problem posed

$$u(x, t) = \left(\frac{B - A}{L}\right)x + A + \frac{2}{\pi} \sum_{n=1}^{\infty} \frac{B(-1)^n - A}{n} \exp\left[-n^2 \left(\frac{\pi a}{L}\right)^2 t \right] \sin \frac{n\pi x}{L}$$

By the same reasoning as in the previous example we see that $u(x, t)$ will tend to $u_*(x)$ as t tends to infinity.

Verification of Solutions

We now turn to the problem of verifying that method (4.5) furnishes us with a rigorous solution to problem (4.1). For convenience of notation in the arguments to follow we shall assume that L equals π and that a^2 equals 1; those assumptions have no effect on the generality of our discussion and the reader may easily modify our notation to obtain proofs for the more general case. The series in (4.5) has the form

(4.10) $$u(x, t) = \sum_{n=1}^{\infty} B_n e^{-n^2 t} \sin nx \qquad \left(B_n = \frac{2}{\pi} \int_0^{\pi} f(x) \sin nx \, dx \right)$$

That the series above satisfies the boundary conditions in (4.1) is obvious; we now prove that it also satisfies the heat equation.

If we bring the absolute value sign inside the integral defining B_n we get

$$|B_n| \leq \frac{2}{\pi} \int_0^{\pi} |f(x)| \, dx$$

Letting M denote the value of the quantity on the right side above, we have for all n, $|B_n| \leq M$. Therefore, for each n

$$|B_n e^{-n^2 t} \sin nx| \leq M e^{-n^2 t}$$

Furthermore, if δ is some positive constant, then for all $t \geq \delta$ we have

$$M e^{-n^2 t} \leq M e^{-n^2 \delta} \qquad (t \geq \delta)$$

since e^{-y} is a decreasing function of y. Combining these last two inequalities yields

(4.11) $$|B_n e^{-n^2 t} \sin nx| \leq M e^{-n^2 \delta} \qquad (t \geq \delta \geq 0)$$

We will now show that the numerical series $\sum_{n=1}^{\infty} M e^{-n^2 \delta}$ converges; this will prove that the series for u converges for all values of $t \geq \delta$ and of x. The convergence of that numerical series can be shown by applying the Ratio Test. For purposes of later applications, however, we prefer to use a different argument.

The numerical series $\sum_{n=1}^{\infty} M e^{-n^2 \delta}$ satisfies the following relations:

$$\sum_{n=1}^{\infty} M e^{-n^2 \delta} = M e^{-\delta}[1 + e^{-3\delta} + e^{-8\delta} + e^{-15\delta} + e^{-24\delta} + \cdots]$$

$$< M e^{-\delta}[1 + e^{-3\delta} + (e^{-3\delta})^2 + (e^{-3\delta})^3 + (e^{-3\delta})^4 + \cdots]$$

$$= M e^{-\delta}\left[\sum_{n=1}^{\infty} (e^{-3\delta})^{n-1} \right]$$

Since $\sum_{n=1}^{\infty} (e^{-3\delta})^{n-1}$, is a geometric series, we find that its sum is $1/(1 - e^{-3\delta})$. Therefore,

(4.12) $$\sum_{n=1}^{\infty} M e^{-n^2 \delta} < M e^{-\delta}/(1 - e^{-3\delta})$$

Combining this last result with (4.11) we deduce from Weierstrass' M-test and Theorem (5.4), Chapter 1, that $u(x, t)$ is a continuous function of x and t for all $t \geq \delta$ and all x. Letting δ pass toward zero, this last result holds for all $t > 0$ and all x.

If we partially differentiate the series for u term by term with respect to t we obtain

$$\sum_{n=1}^{\infty} B_n(-n^2)e^{-n^2t} \sin nx$$

Analyzing that series as we did for u above, we find that it is dominated (majorized) for $t \geq \delta > 0$ by the numerical series

$$\sum_{n=1}^{\infty} Mn^2 e^{-\delta}(e^{-3\delta})^{n-1}$$

which converges by the Ratio Test. Applying Theorem (5.5), Chapter 1, for the variable t *with x held fixed*, we conclude that for $t \geq \delta$ the partial derivative u_t is given by

(4.13)
$$u_t(x, t) = \sum_{n=1}^{\infty} B_n(-n^2)e^{-n^2t} \sin nx$$

Letting δ tend to zero, we obtain (4.13) for all $t > 0$. A similar analysis shows that u_x and u_{xx} exist for all $t > 0$ and

(4.14)
$$u_{xx}(x, t) = \sum_{n=1}^{\infty} B_n(-n^2)e^{-n^2t} \sin nx$$

Comparing (4.13) and (4.14) we see that the heat equation $u_t = u_{xx}$ is satisfied for all $t > 0$ and all x.

(4.15) Remark. The main fact underlying the discussion above was that $|B_n| \leq M$ for all n. Therefore, it follows that every series

$$v(x, t) = \sum_{n=1}^{\infty} C_n \exp\left[-n^2\left(\frac{\pi a}{L}\right)^2 t\right] \sin \frac{n\pi x}{L}$$

for which $|C_n| \leq A$ where A is some positive constant satisfies the heat equation $v_t = a^2 v_{xx}$. For instance, if $C_n = 1$ for each n, then the series v is a solution of the heat equation which is *not* generated by an initial temperature function f. [See Exercise (4.27).]

To prove that the initial temperature condition in (4.1) is satisfied by the series in (4.10) we need to make some assumptions concerning the function f. *Suppose that f is piecewise smooth on the interval $(0, \pi)$.* If x_0 is one of the finite number of points of discontinuity of f in the open interval $(0, \pi)$, then we (re)define $f(x_0)$ to be

(4.16)
$$f(x_0) = \tfrac{1}{2}[f(x_0+) + f(x_0-)]$$

Also, if necessary, we (re)define $f(0)$ and $f(\pi)$ to be zero. Then Theorem (4.5), Chapter 1, assures us that[16]

$$u(x, 0) = \sum_{n=1}^{\infty} B_n \sin nx = f(x) \qquad (0 \leq x \leq \pi)$$

This last result finishes the verification that u rigorously solves (4.1).

There is, however, one remaining point. We proved above that $u(x, 0) = f(x)$ and u is continuous for all real x and all $t > 0$. For completeness, we should also show that

(4.17) $\lim_{(x,t)\to(x_0,0+)} u(x, t) = f(x_0) \qquad (0 \leq x_0 \leq \pi)$

Formula (4.17) shows that the temperature u evolves continuously in time from the initial temperature f. Therefore, for physical reasons, it is important to establish (4.17).

To establish (4.17) we will, for now, *assume that f has an odd, periodic extension that is continuous and piecewise smooth*. But then the Fourier series for that extension is just the sine series for f, namely $\sum_{n=1}^{\infty} B_n \sin nx$. Hence, as we showed in the proof of Theorem (4.4) in Chapter 2, $\sum_{n=1}^{\infty} |B_n|$ converges. Then, for all $t \geq 0$ we have

$$|B_n e^{-n^2 t} \sin nx| \leq |B_n|$$

and Weierstrass' M-test implies that the series $\sum_{n=1}^{\infty} B_n e^{-n^2 t} \sin nx$ converges uniformly for all $t \geq 0$ and all x values. Formula (4.17) then holds due to continuity of the uniformly convergent series.

In Chapter 4, §7, we will show that (4.17) holds as long as $\int_0^{\pi} |f(x)| \, dx$ is finite (converges) and x_0 is a point of continuity of the odd periodic extension of f. In particular, (4.17) *holds whether or not* $\sum_{n=1}^{\infty} B_n \sin nx$ *converges to* $f(x)$. For instance, if f is one of those pathological continuous functions whose sine series diverges for some x values, then (4.17) will still be valid even for those x values where the sine series diverges. Furthermore, the discussion of the heat equation given above shows that $\int_0^{\pi} |f(x)| \, dx$ being finite is enough to ensure that the series for u solves the heat equation for $t > 0$. *Thus, once we have established* (4.17) *we will have completely proved the following theorem.*

(4.18) Theorem.   If f is continuous on $[0, \pi]$ and satisfies $f(0) = f(\pi) = 0$ then the function u, defined by (4.5) for $t > 0$ and equalling f for $t = 0$, solves problem (4.1) and is continuous for all $t \geq 0$ and $0 \leq x \leq \pi$.

We close this section with an illustration of an approximation problem.

(4.19) Example.   Consider the solution to the problem posed in Example (4.6). Suppose $T = 100°C$, $L = 10$ cm, and $a^2 = 1.14$ cm^2/sec (copper as

---

16. Changing the values of f at a finite number of points has no effect on the integrals that define the sine coefficients B_n. Theorem (4.5), Chapter 1, is applied to the odd periodic extension of f which reduces to f on $[0, \pi]$.

material). Find a small enough value of δ so that $t \geq \delta > 0$ implies that u is approximated to within ± 0.0001 by its first term. How many terms are sufficient for approximating u to within ± 0.0001 if $t \geq 1$ sec?

Solution.   Using the same kind of reasoning as in the verification that u satisfies the heat equation, we obtain [*Note:* $4T/\pi \doteq 127.3$ and $(\pi a/L)^2 \doteq 0.1125$] for $t \geq \delta > 0$

$$\left| u(x, t) - 127.3 e^{-0.1125t} \sin \frac{\pi x}{10} \right| \leq 127.3 \sum_{k=1}^{\infty} \frac{1}{2k+1} e^{-(2k+1)^2 0.1125t}$$

$$\leq 127.3 \sum_{k=1}^{\infty} \frac{1}{2k+1} e^{-(2k+1)^2 0.1125\delta}$$

$$\leq \frac{127.3}{3} e^{-9(0.1125)\delta} / [1 - e^{-16(0.1125)\delta}]$$

where we have used $[1/(2k+1)] \leq \frac{1}{3}$ in the last inequality. Setting the last quantity in the equalities above to be less than 0.0001 and *assuming* that δ is large enough that the denominator $1 - e^{-16(0.1125)\delta}$ is greater than $\frac{1}{2}$ we solve

$$(127.3)\tfrac{2}{3} e^{-9(0.1125)\delta} < 0.0001$$

which yields $\delta > 13.48$. Trying $\delta = 13$ we get that the value of the last quantity in the set of inequalities above is 0.000081527 which is less than 0.0001. Thus, after 13 sec the first term of u is an adequate approximation of u to within ± 0.0001. If $t \geq 1$, then

$$\left| u(x, t) - (127.3) \sum_{k=0}^{N} \frac{e^{-(2k+1)^2 0.1125t}}{2k+1} \sin \frac{(2k+1)\pi x}{10} \right|$$

$$\leq \frac{127.3}{2N+3} e^{-(2N+3)^2(0.1125)} / [1 - e^{-(8N+16)(0.1125)}]$$

Trying different values of N we find that $N = 4$ is sufficient to make the right side of the inequality above less than 0.0001. Therefore, five terms are sufficient for the desired approximation if $t \geq 1$.

Exercises

(4.20) Suppose that a cylindrical rod of iron ($a^2 = 0.12 \text{ cm}^2/\text{sec}$) that is 10 cm long is raised to a constant temperature of 100°C. If the lateral surface of the rod is insulated and the ends are held at 0°C, show that it takes slightly more than 2.5 min for the rod to cool to a temperature of less than 20°C throughout. Show that if the rod is made from concrete ($a^2 = 0.005 \text{ cm}^2/\text{sec}$) then it takes about 62.5 min.

(4.21) Prove that $u(x, t)$ as given in (4.5) tends to zero as t tends to infinity. Also, prove that

$$\lim_{t \to \infty} u(x, t) - B_1 \exp\left[-\left(\frac{\pi a}{L}\right)^2 t \right] \sin \frac{\pi x}{L} = 0$$

Hence $u(x, t)$ can be approximated by its first term for large enough t.

(4.22) Consider the solution to the heat conduction problem found for Example (4.6). Suppose that $T = 50°C$, $L = 10$ cm, and $a^2 = 0.005$ (concrete as material). Solve the following problems.

(a) How many terms are needed to approximate $u(x, t)$ to within ± 0.0001 for all x values if $t \geq 1$?
(b) If $t \geq \delta$, how small a value of δ is sufficient for u to be approximated by its first term to within ± 0.0001?
(c) Answer (a) and (b) if $a^2 = 1.14$ (copper as material).

(4.23) Suppose that the faces of a rod at $x = 0$ and $x = L$ are insulated $[u_x(0, t) = u_x(L, t) = 0]$. If the rod has an initial temperature of f, then derive the temperature formula

$$u(x, t) = \tfrac{1}{2}A_0 + \sum_{n=1}^{\infty} A_n \exp\left[-n^2\left(\frac{\pi a}{L}\right)^2 t\right] \cos\frac{n\pi x}{L}$$

$$A_n = \frac{2}{L} \int_0^L f(x) \cos\frac{n\pi x}{L} \, dx$$

Explain why, and then prove that $\lim_{t\to\infty} u(x, t) = \tfrac{1}{L}\int_0^L f(x)\, dx$. Interpret that limit physically.

(4.24) Verify rigorously that the series for u given in (4.23) does satisfy

$$u_t = a^2 u_{xx} \qquad u_x(0, t) = u_x(L, t) = 0 \qquad u(x, 0) = f(x)$$

and check that $\lim_{t\to 0+} u(x, t) = f(x)$ (under certain conditions).

(4.25) Suppose a rod is initially at a temperature f and its faces at $x = 0$ and $x = \pi$ are kept at temperature zero. Suppose also that *heat is generated at a parabolic rate throughout the rod*, that is, in (1.23) set $(c\rho A)^{-1}Q(x, t)$ equal to $Cx(x - \pi)$ where C is a constant. Solve the following problems.

(a) Using (1.23) show that the temperature u has the form

$$u(x, t) = \frac{-C}{12a^2}(x^4 - 2\pi x^3 + \pi^3 x) + \sum_{n=1}^{\infty} B_n e^{-n^2 a^2 t} \sin nx$$

where

$$B_n = \frac{2}{\pi} \int_0^\pi \left\{ \frac{C}{12a^2}(x^4 - 2\pi x^3 + \pi^3 x) + f(x)\right\} \sin nx \, dx$$

(b) Find $u(x, t)$ when $f(x) = \tfrac{1}{2}\pi - |x - \tfrac{1}{2}\pi|$ and when $f = 0$.
(c) Explain why for every initial temperature f we always have

$$\lim_{t\to +\infty} u(x, t) = \frac{-C}{12a^2}(x^4 - 2\pi x^3 + \pi^3 x)$$

and interpret this limit physically. Prove rigorously that the limit holds (under certain conditions on f) and that the series for u is a rigorous solution to the given problem.

(d) Generalize the results above to the case where $(c\rho A)^{-1}Q(x, t)$ is a continuous function $g(x)$ dependent on x and not t.

(4.26) (a) Let $v(x, t)$ denote the temperature of a rod as described in (1.20). Suppose that the ends of the rod at $x = 0$ and $x = L$ are insulated and the initial temperature is given by the function f. Show that $v(x, t) = T + e^{-ht}u(x, t)$ where u is the temperature function found in (4.23). (b) If the ends of the rod in part (a) are not insulated, but are kept at temperature T, find the temperature function. [*Hint*: For both (a) and (b) the substitution $w(x, t) = v(x, t) - T$ is helpful.]

*(4.27) Show that $v(x, t) = \sum_{n=1}^{\infty} e^{-n^2 t} \sin nx$ solves the heat equation $v_t = v_{xx}$ as well as $v(0, t) = v(\pi, t) = 0$, but v is *not* generated by an initial temperature f.

*Theta Function and Heat Conduction

The following three problems are needed for Chapter 4, §7.

(4.28) Consider the solution (4.5) to the heat conduction problem (4.1). Let F denote the odd periodic extension of f. Prove that u can be written in the *complex form*

$$u(x, t) = \sum_{n=-\infty}^{\infty} c_n e^{i(n\pi/L)x} e^{-n^2(\pi a/L)^2 t}$$

$$c_n = \frac{1}{2L} \int_{-L}^{L} F(s) e^{-i(n\pi/L)s} \, ds$$

(4.29) Substitute the integral form of c_n into the series for u above and obtain, in a formal way,

$$u(x, t) = \frac{1}{2L} \int_{-L}^{L} F(s)\theta(x - s, t) \, ds$$

where

$$\theta(x, t) = \sum_{n=-\infty}^{\infty} e^{i(n\pi x/L)} e^{-n^2(\pi a/L)^2 t}$$

Justify this derivation, for $t > 0$. [*Hint*: Use Weierstrass' M-test for $t \geq \delta$, then let δ tend to 0.]

(4.30) Using periodicity and (4.29), prove that

$$u(x, t) = \frac{1}{2L} \int_{-L}^{L} F(x - s)\theta(s, t) \, ds.$$

The function θ is known as a *theta function*; we will discuss some of its basic properties in §7 of Chapter 4.

§5. The Applied Force Problem for the Vibrating String

We now return to problems involving the vibration of elastic strings. We shall solve the applied force problem for the vibrating string: Equation (1.8) subject to the conditions in (1.1). As a side result, we will show that our proposed solution is the only possible one if we demand a rigorous solution. The reader who is interested in a rigorous verification of our proposed solution is invited to solve Exercises (5.32) through (5.35).

The problem to be solved is the following one

(5.1)
$$\begin{cases} c^2 y_{xx} + F(x, t)/\rho = y_{tt} & \text{(inhomogeneous wave equation)} \\ y(0, t) = 0 \quad y(L, t) = 0 & \text{(boundary conditions)} \\ y(x, 0) = f(x) & \text{(initial position)} \\ y_t(x, 0) = g(x) & \text{(initial velocity)} \end{cases}$$

Suppose that the function y satisfies (5.1) in the rigorous sense.[17] Then, holding t fixed, y has a sine series that converges to it for $0 \le x \le L$. Thus, for $t \ge 0$

(5.2) $$y(x, t) = \sum_{n=1}^{\infty} B_n(t) \sin \frac{n\pi x}{L} \qquad B_n(t) = \frac{2}{L} \int_0^L y(x, t) \sin \frac{n\pi x}{L} \, dx$$

We have labeled the sine coefficients by $B_n(t)$ since they clearly depend on what value of t is held fixed. Performing an integration by parts and using the boundary conditions, we get

$$B_n(t) = -\frac{2}{n\pi} y(x, t) \cos \frac{n\pi x}{L} \Big|_{x=0}^{x=L} + \frac{2}{n\pi} \int_0^L y_x(x, t) \cos \frac{n\pi x}{L} \, dx$$

$$= \frac{2}{n\pi} \int_0^L y_x(x, t) \cos \frac{n\pi x}{L} \, dx$$

Performing another integration by parts yields

$$B_n(t) = -\frac{2L}{n^2\pi^2} \int_0^L y_{xx}(x, t) \sin \frac{n\pi x}{L} \, dx$$

Since y satisfies the inhomogeneous wave equation, we use that equation to solve for y_{xx} and then substitute into the equation above, obtaining

$$B_n(t) = -\frac{2L}{n^2\pi^2} \int_0^L \left[\frac{y_{tt}(x, t)}{c^2} - \frac{F(x, t)}{c^2\rho} \right] \sin \frac{n\pi x}{L} \, dx$$

With a little algebra we obtain the more suggestive form

(5.3) $$B_n(t) = -\left(\frac{L}{nc\pi}\right)^2 \left[\frac{2}{L} \int_0^L y_{tt}(x, t) \sin \frac{n\pi x}{L} \, dx - \frac{2}{L} \int_0^L \frac{F(x, t)}{\rho} \sin \frac{n\pi x}{L} \, dx \right]$$

---

17. See the discussion preceding Theorem (3.9); here we also assume that F is continuous for $0 \le x \le L$ and $t \ge 0$.

If we define $F_n(t)$ by

(5.4)
$$F_n(t) = \frac{2}{L} \int_0^L \frac{F(x, t)}{\rho} \sin \frac{n\pi x}{L} \, dx$$

then Eq. (5.3) takes the form

(5.5)
$$B_n(t) = -\left(\frac{L}{nc\pi}\right)^2 \left[\frac{2}{L} \int_0^L y_{tt}(x, t) \sin \frac{n\pi x}{L} \, dx - F_n(t)\right]$$

To proceed further we need the following well known theorem of advanced calculus.

(5.6) Theorem. If the function $h(x, t)$ is continuous on the rectangle $R = \{(x, t): a \le x \le b \text{ and } c \le t \le d\}$ then $\int_a^b h(x, t) \, dx$ is a continuous function of t on the interval $[c, d]$. Moreover, we have

$$\int_c^d \left[\int_a^b h(x, t) \, dx\right] dt = \int_a^b \left[\int_c^d h(x, t) \, dt\right] dx$$

Furthermore, if $h_t(x, t)$ is continuous on R then $\int_a^b h(x, t) \, dx$ is continuously differentiable and

$$\frac{d}{dt} \int_a^b h(x, t) \, dx = \int_a^b h_t(x, t) \, dx$$

We apply Theorem (5.6) to formula (5.5), with $h(x, t) = y(x, t) \sin(n\pi x/L)$ and $R = \{(x, t): \quad 0 \le x \le L \text{ and } 0 \le t \le K\}$ where K is *any* positive constant. Applying the last statement of that theorem in *reverse*, we bring the partial differentiation by t outside the integral sign in (5.5), hence

$$B_n(t) = -\left(\frac{L}{nc\pi}\right)^2 \left[\frac{2}{L} \frac{d}{dt} \int_0^L y_t(x, t) \sin \frac{n\pi x}{L} \, dx - F_n(t)\right]$$

$$= -\left(\frac{L}{nc\pi}\right)^2 \left[\frac{2}{L} \frac{d^2}{dt^2} \int_0^L y(x, t) \sin \frac{n\pi x}{L} \, dx - F_n(t)\right]$$

$$= -\left(\frac{L}{nc\pi}\right)^2 \left[\frac{d^2}{dt^2} \left(\frac{2}{L} \int_0^L y(x, t) \sin \frac{n\pi x}{L} \, dx\right) - F_n(t)\right]$$

Using the original definition of $B_n(t)$ from (5.2) yields, after some algebra, the following differential equation

(5.7)
$$B_n''(t) + \left(\frac{nc\pi}{L}\right)^2 B_n(t) = F_n(t) \qquad (n = 1, 2, 3, \ldots)$$

Equation (5.7) is a differential equation for the unknown function $B_n(t)$; the function $F_n(t)$ is assumed to be known since $F(x, t)/\rho$ is assumed to be known. Since K was an arbitrary positive constant, Eq. (5.7) holds for all $t \ge 0$. The differential equation for B_n becomes an initial value problem if we note that

$$B_n(0) = \frac{2}{L} \int_0^L y(x, 0) \sin \frac{n\pi x}{L} dx = A_n$$

where

$$A_n = \frac{2}{L} \int_0^L f(x) \sin \frac{n\pi x}{L} dx$$

and

$$B_n'(0) = \frac{2}{L} \int_0^L y_t(x, 0) \sin \frac{n\pi x}{L} dx = b_n$$

where

$$b_n = \frac{2}{L} \int_0^L g(x) \sin \frac{n\pi x}{L} dx$$

Thus, we have obtained the following result.

Problem (5.1) is solved by the following procedure:
(a) Compute

$$F_n(t) = \frac{2}{L} \int_0^L \frac{F(x, t)}{\rho} \sin \frac{n\pi x}{L} dx$$

$$A_n = \frac{2}{L} \int_0^L f(x) \sin \frac{n\pi x}{L} dx$$

and

$$b_n = \frac{2}{L} \int_0^L g(x) \sin \frac{n\pi x}{L} dx$$

(5.8)

(b) Solve the initial value problem for B_n

$$B_n''(t) + \left(\frac{nc\pi}{L}\right)^2 B_n(t) = F_n(t)$$

$$B_n(0) = A_n \qquad B_n'(0) = b_n$$

(c) The solution to (5.1) is given by the following series

$$y(x, t) = \sum_{n=1}^\infty B_n(t) \sin \frac{n\pi x}{L}$$

(5.9) Remark.   The set of functions $\{B_n(t)\}_{n=1}^\infty$ defined above is called the *finite sine transform* of y. Likewise, $\{F_n(t)\}$ is the finite sine transform of

$F(x, t)/\rho$. Sometimes, for simplicity, we might say $B_n(t)$ is the finite sine transform of y. For some problems, finite cosine transforms might be needed. The *finite cosine transform* of y is defined as

$$A_n(t) = \frac{2}{L} \int_0^L y(x, t) \cos \frac{n\pi x}{L} dx \qquad (n = 0, 1, 2, \ldots)$$

We will now examine a couple of sample problems. *Throughout the discussion we shall use the convenient notation* $w_n = nc\pi/L$ *for* $n = 1, 2, 3, \ldots$. The reader may recall that w_n is the frequency of an nth order harmonic for the string (see §2). Each such frequency is called a *natural frequency* for the string.

(5.10) Example: Constant load.   Suppose that the applied force is given by

$$F(x, t) = \begin{cases} 0 & \text{for } 0 \leq x < a \\ C\rho & \text{for } a < x < b \quad (C \text{ constant}) \\ 0 & \text{for } b < x < L \end{cases}$$

Find the subsequent motion of the string if the string is initially in a horizontal position with zero initial velocity.

Solution.   For this problem, the method of (5.8) yields $A_n = 0$, $b_n = 0$, and

$$F_n(t) = \frac{2}{L} \int_a^b C \sin \frac{n\pi x}{L} dx = \frac{2C}{n\pi} \left(\cos \frac{n\pi a}{L} - \cos \frac{n\pi b}{L} \right)$$

Each function $F_n(t)$ is a constant; so let's use the symbol K_n to denote these constants. We must solve

(5.11) $B_n''(t) + w_n^2 B_n(t) = K_n \qquad B_n(0) = 0 \qquad B_n'(0) = 0$

where we have used w_n in place of $nc\pi/L$. The solution to the differential equation in (5.11) is

$$B_n(t) = K_n/w_n^2 + C_n \cos w_n t + D_n \sin w_n t$$

where C_n and D_n are arbitrary constants that we determine from the initial conditions in (5.11). The result is

$$B_n(t) = (K_n/w_n^2)[1 - \cos w_n t]$$

Therefore, our solution is

$$y(x, t) = \sum_{n=1}^\infty (K_n/w_n^2)[1 - \cos w_n t] \sin \frac{n\pi x}{L}$$

$$= \frac{2C}{\pi} \sum_{n=1}^\infty \left[\frac{\cos(w_n a/c) - \cos(w_n b/c)}{n w_n^2} \right] (1 - \cos w_n t) \sin(w_n x/c)$$

(5.12) Example: Resonance.   Suppose that an elastic string with ends fixed is at rest in a horizontal position and the section of the string between $x = a$

and $x = b$ $(0 < a < b < L)$ is acted on by a periodic force $F(x, t) = C\rho \sin wt$ (C and w constants). Find the subsequent motion of the string.

Solution. For this problem, the method of (5.8) yields $A_n = 0$, $b_n = 0$, and $F_n(t) = 2C(n\pi)^{-1}[\cos(n\pi a/L) - \cos(n\pi b/L)] \sin wt$. If we let the constant factor on $\sin wt$ be denoted by K_n then $F_n(t) = K_n \sin wt$. Hence, we must solve

(5.13) $B_n''(t) + w_n^2 B_n(t) = K_n \sin wt$ $B_n(0) = 0$ $B_n'(0) = 0$

If $w \neq w_n$, then the method of undetermined coefficients tells us that the general solution to the differential equation in (5.12) is

$$B_n(t) = K_n \frac{\sin wt}{w_n^2 - w} + C_n \cos w_n t + D_n \sin w_n t$$

where C_n and D_n are found by using the initial conditions in (5.13). Thus,

$$0 = C_n \quad \text{and} \quad 0 = K_n \frac{w}{w_n^2 - w^2} + D_n w_n$$

Hence solving for D_n and substituting back into the equation for $B_n(t)$ above yields, after some algebra,

$$B_n(t) = K_n \left[\frac{w_n \sin wt - w \sin w_n t}{w_n(w_n^2 - w^2)} \right]$$

If $w \neq w_n$ for all integers n, then our solution is

$$y(x, t) = \sum_{n=1}^{\infty} K_n \left[\frac{w_n \sin wt - w \sin w_n t}{w_n(w_n^2 - w^2)} \right] \sin(w_n x/c)$$

where $w_n = nc\pi/L$ and $K_n = 2C(n\pi)^{-1}[\cos(w_n a/c) - \cos(w_n b/c)]$.

To obtain the solution above we assumed that the time frequency w of the applied force was not equal to any of the natural frequencies w_n of the string. If, however, $w = w_m$ for some integer m, then (5.13) becomes for $n = m$

(5.13') $B_m''(t) + w_m^2 B_m(t) = K_m \sin w_m t$ $B_m(0) = 0$ $B_m'(0) = 0$

In this case, using $At \sin w_m t + Bt \cos w_m t$ yields, by the method of undetermined coefficients, a particular solution to the differential equation in (5.13'). Hence that equation's general solution is found to be

$$B_m(t) = -K_m(2w_m)^{-1} t \cos w_m t + C_m \cos w_m t + D_m \sin w_m t$$

The initial conditions in (5.13') then yield

$$B_m(t) = -K_m(2w_m)^{-1} t \cos w_m t + K_m(2w_m^2)^{-1} \sin w_m t$$

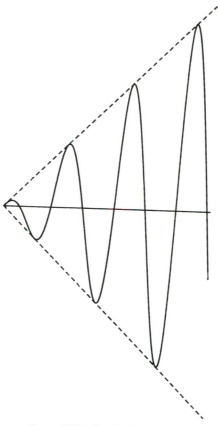

Figure 3.11 Graph of $t \cos w_m t$.

Our solution in this case is[18]

$$y(x, t) = - K_m(2w_m)^{-1} t \cos w_m t \sin(w_m x/c)$$
$$+ K_m(2w_m^2)^{-1} \sin w_m t \sin(w_m x/c)$$
$$+ \sum_{\substack{n=1 \\ (n \neq m)}}^{\infty} K_n \left[\frac{w_n \sin w_m t - w_m \sin w_n t}{w_n(w_n^2 - w_m^2)} \right] \sin(w_n x/c)$$

The harmonic $- K_m(2w_m)^{-1} t \cos w_m t \sin(w_m x/c)$ has a time-dependent amplitude equal to a constant $-K_m(2w_m)^{-1}$ times

$$t \cos w_m t$$

which oscillates in an unbounded manner as t tends to infinity. The graph of $t \cos w_m t$ looks like the graph in Figure 3.11. We say that this harmonic *resonates*.

---

18. The particular (resonance) term $-K_m(2w_m)^{-1} t \cos(w_m t)\sin(w_m x/c) + K_m(2w_m^2)^{-1} \sin(w_m t)$ $\sin(w_m x/c)$ can also be found by applying l'Hôpital's Rule to

$$\lim_{w \to w_m} K_m \left[\frac{w_m \sin wt - w \sin w_m t}{w_m(w_m^2 - w^2)} \right] \sin(w_m x/c)$$

It can be shown [see Exercise (5.22)] that resonance occurs whenever a force of time frequency w equal to one of the natural frequencies for the string is applied to that string.

(5.14) Remark.   In both examples above, we assumed that $y(x, 0) = 0$ and $y_t(x, 0) = 0$. We did that because of the following fact [see Exercise (5.24)]:

If y is a solution of problem (5.1), *then $y = v + w$ where v solves the applied force problem with the same force but $v(x, 0) = 0 = v_t(x, 0)$ and w solves the force free problem* (2.1) *with $w(x, 0) = f(x)$ and $w_t(x, 0) = g(x)$.*

We conclude this section by demonstrating that when a solution to (5.1) exists, it is unique. In particular, when a solution exists, then the series given by method (5.8) does converge and equals that solution. These results are useful from a physical standpoint; for physical reasons, we assume that a solution to (5.1) exists, our theorem tells us that the series given by (5.8) equals that solution.

(5.15) Theorem: Uniqueness of Solutions.   Suppose that y rigorously solves (5.1), then y is unique. Moreover, the series given by method (5.8) converges for all $t > 0$ and equals y.

Proof.   Method (5.8) shows that the finite sine transform $\{B_n(t)\}$ of y is given by the solutions of the initial value problems

(5.16) $$B_n''(t) + \left(\frac{n c \pi}{L}\right)^2 B_n(t) = F_n(t) \qquad B_n(0) = A_n \qquad B_n'(0) = b_n$$

The reader may recall from the theory of ordinary differential equations that solutions to initial value problems such as (5.16) are unique.[19] If we fix a value of $t > 0$, say t_0, then $B_n(t_0)$ are the sine coefficients for the sine series of $y(x, t_0)$. Moreover, since the function $y(x, t_0)$ is continuous for $0 \leq x \leq L$ and has a derivative $y_x(x, t_0)$ for $0 < x < L$, we know that its sine series converges to it. Thus, for *each* $t > 0$

(5.17) $$y(x, t) = \sum_{n=1}^{\infty} B_n(t) \sin \frac{n \pi x}{L} \qquad (0 \leq x \leq L)$$

If $y^\#$ were some other solution to the same problem (5.1), then for some point (x_0, t_0) we would have $y(x_0, t_0) \neq y^\#(x_0, t_0)$. Hence, by the Uniqueness Theorem (6.7), Chapter 2, we conclude that some sine coefficient $B_m^\#(t_0)$ for $y^\#(x, t_0)$ is different from $B_m(t_0)$. But, we would then have two different solutions $B_m^\#$ and B_m to (5.16) for $n = m$, contrary to the uniqueness property mentioned after that equation. This contradiction establishes the uniqueness of y. And, Eq. (5.17) demonstrates the other part of our theorem. ∎

---

19. There are two ways to see this: (1) consult a text in which such a general theorem is stated, for example, Boyce and DiPrima (1977), p. 84; (2) see Exercises (5.29) and (5.30) in which uniqueness is proved for equations such as those in (5.16).

Setting $F = 0$ in the theorem just proved establishes the following corollary which we mentioned at the end of §3.

(5.18) Corollary. A rigorous solution to a force-free vibrating string problem is unique and must equal the series shown in (2.12).

Remark.   The theory of resonance has many important applications. For example, the reader will find a very detailed yet elementary discussion of how Fourier analysis enables us to understand the structure of music, musical instruments, and human hearing in Chapters 5 through 7 of Olson (1967).

Exercises

(5.19) Suppose a force $F(x, t) = C\rho \sin wt$ is applied to an elastic string with ends fastened at $x = 0$ and $x = L$. Determine the subsequent motion of the string. Does resonance occur?

(5.20) Suppose a force $F(x, t) = F(x) \cos wt$ where $F(x) = x$ if $a < x < b$ and 0 elsewhere $(0 < a < b < L)$ is applied to an elastic string of length L with fixed ends. Determine the subsequent motion of the string. Does resonance occur?

*(5.21) Let a concentrated force $F(x, t) = A \sin wt$ act upon the *point $x = b$* of an elastic string of length L with fixed ends. Find the subsequent motion of the string. Does resonance occur? [*Hint*: Use a Dirac δ function.]

(5.22) Prove that if $F(x, t) = F(x) \sin wt$, or $F(x, t) = F(x) \cos wt$, then resonance occurs if such a force is applied to an elastic string with fixed ends *when w equals one of the natural frequencies of that string.*

(5.23) Using Fourier analysis, generalize the result of (5.22) as far as you can.

(5.24) Verify the result stated in Remark (5.14).

(5.25) Suppose an elastic string of length L is fastened at $x = 0$ and the right end of the string is forced to undergo the motion described by $y(L, t) = \cos wt$. (See Figure 3.12.) If the string has initial position $y(x, 0) = x/L$ and initial velocity $y_t(x, 0) = 0$, then determine the subsequent motion of the string. [*Hint*: Letting $y = v + (x/L) \cos wt$ converts the force free problem for y into an applied force problem for v.]

(5.26) Using the results of (5.25), explain the following frequently observed occurrence: A child is playing on a swing; the child soon discovers (to its excitement, and peril) that by swinging in a certain way the swing moves in a very quick and exciting way.

(5.27) *Effect of Frictional Forces.*   Consider problem (2.24). Suppose a force $F(x, t) = C\rho \sin wt$ is applied to the string: (a) Determine the subsequent motion of the string. (b) Explain why $y(x, t)$ tends, as t tends to

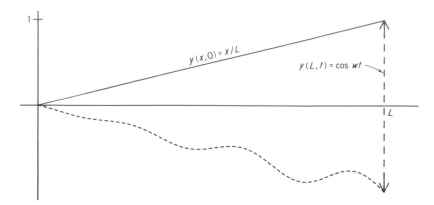

Figure 3.12   Illustration of the *whipped string* described in Exercise (5.25).

infinity, to an equilibrium motion $y_*(x)$ and determine that equilibrium motion. (c) Does resonance occur?

(5.28)   Solve the following problem by either separation of variables, or by a finite sine transform

$$u_{tt} + 2au_t + bu = c^2 u_{xx} \qquad (0 < x < L, t > 0)$$

$$u(0, t) = u(L, t) = 0 \qquad u(x, 0) = 0 \qquad u_t(x, 0) = g(x)$$

[This problem arises in the transmission of electrical impulses in a long wire with distributed capacitance, inductance, and resistance. It is called the *telegrapher's equation.*]

(5.29)   (a)   Show that the differential equation $y''(t) + k^2 y(t) = 0$ $(k > 0)$ can be expressed as the system

$$(*) \quad y'(t) + iky(t) = w(t) \qquad w'(t) - ikw(t) = 0$$

Then show, by the method of integrating factors, that all solutions of $(*)$ have the form

$$w = ce^{ikt} \qquad y = Ce^{ikt} + De^{-ikt}$$

where c, C, and D are complex constants.

(b)   Using (a) show that the *unique* solution to the initial value problem $(k > 0)$

$$y'' + k^2 y = 0 \qquad y(0) = a \qquad y'(0) = b$$

is $y = a \cos kt + (b/k) \sin kt$.

(c)   Generalize this last result to the problem $y'' + Ay' + By = 0$, $y(0) = a$, $y'(0) = b$.

(5.30)   Suppose B_n and $B_n^{\#}$ both solve (5.16). Prove that $B_n = B_n^{\#}$. [*Hint:* Consider $B_n - B_n^{\#}$ and use (5.29b).]

(5.31) Prove that the heat conduction problem (4.1) has a unique solution.

In this set of three exercises we outline the verification that (5.8) solves (5.1).

(5.32) Show that $B_n(t)$ as defined in (5.8) is given by

$$B_n(t) = \int_0^t \frac{\sin w_n(t-s)}{w_n} F_n(s)\, ds$$

provided $f = g = 0$. [*Hint*: Differentiating the integral requires the generalized Leibniz Rule.[20]]

(5.33) Suppose $F(x, t)$ has continuous partial derivatives in x and t and $F(0, t) = F(L, t) = 0$. Proceeding in a formal way, show that by substitution of the integral form for B_n given in (5.32) into the series for y in method (5.8) we obtain (provided again that $f = g = 0$)

$$y(x, t) = \frac{1}{2c} \int_0^t \int_{x-c(t-s)}^{x+c(t-s)} \frac{F(r, s)}{\rho}\, dr\, ds$$

(5.34) Verify that the form for y given in (5.33) solves (5.1) for $f = g = 0$, then prove that method (5.8) does furnish the solution to (5.1) provided that F satisfies the hypotheses given in (5.33) and f and g satisfy the hypotheses given in Theorem (3.9).

References

Further discussion of partial differential equations can be found in Sommerfeld (1949), Weinberger (1965), Berg and McGregor (1966), and Copson (1975). Applications to vibration and sound are discussed in Morse (1948), Helmholtz (1954), and Rayleigh (1945). See also Olson (1967) for musical aspects of sound engineering. Heat conduction and diffusion is discussed in Carslaw and Jaeger (1959), Widder (1975), Fourier (1955) and Crank (1956).

---

20. See Kaplan (1984), p. 258, for a discussion of this rule. The integral form for B_n can be derived by using Laplace transforms, see Chapter 9, §3.

4

Some Harmonic Function Theory

Almost every major area of mathematical physics finds some use for the theory of harmonic functions. Such a ubiquitous mathematical theory has consequently over the course of two centuries been developed by many great mathematicians into a grand, beautiful structure. In this chapter, we shall discuss the principal aspects of the theory of harmonic functions of two variables. Our main tool will be Poisson's Integral which we shall derive using Fourier series. Once Poisson's Integral is in hand, we can then explore some of the many beautiful properties of harmonic functions, such as the maximum–minimum principle, Gauss' mean value theorem, and Harnack's convergence theorems. We shall also briefly treat the applications of Gauss' divergence theorem. In the next to last section we will describe some interesting parallels between harmonic functions and solutions to the heat equation. The final section treats further properties of harmonic functions on connected open sets, subharmonic functions, and Perron's elegant proof of the existence of solutions to Dirichlet's problem. Perron's proof is difficult and should be regarded as optional.

Although the theory of analytic functions of a complex variable is most often used for discussing harmonic functions of two variables, in this chapter we will use methods that generalize to harmonic functions of three variables. Such harmonic functions will be analyzed in Chapter 8.

§1. Introduction

In this section we shall introduce some notation, provide some elementary examples, and briefly discuss the physical applications of harmonic functions.

We shall denote the Euclidean plane by $\mathbb{R}^2$. Every point in $\mathbb{R}^2$ can be described by a pair of Cartesian coordinates (x, y) relative to a fixed pair of perpendicular axes. The *open disk* $B_r(a, b)$ of radius r centered at the point

(a, b) consists of all the points (x, y) that satisfy

$$[(x - a)^2 + (y - b)^2]^{1/2} < r$$

Recall that an *open set* U in $\mathbb{R}^2$ is a set of points having the property that every point in U is contained in some open disk which is a subset of U. (See Figure 4.1.) A set F in $\mathbb{R}^2$ is called *closed* if its complement

$$\mathbb{R}^2 - F = \{(x, y): (x, y) \text{ not in } F\}$$

is open. The reader should be familiar with the following examples of open and closed sets in $\mathbb{R}^2$.

(a) The open disk $B_r(a, b)$ is, as its name indicates, an open set.
(b) The *closed disk* $\bar{B}_r(a, b) = \{(x, y): [(x - a)^2 + (y - b)^2]^{1/2} \le r\}$ is a closed set.
(c) The circle $\partial B_r(a, b) = \{(x, y): [(x - a)^2 + (y - b)^2]^{1/2} = r\}$ is a closed set.
(d) The open rectangle $\{(x, y): a < x < b \text{ and } c < y < d\}$ is an open set.

Let U be some open set in $\mathbb{R}^2$. A real valued function Ψ defined on U is called *harmonic* if Ψ and its first and second partial derivatives are continuous functions on U and Ψ satisfies *Laplace's equation*

$$\Delta \Psi = \Psi_{xx} + \Psi_{yy} = 0$$

on all of U. Here are some examples of harmonic functions.

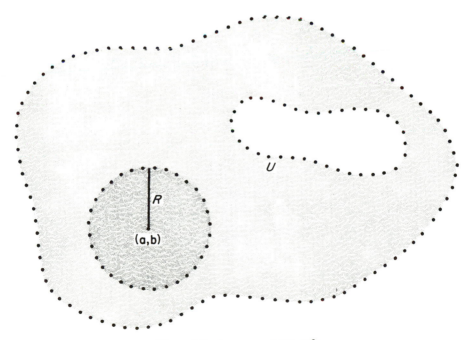

Figure 4.1   An open set U in $\mathbb{R}^2$.

(1.1) Example.

(a)   The functions $\Psi_n(x, y) = \sin nx \sinh ny$ $(n = 1, 2, 3, \ldots)$ are harmonic on $\mathbb{R}^2$.

(b)   The functions $g_n(x, y) = \sin nx e^{-ny}$ $(n = 1, 2, 3, \ldots)$ are harmonic on $\mathbb{R}^2$.

(c)   Let (a, b) be a fixed point in $\mathbb{R}^2$. The function p defined by

$$p(x, y) = \frac{(a^2 + b^2) - (x^2 + y^2)}{(x - a)^2 + (y - b)^2}$$

is harmonic on the complement $\mathbb{R}^2 - \{(a, b)\}$.

(d)   The function f defined by $f(x, y) = \log[x^2 + y^2]^{1/2}$ is harmonic on the complement $\mathbb{R}^2 - \{(0, 0)\}$.

Example (1.1c) is especially important for this chapter [see Exercise (1.7)].

The concepts of open and closed sets in Euclidean space $\mathbb{R}^3$ have definitions similar to those given above for the plane. The *open ball* $B_r(a, b, c)$ of radius r centered at the point (a, b, c) consists of all the points (x, y, z) in $\mathbb{R}^3$ that satisfy $[(x - a)^2 + (y - b)^2 + (z - c)^2]^{1/2} < r$. Open sets in $\mathbb{R}^3$ are those sets U for which each point in U is contained in some open ball that is a subset of U; the closed sets are the complements of the open sets. A real valued function Ψ defined on an open set U in $\mathbb{R}^3$ is called *harmonic* if Ψ and its first and second partial derivatives are continuous functions on U and Ψ satisfies *Laplace's equation*

$$\Delta \Psi = \Psi_{xx} + \Psi_{yy} + \Psi_{zz} = 0$$

on all of U. Here are a few examples of harmonic functions of three variables.

(1.2) Example.

(a)   For every pair of positive integers m and n the function $\Psi_{mn}(x, y, z) = \sin mx \sin ny \sinh[m^2 + n^2]^{1/2}z$ is harmonic on $\mathbb{R}^3$.

(b)   Let (a, b, c) be a fixed point in $\mathbb{R}^3$. The function p defined by

$$p(x, y, z) = \frac{(a^2 + b^2 + c^2) - (x^2 + y^2 + z^2)}{[(x - a)^2 + (y - b)^2 + (z - c)^2]^{3/2}}$$

is harmonic on the complement $\mathbb{R}^3 - \{(a, b, c)\}$.

(c)   The function f defined by $f(x, y, z) = [x^2 + y^2 + z^2]^{-1/2}$ is harmonic on the complement $\mathbb{R}^3 - \{(0, 0, 0)\}$.

Although we will concentrate on harmonic functions of two variables in this chapter, we introduced the concept of harmonic functions of three variables because they are the principal kind of harmonic functions that occur in physical applications.

Applications in Physics

In lieu of a detailed treatment of physical applications of harmonic functions, we briefly summarize some of the areas of applications and give some references. Harmonic functions can be interpreted physically in a variety of ways:

- The gravitational potential Ψ in regions not occupied by attracting matter [e.g., Kellogg (1928)].
- The electrostatic potential Ψ in a uniform dielectric medium [e.g., Jackson (1962) or Slater and Frank (1947)].
- The magnetostatic potential Ψ in free space [e.g., Jackson (1962), Slater and Frank (1947), or Reitz, Milford, and Christy (1979)].
- The velocity potential Ψ at points of a homogeneous liquid moving irrotationally [e.g., Meyer (1971)].
- The temperature Ψ in the theory of thermal equilibrium in solids [e.g., Carslaw and Jaeger (1959) and Fourier (1955)].

The last example is particularly important because certain properties of harmonic functions have natural interpretations as properties of equilibrium (steady-state) temperatures. For example, suppose that U is a solid body enclosed by a surface (which we denote by ∂U). If f is a fixed (in time) temperature distribution on ∂U, for example, f is a given continuous function on ∂U, then in the *absence of heat sources* the temperature in the body U will tend to an equilibrium (steady-state) temperature Ψ. Since Ψ does not change in time we expect that Ψ *has no maxima or minima inside U* (otherwise, the temperature Ψ would change due to heat flow). This lack of maxima or minima for a harmonic function on an open set U is known as the *Maximum–Minimum Principle*. We will discuss this principle further in §§4 and 8.

Furthermore, if Ψ is harmonic in U, is equal to a continuous function f on ∂U, and is the equilibrium temperature for U, then it is reasonable to expect that Ψ is a continuous function on the union $U \cup \partial U$. Or, in other words, that there are no discontinuities in temperature in passing from the surface ∂U into the body U. Mathematically, we are saying that Ψ is a solution to the classic *Dirichlet problem*: Find a function Ψ which is harmonic in U, continuous on $U \cup \partial U$, and which equals a given continuous function f on ∂U. We will examine Dirichlet's problem at various times during the course of this chapter; our investigations will culminate in the last section where we shall prove that for open sets U which are typically encountered in physical applications Dirichlet's problem does have a solution. Besides the theory of heat, Dirichlet's problem is also encountered in electrostatic theory [see Jackson (1962) or Slater and Frank (1947)].

Exercises

(1.3) Verify Examples (1.1a, b, and d).

(1.4) Prove that if $\Psi_1, \Psi_2, \ldots, \Psi_m$ are harmonic on an open set U then so is $a_1 \Psi_1 + a_2 \Psi_2 + \cdots + a_m \Psi_m$ for each set of real numbers $a_1, a_2, \ldots, a_m$.

*(1.5) Let (a, b) be a fixed point in $\mathbb{R}^2$ and c a positive constant. Prove that if Ψ is harmonic on the open set U in $\mathbb{R}^2$ then $\Psi(x - a, y - b)$ is harmonic on the open set $U + (a, b)$ that consists of all points (x, y) such that $(x - a, y - b)$ is in U. Moreover, prove that $\Psi(x/c, y/c)$ is harmonic on cU the open set consisting of all the points (x, y) such that $(x/c, y/c)$ is in U.

(1.6) Prove that if Ψ has three continuous partials and Ψ is harmonic, then $\partial \Psi/\partial x$, $\partial \Psi/\partial y$ (and $\partial \Psi/\partial z$) are harmonic.

*(1.7) Verify (1.1c). [*Hint*: Use (1.5), (1.1d), and (1.6).]

(1.8) Verify the examples in (1.2).

*In the remaining exercises we consider the use of Fourier series to exhibit *harmonic functions in rectangular regions in* $\mathbb{R}^2$.

(1.9) Using separation of variables, find a series solution to the following problem

$$\begin{cases} \Delta \Psi = \Psi_{xx} + \Psi_{yy} = 0 & [0 < x < \pi \quad 0 < y < L] \\ \Psi(0, y) = \Psi(\pi, y) = 0 & [0 \leq y \leq L] \\ \Psi(x, 0) = g(x) \quad \Psi(x, L) = 0 & [0 \leq x \leq \pi] \end{cases}$$

$$\left(Answer: \qquad \Psi = \sum_{n=1}^{\infty} \left[\frac{2}{\pi} \int_0^\pi g(u) \sin nu \, du \right] \frac{\sinh n(L - y)}{\sinh nL} \sin nx \right)$$

(1.10) Give an explicit series solution for problem (1.9) if $L = 4$ and g is the constant function $g(x) = 1$.

(1.11) Give an explicit series solution for problem (1.9) if $L = 4$ and g is the function defined by

$$g(x) = \begin{cases} 1 & \text{if } c < x < d \\ 0 & \text{if } 0 < x < c \quad \text{or} \quad d < x < \pi \end{cases}$$

(1.12) Find a series solution to the following problem

$$\begin{cases} \Delta \Psi = 0 & (0 < x < \pi \quad 0 < y < L) \\ \Psi(0, y) = 0 \quad \Psi(\pi, y) = h(y) & (0 \leq y \leq L) \\ \Psi(x, 0) = \Psi(x, L) = 0 & (0 \leq x \leq \pi) \end{cases}$$

$$\left(Answer: \qquad \Psi = \sum_{n=1}^{\infty} \left[\frac{2}{L} \int_0^L h(u) \sin \frac{n\pi u}{L} \, du \right] \frac{\sinh(n\pi x/L)}{\sinh(n\pi^2/L)} \sin \frac{n\pi y}{L} \right)$$

(1.13) Suppose that f is a function given on the boundary ∂R of the rectangle $R = \{(x, y): 0 < x < \pi \text{ and } 0 < y < L\}$ in $\mathbb{R}^2$. (See Figure 4.2.) Show that the problem

$$\Delta \Psi = 0 \quad \text{on} \quad R \qquad \Psi = f \quad \text{on} \quad \partial R$$

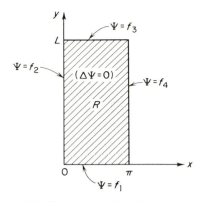

Figure 4.2 The function f on the boundary ∂R of the rectangle consists of four functions f_1, f_2, f_3, and f_4 on each of the four sides of the rectangle R. (We agree to set $f_j = \frac{1}{2}f$ at each corner of ∂R.)

is solved by setting $\Psi = \Psi_1 + \Psi_2 + \Psi_3 + \Psi_4$ where

$$\Psi_1 = \sum_{n=1}^{\infty} \left[\frac{2}{\pi} \int_0^\pi f_1(u) \sin nu \right] \frac{\sinh n(L-y)}{\sinh nL} \sin nx$$

$$\Psi_2 = \sum_{n=1}^{\infty} \left[\frac{2}{L} \int_0^L f_2(u) \sin \frac{n\pi u}{L} \, du \right] \frac{\sinh n\pi(\pi - x)/L}{\sinh(n\pi^2/L)} \sin \frac{n\pi y}{L}$$

$$\Psi_3 = \sum_{n=1}^{\infty} \left[\frac{2}{\pi} \int_0^\pi f_3(u) \sin nu \, du \right] \frac{\sinh ny}{\sinh nL} \sin nx$$

$$\Psi_4 = \sum_{n=1}^{\infty} \left[\frac{2}{L} \int_0^L f_4(u) \sin \frac{n\pi u}{L} \, du \right] \frac{\sinh(n\pi x/L)}{\sinh(n\pi^2/L)} \sin \frac{n\pi y}{L}$$

(1.14)   Check that all the partial sums for the series in (1.13) are harmonic on $\mathbb{R}^2$.

(1.15)   Solve the following two problems

(a) $\begin{cases} \Delta \psi = 0 & (0 < x < \pi \quad 0 < y < 5) \\ \Psi(0, y) = \Psi(\pi, y) = 0 & \Psi(x, 0) = x(x - \pi) \quad \Psi(x, 5) = 0 \end{cases}$

(b) $\begin{cases} \Delta \Psi = 0 & (0 < x < 7 \quad 0 < y < \pi) \\ \Psi(0, y) = 1 \quad \Psi(7, y) = 0 & \Psi(x, 0) = 0 \quad \Psi(x, \pi) = 2 \end{cases}$

(1.16)   Suppose that $\delta > 0$. Prove that if $\delta \le y \le L$ then the series for Ψ in (1.9) converges uniformly.[1] [By Harnack's theorem (5.5) that result will imply that Ψ is harmonic on the open rectangle R.]

(1.17)   Can you prove that the series for Ψ found in (1.9) is harmonic in the open rectangle R (i.e., without recourse to Harnack's theorem)?

---

1. You may assume that g in (1.9) is piecewise continuous.

§2. Harmonic Functions on a Disk; Poisson's Integral

Let's consider the problem of finding functions that are harmonic on an open disk in $\mathbb{R}^2$. The outcome of our investigation will be Poisson's integral, which will provide us with a powerful tool for analyzing harmonic functions.

We begin by finding a function Ψ that is harmonic on the open disk $B_R(0, 0)$ and that equals a given function f defined on the circle $\partial B_R(0, 0)$ that encloses the disk $B_R(0, 0)$. Hence we aim to solve

(2.1) $\Delta\Psi = 0$ on $B_R(0, 0)$ $\Psi = f$ on $\partial B_R(0, 0)$

Since our domain is circular in nature, we begin by expressing the *Laplacian* $\Delta\Psi$ in polar coordinates. Polar coordinates are defined by

(2.2) $x = r\cos\phi$ $y = r\sin\phi$ $(0 < r < +\infty)$

where r is the radial coordinate and ϕ is the angular coordinate (measured counterclockwise from the positive x axis). Using (2.2) we can substitute for x and y in the functions Ψ and f, obtaining

(2.3) $\Psi(x, y) = \Psi(r\cos\phi, r\sin\phi)$ $f(x, y) = f(R\cos\phi, R\sin\phi)$

In light of (2.3) we can view Ψ as a function of r and ϕ, and f as a function of ϕ (R is a constant). *Sometimes, in order to avoid cumbersome notation, we may write $\Psi(r, \phi)$ in place of $\Psi(r\cos\phi, r\sin\phi)$ and $f(\phi)$ in place of $f(R\cos\phi, R\sin\phi)$.* Applying the Chain Rule, we have

$$\frac{\partial\Psi}{\partial r} = \left(\frac{\partial\Psi}{\partial x}\right)\frac{\partial x}{\partial r} + \left(\frac{\partial\Psi}{\partial y}\right)\frac{\partial y}{\partial r} = \cos\phi\,\frac{\partial\Psi}{\partial x} + \sin\phi\,\frac{\partial\Psi}{\partial y}$$

$$\frac{\partial\Psi}{\partial\phi} = -r\sin\phi\,\frac{\partial\Psi}{\partial x} + r\cos\phi\,\frac{\partial\Psi}{\partial y}$$

Solving the equations above for $\partial\Psi/\partial x$ and $\partial\Psi/\partial y$ we get

(2.4) $\dfrac{\partial\Psi}{\partial x} = \cos\phi\,\dfrac{\partial\Psi}{\partial r} - \dfrac{\sin\phi}{r}\dfrac{\partial\Psi}{\partial\phi}$ $\dfrac{\partial\Psi}{\partial y} = \sin\phi\,\dfrac{\partial\Psi}{\partial r} + \dfrac{\cos\phi}{r}\dfrac{\partial\Psi}{\partial\phi}$

If we apply (2.4) to $\partial\Psi/\partial x$ and $\partial\Psi/\partial y$, instead of Ψ, we obtain

(2.4′)
$$\Psi_{xx} = \cos\phi\,\frac{\partial}{\partial r}\left(\frac{\partial\Psi}{\partial x}\right) - \frac{\sin\phi}{r}\frac{\partial}{\partial\phi}\left(\frac{\partial\Psi}{\partial x}\right)$$

$$\Psi_{yy} = \sin\phi\,\frac{\partial}{\partial r}\left(\frac{\partial\Psi}{\partial y}\right) + \frac{\cos\phi}{r}\frac{\partial}{\partial\phi}\left(\frac{\partial\Psi}{\partial y}\right)$$

Substituting the equations from (2.4) into (2.4′) and carrying out the necessary differentiations, we have

(2.5) $\Delta\Psi = \Psi_{rr} + (1/r)\Psi_r + (1/r^2)\Psi_{\phi\phi}$ $(0 < r < +\infty)$

Equation (2.5) defines the *polar coordinate form* for the two-dimensional Laplacian. Note that (2.5) only holds for $r > 0$ which makes sense because polar coordinates are not well defined when $r = 0$. Using (2.5), problem (2.1) assumes the following polar coordinate form:

(2.6)
$$\begin{cases} \Psi_{rr} + (1/r)\Psi_r + (1/r^2)\Psi_{\phi\phi} = 0 & (0 < r < R) \\ \Psi(R \cos \phi, R \sin \phi) = f(R \cos \phi, R \sin \phi) \end{cases}$$

Since the function f is defined on the circle of radius R about the origin it follows that $f(R \cos \phi, R \sin \phi)$ is a periodic function of ϕ. Therefore, we have the following Fourier series expansion

(2.7)
$$\begin{cases} f \sim \tfrac{1}{2}A_0 + \sum_{n=1}^{\infty} (A_n \cos n\phi + B_n \sin n\phi) \\ A_n = \dfrac{1}{\pi} \int_{-\pi}^{\pi} f(R \cos \phi, R \sin \phi) \cos n\phi \, d\phi \\ B_n = \dfrac{1}{\pi} \int_{-\pi}^{\pi} f(R \cos \phi, R \sin \phi) \sin n\phi \, d\phi \end{cases}$$

The function Ψ will be defined on the circular disk $B_R(0,0)$ hence $\Psi(r \cos \phi, r \sin \phi)$ will be periodic in the angular variable ϕ. Therefore, we apply the *finite Fourier transform* method to find Ψ. For each value of r, the function Ψ has a Fourier series expansion in ϕ whose coefficients depend on r

(2.8)
$$\begin{cases} \Psi \sim \tfrac{1}{2}C_0(r) + \sum_{n=1}^{\infty} [C_n(r) \cos n\phi + D_n(r) \sin n\phi] \\ C_n(r) = \dfrac{1}{\pi} \int_{-\pi}^{\pi} \Psi(r, \phi) \cos n\phi \, d\phi \qquad D_n(r) = \dfrac{1}{\pi} \int_{-\pi}^{\pi} \Psi(r, \phi) \sin n\phi \, d\phi \end{cases}$$

Performing an integration by parts on the integral that defines $C_n(r)$ and using the periodicity of Ψ we get (for $n \neq 0$)

(2.9)
$$C_n(r) = -\frac{1}{n\pi} \int_{-\pi}^{\pi} \Psi_\phi(r, \phi) \sin n\phi \, d\phi$$
$$= -\frac{1}{n^2\pi} \int_{-\pi}^{\pi} \Psi_{\phi\phi}(r, \phi) \cos n\phi \, d\phi$$

Since $\Psi_{\phi\phi} = -r^2\Psi_{rr} - r\Psi_r$ we have

(2.10)
$$C_n(r) = \frac{1}{n^2} \left[\frac{r^2}{\pi} \int_{-\pi}^{\pi} \Psi_{rr}(r, \phi) \cos n\phi \, d\phi + \frac{r}{\pi} \int_{-\pi}^{\pi} \Psi_r(r, \phi) \cos n\phi \, d\phi \right]$$
$$= \frac{r^2}{n^2} C_n''(r) + \frac{r}{n^2} C_n'(r)$$

Thus, for $n \neq 0$ we have obtained

(2.11)
$$r^2 C_n''(r) + r C_n'(r) - n^2 C_n(r) = 0$$

Equation (2.11) also holds for $n = 0$ [see Exercise (2.20)]. If we try a solution to (2.11) of the form $C_n(r) = r^p$ we get

(2.12) $0 = [p(p-1) + p - n^2]r^p = [p^2 - n^2]r^p$

Hence $p = \pm n$, so $C_n(r) = a_n r^n + b_n r^{-n}$ for $n = 1, 2, 3, \ldots$. If $n = 0$, we get $C_0(r) = a_0 + b_0 \log r$. Since we want Ψ to be harmonic in the disk $B_R(0, 0)$ it should also be continuous there. In particular, Ψ should be continuous at $r = 0$. Therefore, we set $b_n = 0$ for each n. Thus, $C_n(r) = a_n r^n$ and since $\Psi(R, \phi) = f(R \cos \phi, R \sin \phi)$ we observe that $C_n(R) = A_n$. It follows that $C_n(r) = A_n(r/R)^n$. A similar argument yields $D_n(r) = B_n(r/R)^n$. Hence our proposed solution to (2.6) is as follows.

For $A_n = \frac{1}{\pi}\int_{-\pi}^{\pi} f(R \cos \phi, R \sin \phi) \cos n\phi \, d\phi$
and $B_n = \frac{1}{\pi}\int_{-\pi}^{\pi} f(R \cos \phi, R \sin \phi) \sin n\phi \, d\phi$
we define Ψ as

(2.13) $\Psi(r \cos \phi, r \sin \phi) = \frac{1}{2}A_0$

$$+ \sum_{n=1}^{\infty} (A_n \cos n\phi + B_n \sin n\phi)\left(\frac{r}{R}\right)^n \qquad (0 \le r < R)$$

while if $r = R$ we set $\Psi(R \cos \phi, R \sin \phi) = f(R \cos \phi, R \sin \phi)$.

Or, if we use (2.2) we can express (2.13) in a slightly different form.

For A_n and B_n as in (2.13) we have for (x, y) in the disk $B_R(0, 0)$

$$\textbf{(2.13')} \quad \Psi(x, y) = \tfrac{1}{2}A_0 + \sum_{n=1}^{\infty} (A_n \cos n\phi + B_n \sin n\phi)\left(\frac{r}{R}\right)^n$$

while if (x, y) is on the circle $\partial B_R(0, 0)$ we set $\Psi(x, y) = f(x, y)$.

In Exercise (2.26), we shall see how the series in (2.13') can be put into *Cartesian form*, that is, expressed directly in terms of x and y.

Poisson's Integral

The series form for Ψ shown in (2.13') can be put into a more compact form known as *Poisson's Integral*. Substituting the integral expressions for A_n and B_n into the series form of Ψ in (2.13') we obtain, after changing the variable of integration from ϕ to α and applying the cosine subtraction formula

$$\Psi(x, y) = \frac{1}{\pi} \int_{-\pi}^{\pi} \tfrac{1}{2}f(R\cos\alpha, R\sin\alpha)\, d\alpha$$

$$+ \sum_{n=1}^{\infty} \left\{ \frac{1}{\pi} \int_{-\pi}^{\pi} f(R\cos\alpha, R\sin\alpha)[\cos n\alpha \cos n\phi \right.$$

$$\left. + \sin n\alpha \sin n\phi](r/R)^n\, d\alpha \right\}$$

$$= \frac{1}{\pi} \int_{-\pi}^{\pi} \tfrac{1}{2}f(R\cos\alpha, R\sin\alpha)\, d\alpha$$

$$+ \sum_{n=1}^{\infty} \left\{ \frac{1}{\pi} \int_{-\pi}^{\pi} f(R\cos\alpha, R\sin\alpha)\cos n(\phi - \alpha)(r/R)^n\, d\alpha \right\}$$

Let's assume for the moment that f is continuous. If we let M equal the maximum for $|f|$ on the circle $\partial B_R(0, 0)$ then

$$|f(R\cos\alpha, R\sin\alpha)\cos n(\phi - \alpha)(r/R)^n| \le M(r/R)^n$$

and the series $\sum_{n=1}^{\infty} M(r/R)^n$ converges for each fixed r such that $0 \le r < R$. It follows that we may interchange the summation and integration in the last line of integrals above. Thus, for $0 \le r < R$

(2.14) $$\Psi(x, y) = \frac{1}{\pi} \int_{-\pi}^{\pi} f(R\cos\alpha, R\sin\alpha)\left[\tfrac{1}{2} + \sum_{n=1}^{\infty} (r/R)^n \cos n(\phi - \alpha)\right] d\alpha$$

where $x = r\cos\phi$ and $y = r\sin\phi$.

(2.15) Remark. To get (2.14) we assumed that f was continuous. If we only assume, however, that $f(R\cos\alpha, R\sin\alpha)$ is *absolutely integrable*, then (2.14) will still hold. See Exercise (2.33).

We now evaluate the series in brackets in (2.14). Our argument involves complex numbers; an alternate approach is described in Exercise (2.34). Let $\beta = \phi - \alpha$ and $\rho = r/R$, then

$$\tfrac{1}{2} + \sum_{n=1}^{\infty} \rho^n \cos n\beta = \tfrac{1}{2} + \sum_{n=1}^{\infty} \text{Re}[(\rho e^{i\beta})^n]$$

$$= \tfrac{1}{2} + \text{Re}\left[\sum_{n=1}^{\infty} (\rho e^{i\beta})^n\right]$$

where Re denotes taking the real part of a complex number. Since $|\rho e^{i\beta}| < 1$, the geometric series $\sum_{n=1}^{\infty} (\rho e^{i\beta})^n$ converges to $\rho e^{i\beta}/(1 - \rho e^{i\beta})$. Thus,

$$\tfrac{1}{2} + \sum_{n=1}^{\infty} \rho^n \cos n\beta = \frac{1}{2} + \text{Re}\left(\frac{\rho e^{i\beta}}{1 - \rho e^{i\beta}}\right)$$

$$= \frac{1}{2} \frac{1 - \rho^2}{1 + \rho^2 - 2\rho \cos \beta}$$

Substituting $\rho = r/R$ and $\beta = \phi - \alpha$ into the equalities above, we obtain after a little algebra

$$\tfrac{1}{2} + \sum_{n=1}^{\infty} (r/R)^n \cos n(\phi - \alpha) = \frac{1}{2}\frac{R^2 - r^2}{R^2 + r^2 - 2Rr\cos(\phi - \alpha)} \qquad (0 \le r < R)$$

We will use the expression $P(r; \phi - \alpha)$ for *twice* the quotient on the right above; thus

$$P(r; \phi - \alpha) = \frac{R^2 - r^2}{R^2 + r^2 - 2Rr\cos(\phi - \alpha)} \qquad (0 \le r < R)$$

Using the two equalities above, we transform (2.14) into the following form.

For (x, y) in the disk $B_R(0, 0)$ we have $(x, y) = (r \cos \phi, r \sin \phi)$ and

(2.16a) $\Psi(x, y) = \dfrac{1}{2\pi} \displaystyle\int_{-\pi}^{\pi} f(R \cos \alpha, R \sin \alpha) P(r; \phi - \alpha)\, d\alpha$

while for (x, y) on the circle $\partial B_R(0, 0)$ we set $\Psi(x, y) = f(x, y)$.

By a change of variables and periodicity, (2.16a) can be rewritten as follows.

For (x, y) in the disk $B_r(0, 0)$ we have $(x, y) = (r \cos \phi, r \sin \phi)$ and

(2.16b) $\Psi(x, y) = \dfrac{1}{2\pi} \displaystyle\int_{-\pi}^{\pi} f[R \cos(\phi - \alpha), R \sin(\phi - \alpha)] P(r; \alpha)\, d\alpha$

while for (x, y) on the circle $\partial B_R(0, 0)$ we set $\Psi(x, y) = f(x, y)$.

The function $P(r; \alpha)$ is called *Poisson's kernel*, in polar coordinates, for the disk $B_R(0, 0)$. Note that as a series we have

(2.17) $P(r; \alpha) = \dfrac{R^2 - r^2}{R^2 + r^2 - 2Rr\cos \alpha} = 1 + 2\sum_{n=1}^{\infty} (r/R)^n \cos n\alpha$

We now express Ψ completely in Cartesian coordinates. Suppose that A is a point on the circle $\partial B_R(0, 0)$ which has Cartesian coordinates (x_0, y_0), while B is a point in the disk $B_R(0, 0)$ which has Cartesian coordinates (x, y). (See Figure 4.3.) Then for some angles α and ϕ we have

A: $(x_0, y_0) = (R \cos \alpha, R \sin \alpha)$ B: $(x, y) = (r \cos \phi, r \sin \phi)$

Letting $\overline{AB}$ denote the length of the line segment between A and B, we obtain by the Law of Cosines [or by simplifying $\overline{AB}^2 = (x - x_0)^2 + (y - y_0)^2$]

$$\overline{AB}^2 = R^2 + r^2 - 2Rr\cos(\phi - \alpha)$$

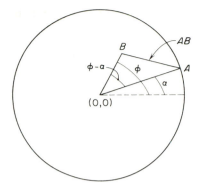

Figure 4.3 Angles between a point A on a circle and a point B inside that circle.

It follows that $P(r; \phi - \alpha)$ can be expressed in the Cartesian form

$$P(r; \phi - \alpha) = \frac{R^2 - r^2}{AB^2} = \frac{R^2 - (x^2 + y^2)}{(x - R \cos \alpha)^2 + (y - R \sin \alpha)^2}$$

Based on the equalities above, if we define $p(x, y; \alpha)$ to be

$$p(x, y; \alpha) = \frac{R^2 - (x^2 + y^2)}{(x - R \cos \alpha)^2 + (y - R \sin \alpha)^2}$$

then we can write (2.16a) as follows.

For (x, y) in the disk $B_R(0, 0)$ we have

(2.18) $\quad \Psi(x, y) = \dfrac{1}{2\pi} \displaystyle\int_{-\pi}^{\pi} f(R \cos \alpha, R \sin \alpha) p(x, y; \alpha)\, d\alpha$

while for (x, y) on the circle $\partial B_R(0, 0)$ we set $\Psi(x, y) = f(x, y)$.

The function $p(x, y; \alpha)$ is called *Poisson's kernel*, in Cartesian coordinates, for the disk $B_R(0, 0)$.

(2.19) Notation. The integral in (2.18) will be denoted by P.I.$[f]$ for "Poisson integral of f;" thus (2.18) can be written as $\Psi = $ P.I.$[f]$. If we let $F(\phi) = f(R \cos \phi, R \sin \phi)$ then we will sometimes write (2.16a) and (2.16b) as $\Psi(x, y) = F *_r P(\phi) = {}_r P * F(\phi)$ where ${}_r P(\phi) = P(r; \phi)$ is Poisson's kernel in polar coordinates.

Exercises

(2.20) Check (2.5) and show that (2.11) holds for $n = 0$. Verify that $\log r$ does satisfy (2.11) for $n = 0$. Express the function $g(r) = \log r$ in Cartesian coordinates and compare with Example (1.1d).

(2.21) Using (2.13), find series solutions to the following problems.

(a) $\qquad \Delta \Psi = 0 \qquad (0 < r < 4)$

$\qquad \Psi(4, \phi) = |\sin \phi|$

(b) $\qquad \Delta \Psi = 0 \qquad (0 < r < 3)$

$$\Psi(3, \phi) = \begin{cases} 1 & \text{if } 0 < \phi < \pi \\ 0 & \text{if } -\pi < \phi < 0 \end{cases}$$

(c) $\qquad \Delta \Psi = 0 \qquad (0 < r < 6)$

$\qquad \Psi(6, \phi) = \phi^2 \qquad (-\pi \le \phi \le \pi)$

(d) $\qquad \Delta \Psi = 0 \qquad (0 < r < 7)$

$$\Psi(7, \phi) = \begin{cases} 2 & \text{if } 0 < \phi < \pi \\ 3 & \text{if } -\pi < \phi < 0 \end{cases}$$

(2.22) Using (2.13'), find series solutions to the following problems.

(a) $\qquad \Delta \Psi = 0 \qquad \text{in } B_2(0, 0)$

$\qquad \Psi(x, y) = |x| \qquad \text{on } \partial B_2(0, 0)$

(b) $\qquad \Delta \Psi = 0 \qquad \text{in } B_4(0, 0)$

$\qquad \Psi(x, y) = y^2 \qquad \text{on } \partial B_4(0, 0)$

(c) $\qquad \Delta \Psi = 0 \qquad \text{in } B_5(0, 0)$

$$\Psi(x, y) = \begin{cases} 1 & \text{if } y > 0 \\ 0 & \text{if } y < 0 \end{cases} \quad \text{on } \partial B_5(0, 0)$$

*(2.23) Prove that $r^n \cos n\phi$ and $r^n \sin n\phi$ are both polynomials in x and y ($n = 0, 1, 2, \ldots$). Prove also that those polynomials have the following two properties:

(1) They are harmonic on $\mathbb{R}^2$.
(2) The sum of the powers of x and y in each term is n.

Such polynomials are called *homogeneous harmonic polynomials of degree n*. [*Hint*: Use mathematical induction.]

*(2.24) Make a list of the polynomials in (2.23) for $n = 0, 1, 2, 3,$ and 4.

*(2.25) Use the list from (2.24) to express the first five terms of each of the series found in (2.22) as explicit functions of x and y.

*(2.26) Using (2.23) express the series in (2.13') in Cartesian form. Then prove that any partial sum of that series satisfies Laplace's equation (in its original Cartesian form) on $\mathbb{R}^2$.

(2.27) Suppose that $\delta > 0$. Prove that if $0 \le r \le R - \delta$ then the series for Ψ in (2.13) and (2.13') converge uniformly.

*(2.28) Let $q(x, y)$ denote a homogeneous harmonic polynomial of degree n [i.e., q satisfies (1) and (2) in (2.23)]. Prove that q is a linear combination of the polynomials found in (2.23) for $r^n \cos n\phi$ and $r^n \sin n\phi$.

(2.29)   Express the solutions to (2.22) in Poisson integral form, P.I.$[f]$.

(2.30)   Using (2.18), solve the following two problems.

(a)
$$\Delta\Psi = 0 \quad \text{in } B_1(0, 0)$$
$$\Psi(x, y) = 1 + x \quad \text{on } \partial B_1(0, 0)$$

(b)
$$\Delta\Psi = 0 \quad \text{in } B_1(0, 0)$$
$$\Psi(x, y) = x + y^2 \quad \text{on } \partial B_1(0, 0)$$

(2.31)   Using the list from (2.24), express the solutions to the problems in (2.30) as polynomials in x and y.

(2.32)   A rotation of coordinate axes in $\mathbb{R}^2$ is accomplished by the following definitions of new coordinates x' and y'

$$x' = x \cos\theta + y \sin\theta \qquad y' = -x \sin\theta + y \cos\theta$$

for some angle θ. Show that $\Psi_{xx} + \Psi_{yy} = 0$ transforms into $\Psi_{x'x'} + \Psi_{y'y'} = 0$ under this change of coordinates (i.e., Laplace's equation is invariant under rotations of $\mathbb{R}^2$ about the origin).

(2.33)   Formula (2.16a) actually holds if $f(R \cos\alpha, R \sin\alpha)$ is merely absolutely integrable over $[-\pi, \pi]$. To prove it, consider

$$\frac{1}{2\pi} \int_{-\pi}^{\pi} f(R \cos\alpha, R \sin\alpha) P(r; \phi - \alpha) \, d\alpha$$

$$- \left\{ \tfrac{1}{2}A_0 + \sum_{n=1}^{N} [A_n \cos n\phi + B_n \sin n\phi](r/R)^n \right\}$$

for fixed r and show that for N sufficiently large the difference above has negligible magnitude. From (2.16a) we then obtain (2.16b) and (2.18) as in the text. [*Hint*: Express the difference above as one integral.]

(2.34)   Prove that

$$[1 + \rho^2 - 2\rho \cos\beta]\left[\tfrac{1}{2} + \sum_{n=1}^{\infty} \rho^n \cos n\beta \right] = 1 - \rho^2 \qquad (|\rho| < 1)$$

hence obtaining (2.16a) without using complex variables. [*Hint*: use the trigonometric identity $\cos\beta \cos n\beta = \tfrac{1}{2}\cos(n+1)\beta + \tfrac{1}{2}\cos(n-1)\beta$.]

§3. Poisson's Integral

In this section we will show that the function Ψ, described in (2.18), provides a solution to Dirichlet's problem for the disk $B_R(0, 0)$. To state Dirichlet's problem in a mathematically precise form we need to make some preliminary observations. Let A be a subset of $\mathbb{R}^2$. The points of $\mathbb{R}^2$ fall into three mutually exclusive categories (see Figure 4.4 below):

(1)   The points in the *interior* of A, denoted by int(A), consisting of all those points in A that are contained in some open disk that lies entirely in A.

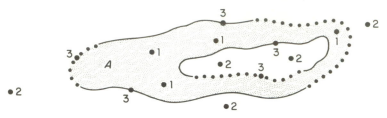

Figure 4.4 Interior, exterior, and boundary points for a set A in $\mathbb{R}^2$.

(2) The points in the *exterior* of A, denoted by ext(A); the exterior of A is the interior of the complement of A in $\mathbb{R}^2$. Thus, ext(A) = int($\mathbb{R}^2 - A$).

(3) The points on the *boundary of A*, denoted by ∂A, consisting of all those points that are in neither the interior nor the exterior of A. Or, put another way, those points (x, y) for which every open ball $B_r(x, y)$ contains points belonging to A and $\mathbb{R}^2 - A$.

For example, the boundary of the disk $B_R(0, 0)$ is the set of all points (x, y) such that $[x^2 + y^2]^{1/2} = R$. Hence $\partial B_r(0, 0)$ is the circle of radius R centered at the origin $(0, 0)$ (so our new notation does not conflict with our original designation of that set). It is easily shown that a set A in $\mathbb{R}^2$ is an open set if and only if int(A) = A. Moreover, we clearly have that A is always a subset of the union int(A) $\cup \partial A$. We define the *closure* of a set A to be the union int(A) $\cup \partial A$ and denote it by $\bar{A}$. For example, the closed disk $\bar{B}_R(0, 0)$ consisting of all points (x, y) such that $[x^2 + y^2]^{1/2} \leq R$ is the closure of the disk $B_R(0, 0)$. An easy exercise consists in showing that a set A in $\mathbb{R}^2$ is closed if and only if $\bar{A} = A$.

We are now prepared to state Dirichlet's problem in a precise way.

(3.1) Dirichlet's Problem.   Let U be an open set in $\mathbb{R}^2$ with boundary ∂U. Given a continuous function f on ∂U, find a function Ψ which is harmonic in U, continuous on the closure $\bar{U}$, and equal to f on the boundary ∂U.

In §8, we will show that Dirichlet's problem is solvable for a rather large class of open sets. Moreover, in the next section we will prove that if the closure $\bar{U}$ is a compact set, then there can exist at most one solution to Dirichlet's problem for each continuous function f (i.e., if a solution exists, it is unique). We shall now prove that the function Ψ, described in (2.18), solves Dirichlet's problem for the open disk $B_R(0, 0)$.

To prove that Ψ is harmonic in $B_R(0, 0)$ we just differentiate under the integral sign in (2.18); thus

$$\Delta\Psi(x, y) = \frac{1}{2\pi} \int_{-\pi}^{\pi} f(R \cos \alpha, R \sin \alpha)\Delta p(x, y; \alpha) \, d\alpha$$

$$= \frac{1}{2\pi} \int_{-\pi}^{\pi} f(R \cos \alpha, R \sin \alpha) 0 \, d\alpha = 0$$

since $\Delta p(x, y; \alpha) = p_{xx} + p_{yy} = 0$ by Exercise (1.7). Of course, we must justify differentiating under the integral sign. To do that we observe that

$p_x(x, y; \alpha)$
$$= \frac{-2x[(x - R \cos \alpha)^2 + (y - R \sin \alpha)^2] - 2[R^2 - (x^2 + y^2)](x - R \cos \alpha)}{[(x - R \cos \alpha)^2 + (y - R \sin \alpha)^2]^2}$$

which is a continuous function of the point $(x, y; \alpha)$ in $\mathbb{R}^3$ provided (x, y) belongs to the closed disk $\bar{B}_{R-\epsilon}(0, 0)$ for some $\epsilon > 0$. Therefore, *since f is continuous on* $\partial B_R(0, 0)$, if we *hold y fixed*, then we may apply Theorem (5.6), Chapter 3,[2] to obtain

$$\Psi_x = \frac{1}{2\pi} \int_{-\pi}^{\pi} \frac{\partial}{\partial x} [f(R \cos \alpha, R \sin \alpha)p(x, y; \alpha)] \, d\alpha$$

$$= \frac{1}{2\pi} \int_{-\pi}^{\pi} f(R \cos \alpha, R \sin \alpha)p_x(x, y; \alpha) \, d\alpha$$

This justifies differentiating $\Psi = $ P.I.$[f]$ under the integral sign with respect to x, or y by a similar argument, in all of $B_R(0, 0)$ (when we let ϵ tend to zero). In the same manner, we can justify differentiating $\Psi = $ P.I.$[f]$ under the integral sign *any number of times*; in particular, our demonstration above that Ψ is harmonic in $B_R(0, 0)$ is justified.

(3.2) Remark. The Poisson integral P.I.$[f]$ is still harmonic even if $f(R \cos \alpha, R \sin \alpha)$ is only absolutely integrable. The proof, however, is more technical than the one above and not germane to our present discussion, so we will wait until the end of this section to give it.

In (2.18) we defined $\Psi(x, y) = f(x, y)$ for (x, y) on the boundary circle $\partial B_R(0, 0)$; therefore, all that remains for us to show is that Ψ is continuous on the closed disk $\bar{B}_R(0, 0)$. Since f is continuous on $\partial B_R(0, 0)$ we only have to prove that for each (x_0, y_0) on $\partial B_R(0, 0)$ we have for (x, y) in $B_R(0, 0)$

(3.3)
$$\lim_{(x, y) \to (x_0, y_0)} \Psi(x, y) = f(x_0, y_0)$$

Our proof of (3.3) will use the fact that Poisson's kernel is a *summation kernel*. This is most easily seen if we look at the polar form $P(r; \alpha)$ of Poisson's kernel.

(3.4) Lemma. Poisson's kernel $P(r; \alpha)$ satisfies

(A$_1$) $\dfrac{1}{2\pi} \displaystyle\int_{-\pi}^{\pi} P(r; \alpha) \, d\alpha = 1$ for each r such that $0 \leq r < R$

(A$_2$) $P(r; \alpha) \geq 0$ for $0 \leq r < R$ and all α

(A$_3$) For every $\epsilon > 0$ and $\delta > 0$ we have

$$0 \leq P(r; \alpha) < \epsilon$$

provided $\pi \geq |\alpha| \geq \delta$ and r is taken close enough to R.

---

2. To the function $h(x, \alpha) = f(R \cos \alpha, R \sin \alpha)p(x, y; \alpha)$ on the rectangle $R = \{(x, \alpha): |x| \leq [(R - \epsilon)^2 - y^2]^{1/2}$ and $|\alpha| \leq \pi\}$. Note: y is *fixed*.

Proof. If we fix a value of $r < R$, then the series shown in (2.17) converges uniformly to $P(r; \alpha)$ for $-\pi \le \alpha \le \pi$. Hence, integrating term by term,

$$\frac{1}{2\pi} \int_{-\pi}^{\pi} P(r; \alpha) \, d\alpha = \frac{1}{2\pi} \int_{-\pi}^{\pi} 1 \, d\alpha + \frac{1}{\pi} \sum_{n=1}^{\infty} \left[(r/R)^n \int_{-\pi}^{\pi} \cos n\alpha \, d\alpha \right]$$

$$= 1$$

Thus, (A_1) holds. Property (A_2) holds because $r^2 < R^2$ and $(R - r)^2 \le R^2 + r^2 - 2Rr \cos \alpha$. If $\pi \ge |\alpha| \ge \delta$, then $-1 \le \cos \alpha \le 1 - \eta$ for some $\eta > 0$. Hence

$$(R - r)^2 + 2Rr\eta \le R^2 + r^2 - 2Rr \cos \alpha \le (R + r)^2$$

Therefore,

$$0 \le P(r; \alpha) \le \frac{R^2 - r^2}{(R - r)^2 + 2Rr\eta}$$

Letting r tend to R, the quantity on the right side of the inequality above tends to zero. Therefore, given some $\epsilon > 0$, we are ensured that for r close enough to R

$$\frac{R^2 - r^2}{(R - r)^2 + 2Rr\eta} < \epsilon$$

and (A_3) follows immediately. ∎

Comparing Lemma (3.4) with Definition (8.1) of Chapter 2, we see that Poisson's kernel is a summation kernel on $(-\pi, \pi)$, parameterized by $r \to R-$ instead of $\tau \to 0+$. Therefore, if F is an absolutely integrable function over $[-\pi, \pi]$ we have

(3.5) $$\lim_{(\phi, r) \to (\phi_0, R-)} \frac{1}{2\pi} \int_{-\pi}^{\pi} F(\phi - \alpha) P(r; \alpha) \, d\alpha = F(\phi_0)$$

provided ϕ_0 is a point of continuity for F. [See Theorem (8.4), Chapter 2.]

We can now prove a theorem that yields as a consequence the validity of the limit in (3.3) [since $\psi = \text{P.I.}[f]$ in the disk $B_R(0, 0)$].

(3.6) Theorem.   Suppose that $f(R \cos \alpha, R \sin \alpha)$ is absolutely integrable over the interval $[-\pi, \pi]$. Then if (x_0, y_0) is a point of continuity for f on the boundary circle $\partial B_R(0, 0)$ we have for (x, y) in the disk $B_R(0, 0)$

$$\lim_{(x, y) \to (x_0, y_0)} \text{P.I.}[f](x, y) = f(x_0, y_0)$$

Proof. The desired limit is easier to prove if we express it in polar coordinate form. Let $(x, y) = (r \cos \phi, r \sin \phi)$ and $(x_0, y_0) = (R \cos \phi_0, R \sin \phi_0)$ then clearly

(3.7) $(x, y) \to (x_0, y_0)$ if and only if $(\phi, r) \to (\phi_0, R-)$

Moreover, if we define the function F by $F(\alpha) = f(R \cos \alpha, R \sin \alpha)$ then *the polar coordinate form of* P.I.$[f]$, *which is given in* (2.16b), *can be written as*

$$\frac{1}{2\pi} \int_{-\pi}^{\pi} F(\phi - \alpha) P(r; \alpha) \, d\alpha$$

Because of (3.5) we have

$$\frac{1}{2\pi} \int_{-\pi}^{\pi} F(\phi - \alpha) P(r; \alpha) \, d\alpha \to F(\phi_0) \qquad \text{as} \qquad (\phi, r) \to (\phi_0, R-)$$

Hence, in Cartesian coordinates

$$\text{P.I.}[f](x, y) \to f(x_0, y_0) \qquad \text{as} \qquad (x, y) \to (x_0, y_0)$$

and our theorem is proved. ∎

Since we have been assuming that f is a continuous function on the circle $\partial B_R(0, 0)$, Theorem (3.6) along with our proof above that $\Psi = \text{P.I.}[f]$ is harmonic in the disk $B_R(0, 0)$ tells us that *the function Ψ described in* (2.18) *solves Dirichlet's problem for the disk* $B_R(0, 0)$. The following generalization is left to the reader as an exercise.

(3.8) Theorem. Let f be a continuous function on the circle $\partial B_R(a, b)$. The function Ψ defined by

$$\Psi(x, y) = \frac{1}{2\pi} \int_{-\pi}^{\pi} f(a + R \cos \alpha, b + R \sin \alpha) p(x - a, y - b; \alpha) \, d\alpha$$

for (x, y) in the disk $B_R(a, b)$, and $\Psi(x, y) = f(x, y)$ for (x, y) on the circle $\partial B_R(a, b)$, solves Dirichlet's problem for the disk $B_R(a, b)$. ∎

We shall designate the integral $(2\pi)^{-1} \int_{-\pi}^{\pi} f(a + R \cos \alpha, b + R \sin \alpha) p(x - a, y - b; \alpha) \, d\alpha$ by P.I.$[f]$. This notation will require that we always specify precisely which disk the Poisson integral is defined on.

We conclude this section with a proof that P.I.$[f]$ is harmonic in the disk $B_R(0, 0)$ whenever $f(R \cos \alpha, R \sin \alpha)$ is absolutely integrable [see Remark (3.2)].

(3.9) Theorem. Suppose that $F(\alpha) = f(R \cos \alpha, R \sin \alpha)$ is absolutely integrable over the interval $[-\pi, \pi]$, then $\Psi = \text{P.I.}[f]$ is harmonic in the disk $B_R(0, 0)$.

Proof. Consider the quantity I defined (for $h \neq 0$) by

$$I = \frac{1}{h} [\Psi(x + h, y) - \Psi(x, y)] - \frac{1}{2\pi} \int_{-\pi}^{\pi} f(R \cos \alpha, R \sin \alpha) p_x(x, y; \alpha) \, d\alpha$$

Writing out $\Psi = \text{P.I.}[f]$ explicitly and combining terms under the integral sign we find that

$$|I| \leq \frac{1}{2\pi} \int_{-\pi}^{\pi} |f(R \cos \alpha, R \sin \alpha)|$$

$$\cdot \left| \frac{p(x + h, y; \alpha) - p(x, y; \alpha)}{h} - p_x(x, y; \alpha) \right| d\alpha$$

Since partial differentiation is just differentiation with respect to one variable while the other variables are held fixed, we may apply the Mean Value Theorem to conclude that

$$\frac{p(x + h, y; \alpha) - p(x, y; \alpha)}{h} = p_x(c, y, \alpha)$$

where c lies between x and $x + h$. If we restrict (x, y) to lying in the closed disk $\bar{B}_{R-\delta}(0, 0)$ for some $\delta > 0$, then for $|h|$ sufficiently small we will have that both (x, y) and (c, y) are in the closed disk $\bar{B}_{R-\frac{1}{2}\delta}(0, 0)$. Since $p_x(x, y; \alpha)$ is uniformly continuous on the compact cylinder $\bar{B}_{R-\frac{1}{2}\delta}(0, 0) \times [-\pi, \pi]$ it follows that for $|h|$ sufficiently small (because c lies between x and $x + h$)

$$|p_x(c, y; \alpha) - p_x(x, y; \alpha)| < \epsilon$$

(where ϵ is some preassigned small number). Hence

$$|I| \leq \frac{\epsilon}{2\pi} \int_{-\pi}^{\pi} |f(R \cos \alpha, R \sin \alpha)| \, d\alpha$$

Because $\epsilon > 0$ can be taken arbitrarily small it follows that Ψ_x exists *and*

$$\Psi_x = \frac{1}{2\pi} \int_{-\pi}^{\pi} f(R \cos \alpha, R \sin \alpha) p_x(x, y; \alpha) \, d\alpha$$

Thus, we have shown that we may partially differentiate $\Psi = \text{P.I.}[f]$ under the integral sign with respect to x in the disk $B_R(0, 0)$ (letting δ pass to zero). The main idea involved in the argument above was that $p_x(x, y; \alpha)$ is continuous on the disk $B_R(0, 0)$. Therefore, since $p(x, y; \alpha)$ can be continuously differentiated any number of times for (x, y) in $B_R(0, 0)$, it follows that P.I.$[f]$ can be *partially differentiated (under the integral sign) any number of times with respect to x and y*. In particular, since $\Delta p(x, y; \alpha) = 0$ we get that $\Psi = \text{P.I.}[f]$ is harmonic in the disk $B_R(0, 0)$. ∎

Of course Theorem (3.9) generalizes to Poisson integrals P.I.$[f]$ defined on the disk $B_R(a, b)$; the details are left as an exercise.

Exercises

(3.10)   Suppose that $F(\phi) = f(R \cos \phi, R \sin \phi)$ is a piecewise continuous function of ϕ. Prove that if ϕ_0 is a point of discontinuity of F then

$$\lim_{(x, y) \to (x_0, y_0)} \text{P.I.}[f](x, y)$$

does *not* exist when $(x_0, y_0) = (R \cos \phi_0, R \sin \phi_0)$.

(3.11)   Let f, F, and ϕ_0 be as in (3.10). Prove that the following *radial limit* (see Figure 4.5) holds

$$\lim_{r \to R-} \text{P.I.}[f](x, y) = \tfrac{1}{2}[F(\phi_0+) + F(\phi_0-)]$$

where $(x, y) = (r \cos \phi_0, r \sin \phi_0)$.

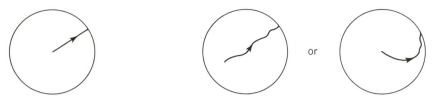

Figure 4.5 A radial limit path versus a general limit path.

(3.12) Suppose that $F(\phi)$ is k times continuously differentiable. Prove that

$$\lim_{(\phi,\,r)\to(\phi_0,\,R-)} \frac{\partial^j}{\partial\phi^j}[_rP * F(\phi)] = \frac{d^j F}{d\phi^j}(\phi_0) \qquad (j = 1, 2, \ldots, k)$$

(3.13) Suppose that K_τ is a summation kernel over $(-a, a)$ [see (8.1), Chapter 2]. Suppose also that $K_\tau(u)$ is k times continuously differentiable with respect to u, and has period $2a$, for each $\tau > 0$. Prove that $f * K_\tau(u)$ is k times continuously differentiable with respect to u whenever f is a continuous (absolutely integrable) function with period $2a$.

(3.14) Prove Theorem (3.8). [*Hint*: Use Exercise (1.5).]

(3.15) Show that $P(r; 0)$ tends to $+\infty$ as r tends to R, while $P(r; \alpha)$ tends to 0 if $\pi \geq |\alpha| > 0$.

★(3.16) Prove that all partial derivatives of a Poisson integral P.I.$[f]$ are harmonic functions on the disk $B_R(a, b)$ where P.I.$[f]$ is defined.

(3.17) Solve the following problems using Poisson integrals

(a)
$$\Delta\Psi = 0 \qquad \text{on } B_2(3, 4)$$
$$\Psi(x, y) = x \qquad \text{on } \partial B_2(3, 4)$$

(b)
$$\Delta\Psi = 0 \qquad \text{on } B_5(-1, 2)$$
$$\Psi(x, y) = |x| \qquad \text{on } \partial B_5(-1, 2)$$

(c)
$$\Delta\Psi = 0 \qquad \text{on } B_3(-4, -1)$$
$$\Psi(x, y) = 1 \qquad \text{on } \partial B_3(-4, -1)$$

(d)
$$\Delta\Psi = 0 \qquad \text{on } B_6(1, -3)$$
$$\Psi(x, y) = |y| \qquad \text{on } \partial B_6(1, -3)$$

(3.18) Express the solutions for (a) through (c) in (3.17) in series form, using polar coordinates.

(3.19) Using the polynomials found in Exercise (2.24), express the first five terms of the series found in (3.18) as explicit functions of x and y.

★(3.20) Prove that $(2\pi)^{-1}\int_{-\pi}^{\pi} p(x - a, y - b; \alpha)\, d\alpha = 1$ for all points (x, y) in $B_R(a, b)$ and that $(2\pi)^{-1}\int_{-\pi}^{\pi} p(x - a, y - b; \alpha)\, d\alpha = -1$ for all points (x, y) in the complement $\mathbb{R}^2 - \bar{B}_R(a, b)$. [*Hint*: Use polar coordinates centered at (a, b).]

§4. The Maximum–Minimum Principle

Harmonic functions possess an interesting property known as the maximum–minimum principle. We briefly discussed the physical motivation for this principle in §1. In this section we will prove one of the two alternative forms for this principle, called the *weak* form of the maximum–minimum principle. This form is called weak because it gives slightly less information than the *strong* form (which we shall prove in §8).

We recall that the *compact* sets in $\mathbb{R}^2$ are, by the Heine–Borel theorem, precisely those sets that are closed and bounded. [A set is bounded if it is contained in some disk $B_R(0, 0)$.]

(4.1) Theorem: Maximum–Minimum Principle, Weak Form.   Let U be an open set in $\mathbb{R}^2$ with compact closure $\bar{U}$. If Ψ is harmonic on U and continuous on $\bar{U}$, then the maximum (minimum) for Ψ on $\bar{U}$ is attained at some point of the boundary ∂U. In other words,

$$\max_{\bar{U}} \Psi = \max_{\partial U} \Psi \qquad \min_{\bar{U}} \Psi = \min_{\partial U} \Psi$$

Proof.   The closed set ∂U is contained in the compact set $\bar{U}$ so ∂U is compact. Thus, $\bar{U}$ and ∂U are both compact hence the continuous function Ψ attains the maxima and minima required by the theorem. Since ∂U is contained in $\bar{U}$ we clearly have

$$\max_{\bar{U}} \Psi \geq \max_{\partial U} \Psi$$

Suppose that

(4.2)
$$\max_{\bar{U}} \Psi > \max_{\partial U} \Psi$$

we will derive a contradiction. Inequality (4.2) implies that for some point (a, b) in U we have

$$\Psi(a, b) = \max_{\bar{U}} \Psi > \max_{\partial U} \Psi$$

Consider the function $f(x, y) = \Psi(x, y) + \delta[(x - a)^2 + (y - b)^2]$ for some $\delta > 0$. Since the continuous function g defined by $g(x, y) = [(x - a)^2 + (y - b)^2]$ attains its maximum on ∂U, it follows that by taking δ sufficiently small we can have[3]

$$f(a, b) = \Psi(a, b) > \max_{\partial U} f$$

Therefore, the maximum of f on $\bar{U}$ must be attained at some point (c, d) in U [possibly, but not neccessarily, at $(c, d) = (a, b)$]. At that point (c, d)

(4.3) $f_x(c, d) = 0 \qquad f_y(c, d) = 0 \qquad f_{xx}(c, d) \leq 0 \qquad f_{yy}(c, d) \leq 0$

---

3. Since $\max_{\partial U} f \leq \max_{\partial U} \Psi + \delta \max_{\partial U} g < \Psi(a, b)$ provided δ is small enough.

[If $f_{xx}(c, d) > 0$ there would be a local minimum for f at (c, d) along the x direction; a similar contradiction would obtain if $f_{yy}(c, d)$ were greater than zero.] From (4.3) we conclude that $\Delta f \leq 0$ at the point (c, d). However,

$$\Delta f(c, d) = \Delta \Psi(c, d) + 4\delta = 4\delta > 0$$

which contradicts $\Delta f \leq 0$ at the point (c, d). Thus inequality (4.2) must be false; hence

$$\max_{\bar{U}} \Psi = \max_{\partial U} \Psi$$

Applying this last result to $-\Psi$ we conclude that

$$\min_{\bar{U}} \Psi = \min_{\partial U} \Psi \qquad \blacksquare$$

The Maximum–Minimum Principle is easily generalized to harmonic functions of three variables [see Exercise (4.9)]. From Theorem (4.1) we get the following useful corollary.

(4.4) Corollary: Uniqueness Theorem.   Suppose that U is an open set in $\mathbb{R}^2$ with compact closure $\bar{U}$. If Ψ_1 and Ψ_2 are both harmonic in U, continuous on $\bar{U}$, and equal to each other on ∂U, then they are equal to each other on all of $\bar{U}$.

Proof.   Apply Theorem (4.1) to $\Psi_1 - \Psi_2$. $\blacksquare$

It follows from Corollary (4.4) that if Ψ is a solution to Dirichlet's problem (3.1) *for U having compact closure $\bar{U}$*, then Ψ is the unique solution. In particular, the solution Ψ to Dirichlet's problem for the disk $B_R(a, b)$ given in Theorem (3.8) is unique. This latter fact leads to the following theorem, which will play a major role in the next section when we investigate the general properties of harmonic functions.

(4.5) Theorem.   Suppose that Ψ is a harmonic function on an open set U in $\mathbb{R}^2$. For each closed disk $\bar{B}_R(a, b)$ contained in U, the Poisson integral P.I.$[\Psi]$ is equal to Ψ in the open disk $B_R(a, b)$.

Proof.   Since the closed disk $\bar{B}_R(a, b)$ is contained in U, it follows that Ψ is harmonic and, therefore, also continuous on $\bar{B}_R(a, b)$. In particular, Ψ is continuous on the circle $\partial B_R(a, b)$. This last fact allows us to define a function $\Psi^\#$ by applying Theorem (3.8) to Ψ. Thus we define $\Psi^\#(x, y)$ to equal P.I.$[\Psi](x, y)$ for (x, y) in the disk $B_R(a, b)$ and $\Psi^\#(x, y) = \Psi(x, y)$ for (x, y) on the circle $\partial B_R(a, b)$. Theorem (3.8) tells us that $\Psi^\#$ solves Dirichlet's problem for the disk $B_R(a, b)$ with Ψ as the continuous function on $\partial B_R(a, b)$. But Ψ itself also solves that particular version of Dirichlet's problem. Therefore, we conclude from the uniqueness of solutions to Dirichlet's problem that $\Psi = \Psi^\#$ on the closed disk $\bar{B}_r(a, b)$. In particular, $\Psi = $ P.I.$[\Psi]$ in the open disk $B_R(a, b)$. $\blacksquare$

A concise way of phrasing Theorem (4.5) is that "Every harmonic function is locally a Poisson integral" (of itself!).

We close this section with a generalization of the maximum–minimum principle. Recall that a real valued function f defined on a set A contained in $\mathbb{R}^2$ is called *bounded above* if for some constant b we have $f(x, y) \leq b$ for all points (x, y) in A. The constant b is called an *upper bound* for f (which must, therefore, have a *least* upper bound). If for some constant a we have $a \leq f(x, y)$ for all points (x, y) in A then we say that f is *bounded below*. The constant a is called a *lower bound* for f (which must, therefore, have a *greatest* lower bound).

(4.6) Theorem: Phragmèn-Lindelöf. If Ψ is harmonic in an open set U in $\mathbb{R}^2$ and continuous on the closure $\bar{U}$, except for a finite number of points on the boundary ∂U, and if Ψ is bounded on $\bar{U}$, then Ψ is bounded above by the least upper bound and bounded below by the greatest lower bound of its values at its points of continuity on ∂U.

The proof of this Phragmèn–Lindelöf Theorem is beyond the scope of this text; the interested reader might consult Nevanlinna and Paatero (1982), p. 194. Using the Phragmèn–Lindelöf theorem, the uniqueness theorem for Dirichlet's problem can be extended to include functions f on the boundary ∂U that are bounded, and continuous except for a finite number of points.

Exercises

*(4.7) Prove that each harmonic function of two variables is infinitely partially differentiable, and that all its partial derivatives are harmonic functions.

*(4.8) A real valued function f is called *real analytic* at a point (a, b) in $\mathbb{R}^2$ if on some open disk $B = B_R(a, b)$ containing (a, b) we have

$$f(x, y) = \sum_{n,m=0}^{\infty} a_{mn}(x - a)^m (y - b)^n$$

where the series on the right converges for all points (x, y) in B. Prove that each harmonic function of two variables is real analytic wherever it is defined. [*Hint*: Use Theorem (4.5) and Exercise (2.26).]

(4.9) Generalize (and prove) the maximum–minimum principle to include harmonic functions of three variables. What about harmonic functions of more than three variables?

(4.10) Show that the solutions found for Exercises (2.22), (2.30), and (3.17) are unique. Are the solutions found for Exercise (2.21) unique?

(4.11) Let U be an open set in $\mathbb{R}^2$ with compact closure $\bar{U}$. Suppose that $\Psi = C$, a constant, on ∂U. Prove that $\Psi = C$ on all of $\bar{U}$.

(4.12) Suppose that U is a set of points in $\mathbb{R}^2$ which lie inside of an n-sided regular polygon, that polygon being ∂U. Let Ψ be harmonic in U, and let

$\Psi = C_j$, a constant, on each of the n sides of the polygon ($j = 1, 2, 3, \ldots, n$). Prove that the value of Ψ at the center of the polygon equals the average of the constants C_j.

§5. Properties of Harmonic Functions

The fact that every harmonic function is locally a Poisson integral can be used to derive a large number of properties of harmonic functions. Our principal tools for the following discussion will be Theorems (3.8) and (4.5).

We begin with a result known as *Gauss' Mean Value Theorem*.

(5.1) Theorem: Mean Value Theorem. Let Ψ be a harmonic function on an open set U. If the closed disk $\bar{B}_R(a, b)$ is contained in U then

$$\Psi(a, b) = \frac{1}{2\pi} \int_{-\pi}^{\pi} \Psi(a + R \cos \alpha, b + R \sin \alpha) \, d\alpha$$

Proof. By Theorem (4.5), Ψ equals P.I.$[\Psi]$ on the disk $B_R(a, b)$. Thus, for each point (x, y) in $B_R(a, b)$

$$\Psi(x, y) = \frac{1}{2\pi} \int_{-\pi}^{\pi} \Psi(a + R \cos \alpha, b + R \sin \alpha) p(x - a, y - b; \alpha) \, d\alpha$$

Putting (x, y) equal to (a, b), and observing that $p(0, 0; \alpha) = 1$, we obtain the desired result. ∎

(5.2) Corollary: Harnack's Inequality. Let Ψ be a harmonic function on an open set U. Suppose that Ψ has only nonnegative values. If the closed disk $\bar{B}_R(a, b)$ is contained in U, then for a point (x, y) in the open disk $B_R(a, b)$ which lies at a distance r from (a, b) we have

$$\frac{R - r}{R + r} \Psi(a, b) \le \Psi(x, y) \le \frac{R + r}{R - r} \Psi(a, b)$$

Proof. Poisson's kernel satisfies

$$p(x - a, y - b; \alpha) = \frac{R^2 - [(x - a)^2 + (y - b)^2]}{(x - a - R \cos \alpha)^2 + (y - b - R \sin \alpha)^2}$$

Let $(x, y) = (a + r \cos \phi, b + r \sin \phi)$. Then

$$p(x - a, y - b; \alpha) = \frac{R^2 - r^2}{R^2 + r^2 - 2Rr \cos(\phi - \alpha)}.$$

Since $-1 \le \cos(\phi - \alpha) \le 1$ we find that

$$\frac{R - r}{R + r} \le p(x - a, y - b; \alpha) \le \frac{R + r}{R - r}$$

Using the Mean Value Theorem (5.1) and the fact that $\Psi = \text{P.I.}[\Psi]$ on $B_R(a, b)$, when we multiply the inequalities above by the nonnegative quantity $(2\pi)^{-1}$ times $\Psi(a + R\cos\alpha, b + R\sin\alpha)$ and then integrate with respect to α from $-\pi$ to π we get

$$\frac{R-r}{R+r}\Psi(a, b) \le \Psi(x, y) \le \frac{R+r}{R-r}\Psi(a, b) \quad \blacksquare$$

(5.3) Remark. Note the pleasant symmetry between the two end members of Harnack's Inequality. It is not hard to show that for $0 \le r' \le r$

$$\frac{R-r}{R+r} \le \frac{R-r'}{R+r'} \quad \text{and} \quad \frac{R+r'}{R-r'} \le \frac{R+r}{R-r}$$

hence Harnack's Inequality holds for all points (x, y) in the disk $B_R(a, b)$ which are at a distance *less than or equal to* r from (a, b).

Harnack's Inequality leads to a quite interesting result due to Picard.

(5.4) Theorem: Picard. Suppose that Ψ is harmonic on all of $\mathbb{R}^2$. If Ψ is bounded below (above) by a constant, then Ψ is a constant function.

Proof. First, suppose that Ψ is nonnegative (i.e., Ψ is bounded below by 0). Let (x, y) be some point in $\mathbb{R}^2$ and suppose that (x, y) lies a distance r from the origin. If $R > r$, then applying Harnack's inequality for the disk $\bar{B}_R(0, 0)$ we obtain

$$\frac{R-r}{R+r}\Psi(0, 0) \le \Psi(x, y) \le \frac{R+r}{R-r}\Psi(0, 0)$$

Letting R tend to infinity we find that $\Psi(x, y) = \Psi(0, 0)$. Therefore, Ψ is a constant function.

Now, if $\Psi(x, y) \ge C$, a constant, for all points (x, y) then $\Psi - C \ge 0$. Hence, $\Psi - C$ is a constant function, which implies the same thing for Ψ. If Ψ is bounded above by a constant, then we apply the last result to $-\Psi$. $\blacksquare$

We shall apply our next theorem to some of the harmonic functions on rectangles which were discussed in the exercises for §1; this is just one of this theorem's many uses.

(5.5) Theorem: Harnack's First Convergence Theorem. Let $\{\Psi_n\}$ be a sequence of harmonic functions on an open set U. If $\{\Psi_n\}$ converges uniformly to the function g on every compact subset of U, then g is harmonic in U.

Proof. Choose a point (c, d) in U and suppose that the closed disk $\bar{B}_R(c, d)$ is contained in the open set U; we will prove that g is harmonic in $B_R(c, d)$. Because of the first part of Exercise (1.5), we may assume *without loss of generality* that $(c, d) = (0, 0)$. From Theorem (4.5), we have

(5.6) $$\Psi_m = \text{P.I.}[\Psi_m]$$

on the disk $B_R(0, 0)$ for each m. When m tends to infinity, the left side of (5.6) tends to g. We now show that the right side of (5.6) tends to P.I.$[g]$.

Given $\epsilon > 0$, we can find a constant M so large that for all $m \geq M$

$$|g(x, y) - \Psi_m(x, y)| < \epsilon$$

for all points (x, y) in the compact disk $\bar{B}_R(0, 0)$. Hence for all $m \geq M$

$$|\text{P.I.}[g](x, y) - \text{P.I.}[\Psi_m](x, y)| \leq \frac{1}{2\pi} \int_{-\pi}^{\pi} |g(R \cos \alpha, R \sin \alpha)$$
$$- \Psi_m(R \cos \alpha, R \sin \alpha)| \, p(x, y; \alpha) \, d\alpha$$
$$< \frac{\epsilon}{2\pi} \int_{-\pi}^{\pi} p(x, y; \alpha) \, d\alpha$$
$$= \epsilon$$

Thus the right side of (5.6) does tend to P.I.$[g]$ on $B_R(0, 0)$. Hence $g = \text{P.I.}[g]$ on $B_R(0, 0)$ which proves that g is harmonic on that disk. Our theorem then follows. ■

(5.7) Example. [See Exercises (1.9) through (1.16).]

(a)   Consider the formal solution

$$\Psi = -\frac{8}{\pi} \sum_{k=0}^{\infty} \frac{1}{(2k + 1)^3} \frac{\sinh(2k + 1)[5 - y]}{\sinh(2k + 1)5} \sin(2k + 1)x$$

to Exercise (1.15a). When $0 \leq y \leq 5$ we have

$$\left| \frac{1}{(2k + 1)^3} \frac{\sinh(2k + 1)[5 - y]}{\sinh(2k + 1)5} \sin(2k + 1)x \right| \leq \frac{1}{(2k + 1)^3}$$

hence Weierstrass' M-Test implies that the series for Ψ converges uniformly to Ψ on the closed rectangle $\bar{R} = \{(x, y): 0 \leq x \leq \pi, 0 \leq y \leq 5\}$. Since Ψ is the uniform limit of continuous functions (its partial sums), Ψ is continuous on $\bar{R}$. Furthermore, Theorem (5.5) tells us that Ψ is harmonic on the open rectangle $R = \{(x, y): 0 < x < \pi, 0 < y < 5\}$. Therefore, Ψ solves Dirichlet's problem on R where the given boundary function f equals 0 on three sides of ∂R and equals $x(x - \pi)$ on the fourth side of ∂R. Since f is continuous on R, we know that Ψ is the unique solution to Dirichlet's problem.

(b)   Consider the formal solution

$$\Psi = \sum_{n=1}^{\infty} \left[\frac{2}{\pi} \int_0^{\pi} g(u) \sin nu \, du \right] \frac{\sinh n(L - y)}{\sinh nL} \sin nx$$

to Exercise (1.9). Exercise (1.16) tells us that the series converges uniformly to Ψ on the closed rectangle $\bar{R}_\delta = \{(x, y): 0 \leq x \leq \pi, \delta \leq y \leq L\}$. It follows from Theorem (5.5) that Ψ is harmonic in the open rectangle $R_\delta = \{(x, y): 0 < x < \pi, \delta < y < L\}$. Letting δ pass to zero, we see that Ψ is harmonic on the open rectangle R.

Actually, all that is needed for a series

$$\sum_{n=1}^{\infty} a_n \frac{\sinh n(L-y)}{\sinh nL} \sin nx$$

to converge to a harmonic function Ψ on R is that $|a_n| \le A$ a constant. [See Exercise (5.15).]

(c)   Suppose that the function g in Exercise (1.9) is continuous on the interval $[0, \pi]$, piecewise smooth on the interval $(0, \pi)$, and satisfies $g(0) = g(\pi) = 0$. Setting $B_n = 2/\pi \int_0^\pi g(u) \sin nu \, du$, we see that for

$$\Psi = \sum_{n=1}^{\infty} B_n \frac{\sinh n(L-y)}{\sinh nL} \sin nx$$

we have

$$\left| B_n \frac{\sinh n(L-y)}{\sinh nL} \sin nx \right| \le |B_n|$$

Since $\sum_{n=1}^{\infty} B_n \sin nx$ is the Fourier series for the continuous and piecewise smooth odd periodic extension of g, it follows that[4] $\sum_{n=1}^{\infty} |B_n|$ converges. Hence, by the same reasoning as in (a) above, we see that Ψ solves Dirichlet's problem for the rectangle R with the boundary function f equal to 0 on three sides of ∂R and equal to g on the remaining side.

The following Corollary to Theorem (5.5) is also sometimes called Harnack's first convergence theorem; it follows from (5.5) by an application of the maximum–minimum principle. We leave its proof to the reader as Exercise (5.16).

(5.8) Corollary.   Suppose that U is an open set with compact closure $\bar{U}$. If $\{\Psi_j\}$ are harmonic functions on U which are all continuous on $\bar{U}$ and if the series $\sum_{j=1}^{\infty} \Psi_j$ converges uniformly on the boundary ∂U, then that series is uniformly convergent in $\bar{U}$ and its sum $\Psi = \sum_{j=1}^{\infty} \Psi_j$ is harmonic in U and continuous on $\bar{U}$.

Besides Theorem (5.5) and Corollary (5.8), which deal with uniform convergence of harmonic functions, Harnack discovered another theorem that deals with *monotone convergence* of harmonic functions. Unlike his two previous results, *Harnack's second convergence theorem* is not valid for general open sets but only for those that are *connected*. We will discuss connected open sets and Harnack's second convergence theorem in §8.

Our next theorem will be used in Chapter 7 for some uniqueness results, but it is convenient to prove the theorem now. We use the notation $\mathbb{R}^2_+$ to denote the *upper half plane* consisting of all points (x, y) such that $y > 0$, and $\mathbb{R}^2_-$ to denote the *lower half plane* consisting of all points (x, y) such that $y < 0$.

---

4. This was shown in the proof of Theorem (4.4), Chapter 2.

(5.9) Theorem: Schwarz's Reflection Principle. Suppose that Ψ is harmonic in the upper half plane $\mathbb{R}^2_+$, and continuous on the closure $\bar{\mathbb{R}}^2_+ = \{(x, y): y \geq 0\}$, and $\Psi = 0$ on the x axis $\partial\mathbb{R}^2_+$. Then the function $\Psi^\#$ defined by

$$\Psi^\#(x, y) = \begin{cases} \Psi(x, y) & \text{if } y \geq 0 \\ -\Psi(x, -y) & \text{if } y < 0 \end{cases}$$

is harmonic on $\mathbb{R}^2$.

Proof. Let the function w be defined on a closed disk $\bar{B}_R(0, 0)$ by

$$w(x, y) = \begin{cases} \Psi^\#(x, y) & \text{for } (x, y) \text{ on } \partial B_R(0, 0) \\ \text{P.I.}[\Psi^\#](x, y) & \text{for } (x, y) \text{ in } B_R(0, 0) \end{cases}$$

Then $w = \Psi^\#$ on the *lower semicircle* $\partial B_R(0, 0) \cap \mathbb{R}^2_-$. Moreover, for $(x, 0)$ in $B_R(0, 0)$ we have $p(x, 0; -\alpha) = p(x, 0; \alpha)$ hence

$$\text{P.I.}[\Psi^\#](x, 0) = \frac{1}{2\pi}\int_{-\pi}^{\pi} \Psi^\#(R\cos\alpha, R\sin\alpha)p(x, 0; \alpha)\,d\alpha$$

$$= \frac{1}{2\pi}\int_0^{\pi} -\Psi(R\cos\alpha, R\sin\alpha)p(x, 0; \alpha)\,d\alpha$$

$$+ \frac{1}{2\pi}\int_0^{\pi} \Psi(R\cos\alpha, R\sin\alpha)p(x, 0; \alpha)\,d\alpha$$

$$= 0$$

It follows that $\Psi^\#$ and w agree on the boundary of the *lower hemisphere* $B_R(0, 0) \cap \mathbb{R}^2_-$, whereas $\Psi^\#$ is harmonic on that lower hemisphere.[5] Therefore, by uniqueness of solutions to Dirichlet's problem (with $\Psi^\#$ as boundary function), we conclude that $\Psi^\# = w$ on the lower hemisphere and on the x axis. Similarly, $\Psi^\# = w$ on the *upper hemisphere* $B_R(0, 0) \cap \mathbb{R}^2_+$. Hence $\Psi^\# = w = \text{P.I.}[\Psi^\#]$ in the disk $B_R(0, 0)$ from which we conclude that $\Psi^\#$ is harmonic in $B_R(0, 0)$; therefore in the whole plane (letting R tend to infinity). ∎

(5.10) Corollary. If Ψ_1 and Ψ_2 agree on the x axis, are harmonic and bounded on the upper half plane, and continuous on its closure, then $\Psi_1 = \Psi_2$.

We close this section with a converse to Gauss' mean value theorem (5.1).

(5.11) Definition. A continuous function g defined on an open set U is said to have the *Mean Value Property* if for each closed disk $\bar{B}_R(a, b)$ contained in U

$$g(a, b) = \frac{1}{2\pi}\int_{-\pi}^{\pi} g(a + R\cos\alpha, b + R\sin\alpha)\,d\alpha \qquad ∎$$

---

5. Since $\Delta\Psi^\#(x, y) = -\Delta\Psi(x, -y) = -0 = 0$.

We proved in Theorem (5.1) that every harmonic function has the mean value property; we will now prove the converse.

(5.12) Theorem. Suppose that g is a continuous function on an open set U. If g has the mean value property, then g is harmonic in U.

Proof. Let (c, d) be some point in U and suppose that the closed disk $\bar{B}_R(c, d)$ is contained in U. Consider the function h defined on $\bar{B}_R(c, d)$ by

$$h(x, y) = \begin{cases} g(x, y) - \text{P.I.}[g](x, y) & \text{if } (x, y) \text{ is in } B_R(c, d) \\ 0 & \text{if } (x, y) \text{ is on } \partial B_R(c, d) \end{cases}$$

By Theorem (3.8) we know that h is continuous on $\bar{B}_R(c, d)$. Moreover, by our hypothesis and Theorem (5.1) we know that h satisfies the mean value property in $B_R(c, d)$. Let M be the maximum for h on the compact disk $\bar{B}_R(c, d)$. Since $h = 0$ on the circle $\partial B_R(c, d)$ we must have $M \geq 0$.

Suppose that $M > 0$; we will obtain a contradiction. Let $h(a, b) = M > 0$ where (a, b) is in $B_R(c, d)$. Then for r sufficiently small the closed disk $\bar{B}_r(a, b)$ is contained in $B_R(c, d)$, hence by the mean value property

(5.13) $$h(a, b) = M = \frac{1}{2\pi} \int_{-\pi}^{\pi} h(a + r \cos \alpha, b + r \sin \alpha) \, d\alpha$$

Each point $(a + r \cos \alpha, b + r \sin \alpha)$ lies on $\partial B_r(a, b)$, therefore, from the continuity of h and the fact that $h \leq M$ we conclude that $h = M$ on all of $\partial B_r(a, b)$ [otherwise (5.13) would not hold].[6] Letting r increase, when $\partial B_r(a, b)$ intersects $\partial B_R(c, d)$ we obtain a contradiction.

Therefore $M = 0$ hence $h \leq 0$ on all of $\bar{B}_R(c, d)$. Applying the same argument to $-h$ we get $h = 0$ on all of $\bar{B}_R(c, d)$. Thus, $g = \text{P.I.}[g]$ on $B_R(c, d)$ hence g is harmonic on that disk. Since the choice of (c, d) was arbitrary g is harmonic on U. ∎

Exercises

(5.14) Verify the results in Remark (5.3) and prove Corollary (5.10).

(5.15) Suppose $|a_n| \leq A$ a constant, prove that

$$\Psi = \sum_{n=1}^{\infty} a_n \frac{\sinh n(L - y)}{\sinh nL} \sin nx$$

is harmonic in $R = \{(x, y): 0 < x < \pi \text{ and } 0 < y < L\}$.

(5.16) Prove Corollary (5.8). [*Hint*: Consider $|\sum_{j=1}^{m} \Psi_j(x, y) - \sum_{j=1}^{n} \Psi_j(x, y)|$ for $m > n \geq N$, in the light of the maximum–minimum principle.]

(5.17) Show that the formal solution to Exercise (1.9), which was

---

6. In fact $h < M$ on an interval implies that $(2\pi)^{-1} \int_{-\pi}^{\pi} h < M$. See also the proof of (8.6).

discussed in (5.7b) above, can be expressed as

(a) $\Psi(x, y) = \int_0^\pi K_1(x, y; u)g(u)\, du$

where, for each fixed u, K_1 is a harmonic function of (x, y) in R. Using (a), give a second proof that Ψ is harmonic in R.

(5.18)   Let $\{\Psi_n\}$ converge to g over U as in Harnack's theorem (5.5). Prove that $\partial\Psi_n/\partial x$ $(\partial\Psi_n/\partial y)$ converges uniformly to $\partial\Psi/\partial x$ $(\partial\Psi/\partial y)$ on each compact subset of U.

(5.19)   Show that the formal solutions given in Exercise (1.13) can be expressed as

$$\Psi_j(x, y) = \int_0^B K_j(x, y; u)f_j(u)\, du \qquad (B = \pi \text{ or } L)$$

where each K_j is a harmonic function of (x, y) in R. Then prove that each Ψ_j is harmonic in R.

★(5.20)**   Suppose that Ψ is harmonic in an open disk $B_R(a, b)$ and Ψ has only nonnegative values. Prove that if $\Psi(a, b) = 0$ then $\Psi = 0$ on all of $B_R(a, b)$.

★(5.21)**   Suppose that Ψ is harmonic in an open set U and Ψ achieves a local minimum (maximum) at a point (a, b) in U. Prove that Ψ is *constant in value* on some open disk $B_R(a, b)$ contained in U.

★(5.22)**   (a) Suppose that Ψ is harmonic on the disk $B_R(a, b)$ and is identically zero on the smaller disk $B_S(a, b)$ where $S < R$. Prove that Ψ is identically zero on the whole disk $B_R(a, b)$. [*Hint*: Assume $(a, b) = (0, 0)$ and use (2.13).] (b) Show that the function Ψ in (5.21) is constant in value on *every* open disk $B_R(a, b)$ contained in U.

§6. Some Applications of the Divergence Theorem

In this section, we shall briefly consider some applications of Gauss' divergence theorem to harmonic functions of three variables. First, however, let's look again at Gauss' mean value theorem for a harmonic function Ψ of two variables:

(6.1) $\Psi(a, b) = \dfrac{1}{2\pi} \int_{-\pi}^\pi \Psi(a + R \cos \alpha, b + R \sin \alpha)\, d\alpha$

The point $(a + R \cos \alpha, b + R \sin \alpha)$ travels around the circle $\partial B_R(a, b)$ in the counterclockwise sense as α goes from $-\pi$ to π. The quantity $Rd\alpha$ is the element of arclength ds on the circle $\partial B_R(a, b)$, hence the integral in (6.1) can be expressed as a *line integral* over $\partial B_R(a, b)$

$$\frac{1}{2\pi} \int_{-\pi}^{\pi} \Psi(a + R \cos \alpha, b + R \sin \alpha) \, d\alpha$$

$$= \frac{1}{2\pi R} \int_{-\pi}^{\pi} \Psi(a + R \cos \alpha, b + R \sin \alpha) \, R d\alpha$$

$$= \frac{1}{2\pi R} \oint_{\partial B_R(a, b)} \Psi \, ds$$

Thus (6.1) becomes

(6.1') $$\Psi(a, b) = \frac{1}{2\pi R} \oint_{\partial B_R(a, b)} \Psi \, ds$$

Since $2\pi R$ is the circumference of the circle $\partial B_R(a, b)$, the right side of (6.1') is called the *mean value* of Ψ over the circle $\partial B_R(a, b)$. By deriving (6.1') we have shown that Gauss' mean value theorem states that the value of a harmonic function at the center of a disk is determined by the mean value of its values on the boundary of that disk.

Poisson's integral can also be written as a line integral over $\partial B_R(a, b)$. The reader may show that

(6.2) $$\Psi(x, y) = \frac{1}{2\pi R} \oint_{\partial B_R(a, b)} \Psi(u, v) \frac{R^2 - (x^2 + y^2)}{(x - u)^2 - (y - v)^2} \, ds_{(u, v)}$$

given that Ψ is harmonic in an open set U which contains the closed disk $\bar{B}_R(a, b)$. From (6.2) we see that the values of Ψ in the open disk $B_R(a, b)$ are determined by its values on the boundary $\partial B_R(a, b)$.

The discussion above shows that fundamental results for harmonic functions of two variables, especially (6.2), can be phrased using line integrals. Furthermore, all of the results of the previous three sections could have been worked out by first obtaining (6.2) and then (6.1'). For harmonic functions of three variables this latter approach, using *surface integrals*, plays an important role as we shall see in Chapter 8. We shall limit ourselves, for now, to deriving a three-dimensional analogue of (6.1').

Our starting point is the following theorem, due to Gauss,[7] which is adequately discussed in any advanced calculus textbook. (See Figure 4.6.)

(6.3) Theorem: Divergence Theorem: Let $\mathbf{W} = (X, Y, Z)$ be a continuously differentiable vector field[8] on an open set U in $\mathbb{R}^3$. Suppose that U contains a smooth surface $\partial \mathcal{R}$ that surrounds an open set (region) $\mathcal{R}$. If $\boldsymbol{\eta} = (\alpha, \beta, \gamma)$ denotes the outward pointing unit normal vector field on $\partial \mathcal{R}$, then

$$\int_{\mathcal{R}} div \, \mathbf{W} \, dV = \int_{\partial \mathcal{R}} \mathbf{W} \cdot \boldsymbol{\eta} \, dA$$

---

7. Theorem (6.3) is sometimes called *Gauss' theorem*.

8. That is, each function X, Y, and Z is a continuously differentiable function of x, y, and z over U.

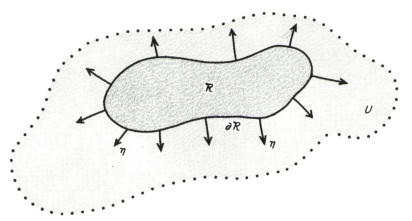

Figure 4.6 By smoothness of $\partial \mathcal{R}$ we mean that $\boldsymbol{\eta}$, the outward pointing unit normal, is a continuously differentiable vector field on $\partial \mathcal{R}$.

or, more explicitly,

$$\int_{\mathcal{R}} \left(\frac{\partial X}{\partial x} + \frac{\partial Y}{\partial y} + \frac{\partial Z}{\partial z} \right) dV = \int_{\partial \mathcal{R}} (X\alpha + Y\beta + Z\gamma) \, dA$$

While the Divergence Theorem takes some doing to prove with any degree of generality, we shall need it only for quite simple cases where $\mathcal{R}$ might be one of the following: an open ball, the region between two concentric spheres, or the region inside of a ball lying to one side of a plane (see Figure 4.7.).

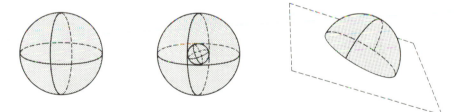

Figure 4.7 Some simple cases of $\mathcal{R}$.

For such simple cases, the divergence theorem is adequately discussed in some elementary calculus texts.[9]

Two simple consequences of the divergence theorem are *Green's identities*

(6.4a) $$\int_{\mathcal{R}} v \, \Delta u \, dV + \int_{\mathcal{R}} (\text{grad } u) \cdot (\text{grad } v) \, dV = \int_{\partial \mathcal{R}} \frac{\partial u}{\partial \boldsymbol{\eta}} v \, dA$$

(6.4b) $$\int_{\mathcal{R}} [v \, \Delta u - u \, \Delta v] \, dV = \int_{\partial \mathcal{R}} \left[\frac{\partial u}{\partial \boldsymbol{\eta}} v - u \frac{\partial v}{\partial \boldsymbol{\eta}} \right] dA$$

9. See, e.g., Thomas and Finney (1984), Chap. 17.

where u and v are twice continuously differentiable functions on the open set U (which contains $\bar{\mathscr{R}} = \mathscr{R} \cup \partial \mathscr{R}$). To prove (6.4a), set

$$\mathbf{W} = \left(v \frac{\partial u}{\partial x}, v \frac{\partial u}{\partial y}, v \frac{\partial u}{\partial z} \right)$$

in the divergence theorem, obtaining

$$\int_{\mathscr{R}} [v \, \Delta u + (\text{grad } u) \cdot (\text{grad } v)] \, dV = \int_{\partial \mathscr{R}} v \left(\frac{\partial u}{\partial x}, \frac{\partial u}{\partial y}, \frac{\partial u}{\partial z} \right) \cdot \boldsymbol{\eta} \, dA$$

or, in other words, (6.4a). Interchanging the roles of u and v in (6.4a) and then subtracting yields (6.4b).

If we let $u = \Psi$, a harmonic function on U, and $v = 1$, a constant function, then (6.4a) reduces to

(6.5)
$$\int_{\partial \mathscr{R}} \frac{\partial \Psi}{\partial \eta} \, dA = 0$$

We can now prove Gauss' mean value theorem.

(6.6) Theorem: Mean Value Theorem. Let Ψ be a harmonic function on an open set U in $\mathbb{R}^3$. If the closed ball $\bar{B}_R(a, b, c)$ is contained in U, then

$$\Psi(a, b, c) = \frac{1}{4\pi R^2} \int_{\partial B_R(a, b, c)} \Psi \, dA$$

[*Remark*: The right side of the equation above is the surface integral of Ψ over the sphere $\partial B_R(a, b, c)$ divided by the surface area $4\pi R^2$ of $\partial B_R(a, b, c)$. Hence the right side of the equation above is called the *mean value* of Ψ over $\partial B_R(a, b, c)$.]

Proof. Let ∂B_R stand for the sphere $\partial B_R(a, b, c)$. Also let $0 < \epsilon < R$. If we define u by $u(x, y, z) = [(x - a)^2 + (y - b)^2 + (z - c)^2]^{-1/2}$ then u is harmonic away from (a, b, c) [see Exercise (6.9)]. Letting v equal Ψ and applying (6.4b) to the region $\mathscr{R}$ between the two spheres ∂B_R and ∂B_ϵ we get

$$0 = \int_{\partial B_R \cup \partial B_\epsilon} \left(\Psi \frac{\partial u}{\partial \eta} - \frac{\partial \Psi}{\partial \eta} u \right) dA$$

$$= \int_{\partial B_R} \Psi \frac{\partial}{\partial r} \left(\frac{1}{r} \right) dA - \int_{\partial B_\epsilon} \Psi \frac{\partial}{\partial r} \left(\frac{1}{r} \right) dA - \int_{\partial B_R} \frac{\partial \Psi}{\partial \eta} \frac{1}{R} \, dA + \int_{\partial B_\epsilon} \frac{\partial \Psi}{\partial \eta} \frac{1}{\epsilon} \, dA$$

where $r = [(x - a)^2 + (y - b)^2 + (z - c)^2]^{1/2}$ is the radial coordinate which describes the normal direction to spheres centered at (a, b, c). Applying (6.5) we have

$$0 = \frac{1}{\epsilon^2} \int_{\partial B_\epsilon} \Psi \, dA - \frac{1}{R^2} \int_{\partial B_R} \Psi \, dA$$

Hence,

$$\frac{1}{4\pi R^2} \int_{\partial B_R} \Psi \, dA = \frac{1}{4\pi \epsilon^2} \int_{\partial B_\epsilon} \Psi \, dA$$

Making ϵ tend to zero, since Ψ is continuous at (a, b, c), we clearly have

$$\frac{1}{4\pi R^2} \int_{\partial B_R} \Psi \, dA = \Psi(a, b, c) \quad \blacksquare$$

Exercises

(6.7) Derive (6.2).

(6.8) Show that the solution $\Psi = \sum_{j=1}^{4} \Psi_j$ in Exercise (1.13) can be expressed as the following line integral over the boundary ∂R of the rectangle R

$$\Psi(x, y) = \oint_{\partial R} K(x, y; u, v) f(u, v) \, ds_{(u, v)}$$

[where (u, v) denotes a point on ∂R].

(6.9) Prove that $u(x, y, z) = [(x - a)^2 + (y - b)^2 + (z - c)^2]^{-1/2}$ is harmonic for all $(x, y, z) \neq (a, b, c)$.

★(6.10) Show that for the function p defined in Example (1.2b) we have for $R^2 = a^2 + b^2 + c^2$ and $r^2 = x^2 + y^2 + z^2$ (where $r < R$)

$$\frac{R}{4\pi r^2} \int_{\partial B_r(0, 0, 0)} p(x, y, z) \, dA = 1$$

(6.11) Suppose that Ψ is harmonic in an open set U in $\mathbb{R}^3$ and continuous on the compact closure $\bar{U}$. Prove that if Ψ attains a local maximum M at a point (a, b, c) in U, then on some ball $B_R(a, b, c)$ we have $\Psi = M$, and a similar result holds for a local minimum of Ψ.

§7. The Heat Equation and the Theta Function Kernel

There are many interesting parallels between the properties of solutions to the heat equation and the properties of solutions to Laplace's equation. We will limit ourselves to discussing just two of these parallels, namely, the maximum–minimum principle, and the theta function kernel (which is analogous to the Poisson kernel).

Let $U = (c, d)$ be an open interval in $\mathbb{R}$. For a fixed time $T > 0$ we form the open rectangle U_T in $\mathbb{R}^2$ with base U and height T

$$U_T = \{(x, t): \quad c < x < d \quad \text{and} \quad 0 < t < T\}$$

The boundary ∂U_T consists of two separate portions, a lower boundary

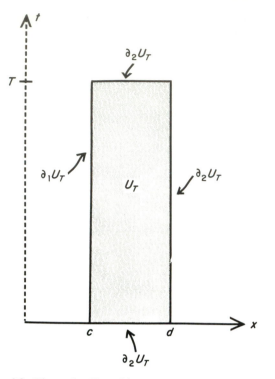

Figure 4.8   The region U_T and its boundary parts $\partial_1 U_T$ and $\partial_2 U_T$.

$\partial_1 U_T$ and an upper boundary $\partial_2 U_T$. (See Figure 4.8.):

$$\partial_1 U_T = \{(x, t): \quad x = c \quad \text{and} \quad 0 \le t \le T \quad \text{or } x = d \quad \text{and} \quad 0 \le t \le T$$
$$\text{or } c \le x \le d \quad \text{and} \quad t = 0\}$$
$$\partial_2 U_T = \{(x, t): \quad c < x < d \quad \text{and} \quad t = T\}$$

As in the case of Laplace's equation (harmonic functions), the maximum, or minimum, of a solution to the heat equation is attained on the boundary ∂U_T. However, for the heat equation we can say a little more.

(7.1) Theorem: Maximum–Minimum Principle.   Let u be continuous on $\bar{U}_T$, and u_t, u_{xx} exist and be continuous for $c < x < d$ and $t > 0$, and satisfy $u_t - a^2 u_{xx} = 0$ (for $a^2 > 0$). Then

$$\max_{\bar{U}_T} u = \max_{\partial_1 U_T} u \qquad \min_{\bar{U}_T} u = \min_{\partial_1 U_T} u$$

Proof.   Let M equal the maximum of u over the compact set $\partial_1 U_T$. Clearly, we have

$$\max_{\bar{U}_T} u \ge M$$

Let $\delta > 0$ and let v be the function defined by $v(x, t) = u(x, t) + \delta x^2$. Then $v_t - a^2 v_{xx} = -2a^2 \delta < 0$.

Since v is continuous, it attains its maximum at some point (x_0, t_0) in the compact set $\overline{U}_T$. Suppose that (x_0, t_0) belonged to $U_T \cup \partial_2 U_T$; we will obtain a contradiction. If (x_0, t_0) lies in U_T, then[10]

(7.2) $$v_t(x_0, t_0) = 0 \qquad v_{xx}(x_0, t_0) \leq 0$$

while if (x_0, t_0) lies in $\partial_2 U_T$ then

(7.2') $$v_t(x_0, t_0) \geq 0 \qquad v_{xx}(x_0, t_0) \leq 0$$

[The first inequality in (7.2') holds because $v(x_0, t)$ must be increasing as t increases up to $t_0 = T$.] In any case, we have $v_t(x_0, t_0) \geq 0$ and $v_{xx}(x_0, t_0) \leq 0$ which contradicts $v_t - a^2 v_{xx} < 0$.

Therefore, v assumes its maximum over $\overline{U}_T$ at a point belonging to $\partial_1 U_T$ hence $u \leq M$ at that point. Therefore, on $\overline{U}_T$

$$v \leq M + \delta \underset{c \leq x \leq d}{\text{maximum}} (x^2)$$

Since $u \leq v$ and δ can be made arbitrarily small, it follows that $u \leq M$ on $\overline{U}_T$. Hence

$$\max_{\overline{U}_T} u \leq M$$

The reverse inequality, stated at the beginning, implies that

$$\max_{\overline{U}_T} u = M = \max_{\partial_1 U_T} u$$

We get the analogous equality for minimums by applying the equality just obtained to $-u$. ∎

(7.3) Corollary. Let u satisfy the hypotheses of Theorem (7.1). Then u is determined uniquely in $\overline{U}_T$ by its values on $\partial_1 U_T$.

Proof. Exercise (7.10). ∎

Corollary (7.3) implies that the heat conduction problem (4.1), Chapter 3, has a unique solution. Furthermore, our results above generalize to the heat equation in higher dimensions [see Exercise (7.12)].

We now describe the theta function kernel for the heat equation. The results of Exercises (4.28) through (4.30) from Chapter 3 will be used below. In those Exercises it was found that the solution (4.5) to the heat conduction problem (4.1), in Chapter 3, could be expressed in the form:

(7.4) $$u(x, t) = \frac{1}{2L} \int_{-L}^{L} F(x - s)\theta(s, t)\, ds$$

where

(7.4') $$\theta(s, t) = \sum_{n=-\infty}^{+\infty} e^{i(n\pi s/L)} e^{-n^2(\pi a/L)^2 t}$$

---

10. We get (7.2) in the same way as (4.3).

and F is the odd periodic extension of f. The function θ is called (Jacobi's) theta function. For convenience, we shall henceforth assume that $L = \frac{1}{2}$ and $a^2 = \frac{1}{2}$. *This results in no loss of generality.* We now have

(7.4") $$\theta(s, t) = \sum_{n=-\infty}^{+\infty} e^{i(2n\pi s)} e^{-2n^2\pi^2 t}$$

The theta function satisfies the following identity, discovered by Jacobi,

(7.5) $$\theta(s, t) = \frac{1}{\sqrt{2\pi t}} \sum_{k=-\infty}^{+\infty} e^{-(s-k)^2/2t} \qquad (t > 0)$$

To prove (7.5) we shall need to borrow the following result from Chapter 6 [see formula (3.3) from that chapter].

(7.6) $$\int_{-\infty}^{+\infty} e^{-cx^2} e^{-i2\pi u x} \, dx = (\pi/c)^{1/2} e^{-\pi^2 u^2/c} \qquad (c > 0)$$

Fix a value of $t > 0$. Consider the function g defined by $g(s) = \sum_{k=-\infty}^{+\infty} e^{-(s-k)^2/2t}$. The function g is periodic with period 1. If we compute its complex Fourier series

$$g \sim \sum_{n=-\infty}^{+\infty} c_n e^{i(2n\pi s)}$$

then

$$c_n = \int_{-1/2}^{1/2} g(s) e^{-i(2n\pi s)} \, ds$$

$$= \int_{-1/2}^{1/2} \left[\sum_{k=-\infty}^{+\infty} e^{-(s-k)^2/2t} \right] e^{-i(2n\pi s)} \, ds$$

It is not hard to verify that $\sum_{k=-\infty}^{+\infty} e^{-(s-k)^2/2t} e^{-i(2n\pi s)}$ converges uniformly (and absolutely) for $-\frac{1}{2} \leq s \leq \frac{1}{2}$, from which

$$c_n = \sum_{k=-\infty}^{+\infty} \int_{-1/2}^{1/2} e^{-(s-k)^2/2t} e^{-i(2n\pi s)} \, ds$$

$$= \int_{-\infty}^{+\infty} e^{-s^2/2t} e^{-i(2n\pi s)} \, ds$$

where the last integral was obtained by substituting $s + k$ for s and using periodicity. Applying (7.6),[11] we obtain

$$c_n = \sqrt{2\pi t} e^{-2\pi^2 n^2 t}$$

Therefore, since $\sum_{n=-\infty}^{+\infty} |c_n|$ converges, we have

$$\sum_{k=-\infty}^{+\infty} e^{-(s-k)^2/2t} = g(s) = \sqrt{2\pi t} \sum_{n=-\infty}^{+\infty} e^{-2\pi^2 n^2 t} e^{i(2n\pi s)}$$

Dividing the equality above by $\sqrt{2\pi t}$ we get (7.5).

---

11. For $s = x$, $n = u$, and $1/2t = c$.

Using Jacobi's identity (7.5) we can show that the theta function is a summation kernel over $(-\frac{1}{2}, \frac{1}{2})$, parameterized by $t > 0$.

(7.7) Lemma.   The theta function $\theta(s, t)$ satisfies

(A$_1$) $\displaystyle\int_{-1/2}^{1/2} \theta(s, t)\, ds = 1$ for all $t > 0$

(A$_2$) $\theta(s, t) \geq 0$ for all $t > 0$ and all s

(A$_3$)   For every $\epsilon > 0$ and $\delta > 0$ we have

$$0 \leq \theta(s, t) < \epsilon$$

provided $\frac{1}{2} \geq |s| \geq \delta$ and t is taken close enough to 0.

Proof.   Fix a value of $t > 0$. The series for θ in (7.4″) converges uniformly for all s values. Hence

$$\int_{-1/2}^{1/2} \theta(s, t)\, ds = \sum_{n=-\infty}^{+\infty} \int_{-1/2}^{1/2} e^{i(2n\pi s)}\, ds \cdot e^{-2n^2\pi^2 t} = 1$$

since only the integral with $n = 0$ does not vanish. Thus (A$_1$) holds. Property (A$_2$) follows immediately from (7.5). As for (A$_3$), if $\frac{1}{2} \geq |s| \geq \delta$, then for $|k| \geq 1$ we have

$$k^2/8t \leq (s - k)^2/2t$$

while for $k = 0$ we have $\delta^2/2t \leq s^2/2t$. Therefore (7.5) tells us that

$$0 \leq \theta(s, t) \leq \frac{1}{\sqrt{2\pi t}} e^{-\delta^2/2t} + \frac{2}{\sqrt{2\pi t}} e^{-1/8t} + \frac{2}{\sqrt{2\pi t}} \sum_{k=2}^{\infty} e^{-k^2/8t}$$

$$< \frac{1}{\sqrt{2\pi t}} e^{-\delta^2/2t} + \frac{2}{\sqrt{2\pi t}} e^{-1/8t} + \frac{2}{\sqrt{2\pi t}} \int_{1}^{\infty} e^{-x^2/8t}\, dx$$

Substitute $x = \sqrt{8t}\, r$ and we have

$$0 \leq \theta(s, t) < \frac{1}{\sqrt{2\pi t}} e^{-\delta^2/2t} + \frac{2}{\sqrt{2\pi t}} e^{-1/8t} + \frac{4}{\sqrt{\pi}} \int_{1/\sqrt{8t}}^{\infty} e^{-r^2}\, dr$$

$$< \epsilon$$

if we take t close enough to 0. ∎

Since the theta function is a summation kernel, and u is described by (7.4), we obtain the following theorem as an immediate corollary of Theorem (8.4), Chapter 2.

(7.8) Theorem.   Suppose that F has period 1 and F is absolutely integrable over $[-\frac{1}{2}, \frac{1}{2}]$. Then if x_0 is a point of continuity for F we have

$$\lim_{(x,\, t) \to (x_0,\, 0+)} u(x, t) = F(x_0)$$

(7.9) Remark.   By obtaining Theorem (7.8) we have completed the treatment of the heat equation begun in Chapter 3. In particular, Theorem (4.18), Chapter 3, is now established.

Exercises

(7.10)   Prove Corollary (7.3).

(7.11)   Show *directly from* (7.4) that the function u given there is the unique solution to (4.1), Chapter 3, provided f is continuous and $f(0) = f(L) = 0$.

(7.12)   Formulate and prove a maximum–minimum principle for the two-(three)-dimensional heat equation $u_t = a^2 \Delta u$ where Δ is the two- (three)-dimensional Laplacian.

(7.13)   Suppose that F is k times continuously differentiable, prove that $u(x, t) = \int_{-1/2}^{1/2} F(x - s)\theta(s, t)\, ds$ satisfies

$$\lim_{t \to 0+} \frac{\partial^j u}{\partial x^j}(x, t) = \frac{d^j F}{dx^j}(x) \qquad (j = 1, \ldots, k)$$

(7.14)   Consider (7.4). Prove that if $m \le F(x) \le M$ (or $m \le f(x) \le M$) then $m \le u(x, t) \le M$.

(7.15)   Show that u in (7.4) is infinitely continuously differentiable in x and t, provided that F is continuous (absolutely integrable).

(7.16)   Express u in (7.4) as $u = F * \theta$. Prove that if $\{F_n\}$ is a sequence of continuous functions, with period $2L$, uniformly convergent to F over $[-L, L]$, then $u_n = F_n * \theta$ converges uniformly to u.

*Periodic Solutions to the Heat Equation

In Exercises (7.17) through (7.21) below, when we say that u is a *periodic solution to the heat equation* we mean that $u_t = a^2 u_{xx}$ for $t > 0$ and all x values and that u has period $2L$ in x [$u(x + 2L, t) = u(x, t)$]. Let $\mathbb{R}_+^2$ stand for $\{(x, t): -\infty < x < +\infty$ and $t > 0\}$.

(7.17)   Prove that for $t_0 > 0$ and $t > 0$ a periodic solution u to the heat equation satisfies

$$u(x, t + t_0) = \frac{1}{2L} \int_{-L}^{L} u(x - s, t_0)\theta(s, t)\, ds = \frac{1}{2L} \int_{-L}^{L} u(s, t_0)\theta(x - s, t)\, ds$$

where θ is as in (7.4'). Moreover, prove that u is infinitely continuously differentiable in x and t over $\mathbb{R}_+^2$. [*Hint*: Use $u \sim \sum C_n(t)e^{i(n\pi x/L)}$ and Exercise (8.15c), Chapter 2.]

(7.18)   Suppose that $\{u_n\}$ is a sequence of periodic solutions to the heat equation and $\lim_{n \to \infty} u_n(x, t) = u(x, t)$ uniformly on every compact subset of $\mathbb{R}_+^2$. Prove that u is a periodic solution to the heat equation. Moreover, prove that $\partial u_n / \partial x$ converges uniformly to $\partial u / \partial x$ on every compact subset of $\mathbb{R}_+^2$.

(7.19)   Show that $u = F * \theta$ is a periodic solution to the heat equation [see

(7.16)] when F is piecewise continuous and has period $2L$, and show that not all periodic solutions to the heat equation are of the form $F * \theta$.

(7.20) Prove that if u is a periodic solution to the heat equation, then for each $t_0 > 0$

$$\lim_{t \to \infty} u(x, t) = \frac{1}{2L} \int_{-L}^{L} u(s, t_0) \, ds$$

uniformly for all x values. Conclude that $(1/2L) \int_{-L}^{L} u(x, t) \, dx$ is constant for $t > 0$. Interpret these results physically and give a counterexample when u is not periodic.

⋆(7.21) Suppose that $\{u_n\}$ is a monotone increasing sequence of periodic solutions to the heat equation and $\lim_{n \to \infty} u_n(x, t) = u(x, t)$ for u a continuous (bounded) function on $\mathbb{R}_+^2$. Prove that u is a periodic solution to the heat equation.[12]

⋆(7.22) Generalize (7.17) through (7.21) to solutions of

$$u_t = a^2 u_{xx} \qquad (0 < x < L \qquad t > 0)$$
$$u(0+, t) = u(L-, t) = 0 \qquad (t > 0)$$

(7.23) Show that the problem

$$u_t = a^2 u_{xx} \qquad (c < x < d \qquad t > 0)$$
$$u(x, 0+) = f(x) \qquad (c < x < d)$$

is *not* uniquely solvable for u given a continuous function f.

(7.24) Prove that *each term* $e^{-(x-k)^2/2t}(2\pi t)^{-1/2}$ in Jacobi's series for $\theta(x, t)$ [see (7.5)] is a solution to the heat equation with $a^2 = \frac{1}{2}$.

⋆(7.25) Can you give a physical interpretation to each term $e^{-(x-k)^2/2t}(2\pi t)^{-1/2}$ in Jacobi's series for $\theta(x, t)$? What about the whole series for θ in (7.5)?

§8. Harmonic Functions on Connected Open Sets

In this section we shall examine some of the properties of harmonic functions defined on *connected* open sets. The reader may recall that an open set U is called *connected* if U cannot be expressed as the union of two nonempty disjoint open sets. The following well known theorem allows us to easily determine when a given open set is connected; for a proof, see Buck (1978) or Bartle (1964).

(8.1) Theorem. An open set U is connected if and only if it is path connected.

---

12. *Hint*: For u continuous, the convergence of u_n to u is uniform on compact sets by a theorem of Dini's; see Courant and Hilbert (1953), footnote to p. 57. For u bounded, you need the Lebesgue dominated convergence theorem; see Royden (1968) or Rudin (1974).

By path connected we mean that every two points in U can be joined by a continuous (or even polygonal) path. For example, because of Theorem (8.1) we know that each open disk, open rectangle, or open annulus is a connected open set. Here are a couple of theoretical consequences of assuming that Ψ is harmonic on a connected open set.

(8.2) Theorem: Maximum–Minimum Principle, Strong Form. Let U be a connected open set in $\mathbb{R}^2$ with compact closure $\bar{U}$. The maximum (minimum) value of Ψ over $\bar{U}$ is attained only on ∂U, unless Ψ is a constant function.

Proof. Suppose that Ψ attains its maximum M over $\bar{U}$ at a point (a, b) in U. Then Exercise (5.21) tells us that $\Psi = M$ on some open disk $B_R(a, b)$ contained in U. Therefore, the set $\mathcal{M}$ defined by $\mathcal{M} = \{(x, y): (x, y)$ is in U and $\Psi(x, y) = M\}$ is an open set which includes the point (a, b). Moreover, $U - \mathcal{M} = \{(x, y): (x, y)$ is in U and $\Psi(x, y) \neq M\}$ is an open set because Ψ is a continuous function. Since U is connected and $\mathcal{M}$ is not empty, we conclude that $U = \mathcal{M}$. Thus $\Psi = M$ on all of U, hence, by continuity, on all of $\bar{U}$. Applying the argument above to $-\Psi$ we see that if Ψ attains its minimum m over $\bar{U}$ at a point in U, then $\Psi = m$ on all of $\bar{U}$. ∎

Theorem (8.2) is false if U is not connected [see Exercise (8.15)]. It also generalizes to harmonic functions of three variables.

As we said in §5, Harnack's second convergence theorem depends on U being a connected open set.

(8.3) Theorem: Harnack's Second Convergence Theorem. Let $\{\Psi_n\}$ be a sequence of harmonic functions on a connected open set U. Suppose that $\Psi_n(x, y) \leq \Psi_{n+1}(x, y)$ for every n and every point (x, y) in U. If, for one point (a, b) in U, the sequence $\{\Psi_n(a, b)\}$ is bounded above, then the sequence $\{\Psi_n\}$ converges to a harmonic function Ψ on U. Moreover, the convergence is uniform on every compact subset of U.

Proof. Let $\mathscr{C}$ stand for the set of all points (x, y) in U such that $\{\Psi_n(x, y)\}$ is bounded above. Since (a, b) is in U we know that $\mathscr{C}$ is nonempty. Moreover, for each point (c, d) in $\mathscr{C}$ the sequence $\{\Psi_n(c, d)\}$ is monotone increasing and bounded above, *therefore it converges*. Thus $\{\Psi_n\}$ converges on $\mathscr{C}$ to a function Ψ; we will now show that $\mathscr{C}$ is an open subset of U and Ψ is harmonic on $\mathscr{C}$.

Suppose (c, d) is in $\mathscr{C}$ and $\bar{B}_R(c, d)$ is contained in U. The sequence $\{\Psi_n(c, d)\}$ is convergent, hence for every $\epsilon > 0$ there exists an integer N such that

$$0 \leq \Psi_m(c, d) - \Psi_n(c, d) < \epsilon$$

for all $m > n \geq N$. From Harnack's inequality it follows that for all (x, y) in the closed disk $\bar{B}_r(c, d)$ (where $r < R$)

$$0 \leq \Psi_m(x, y) - \Psi_n(x, y) < \frac{R + r}{R - r} \epsilon$$

provided $m > n \geq N$ [see Remark (5.3)]. Hence for all $m > n > N$

$$0 \leq \Psi_m(x, y) - \Psi_n(x, y) < 3\epsilon$$

when (x, y) is in $\bar{B}_{\frac{1}{2}R}(c, d)$. Moreover, letting $m \to +\infty$ in the last inequality we obtain

$$0 \leq \Psi(x, y) - \Psi_n(x, y) \leq 3\epsilon$$

for all $n \geq N$. Thus $\{\Psi_n\}$ converges uniformly to Ψ on $\bar{B}_{\frac{1}{2}R}(c, d)$. By Theorem (5.5), we conclude that Ψ is harmonic on $B_{\frac{1}{2}R}(c, d)$. Moreover, since convergent sequences are bounded, we have that $\bar{B}_{\frac{1}{2}R}(c, d)$ is contained in $\mathscr{C}$. Thus, we have shown that $\mathscr{C}$ is an open subset of U and Ψ is harmonic on $\mathscr{C}$.

We will now show that $\mathscr{C}$ is all of U. Suppose, on the contrary, that (c, d) is in U but not in $\mathscr{C}$. Then $\Psi_n(c, d)$ tends to $+\infty$. But then if $\bar{B}_R(c, d)$ is contained in U we have for all (x, y) in $\bar{B}_{\frac{1}{2}R}(c, d)$

$$\tfrac{1}{3}[\Psi_n(c, d) - \Psi_1(c, d)] \leq [\Psi_n(x, y) - \Psi_1(x, y)]$$

by Harnack's Inequality. It follows that $\{\Psi_n(x, y)\}$ is unbounded for all (x, y) in $B_{\frac{1}{2}R}(c, d)$. Therefore, because U is connected, we must have $U = \mathscr{C}$, hence $\{\Psi_n\}$ converges to the harmonic function Ψ on all of U. Moreover, suppose K is a compact set in U. Covering each point (c, d) in K by $B_{\frac{1}{2}R}(c, d)$, where some $B_R(c, d)$ is contained in U, and then reducing to a finite subcover, we see that $\{\Psi_n\}$ converges to Ψ uniformly on K. ∎

Subharmonic Functions

Subharmonic functions are an important generalization of harmonic functions. They play a key role in Perron's proof of the existence of solutions to Dirichlet's problem. Moreover, by widening our view, we can learn more about harmonic functions themselves [see Exercises (8.20) through (8.22)].

As we saw in §5, harmonic functions are completely characterized by the fact that they satisfy the mean value property. We mention this in order to motivate the following definition of subharmonic functions.

(8.4) Definition. A continuous function g defined on an open set U is called *subharmonic* if for each point (a, b) in U there exists a radius R [dependent on (a, b)] such that for all radii $r < R$

$$g(a, b) \leq \frac{1}{2\pi} \int_{-\pi}^{\pi} g(a + r \cos \alpha, \, b + r \sin \alpha) \, d\alpha \qquad ∎$$

It is important to note that since Definition (8.4) employs a less than or *equal* sign, *every harmonic function is subharmonic*. The following examples can be used to generate an infinite number of new examples of subharmonic functions.

(8.5) Example.

   (a)   If Ψ is harmonic on U, then $g = |\Psi|$ is subharmonic on U.

   (b)   If $g_1, \ldots, g_m$ are subharmonic on U, then $g = a_1 g_1 + \cdots + a_m g_m$ is subharmonic whenever $a_1, \ldots, a_m$ are positive numbers.

(c)   If g_1, g_2 are subharmonic in U, let $g = \max[g_1, g_2]$ be defined by $g(x, y) = \max[g_1(x, y), g_2(x, y)]$. Then g is subharmonic in U. For, because $\max[g_1, g_2] = \frac{1}{2}[g_1 + g_2 + |g_1 - g_2|]$, we see that g is continuous. Moreover, for r sufficiently small

$$g_j(a, b) \leq \frac{1}{2\pi} \int_{-\pi}^{\pi} g_j(a + r \cos \alpha, b + r \sin \alpha) \, d\alpha$$

$$\leq \frac{1}{2\pi} \int_{-\pi}^{\pi} g(a + r \cos \alpha, b + r \sin \alpha) \, d\alpha$$

where $j = 1, 2$. Hence

$$g(a, b) \leq \frac{1}{2\pi} \int_{-\pi}^{\pi} g(a + r \cos \alpha, b + r \sin \alpha) \, d\alpha$$

which shows that $g = \max[g_1, g_2]$ is subharmonic.

(d)   By induction from (c), we see that if $g_1, \ldots, g_n$ are subharmonic on U, then $g = \max[g_1, \ldots, g_n]$ is subharmonic on U.

Subharmonic functions satisfy a maximum principle on connected open sets.

(8.6) Theorem: Maximum Principle.   Let U be a connected open set with compact closure $\bar{U}$, and let g be subharmonic on U and continuous on $\bar{U}$. If g is not a constant function, then g attains its maximum on $\bar{U}$ only on the boundary ∂U. [*Remark*: Note the absence of any mention of minimum.]

Proof.   Suppose g attains its maximum M on $\bar{U}$ at a point (a, b) in U. Let $\mathcal{M} = \{(x, y): \quad (x, y)$ is in U and $g(x, y) = M\}$. Since (a, b) is in $\mathcal{M}$, we know that $\mathcal{M}$ is nonempty. We now prove that $\mathcal{M}$ is open. Let (c, d) be a point in $\mathcal{M}$ and for $r < R$ suppose that

$$M = g(c, d) \leq \frac{1}{2\pi} \int_{-\pi}^{\pi} g(c + r \cos \alpha, d + r \sin \alpha) \, d\alpha$$

Suppose $g(c + r \cos \beta, d + r \sin \beta) < M$ for some angle β. Because g is continuous, we would then have $g(c + r \cos \alpha, d + r \sin \alpha) < M$ for $|\alpha - \beta| < \delta$ provided $\delta > 0$ is chosen sufficiently small. But, since $g \leq M$ on $\bar{U}$, we would then have

$$M \leq \frac{1}{2\pi} \int_{-\pi}^{\pi} g(c + r \cos \alpha, d + r \sin \alpha) \, d\alpha$$

$$< \frac{1}{2\pi} \int_{|\alpha - \beta| \geq \delta} M \, d\alpha + \frac{1}{2\pi} \int_{|\alpha - \beta| > \delta} M \, d\alpha = M$$

This contradiction proves that $g = M$ on every circle $\partial B_r(c, d)$ for all $r < R$. Thus, $g = M$ on $B_R(c, d)$ which proves that $\mathcal{M}$ is open. Since g is continuous and U is connected, we conclude [as in the proof of (8.2)] that $g = M$ on all of $\bar{U}$.   ∎

The following corollary is an immediate consequence of Theorem (8.6).

(8.7) Corollary.   Under the same hypotheses as (8.6), we have $\max_{\bar{U}} g = \max_{\partial U} g$.

We conclude this brief subsection with the following lemma, which will prove useful in Perron's solution of Dirichlet's problem.

(8.8) Lemma.   Let U be an open set and g a subharmonic function on U. If the open disk $B = B_R(a, b)$ has its closure $\bar{B}$ contained in U, then the function $[g]_B$ defined by

$$[g]_B(x, y) = \begin{cases} g(x, y) & \text{if } (x, y) \text{ is in } U \text{ but not in } B \\ \text{P.I.}[g](x, y) & \text{if } (x, y) \text{ is in } B \end{cases}$$

is subharmonic on U and satisfies $g(x, y) \le [g]_B(x, y)$ for all (x, y) in U. (*Remark*: The function $[g]_B$ is obtained by replacing g on the disk B by P.I.$[g]$.)

Proof.   Since $g(x, y) = [g]_B(x, y)$ for (x, y) in U but not in B, we have $g(x, y) \le [g]_B(x, y)$ for all such points (x, y). Since g is subharmonic on U, it is also subharmonic on B; moreover, so is the harmonic function $-\text{P.I.}[g]$. Therefore, $g - \text{P.I.}[g]$ is subharmonic on B. It follows that $g - [g]_B$ is subharmonic on B, continuous on $\bar{B}$, and equals 0 on ∂B. Hence, by Corollary (8.7), we have $g - [g]_B \le 0$ on B, which along with our initial remarks proves that $g \le [g]_B$ on all of U.

Suppose that (c, d) is in U but not in B. Since $g \le [g]_B$ we have, due to the subharmonicity of g, that

$$[g]_B(c, d) = g(c, d) \le \frac{1}{2\pi} \int_{-\pi}^{\pi} g(c + r \cos \alpha, d + r \sin \alpha) \, d\alpha$$

$$\le \frac{1}{2\pi} \int_{-\pi}^{\pi} [g]_B(c + r \cos \alpha, d + r \sin \alpha) \, d\alpha$$

for all radii r that are sufficiently small. Furthermore, since $[g]_B = \text{P.I.}[g]$ in the disk B and P.I.$[g]$ is harmonic, we know that $[g]_B$ must be subharmonic on B. Therefore, because of these last two results, we conclude that $[g]_B$ is subharmonic on U. ∎

The reader might notice that in the proof above we found that $g \le \text{P.I.}[g]$ on every disk $B_R(a, b)$ for which $\bar{B}_R(a, b)$ is contained in U. That inequality explains why g is called *sub*harmonic.

*Existence of Solutions to Dirichlet's Problem

We are now in a position to discuss Perron's proof that solutions to Dirichlet's problem (3.1) exist. *Throughout this subsection we shall assume that U is a connected open set with compact closure $\bar{U}$.* Let S_f denote the collection of all functions g that are continuous on $\bar{U}$, subharmonic on U, and that satisfy $g \le f$ on ∂U (where f is the given continuous function on ∂U in Dirichlet's problem). Let m equal the minimum and M equal the maximum of the continuous function f over the compact set ∂U. Note that the collection S_f is not empty, because the constant function m belongs to it. Moreover, if g belongs to S_f, then $g \leq M$ by Corollary (8.7). Further, if $\bar{B} = \bar{B}_r(a, b)$ is contained in U, then $[g]_B$ belongs to S_f.[13]

Define Ψ_f by

$$\Psi_f(x, y) = \underset{g \text{ in } S_f}{\text{supremum}} \, g(x, y)$$

We will prove that Ψ_f solves Dirichlet's problem for U with boundary function f, under certain conditions that we shall explain below.

First, we show that Ψ_f is a continuous function on U. Choose a point (a, b) in U and a closed disk $\bar{B} = \bar{B}_R(a, b)$ contained in U. Let $\{(a_n, b_n)\}_{n=0}^{\infty}$ be some sequence of points in $B_R(a, b)$. For each point (a_n, b_n) there exists a sequence $\{g_{mn}\}_{m=0}^{\infty}$ in S_f such that $g_{mn}(a_n, b_n)$ increases up to $\Psi_f(a_n, b_n)$ as m tends to $+\infty$. If we put $h_m = \max[g_{m0}, g_{m1}, \ldots, g_{mm}]$, then $h_m(a_j, b_j)$ also increases up to $\Psi_f(a_j, b_j)$ as m tends to $+\infty$ for each fixed j. Moreover, each function h_m is in S_f. Define $\{w_m\}_{m=0}^{\infty}$ as follows:

$$w_0 = [h_0]_B \qquad w_1 = [\max(h_1, w_0)]_B \cdots w_m = [\max(h_m, w_{m-1})]_B \cdots$$

using the notation of Lemma (8.8). By induction, each function $\max(h_m, w_{m-1})$ is in S_f, hence so is each function w_m. Moreover, each function w_m is harmonic in the disk $B = B_R(a, b)$ and $w_m(a_n, b_n)$ increases up to $\Psi_f(a_n, b_n)$ as m tends to $+\infty$, for every n. Since w_m belongs to S_f for all m, we know that $w_m \leq M$ for each m. Therefore, Theorem (8.3) implies that $\{w_m\}$ converges to a harmonic function Ψ on the disk $B_R(a, b)$. Moreover, $\Psi_f(a_n, b_n) = \Psi(a_n, b_n)$ for all n. Suppose that $(a_0, b_0) = (c, d)$ and $\lim_{n \to \infty}(a_n, b_n) = (c, d)$. Then the continuity of Ψ implies that

$$\lim_{n \to \infty} \Psi(a_n, b_n) = \Psi(c, d)$$

Hence, because $\Psi_f = \Psi$ on $\{(a_n, b_n)\}$, we have that $\lim_{n \to \infty} \Psi_f(a_n, b_n) = \Psi_f(c, d)$ for every point (c, d) in $B_R(a, b)$. Thus Ψ_f is continuous on $B_R(a, b)$, hence on all of U.

Second, we show that Ψ_f is harmonic on U. Let the sequence $\{(a_n, b_n)\}$, considered above, consist of the countably infinite set of points in $B_R(a, b)$ that have rational coordinates. Then $\Psi_f(a_n, b_n) = \Psi(a_n, b_n)$ for all such points. Since every point (c, d) in $B_R(a, b)$ can be approximated arbitrarily closely by points from the sequence $\{(a_n, b_n)\}$, we conclude from the equality of the continuous functions Ψ and Ψ_f that $\Psi_f(c, d) = \Psi(c, d)$ for all points (c, d) in U. Thus Ψ_f equals the harmonic function Ψ in $B_R(a, b)$. Since the choice of (a, b) was arbitrary, we have that Ψ_f is harmonic in U.

It should be noted that since $g \leq f$ on ∂U for all functions g in S_f we must have $\Psi_f \leq f$ on ∂U. Our next task is to show that $\Psi_f = f$ on ∂U and that Ψ_f is continuous on $\bar{U}$ (as well as U). To do that we need to introduce the concept of a *barrier* at a point (a, b) on the boundary ∂U.

---

13. By Lemma (8.8); see Exercise (8.23). Exercise (8.24) will also be used below.

(8.9) Definition. A point (a, b) on ∂U is said to have a *barrier if* there exists a function h which is continuous on $\bar{U}$, subharmonic on U, and satisfies $h(a, b) = 0$ and $h(x, y) < 0$ for all points (x, y) on ∂U such that $(x, y) \neq (a, b)$. We call the function h a *barrier function.* ∎

Suppose that the point (a, b) on ∂U has a barrier function h. We will show that $\lim_{(x, y) \to (a, b)} \Psi_f(x, y) = f(a, b)$. To do that we first show that for (x, y) in $\bar{U}$

(8.10)
$$\liminf_{(x, y) \to (a, b)} \Psi_f(x, y) \geq f(a, b)$$

Consider the function g defined by $g(x, y) = f(a, b) - \delta + Kh(x, y)$ where $\delta > 0$ and $K > 0$. If K is large enough, then we will have $[f(a, b) - \delta + Kh(x, y)] \leq f(x, y)$ for all (x, y) on ∂U, from which it follows that g belongs to S_f. Consequently, for all (x, y) in $\bar{U}$

$$f(a, b) - \delta + Kh(x, y) \leq \Psi_f(x, y)$$

Therefore,

$$\liminf_{(x, y) \to (a, b)} [f(a, b) - \delta + Kh(x, y)] = f(a, b) - \delta$$
$$\leq \liminf_{(x, y) \to (a, b)} \Psi_f(x, y)$$

Letting δ pass to zero we obtain (8.10).

We now prove that for (x, y) in $\bar{U}$

(8.11)
$$\limsup_{(x, y) \to (a, b)} \Psi_f(x, y) \leq f(a, b)$$

Consider the function

$$-\Psi_{-f} = -\sup_{g \text{ in } S_{-f}} g$$

Writing $G = -g$ we have

$$-\Psi_{-f} = \inf_{-G \text{ in } S_{-f}} G$$

Thus, if $g^{\#}$ is in S_f we have $g^{\#} \leq f$ and $-G \leq -f$ on ∂U, hence $g^{\#} - G \leq 0$ on ∂U. Therefore, by Corollary (8.7), we have $g^{\#} - G \leq 0$ on $\bar{U}$. We conclude from the last inequality that

$$\Psi_f = \sup_{g^{\#} \text{ in } S_f} g^{\#} \leq \inf_{-G \text{ in } S_{-f}} G = -\Psi_{-f}$$

on $\bar{U}$. Hence, applying (8.10) to Ψ_{-f} (instead of Ψ_f) we obtain for (x, y) in $\bar{U}$

$$\limsup_{(x, y) \to (a, b)} \Psi_f(x, y) \leq \limsup_{(x, y) \to (a, b)} [-\Psi_{-f}(x, y)]$$
$$= -\liminf_{(x, y) \to (a, b)} \Psi_{-f}(x, y)$$
$$\leq f(a, b)$$

Thus (8.11) holds, which in combination with (8.10) yields

(8.12)
$$\lim_{(x, y)\to(a, b)} \Psi_f(x, y) = f(a, b)$$

at each point (a, b) on ∂U for which a barrier exists. From this last result the following theorem is an immediate consequence.

(8.13) Theorem. Suppose that U is an open connected set with compact closure $\bar{U}$ and that every point of ∂U has a barrier. Then, given a continuous function f on ∂U, Dirichlet's problem is solved by the function Ψ_f.

For example, Dirichlet's problem is solvable for an open rectangle, or an open annulus, or an open rectangle with a closed disk removed, or many other regions, due to the following theorem.

(8.14) Theorem. Let U be an open connected set with compact closure $\bar{U}$. Suppose that for the point (a, b) on ∂U there exists a disk $B_R(c, d)$ such that $\bar{B}_R(c, d)$ and $\bar{U}$ have just the point (a, b) in common. Then (a, b) has a barrier.

Proof. Suppose that $(a, b) = (c + R \cos \phi, \ d + R \sin \phi)$ for $|\phi| \leq \pi$. Define h to be Poisson's kernel for the disk $B_R(c, d)$ evaluated at $\pi + \phi$. Thus $h(x, y) = p(x - c, y - d; \pi + \phi)$. The reader may check that h is a barrier function. ∎

Most open sets that occur in applications do satisfy the conditions demanded by Theorems (8.13) and (8.14); for further discussion, see the References for this chapter.

Exercises

(8.15) Provide counterexamples to Theorems (8.2) and (8.3) if U is an open set which is not connected.

(8.16) Suppose that Ψ is a nonnegative harmonic function on a connected open set U. Prove that if Ψ is not constantly zero, then Ψ is always positive in value. Furthermore, prove that if $\Psi_1 \geq \Psi_2$ on U and $\Psi_1 \neq \Psi_2$, then $\Psi_1 > \Psi_2$ on U.

*★***(8.17)** Prove that a nonconstant harmonic function Ψ has no local maxima or local minima on a connected open set.

(8.18) Verify Examples (8.5a, b, and d).

(8.19) Using (8.5), prove that the following functions are subharmonic

    (a) $g(x, y) = |2x - 3y| + \frac{1}{2}|x + 4y|$
    (b) $h(x, y) = \max[|x^2 - y^2|, \ |x - y|, \ |2x + 6y| + |3x - 4y|]$
    (c) $g(x, y) = |p(x, y; \frac{1}{2}\pi)| + \frac{1}{4}|p(x - 1, y - 2; \frac{1}{4}\pi)|$ where p is the Poisson kernel for the disk $B_8(0, 0)$.

*★***(8.20)** Prove the following: A continuous function g on an open set U is

harmonic if and only if both g and $-g$ are subharmonic. [*Hint*: Use Lemma (8.8).]

⋆(8.21)   Use (8.20) to prove the following: A continuous function g on an open set U is harmonic if and only if for each point (a, b) in U there exists a radius R [dependent upon (a, b)] such that for all radii $r < R$

$$g(a, b) = \frac{1}{2\pi} \int_{-\pi}^{\pi} g(a + r \cos \alpha, b + r \sin \alpha)\, d\alpha$$

[*Remark*: The result above extends Theorem (5.12).]

⋆(8.22)   Use (8.21) to give a different proof of Theorem (5.9).

(8.23)   Prove that if g belongs to S_f, then $[g]_B$ belongs to S_f [where $\bar{B} = \bar{B}_R(a, b)$ is a closed disk contained in U].

(8.24)   Prove that if $g_1, \ldots, g_m$ belong to S_f, then $\max[g_1, \ldots, g_m]$ belongs to S_f.

(8.25)   Prove that if $g_1, \ldots, g_m$ belong to S_f, then $\sum_{j=1}^{m} a_j g_j$ belongs to S_f provided that each constant $a_j \geq 0$ and $\sum_{j=1}^{m} a_j = 1$.

(8.26)   Apply Theorems (8.13) and (8.14) to prove that Dirichlet's problem is solvable for each of the following open sets in $\mathbb{R}^2$

(a)   an open annulus
(b)   an open rectangle $R = \{(x, y): a < x < b$ and $c < y < d\}$
(c)   an open rectangle with one or more closed disks removed
(d)   an open disk with one or more closed disks removed.

(8.27)   Provide a simple example of a connected open set U with compact closure $\bar{U}$ to which the hypothesis of (8.14) does not apply.

(8.28)   Let $U = \{(x, y):\quad 0 < x^2 + y^2 < 1\}$. Show that there is no function Ψ that is continuous on $\bar{U}$, harmonic in U, and equals 0 identically on $\partial B_1(0, 0)$ and 1 at the point $(0, 0)$. [*Remark*: This exercise shows the impossibility of solving Dirichlet's problem for some open sets that are connected and have compact closure.]

(8.29)   Let U be as in (8.28). Show that Dirichlet's problem is solvable if the function f on ∂U satisfies

$$f(0, 0) = \frac{1}{2\pi} \int_{-\pi}^{\pi} f(\cos \alpha, \sin \alpha)\, d\alpha$$

§9. Miscellaneous Exercises

Harmonic Functions on Annuli

(9.1)   Let $a < b$ be two positive constants. Find a series solution to the following problem

$$\Delta \Psi = 0 \qquad (a^2 < x^2 + y^2 < b^2)$$

(9.1') $\Psi(x, y) = f(x, y) \qquad (x^2 + y^2 = a^2)$
$\Psi(x, y) = g(x, y) \qquad (x^2 + y^2 = b^2)$

(9.2) Using the series found in (9.1), solve the following problems

(a) $\Delta\Psi = 0$ $(4 < x^2 + y^2 < 9)$
            $\Psi(x, y) = x$ $(x^2 + y^2 = 4)$ $\Psi(x, y) = y$ $(x^2 + y^2 = 9)$

(b) $\Delta\Psi = 0$ $(1 < x^2 + y^2 < 25)$
            $\Psi(x, y) = |x|$ $(x^2 + y^2 = 1)$ $\Psi(x, y) = x^2$ $(x^2 + y^2 = 25)$

(c) $\Delta\Psi = 0$ $(4 < x^2 + y^2 < 16)$

$$\Psi(x, y) = \begin{cases} 1 & \text{if } x < 0 \\ 0 & \text{if } x > 0 \end{cases} \quad (x^2 + y^2 = 4) \quad \Psi(x, y) = 1 \quad (x^2 + y^2 = 16)$$

(9.3) Discuss Dirichlet's problem for (9.1').

****(9.4)** Discuss the existence of radial limits at the two circular boundaries in problem (9.1') if f or g is not continuous.

Harmonic Functions Outside a Disk

Consider the problem

(9.5)
$$\Delta\Psi = 0 \quad (x^2 + y^2 > R^2)$$
$$\Psi(x, y) = f(x, y) \quad (x^2 + y^2 = R^2)$$

(9.6) If we require a solution Ψ of (9.5) to be bounded for all (x, y) such that $x^2 + y^2 > R^2$, show that Ψ can be expressed as a series

$$\Psi(x, y) = \sum_{n=0}^{\infty} [A_n \cos n\phi + B_n \sin n\phi](R/r)^n \quad (x = r \cos \phi \quad y = r \sin \phi)$$

for appropriate constants A_n and B_n.

(9.7) Solve (9.5) given the following functions f and radii R.

(a) $f(x, y) = x^2$ $R = 3$

(b) $f(x, y) = \begin{cases} 1 & \text{for } x > 0 \\ -1 & \text{for } x < 0 \end{cases}$ $R = 7$

(9.8) Express the series in (9.6) in the form of a Poisson integral, both in polar coordinate and Cartesian coordinate versions.

(9.9) Prove that Dirichlet's problem for $U = \{(x, y): x^2 + y^2 > R^2\}$ is solvable, and uniquely solvable if we require $\lim_{r \to \infty} \Psi(r, \phi) = 0$ uniformly.

Some Complex Methods

Methods of complex function theory are often of great utility in harmonic function theory in $\mathbb{R}^2$; we give just a few simple examples here. Let $z = x + iy$ denote a complex variable. We shall assume the validity of the following power series expansions [see, e.g., Churchill and Brown (1984) or

Copson (1978), Chapter 3]

$$\log(1 + z) = z - z^2/2 + z^3/3 - \cdots + (-1)^n z^{n+1}/(n+1) + \cdots \qquad (|z| < 1)$$

$$-\log(1 - z) = z + z^2/2 + z^3/3 + \cdots + z^n/n + \cdots \qquad (|z| < 1)$$

$$\tan^{-1} z = z - z^3/3 + z^5/5 + \cdots + (-1)^n z^{2n+1}/(2n+1) + \cdots \qquad (|z| < 1)$$

Moreover, if $w = u + iv$ is complex, then

$$\log w = \tfrac{1}{2} \log(u^2 + v^2) + i \tan^{-1}(u/v)$$

$$\tan^{-1} w = \tfrac{1}{2} \tan^{-1}\left[\frac{2u}{1 - (u^2 + v^2)}\right] + \tfrac{1}{4}i \log\left[\frac{(1+v)^2 + u^2}{(1-v)^2 + u^2}\right]$$

Using the results above, the following seven exercises can be solved.

(9.10)   Consider the series

$$\Psi = \frac{4V}{\pi} \sum_{k=0}^{\infty} \frac{\sin[(2k+1)\pi x/L]}{2k+1} e^{-(2k+1)\pi y/L}$$

which solves

$$\Delta\Psi = 0 \qquad [0 < x < L,\ 0 < y < +\infty] \qquad \Psi(x, 0) = V \qquad \text{a constant}$$

Prove that for $0 < x < L$ and $0 < y < +\infty$

$$\Psi(x, y) = \frac{2V}{\pi} \tan^{-1}\left[\frac{\sin(\pi x/L)}{\sinh(\pi y/L)}\right]$$

which is a closed form solution. [*Hint*: Consider $\mathrm{Re}[\tan^{-1}(-iz)]$ for $z = e^{i(\pi x/L)} e^{-\pi y/L}$.]

(9.11)   Find a *closed form* solution to

$$\Delta\Psi = 0 \qquad [0 < x < \pi,\ 0 < y < +\infty] \qquad \Psi(x, 0) = (\pi - x)/2 \qquad \text{for } 0 < x < \pi$$

(9.12)   Let the function f on the circle $\partial B_R(0, 0)$ be defined by $f(x, y) = A$ for $x > 0$ and $f(x, y) = B$ for $x < 0$, where A and B are constants. Show that there is a unique function Ψ which equals f on $B_R(0, 0)$ and is harmonic in the disk $B_R(0, 0)$. Find a closed form expression for Ψ

$$\left(Answer:\ \Psi(x, y) = \tfrac{1}{2}[A + B] + \frac{1}{\pi}(A - B)\tan^{-1}\left[\frac{2Rx}{R^2 - (x^2 + y^2)}\right]\right)$$

(9.13)   A variant of problem (9.12) involves letting f take on the values $\pm A$ alternately on the four quarters of $\partial B_R(0, 0)$ that are marked off by the x–y axes, starting with $+A$ in the first quadrant. Show that Ψ is described inside the disk $B_R(0, 0)$ by the series

$$\Psi(r \cos \phi,\ r \sin \phi) = \frac{4A}{\pi} \sum_{k=0}^{\infty} \left(\frac{r}{R}\right)^{4k+2} \frac{\sin(4k+2)\phi}{2k+1} \qquad (0 \leq r < R)$$

Sum the series above and show that for $0 \le x^2 + y^2 < R^2$

$$\Psi(x, y) = \frac{2A}{\pi} \tan^{-1} \left[\frac{4R^2 xy}{R^4 - (x^2 + y^2)^2} \right]$$

*(9.14)   Find the closed form sum of $\sum_{n=1}^{\infty} n^{-1} \cos nx$ for $0 < x < 2\pi$.

Mean Convergence of Poisson Integrals

(9.15)   Prove that for a continuous function F of period 2π

$$\lim_{r \to R-} {}_r P * F(\phi) = F(\phi)$$

holds uniformly for all ϕ values. [See (2.19).]

(9.16)   Using (9.15), prove that for F continuous over $[-\pi, \pi]$

$$\lim_{r \to R-} \int_{-\pi}^{\pi} |F(\phi) - {}_r P * F(\phi)| \, d\phi = 0$$

*(9.17)   Prove that for a piecewise continuous function F on the interval $(-\pi, \pi)$

$$\lim_{r \to R-} \int_{-\pi}^{\pi} |F(\phi) - {}_r P * F(\phi)| \, d\phi = 0$$

[*Hint*: First approximate F by a continuous function G such that

$$\int_{-\pi}^{\pi} |F(\phi) - G(\phi)| \, d\phi \text{ is small.}]$$

**(9.18)   Prove that for F a piecewise continuous function on the interval $(-\pi, \pi)$

$$\lim_{r \to R-} \|F - {}_r P * F\|_2 = 0$$

References

For further discussion of harmonic functions the reader might consult Kellogg (1928), Copson (1975), Helms (1969), or Doob (1984). For the complex variable approach, see Nevanlinna and Paatero (1982) or Conway (1978). The heat equation and theta function are discussed in Widder (1975), Bellman (1961), or Doob (1984). The theta function has been extensively tabulated; a table of values can be found in Abramowitz and Stegun (1972).

5

Multiple Fourier Series

In this chapter we will discuss multiple Fourier series, concentrating on double Fourier series. Double Fourier series are Fourier series that involve two real variables. First, we will treat the real form of those series, then we will examine the more flexible complex form of double Fourier series. It is that complex form that allows for the most convenient generalization to multiple Fourier series of three, or more, variables. The problems of pointwise and mean square convergence of multiple Fourier series will be discussed for only the simplest cases; pointwise convergence is significantly more difficult to treat rigorously for multiple Fourier series than for single variable Fourier series. We shall discuss some applications of multiple Fourier series to physical problems of vibrations and heat conduction. The treatment of heat conduction allows us to prove completeness (in the mean square sense) of the multiple trigonometric systems.

§1. Double Fourier Series, Real Form

Double Fourier series may be considered as a special case of orthogonal series over a rectangle $[a, b] \times [c, d] = \{(x, y): a \le x \le b \text{ and } c \le y \le d\}$. The definition of such series is formally identical to that given for orthogonal series on an interval (see §1, Chapter 2).

(1.1) Definition. An *orthogonal system (or set) of functions* over $[a, b] \times [c, d]$ is a set of real valued functions $\{g_n(x, y)\}_{n=0}^{\infty}$ such that

$$\int_c^d \int_a^b g_m(x, y)g_n(x, y)\, dx\, dy = 0 \qquad \text{if } m \ne n$$

$$\int_c^d \int_a^b g_n^2(x, y)\, dx\, dy > 0 \text{ for each } n$$

If $\int_c^d \int_a^b g_n^2(x, y) \, dx \, dy = 1$ for each n, then $\{g_n(x, y)\}_{n=0}^\infty$ is called an *orthonormal* system of functions over $[a, b] \times [c, d]$. ∎

If we let $\|g_n\|_2^2$ stand for $\int_c^d \int_a^b g_n^2(x, y) \, dx \, dy$, then the *Fourier series* of a (piecewise) continuous function f relative to $\{g_n\}$ is defined by the formal correspondence

$$f \sim \sum_{n=0}^\infty c_n g_n(x, y) \qquad c_n = \frac{1}{\|g_n\|_2^2} \int_c^d \int_a^b f(x, y) g_n(x, y) \, dx \, dy$$

With the definitions above, the results in Chapter 2 remain valid. In particular, Bessel's inequality, $\lim_{n \to \infty} c_n = 0$, the least squares theorem, and the definition of completeness all remain valid for $\{g_n(x, y)\}$ with the same proofs as those given in Chapter 2.

The following theorem allows for a great many orthogonal sets of functions. We leave its proof to the reader [see Exercise (1.8)].

(1.2) Theorem. Suppose that $\{g_j(x)\}_{j=0}^\infty$ and $\{h_k(y)\}_{k=0}^\infty$ are orthogonal systems of continuous functions over the intervals $[a, b]$ and $[c, d]$, respectively. Then $\{g_j(x) h_k(y)\}_{j,k=0}^\infty$ is an orthogonal set over $[a, b] \times [c, d]$. ∎

For example,

(1.3) $\{\cos mx \cos ny \quad \sin mx \cos ny \quad \cos mx \sin ny \quad \sin mx \sin ny\}_{m,n=0}^\infty$

is an orthogonal set[1] over $[-\pi, \pi] \times [-\pi, \pi]$. We find that

$$\|\cos mx \cos ny\|_2^2 = \pi^2 \qquad \|\cos ny\|_2^2 = \|\cos mx\|_2^2 = 2\pi^2 \qquad \|1\|_2^2 = 4\pi^2$$

$$\|\sin mx \cos ny\|_2^2 = \pi^2 \qquad \|\sin mx\|_2^2 = 2\pi^2$$

$$\|\cos mx \sin ny\|_2^2 = \pi^2 \qquad \|\sin ny\|_2^2 = 2\pi^2$$

$$\|\sin mx \sin ny\|_2^2 = \pi^2$$

Hence, adopting the symbol

$$\lambda_{mn} = \begin{cases} \frac{1}{4} & \text{if } m = 0 \quad n = 0 \\ \frac{1}{2} & \text{if } m = 0 \quad n \neq 0 \quad \text{or } m \neq 0 \quad n = 0 \\ 1 & \text{if } m \neq 0 \quad n \neq 0 \end{cases}$$

we obtain the following double Fourier series expansion

$$f \sim \sum_{m,n=0}^\infty \lambda_{mn}[A_{mn} \cos mx \cos ny + B_{mn} \sin mx \cos ny$$

$$+ C_{mn} \cos mx \sin ny + D_{mn} \sin mx \sin ny]$$

---

1. Here we adopt the convention that $\sin 0x \cos ny$, $\cos mx \sin 0y$, $\sin 0x \sin ny$, and $\sin mx \sin 0y$ are *not* included in the set given in (1.3), since they are identically zero.

where

$$A_{mn} = \frac{1}{\pi^2} \int_{-\pi}^{\pi} \int_{-\pi}^{\pi} f(x, y) \cos mx \cos ny \, dx \, dy$$

$$B_{mn} = \frac{1}{\pi^2} \int_{-\pi}^{\pi} \int_{-\pi}^{\pi} f(x, y) \sin mx \cos ny \, dx \, dy$$

$$C_{mn} = \frac{1}{\pi^2} \int_{-\pi}^{\pi} \int_{-\pi}^{\pi} f(x, y) \cos mx \sin ny \, dx \, dx$$

$$D_{mn} = \frac{1}{\pi^2} \int_{-\pi}^{\pi} \int_{-\pi}^{\pi} f(x, y) \sin mx \sin ny \, dx \, dy$$

The symbol λ_{mn} is needed in order to reconcile the differing values of $\|\cdot\|_2^2$. Perhaps some examples will help to clarify matters.

(1.4) Example. Let $f(x, y) = xy^2$ on $[-\pi, \pi] \times [-\pi, \pi]$. Expand f in a double Fourier series.

Solution. Since x is odd and y^2 is even, it follows that $A_{mn} = 0$, $A_{0n} = 0$, $A_{m0} = 0$, $C_{mn} = 0$, $C_{0n} = 0$, and $D_{mn} = 0$. For instance,

$$A_{mn} = \frac{1}{\pi^2} \int_{-\pi}^{\pi} \int_{-\pi}^{\pi} xy^2 \cos mx \cos ny \, dx \, dy$$

$$= \frac{1}{\pi^2} \int_{-\pi}^{\pi} x \cos mx \, dx \int_{-\pi}^{\pi} y^2 \cos ny \, dy = 0 \int_{-\pi}^{\pi} y^2 \cos ny \, dy = 0$$

While for B_{m0} and B_{mn} we have

$$B_{m0} = \frac{1}{\pi^2} \int_{-\pi}^{\pi} x \sin mx \, dx \int_{-\pi}^{\pi} y^2 \, dy = \frac{4\pi^2(-1)^{m+1}}{3m}$$

$$B_{mn} = \frac{1}{\pi^2} \int_{-\pi}^{\pi} x \sin mx \, dx \int_{-\pi}^{\pi} y^2 \cos ny \, dy = \frac{8(-1)^{m+n+1}}{mn^2}$$

Thus, on $[-\pi, \pi] \times [-\pi, \pi]$

$$xy^2 \sim \frac{2\pi^2}{3} \sum_{m=1}^{\infty} \frac{(-1)^{m+1}}{m} \sin mx - 8 \sum_{m,n=1}^{\infty} \frac{(-1)^{m+n}}{mn^2} \sin mx \cos ny$$

Besides the rectangle $[-\pi, \pi] \times [-\pi, \pi]$, we could just as well have used the rectangle $[0, 2\pi] \times [0, 2\pi]$, as shown by our next example.

(1.5) Example. Let $f(x, y) = xy$ on $[0, 2\pi] \times [0, 2\pi]$. Expand f in a double Fourier series.

Solution. We have

$$C_{mn} = \frac{1}{\pi^2} \int_{0}^{2\pi} x \cos mx \, dx \int_{0}^{2\pi} y \sin ny \, dy = 0 \int_{0}^{2\pi} y \sin ny \, dy = 0$$

and similarly $A_{mn} = A_{m0} = A_{0n} = 0$, $A_{00} = 4\pi^2$, $B_{mn} = 0$, $B_{m0} = -4\pi/m$,

$C_{0n} = -4\pi/n$, $D_{mn} = 4/mn$. Hence on $[0, 2\pi] \times [0, 2\pi]$

$$xy \sim \pi^2 - 2\pi \sum_{m=1}^{\infty} \frac{\sin mx}{m} - 2\pi \sum_{n=1}^{\infty} \frac{\sin ny}{n} + 4 \sum_{m,n=1}^{\infty} \frac{\sin mx \sin ny}{mn}$$

Instead of using rectangles with sides of length 2π, we can use rectangles of length a along the x axis and length b along the y axis. Rather than stating general formulas, we will simply consider the following example. [See Exercise (1.9).]

(1.6) Example. Let $f(x, y) = xy$ on the rectangle $[-2, 2] \times [-3, 3]$. Expand f in a double Fourier series.

Solution. Since f is odd in each variable we have

$$D_{mn} = \frac{1}{2 \cdot 3} \int_{-2}^{2} x \sin \frac{m\pi x}{2} dx \int_{-3}^{3} y \sin \frac{n\pi y}{3} dy$$

$$= \frac{2}{3} \int_{0}^{2} x \sin \frac{m\pi x}{2} dx \int_{0}^{3} y \sin \frac{n\pi y}{3} dy$$

$$= \frac{24(-1)^{m+n}}{m^2 n^2 \pi^2}$$

and all other coefficients are zero. Thus, on $[-2, 2] \times [-3, 3]$

$$xy \sim \frac{24}{\pi^2} \sum_{m,n=1}^{\infty} \frac{(-1)^{m+n}}{m^2 n^2} \sin mx \sin ny$$

This last example is a special case of a *double sine series*, which occurs often in applications.

(1.7) Example. Let f be an odd function in each variable over $[-a, a] \times [-b, b]$. That is,

$$f(-x, y) = -f(x, y) \qquad f(x, -y) = -f(x, y)$$

Therefore

$$f \sim \sum_{m,n=1}^{\infty} D_{mn} \sin \frac{n\pi x}{a} \sin \frac{n\pi y}{b}$$

where

$$D_{mn} = \frac{4}{ab} \int_{0}^{b} \int_{0}^{a} f(x, y) \sin \frac{m\pi x}{a} \sin \frac{n\pi y}{b} dx\, dy$$

Exercises

(1.8) Prove Theorem (1.2) and check Example (1.7).

(1.9) Show   that $\{\cos(n\pi x/a)\cos(n\pi x/b), \sin(m\pi x/a)\cos(n\pi y/b),$

$\cos(m\pi x/a)\sin(n\pi y/b)$, $\sin(m\pi x/a)\sin(n\pi y/b)\}_{m,n=0}^{\infty}$ is an orthogonal set[2] over $[-a, a] \times [-b, b]$, and

$$f \sim \sum_{m,n=0}^{\infty} \lambda_{mn} \left[A_{mn} \cos\frac{m\pi x}{a} \cos\frac{n\pi y}{b} + B_{mn} \sin\frac{m\pi x}{a} \cos\frac{n\pi y}{b} \right.$$

$$\left. + C_{mn} \cos\frac{m\pi x}{a} \sin\frac{n\pi y}{b} + D_{mn} \sin\frac{m\pi x}{a} \sin\frac{n\pi y}{b} \right]$$

for appropriate constants A_{mn}, B_{mn}, C_{mn}, and D_{mn}.

(1.10) Expand the following functions in double Fourier series, using expansions appropriate for the given rectangle.

(a) $f(x, y) = x + y$ on $[-\pi, \pi] \times [-\pi, \pi]$
(b) $f(x, y) = x^3 y^3$ on $[-2, 2] \times [-5, 5]$
(c) $f(x, y) = x^2 y + xy^3$ on $[-2, 2] \times [-4, 4]$
(d) $f(x, y) = |\sin(x + y)|$ on $[-\pi, \pi] \times [-\pi, \pi]$
(e) $f(x, y) = |x| y^2 + x^3 y^3$ on $[-1, 1] \times [-1, 1]$

(1.11) Expand the following functions in double sine series over $(0, a) \times (0, b)$ using the formulas in (1.7). [Or, by generalizing (1.14), Chapter 2.]

(a) $f(x, y) = A$ for A a constant
(b) $f(x, y) = x^2 y^3$
(c) $f(x, y) = \begin{cases} A & \text{for } |x - \frac{1}{2}a| < \frac{1}{4}a \\ 0 & \text{for } |x - \frac{1}{2}a| > \frac{1}{4}a \end{cases}$ and $|y - \frac{1}{2}b| < \frac{1}{4}b$ or $|y - \frac{1}{2}b| > \frac{1}{4}b$

§2. Complex Form for Double Fourier Series, Triple Fourier Series

A function f on $[-\pi, \pi] \times [-\pi, \pi]$ has a *complex form* for its double Fourier series, which is defined by

(2.1)

$$f \sim \sum_{m,n=-\infty}^{+\infty} c_{mn} e^{i(mx+ny)}$$

$$c_{mn} = \frac{1}{4\pi^2} \int_{-\pi}^{\pi} \int_{-\pi}^{\pi} f(x, y) e^{-i(mx+ny)} \, dx \, dy$$

The reader should observe that the definition of c_{mn} is a generalization to two variables of the orthogonality of the complex exponentials [see (7.9) and (7.10), Chapter 1]. We will show that (2.1) defines a formally identical Fourier series to the one defined in the previous section. To that end, let S_{MN} stand for a partial sum of the series in (2.1) given by

$$S_{MN}(x, y) = \sum_{m=-M,n=-N}^{M,N} c_{mn} e^{i(mx+ny)}$$

---

2. Omitting $\sin(0\pi x/a)\cos(n\pi y/b)$, $\cos(m\pi x/a)\sin(0\pi y/b)$, $\sin(0\pi x/a)\sin(n\pi y/b)$, and $\sin(m\pi x/a)\sin(0\pi y/b)$.

We will prove that S_{MN} is identical to the partial sum

$$\sum_{m,n=0}^{M,N} \lambda_{mn}(A_{mn} \cos mx \cos ny + B_{mn} \sin mx \cos ny$$

$$+ C_{mn} \cos mx \sin ny + D_{mn} \sin mx \sin ny)$$

which will prove the desired equivalence of the two formal infinite Fourier series.

Since $e^{-i\theta} = \cos\theta - i\sin\theta$, it follows that

$$c_{mn} = \frac{1}{4\pi^2} \int_{-\pi}^{\pi} \int_{-\pi}^{\pi} f(x, y)[\cos(mx + ny) - i\sin(mx + ny)]\, dx\, dy$$

$$= \frac{1}{4\pi^2} \int_{-\pi}^{\pi} \int_{-\pi}^{\pi} f(x, y)[\cos mx \cos ny - \sin mx \sin ny$$

$$- i\sin mx \cos ny - i\cos mx \sin ny]\, dx\, dy$$

Using the definitions of A_{mn}, B_{mn}, C_{mn}, and D_{mn} from §1, the result above becomes [*Note*: $\cos(-\theta) = \cos\theta$ and $\sin(-\theta) = -\sin\theta$]

$$c_{mn} = \tfrac{1}{4}A_{mn} - \tfrac{1}{4}D_{mn} - i\tfrac{1}{4}B_{mn} - i\tfrac{1}{4}C_{mn}$$

(2.2)
$$c_{m,-n} = \tfrac{1}{4}A_{mn} + \tfrac{1}{4}D_{mn} - i\tfrac{1}{4}B_{mn} + i\tfrac{1}{4}C_{mn}$$

$$c_{-m,n} = \tfrac{1}{4}A_{mn} + \tfrac{1}{4}D_{mn} + i\tfrac{1}{4}B_{mn} - i\tfrac{1}{4}C_{mn}$$

$$c_{-m,-n} = \tfrac{1}{4}A_{mn} - \tfrac{1}{4}D_{mn} + i\tfrac{1}{4}B_{mn} + i\tfrac{1}{4}C_{mn}$$

where we assign B_{0n}, C_{m0}, D_{0n}, and D_{m0} the value of 0 in every case. Therefore, we have

$$S_{MN}(x, y) = \sum_{n=-M,n=-N}^{M,N} c_{mn}[\cos mx \cos ny - \sin mx \sin ny$$

$$+ i\sin mx \cos ny + i\cos mx \sin ny]$$

$$= c_{00} + \sum_{m=1}^{M} [c_{m0} + c_{-m0}]\cos mx + \sum_{m=1}^{M} i[c_{m0} - c_{-m0}]\sin mx$$

$$+ \sum_{n=1}^{N} [c_{0n} + c_{0,-n}]\cos ny + \sum_{n=1}^{N} i[c_{0n} - c_{0,-n}]\sin ny$$

$$+ \sum_{m,n=1}^{M,N} [c_{mn} + c_{m,-n} + c_{-m,n} + c_{-m,-n}]\cos mx \cos ny$$

$$+ \sum_{m,n=1}^{M,N} [c_{m,-n} + c_{-m,n} - c_{mn} - c_{-m,-n}]\sin mx \sin ny$$

$$+ \sum_{m,n=1}^{M,N} i[c_{mn} + c_{m,-n} - c_{-m,n} - c_{-m,-n}]\sin mx \cos ny$$

$$+ \sum_{m,n=1}^{M,N} i[c_{mn} + c_{-m,n} - c_{m,-n} - c_{-m,-n}]\cos mx \sin ny$$

From which we obtain [*Note*: $B_{0n} = C_{m0} = D_{0n} = D_{m0} = 0$]

$$S_{MN}(x, y) = \tfrac{1}{4}A_{00} + \sum_{m=1}^{M} \tfrac{1}{2}A_{m0} \cos mx + \sum_{m=1}^{M} \tfrac{1}{2}B_{m0} \sin mx + \sum_{n=1}^{N} \tfrac{1}{2}A_{0n} \cos ny$$

$$+ \sum_{n=1}^{N} \tfrac{1}{2}C_{0n} \sin ny$$

$$+ \sum_{m,n=1}^{M,N} A_{mn} \cos mx \cos ny + \sum_{m,n=1}^{M,N} B_{mn} \sin mx \cos ny$$

$$+ \sum_{m,n=1}^{M,N} C_{mn} \cos mx \sin ny + \sum_{m,n=1}^{M,N} D_{mn} \sin mx \sin ny$$

$$= \sum_{m,n=0}^{M,N} \lambda_{mn}[A_{mn} \cos mx \cos ny + B_{mn} \sin mx \cos ny$$

$$+ C_{mn} \cos mx \sin ny + D_{mn} \sin mx \sin ny]$$

Thus we have proved the identity of the complex and real forms for the partial sum S_{MN} for all nonnegative integers M and N. Because of this identity of partial sums, we say that the complex form of the Fourier series, shown in (2.1), is identical to the real form of the double Fourier series, discussed in §1. From now on, we will work almost exclusively with the complex form for a double Fourier series. Here are some examples of complex double Fourier series.

(2.3) Example. Let $f(x, y) = xy^2$ on $[-\pi, \pi] \times [-\pi, \pi]$. Expand f in a complex double Fourier series.

Solution. We have for $m \neq 0$ and $n \neq 0$

$$c_{mn} = \frac{1}{4\pi^2} \int_{-\pi}^{\pi} xe^{-imx} dx \int_{-\pi}^{\pi} y^2 e^{-iny} dy$$

$$= \frac{1}{4\pi^2} \left[-i \int_{-\pi}^{\pi} x \sin mx \, dx \right]\left[\int_{-\pi}^{\pi} y^2 \cos ny \, dy \right]$$

$$= \frac{2i(-1)^{m+n}}{mn^2}$$

and

$$c_{m0} = \frac{\pi^2 i(-1)^m}{3m} \qquad c_{0n} = c_{00} = 0$$

Thus, over $[-\pi, \pi] \times [-\pi, \pi]$

$$xy^2 \sim \frac{\pi^2 i}{3} \sum_{m=-\infty}^{+\infty}{}' (-1)^m \frac{e^{imx}}{m} + 2i \sum_{m,n=-\infty}^{+\infty}{}' (-1)^{m+n} \frac{e^{i(mx+ny)}}{mn^2}$$

where $\sum'$ denotes a sum in which all indices involving 0 are omitted.

Instead of using rectangles of length 2π, we can use rectangles of length a along the x axis and length b along the y axis.

(2.4) Example.   Let $f(x, y) = xy$ on $[-1, 1] \times [-3, 3]$. Expand f in a complex double Fourier series.

Solution.   *For $m \neq 0$ and $n \neq 0$*

$$c_{mn} = \frac{1}{2 \cdot 6} \int_{-1}^{1} xe^{-im\pi x} \, dx \int_{-3}^{3} ye^{-in\pi y/3} \, dy$$

$$= \frac{1}{2 \cdot 6} \left[-i \int_{-1}^{1} x \sin m\pi x \right] \left[-i \int_{-3}^{3} y \sin \frac{n\pi y}{3} \, dy \right]$$

$$= -\frac{3(-1)^{m+n}}{mn\pi^2}$$

while $c_{0n} = c_{m0} = 0$. Thus, over $[-1, 1] \times [-3, 3]$

$$xy \sim \sum_{m,n=-\infty}^{+\infty}{}' \; -\frac{3(-1)^{m+n}}{mn\pi^2} \, e^{i[m\pi x + n\pi y/3]}$$

We conclude this section with a brief discussion of *triple Fourier series*. The only sensible way to approach triple Fourier series is through the complex form. Suppose that the function f is defined over the rectangular box

$$[-a, a] \times [-b, b] \times [-c, c]$$
$$= \{(x, y, z): \quad -a \leq x \leq a \quad -b \leq y \leq b \quad -c \leq z \leq c\}$$

then the (complex) triple Fourier series for f is defined by

$$f \sim \sum_{k,m,n=-\infty}^{+\infty} c_{kmn} e^{i[k\pi x/a + m\pi y/b + n\pi z/c]}$$

where

$$c_{kmn} = \frac{1}{8abc} \int_{-c}^{c} \int_{-b}^{b} \int_{-a}^{a} f(x, y, z) e^{-i[k\pi x/a + m\pi y/b + n\pi z/c]} \, dx \, dy \, dz$$

For example, if $f(x, y, z) = xy^2z^3$ on $[-1, 1] \times [-2, 2] \times [-\pi, \pi]$ then for $k \neq 0$, $m \neq 0$, and $n \neq 0$

$$c_{kmn} = \frac{1}{16\pi} \int_{-1}^{1} xe^{-ik\pi x} \, dx \int_{-2}^{2} y^2 e^{-im\pi y/2} \, dy \int_{-\pi}^{\pi} z^3 e^{-inz} \, dz$$

$$= \frac{1}{16\pi} \left[-i \int_{-1}^{1} x \sin k\pi x \, dx \right] \left[\int_{-2}^{2} y^2 \cos \frac{m\pi y}{2} \, dy \right]$$

$$\times \left[-i \int_{-\pi}^{\pi} z^3 \sin nz \, dz \right]$$

$$= \frac{(-1)^{k+m+n} 8[6 - \pi^2 n^2]}{km^2 n^3 \pi^3}$$

while $c_{0mn} = c_{km0} = 0$ and

$$c_{k0n} = \frac{4(-1)^{k+n}[6 - \pi^2 n^2]}{3kn^3\pi}$$

Thus, over $[-1, 1] \times [-2, 2] \times [-\pi, \pi]$

$$xy^2z^3 \sim \sum_{k,n=-\infty}^{+\infty}{}' \frac{4(-1)^{k+n}[6 - \pi^2n^2]}{3kn^3\pi} e^{i[k\pi x + nz]}$$

$$+ \sum_{k,m,n=-\infty}^{+\infty}{}' \frac{(-1)^{k+m+n}8[6 - \pi^2n^2]}{km^2n^3\pi^3} e^{i[k\pi x + (m\pi y/2) + nz]}$$

Remark.   It is often useful to remember that when f is real valued we have that $c_{-m,-n}$ and c_{mn} (or $c_{-k,-m,-n}$ and c_{kmn}) are *complex conjugates.*

Exercises

***(2.5)**   Show that if $f(x, y) = g(x)h(y)$ where $g \sim \sum a_m e^{imx}$ and $h \sim \sum b_n e^{iny}$ then $f \sim \sum a_m b_n e^{i(mx+ny)}$.

(2.6)   Show   that   if $f(x, y) = g(x + y)$ where $g \sim \sum a_n e^{inx}$ then $f \sim \sum a_n e^{i(nx+ny)}$.

(2.7)   Expand the following functions in complex double Fourier series over $[-\pi, \pi] \times [-\pi, \pi]$.

    (a) $f(x, y) = x + y$
    (b) $f(x, y) = (x + y)^2$
    (c) $f(x, y) = |\sin(x + y)| + i\,|x + y|$
    (d) $f(x, y) = |\cos(x + y)| + ix^3y^3$
    (e) $f(x, y) = x^2y^2$
    (f) $f(x, y) = |x^3y^3|$

(2.8)   Compute the complex double Fourier series for the following functions over the given rectangular regions.

    (a) $f(x, y) = |x + y|$ $[-1, 1] \times [-3, 3]$
    (b) $f(x, y) = |x| + i\,|x + y|$ $[-3, 3] \times [-5, 5]$
    (c) $f(x, y) = |\sin(x + y)| + i(x + y)$ $[-\pi, \pi] \times [-\frac{1}{2}\pi, \frac{1}{2}\pi]$

***(2.9)**   Generalize the definition of convolution given in (8.13), Chapter 2, and prove that commutativity, associativity, and the multiplier property [see (8.15), Chapter 2] hold for functions of two (three) variables.

(2.10)   Compute triple Fourier series for the following functions over the given rectangular boxes. [*Hint*: For (a) and (b), generalize (2.6).]

    (a) $f(x, y, z) = (x + y + 3z)^3$ $[-\pi, \pi] \times [-\pi, \pi] \times [-\pi, \pi]$
    (b) $f(x, y, z) = |\sin(x + 2y + 4z)|$ $[-\pi, \pi] \times [-\pi, \pi] \times [-\pi, \pi]$
    (c) $f(x, y, z) = |xyz|$ $[-\pi, \pi] \times [-3, 3] \times [-5, 5]$

§3. Convergence and Completeness

In this section we will discuss some elementary, but quite useful, results mostly concerning pointwise convergence of double Fourier series. We shall also briefly discuss the question of completeness.

The double Fourier series for a function f converges to $f(x, y)$ at the point (x, y) if for

$$S_{MN}(x, y) = \sum_{m=-M, n=-N}^{M, N} c_{mn} e^{i(mx+ny)}$$

we have

$$\lim_{M, N \to \infty} S_{MN}(x, y) = f(x, y)$$

Or, more precisely, given $\epsilon > 0$, we can find positive constants M_0 and N_0 such that for all $M \geq M_0$ and $N \geq N_0$

$$|S_{MN}(x, y) - f(x, y)| < \epsilon$$

The concepts of uniform and absolute convergence are defined in the obvious ways as simple generalizations of the one-dimensional concepts; moreover, Weierstrass' M-test and Theorems (5.4) and (5.5) from Chapter 1 generalize easily. We should also note that our definition of convergence is not meant to apply only to the one type of Fourier series shown above; it applies just as well to Fourier series of the more general form $\sum c_{mn} e^{i(m\pi x/a + n\pi y/b)}$, indeed to all doubly indexed series of functions. We have adopted the form above purely for notational convenience; all of our results below hold just as well for the latter more general Fourier series.

We now prove a quite simple, but important, convergence theorem.

(3.1) Theorem.   Suppose $f(x, y) = g(x)h(y)$ where g and h are piecewise smooth with period 2π. Then the double Fourier series for f converges to $f(x, y)$ at each point (x, y) where g and h are continuous at x and y, respectively. Moreover, if g and h are continuous for all x and y, then the double Fourier series converges uniformly to f.

Proof.   From Exercise (2.5) we have $f \sim \sum a_m b_n e^{i(mx+ny)}$ where $g \sim \sum a_m e^{imx}$ and $h \sim \sum b_n e^{inx}$. Implicit in that result is the fact that

$$S_{MN}(x, y) = S_M(x) S_N^\#(y)$$

where $S_M(x) = \sum_{m=-M}^{M} a_m e^{imx}$ and $S_N^\#(y) = \sum_{n=-N}^{N} b_n e^{iny}$. Because the one-dimensional real and complex Fourier series are identical, we have $\lim_{M \to \infty} S_M(x) = g(x)$ and $\lim_{N \to \infty} S_N^\#(y) = h(y)$ when we apply the convergence theorem (4.5), Chapter 1. Since the limit of a product is the product of the limits, it follows that

$$\lim_{M, N \to \infty} S_{MN}(x, y) = \lim_{M \to \infty} S_M(x) \lim_{N \to \infty} S_N^\#(y)$$

$$= g(x)h(y) = f(x, y)$$

Moreover, if we put

$$A = \underset{-\pi \leq x \leq \pi}{\text{maximum}} |g(x)| \quad \text{and} \quad B = \underset{-\pi \leq y \leq \pi}{\text{maximum}} |h(y)|$$

then A and B are the maximum values for $|g|$ and $|h|$ for all x and y by periodicity. The functions g and h satisfy the conditions for uniform convergence given in Theorem (4.4), Chapter 2. Therefore, given $\epsilon > 0$, we can find positive constants M_0 and N_0 so large that for $M \geq M_0$ and $N \geq N_0$

$$|g(x) - S_M(x)| < \epsilon \qquad \text{and} \qquad |h(y) - S_N^\#(y)| < \epsilon$$

In which case

$$
\begin{aligned}
|f(x, y) &- S_{MN}(x, y)| \\
&= |[g(x)h(y) - S_M(x)h(y)] + [S_M(x)h(y) - S_M(x)S_N^\#(y)]| \\
&\leq |[g(x) - S_M(x)]h(y)| + |S_M(x)[h(y) - S_N^\#(y)]| < \epsilon B + (A + \epsilon)\epsilon
\end{aligned}
$$

Thus for $M \geq M_0$ and $N \geq N_0$ we have $|f(x, y) - S_{MN}(x, y)| < \epsilon(B + A + \epsilon)$ which, because ϵ can be taken arbitrarily small, proves the required uniform convergence. ∎

(3.2) Examples

(a) Let $f(x, y) = xy^2$ on $[-\pi, \pi] \times [-\pi, \pi]$. Then using Example (1.4) and Theorem (3.1) we have for $-\pi < x < \pi$ and $-\pi \leq y \leq \pi$

$$xy^2 = \tfrac{2}{3}\pi^2 \sum_{m=1}^{\infty} \frac{(-1)^{m+1}}{m} \sin mx - 8 \sum_{m,n=1}^{\infty} \frac{(-1)^{m+n}}{mn^2} \sin mx \cos ny$$

(b) By computing a one-dimensional Fourier series, it is easily seen that for $-\pi < x < \pi$ and all y values

$$\tfrac{2}{3}\pi^2 \sum_{m=1}^{\infty} \frac{(-1)^{m+1}}{m} \sin mx = \frac{\pi^2}{3} x$$

Applying that result to (a) we have for $-\pi < x \leq \pi$ and $-\pi \leq y \leq \pi$

$$\sum_{m,n=1}^{\infty} \frac{(-1)^{m+n}}{mn^2} \sin mx \cos ny = \frac{\pi^2}{24} x - \frac{1}{8} xy^2$$

(c) Let $f(x, y) = |xy| + x^4$. Then for $-\pi \leq x \leq \pi$

$$|x| = \frac{\pi}{2} - \frac{2}{\pi} \sum_{m=-\infty}^{+\infty}{}' \frac{1}{(2m - 1)^2} e^{i(2m-1)x}$$

$$x^4 = \frac{\pi^4}{5} + 4 \sum_{m=-\infty}^{+\infty}{}' \frac{(-1)^m(m^2\pi^2 - 6)}{m^4} e^{imx}$$

Hence for $-\pi \leq x \leq \pi$ and $-\pi \leq y \leq \pi$

$$
\begin{aligned}
|xy| + x^4 = {}&\frac{\pi^2}{4} + \frac{4}{\pi^2} \sum_{m,n=-\infty}^{+\infty}{}' \frac{e^{i[(2m-1)x + (2n-1)y]}}{(2m-1)^2(2n-1)^2} - \sum_{m=-\infty}^{+\infty}{}' \frac{e^{i(2m-1)x}}{(2m-1)^2} \\
&- \sum_{n=-\infty}^{+\infty}{}' \frac{e^{i(2n-1)y}}{(2n-1)^2} + \frac{\pi^4}{5} + 4 \sum_{m=-\infty}^{+\infty}{}' \frac{(-1)^m(m^2\pi^2 - 6)}{m^4} e^{imx}
\end{aligned}
$$

We now turn to the question of completeness. In §5 we shall prove the following theorem.

(3.3) Theorem.   The complex exponential system $\{e^{i(mx+ny)}\}$ is complete. That is, for every continuous function f on $[-\pi, \pi] \times [-\pi, \pi]$ we have

$$\lim_{M, N \to +\infty} \left[\int_{-\pi}^{\pi} \int_{-\pi}^{\pi} |f(x, y) - S_{MN}(x, y)|^2 \, dx \, dy \right]^{1/2} = 0$$

Remark.   The complex exponential system may just as well be used to expand complex valued functions as well as real valued ones, so the functions f referred to in (3.3) can be taken as complex valued. Of course, if f is real valued then so is S_{MN} since it is identical to the partial sum of the real form of the Fourier series for f, in that case the limit in (3.3) can be written as

(3.3′)   *Real Case.*

$$\lim_{M, N \to +\infty} \left[\int_{-\pi}^{\pi} \int_{-\pi}^{\pi} [f(x, y) - S_{MN}(x, y)]^2 \, dx \, dy \right]^{1/2} = 0$$

As we noted at the beginning of this chapter, several results from Chapter 2 apply without change to orthogonal systems of functions of two variables. For example, if $\{g_j(x, y)\}_{j=0}^{\infty}$ is an orthonormal system of continuous functions over $[a, b] \times [c, d]$, then Bessel's inequality

$$\sum_{j=0}^{\infty} c_j^2 \leq \int_c^d \int_a^b f^2(x, y) \, dx \, dy$$

is easily seen to hold provided that f is a continuous real valued function.[3] Applying Bessel's inequality to the orthonormal system obtained from (1.3) by dividing by norms, then we obtain

(3.4) $$\sum_{m,n=0}^{+\infty} \lambda_{mn}[A_{mn}^2 + B_{mn}^2 + C_{mn}^2 + D_{mn}^2] \leq \frac{1}{\pi^2} \int_{-\pi}^{\pi} \int_{-\pi}^{\pi} f^2(x, y) \, dx \, dy$$

Using the equalities in (2.2) it is easily shown that (3.4) implies

(3.4′) $$\sum_{m,n=-\infty}^{+\infty} |c_{mn}|^2 \leq \frac{1}{4\pi^2} \int_{-\pi}^{\pi} \int_{-\pi}^{\pi} f^2(x, y) \, dx \, dy$$

where c_{mn} are the coefficients of the real valued (continuous) function f relative to the complex exponential system $\{e^{i(mx+ny)}\}$. Just as with single variable Fourier series, completeness does imply that (3.4) and (3.4′) are *equalities*.

Furthermore, the properties of the norm $\|\cdot\|_2$ described in Theorem (4.5), Chapter 2, easily generalize when we define

$$\|f\|_2 = \left[\int_c^d \int_a^b f^2(x, y) \, dx \, dy \right]^{1/2}$$

---

3. Actually we need only assume that f^2 has a finite integral over $[a, b] \times [c, d]$ in order for Bessel's inequality to hold.

for each real valued continuous function f (or even if f is only square integrable). The proof of this new theorem involves only trivial modifications of the proof of (4.5) in Chapter 2.

We close this section with another pointwise convergence theorem. To prove that theorem we need the following lemma.

(3.5) Lemma.   Suppose that f and g are continuous on $[-\pi, \pi] \times [-\pi, \pi]$ and have the same Fourier coefficients relative to the set $\{e^{i(mx+ny)}\}$, then $f = g$ identically in x and y.

Proof.   Let $h = f - g$. Then the Fourier coefficients c_{mn} of h are all 0. Therefore, performing a double integral as an iterated one, we have

$$0 = c_{mn} = \frac{1}{2\pi} \int_{-\pi}^{\pi} \left[\frac{1}{2\pi} \int_{-\pi}^{\pi} h(x, y)e^{-imx} \, dx \right] e^{-iny} \, dy$$

for all m and n. It follows from the completeness of $\{e^{iny}\}$ over $[-\pi, \pi]$ that for each m

$$0 = \frac{1}{2\pi} \int_{-\pi}^{\pi} h(x, y)e^{-imx} \, dx$$

identically in y.[4] Fix some value of y in the interval $[-\pi, \pi]$. Then, using the completeness of $\{e^{imx}\}$, we have that $h(x, y) = 0$ identically in x for each fixed value of y. Thus $h = 0$ for all points (x, y) in $[-\pi, \pi] \times [-\pi, \pi]$ hence $f = g$ for all such points. ∎

In passing, we note that we proved above that $\{e^{i(mx+ny)}\}$ is *maximally orthogonal* in the sense that if h is a continuous function on $[-\pi, \pi] \times [-\pi, \pi]$ for which $\int_{-\pi}^{\pi} \int_{-\pi}^{\pi} h(x, y)e^{-i(mx+ny)} \, dx \, dy = 0$ for all m and n, then $h = 0$. We can now prove a second pointwise convergence theorem for double Fourier series.

(3.6) Theorem.   If f, f_x, f_y, and f_{xy} are all continuous with period 2π in x and y, then the double Fourier series for f converges uniformly to f.

Proof.   For $m \neq 0$ and $n \neq 0$, let $\bar{c}_{mn}$ denote the m, nth Fourier coefficient of f_{xy}; thus

$$\bar{c}_{mn} = \frac{1}{4\pi^2} \int_{-\pi}^{\pi} \int_{-\pi}^{\pi} f_{xy}(x, y)e^{-i(mx+ny)} \, dx \, dy$$

If we integrate by parts with respect to x, holding y fixed, then by periodicity we obtain

$$\bar{c}_{mn} = \frac{im}{4\pi^2} \int_{-\pi}^{\pi} \int_{-\pi}^{\pi} f_y(x, y)e^{-i(mx+ny)} \, dx \, dy$$

Integrating by parts again yields

$$\bar{c}_{mn} = -mnc_{mn}$$

---

4. Here we are using the fact that, as a function of y, $(1/2\pi) \int_{-\pi}^{\pi} h(x, y)e^{-imx} \, dx$, is continuous on $[-\pi, \pi]$.

where c_{mn} is the m, nth Fourier coefficient of f. If we let $\tilde{c}_{0n}$, for $n \neq 0$, denote the nth Fourier coefficient of f_y, then

$$\tilde{c}_{0n} = \frac{1}{4\pi^2} \int_{-\pi}^{\pi} \int_{-\pi}^{\pi} f_y(x, y)e^{-iny}\, dx\, dy = inc_{0n}$$

and, similarly, for $m \neq 0$

$$\tilde{c}_{m0} = \frac{1}{4\pi^2} \int_{-\pi}^{\pi} \int_{-\pi}^{\pi} f_x(x, y)e^{-imx}\, dx\, dy = imc_{m0}$$

Finally, we define $\tilde{c}_{00}$ to be 0. Bessel's inequality (3.4′) applied to f_x, f_y, and f_{xy} implies that[5]

$$(3.7) \quad \sum_{m,\,n=-\infty}^{+\infty} |\tilde{c}_{mn}|^2 \leq \frac{1}{4\pi^2} \int_{-\pi}^{\pi} \int_{-\pi}^{\pi} [f_x^2(x, y) + f_y^2(x, y) + f_{xy}^2(x, y)]\, dx\, dy$$

From (3.7) we can prove that $\sum |c_{mn}|$ converges. In fact, using the equalities above plus Cauchy's inequality (4.3), Chapter 2, we have

$$\sum |c_{mn}| = |c_{00}| + \sum{}' |\tilde{c}_{mn}| \frac{1}{mn} + \sum{}' |\tilde{c}_{0n}| \frac{1}{n} + \sum{}' |\tilde{c}_{m0}| \frac{1}{m}$$

$$\leq |c_{00}| + \left[\sum{}' |\tilde{c}_{mn}|^2\right]^{1/2} \left[\sum{}' \frac{1}{m^2 n^2}\right]^{1/2} + \left[\sum{}' |\tilde{c}_{0n}|^2\right]^{1/2} \left[\sum{}' \frac{1}{n^2}\right]^{1/2}$$

$$+ \left[\sum{}' |\tilde{c}_{m0}|^2\right]^{1/2} \left[\sum{}' \frac{1}{m^2}\right]^{1/2}$$

Because of (3.7), the last sum above is finite. Thus $\sum |c_{mn}|$ converges, hence by Weierstrass's M-Test $\sum_{m,\,n=-\infty}^{+\infty} c_{mn}e^{i(mx+ny)}$ converges uniformly (and absolutely) to a continuous function, call it g, on $[-\pi, \pi] \times [-\pi, \pi]$. Moreover, the series

$$\sum_{m,\,n=-\infty}^{+\infty} c_{mn}e^{i(mx+ny)}e^{-i(jx+ky)} = g(x, y)e^{-i(jx+ky)}$$

also converges uniformly hence it may be integrated term by term. Thus, we find that g has $\{c_{mn}\}$ as Fourier coefficients. Lemma (3.5) implies that $f = g$ which proves our theorem. ∎

For convenience of notation, we stated all our theorems for double Fourier series. Our theorems are simple enough that they easily generalize to triple Fourier series; we leave that task to the reader. [See Exercise (3.13).]

The theory of pointwise convergence of double (and triple) Fourier series is much more complicated than the single variable case (which is no easy matter either). We have only discussed the most elementary results in this section, more advanced results are the subject of current research. The

---

5. Here we assume that f is real valued, the theorem is also true for f complex valued as we may see by considering real and imaginary parts of f.

interested reader might begin by consulting the article "Multiple Trigonometric Series" by J. M. Ash (1976).

Exercises

(3.8) Discuss the convergence of each of the Fourier series in Exercise (2.7). [*Note*: (2.6) is needed for treating some of those series.] Discuss the convergence of each of the Fourier series in Exercise (2.8).

(3.9) Formulate and prove analogues of Theorems (3.1) and (3.6) for the double sine series of a function f defined on $[0, \pi] \times [0, \pi]$. Then explain why those theorems are also valid for double sine series on $[0, a] \times [0, b]$.

(3.10) Expand the following functions in double sine series over $[0, a] \times [0, b]$ and discuss convergence.

(a) $f(x, y) = xy$
(b) $f(x, y) = x^2 y^2$
(c) $f(x, y) = A$ a constant
(d) $f(x, y) = \begin{cases} A & \text{if } |x - \frac{1}{2}a| < \frac{1}{4}a \quad \text{and} \quad |y - \frac{1}{2}a| < \frac{1}{4}a \\ 0 & \text{if } |x - \frac{1}{2}a| > \frac{1}{4}a \quad \text{or} \quad |y - \frac{1}{2}a| > \frac{1}{4}a \end{cases}$

(3.11) Is $\{\sin mx \sin ny\}_{m,n=1}^{\infty}$ a complete system over $[0, \pi] \times [0, \pi]$? Why?

(3.12) Using (3.4′) prove that $\sum |c_{mn}|^2 \le (4\pi^2)^{-1} \int_{-\pi}^{\pi} \int_{-\pi}^{\pi} |f(x, y)|^2 \, dx \, dy$ holds for every complex valued continuous function.

(3.13) Generalize Theorems (3.1) and (3.6) to triple Fourier series.[6] Discuss convergence for each of the triple Fourier series in Exercise (2.10).

*(3.14) Prove the following generalization of (4.10), Chapter 2.

Theorem. Suppose that f has period $2a$ in x and $2b$ in y and that f has $2k$ continuous derivatives in both x and y. Then for some positive constant A the Fourier coefficients of f satisfy

$$|c_{mn}| \le \frac{A}{m^k n^k} \qquad |c_{m0}| \le \frac{A}{m^k} \qquad |c_{0n}| \le \frac{A}{n^k}$$

for all nonzero integers m and n. Generalize to triple Fourier series.

§4. Some Applications

We will now consider some elementary applications of multiple Fourier series. In the following three subsections we shall use those series to solve the wave equation, Laplace's equation, and the heat equation.

---

6. The obvious generalization of (3.6)—that f has three continuous derivatives in each variable—can be improved. It is sufficient to assume that f has two continuous derivatives in each variable. [See Stein and Weiss (1971), p. 249.]

Wave Equation

Suppose that we have a two-dimensional sheet of elastic material with a constant density σ (per cm^2) and a constant tension $T > 0$. By a balance of forces argument, applied to any rectangular portion of the sheet, we are led to the following *two-dimensional wave equation*

(4.1)
$$c^2[z_{xx} + z_{yy}] = z_{tt} \quad (c^2 = T/\sigma)$$

for $z(x, y, t)$ the height of the sheet above the point (x, y) at time t.

We will solve (4.1) subject to the following conditions. Assume that the elastic sheet is rectangular, that is, $0 \le x \le a$ and $0 \le y \le b$, with edges clamped down. We then have the following *boundary conditions*

(4.2)
$$z(0, y, t) = z(a, y, t) = z(x, 0, t) = z(x, b, t) = 0$$

Moreover, we assume a given initial position f and initial velocity g for the sheet, so we have the following *initial conditions*

(4.3)
$$z(x, y, 0+) = f(x, y) \qquad z_t(x, y, 0+) = g(x, y)$$

Our solution to (4.1) through (4.3) will be found by separation of variables. Substituting $z(x, y, t) = X(x)Y(y)T(t)$ into (4.1) and (4.2) we obtain

$$\frac{X''(x)}{X(x)} + \frac{Y''(y)}{Y(y)} = \frac{T''(t)}{c^2 T(t)} \qquad X(0) = X(a) = 0 \qquad Y(0) = Y(b) = 0$$

Holding x and t fixed and varying y, or holding y and t fixed and varying x, we conclude that

$$X''(x) = \lambda X(x) \qquad Y''(y) = \mu Y(y) \qquad T''(t) = (\lambda + \mu)c^2 T(t)$$
$$X(0) = X(a) = 0 \qquad Y(0) = Y(b) = 0$$

where λ and μ are constants. We obtain as eigenvalues and eigenfunctions

$$\lambda_m = -(m\pi/a)^2 \qquad X_m(x) = \sin(m\pi x/a) \qquad m = 1, 2, 3, \ldots$$
$$\mu_n = -(n\pi/b)^2 \qquad Y_n(y) = \sin(n\pi y/b) \qquad n = 1, 2, 3, \ldots$$

Hence, by superposition, we are led to the following formal series

$$z(x, y, t) = \sum_{m, n=1}^{\infty} [a_{mn} \cos \delta_{mn} ct + b_{mn} \sin \delta_{mn} ct] \sin(m\pi x/a) \sin(n\pi y/b)$$

where $\delta_{mn} = [(m\pi/a)^2 + (n\pi/b)^2]^{1/2}$. To determine a_m and b_{mn} we use double sine series expansions for f and g, based on (4.3); thus our proposed solution to (4.1) through (4.3) is as follows.

$$z(x, y, t) = \sum_{m, n=1}^{\infty} [a_{mn} \cos \delta_{mn} ct$$

$$+ b_{mn} \sin \delta_{mn} ct] \sin(m\pi x/a) \sin(n\pi y/b)$$

$$\delta_{mn} = [(m\pi/a)^2 + (n\pi/b)^2]^{1/2}$$

(4.4)

$$a_{mn} = \frac{4}{ab} \int_0^b \int_0^a f(x, y) \sin(m\pi x/a) \sin(n\pi y/b) \, dx \, dy$$

$$b_{mn} = \frac{4}{ab\delta_{mn}c} \int_0^b \int_0^a g(x, y) \sin(m\pi x/a) \sin(n\pi y/b) \, dx \, dy$$

For example, suppose that f is defined by $f(x, y) = 0.2(x - \pi)x(y - \pi)y$ over $[0, \pi] \times [0, \pi]$ and $g = 0$. Then (4.4) yields the solution

$$z(x, y, t) = \frac{12.8}{\pi^2} \sum_{j, k=0}^{\infty} \frac{\cos([(2j + 1)^2 + (2k + 1)^2]^{1/2} ct)}{(2j + 1)^3 (2k + 1)^3} \sin(2j + 1)x \sin(2k + 1)y$$

We now turn to the question of the validity of (4.4) as a solution, and its uniqueness. Suppose that f is C^4 and g is C^3,[7] and each function is periodic (period 2π) in each variable. Then we can prove that (4.4) provides a rigorous solution to (4.1) through (4.3). For convenience, we are assuming that $a = b = \pi$. Let $\bar{a}_{mn}$ be defined as

$$\bar{a}_{mn} = \frac{4}{\pi^2} \int_0^{\pi} \int_0^{\pi} f_{xxxy}(x, y) \cos mx \cos ny \, dx \, dy$$

Then by integrating by parts and using periodicity, we find that

$$\bar{a}_{mn} = -\frac{4m^3 n}{\pi^2} \int_0^{\pi} \int_0^{\pi} f(x, y) \sin mx \sin ny \, dx \, dy = -m^3 n a_{mn}$$

Hence, applying Bessel's inequality for the system $\{\cos mx \cos ny\}_{m, n=1}^{\infty}$ over $[0, \pi] \times [0, \pi]$ we obtain

$$\sum m^2 |a_{mn}| = \sum \frac{1}{mn} |\bar{a}_{mn}| \leq \left[\sum \frac{1}{m^2 n^2} \right]^{1/2} \left[\sum |\bar{a}_{mn}|^2 \right]^{1/2}$$

$$\leq \frac{\pi^2}{6} \left[\frac{4}{\pi^2} \int_0^{\pi} \int_0^{\pi} f_{xxxy}^2(x, y) \, dx \, dy \right]^{1/2}$$

Similarly, we have

$$\sum n^2 |a_{mn}| \leq \frac{\pi^2}{6} \left[\frac{4}{\pi^2} \int_0^{\pi} \int_0^{\pi} f_{xyyy}^2(x, y) \, dx \, dy \right]^{1/2}$$

---

7. When we say that a function is C^k we mean that it is k times continuously differentiable in all of its variables.

It follows from these last two inequalities that

(4.5)
$$\sum_{m,\,n=1}^{\infty} a_{mn} \cos \delta_{mn} ct \sin mx \sin ny$$

is twice differentiable, term by term, with respect to x, y, or t. In particular, the series in (4.5) defines a solution to the wave equation (4.1). Furthermore, since m/δ_{mn} and n/δ_{mn} tend to 1 as m or n tends to $+\infty$, we only need to assume that g is C^3 to ensure the term by term differentiability of

(4.6)
$$\sum_{m,\,n=1}^{\infty} b_{mn} \sin \delta_{mn} ct \sin mx \sin ny$$

hence (4.6) defines a solution to (4.1). Adding the two solutions in (4.5) and (4.6), we have that the series for z in (4.4) satisfies the wave equation; we leave the verification of (4.2) and (4.3) to the reader. [See Exercise (4.19).]

We now turn to the question of uniqueness. Suppose that z is a solution to (4.1) through (4.3) and that z is C^2 for points (x, y, t) in $[0, \pi] \times [0, \pi] \times [0, +\infty)$. If $z^{\#}$ is another such solution, then $\tilde{z} = z^{\#} - z$ solves

(4.7)
$$c^2[\tilde{z}_{xx} + \tilde{z}_{yy}] = \tilde{z}_{tt}$$
$$\tilde{z}(0, y, t) = \tilde{z}(\pi, y, t) = \tilde{z}(x, 0, t) = \tilde{z}(x, \pi, t) = 0$$
$$\tilde{z}(x, y, 0) = \tilde{z}_t(x, y, 0) = 0$$

and $\tilde{z}$ is C^2. If we expand $\tilde{z}$ in a double sine series

$$\tilde{z}(x, y, t) = \sum_{m,\,n=1}^{\infty} D_{mn}(t) \sin mx \sin ny$$

where

$$D_{mn}(t) = \frac{4}{\pi^2} \int_0^{\pi} \int_0^{\pi} \tilde{z}(x, y, t) \sin mx \sin ny \, dx \, dy$$

then by differentiation under the integral sign[8] plus (4.7), we have

$$D''_{mn}(t) = \frac{4}{\pi^2} \int_0^{\pi} \int_0^{\pi} \tilde{z}_{tt}(x, y, t) \sin mx \sin ny \, dx \, dy$$
$$= \frac{4c^2}{\pi^2} \int_0^{\pi} \int_0^{\pi} [\tilde{z}_{xx}(x, y, t) + \tilde{z}_{yy}(x, y, t)] \sin mx \sin ny \, dx \, dy$$

Integrating by parts twice with respect to x for $\tilde{z}_{xx}$ and y for $\tilde{z}_{yy}$ we get

(4.8)
$$D''_{mn}(t) = -\delta_{mn}^2 c^2 D_{mn}(t) \qquad (t \geq 0)$$

and using the initial conditions $\tilde{z}(x, y, 0) = \tilde{z}_t(x, y, 0) = 0$ we have

(4.8′)
$$D_{mn}(0) = D'_{mn}(0) = 0$$

Therefore, since $D_{mn}(t) = 0$ is the only solution to (4.8) and (4.8′), we have $\tilde{z} = \sum 0 \sin mx \sin ny = 0$. Therefore, $z = z^{\#}$, hence z is unique.

---

8. Here we apply a two-dimensional generalization of Theorem (5.6), Chapter 3.

Uniqueness can also be verified using the Divergence Theorem. See Copson (1975), p. 70 or Weinberger (1965), p. 152.

Laplace's Equation in a Box

We begin by considering Laplace's equation for a cube. In the exercises the reader is asked to treat more general rectangular parallelepipeds (boxes). Let $\mathscr{C}$ stand for the open cube

$$(0, \pi) \times (0, \pi) \times (0, \pi)$$
$$= \{(x, y, z): \quad 0 \le x < \pi \quad 0 < y < \pi \quad 0 < z < \pi\}$$

and let $\partial\mathscr{C}$ stand for the boundary surface of that cube. We will now solve Dirichlet's problem for $\mathscr{C}$. [See Figure 5.1; also (3.1), Chapter 4 *phrased for* $\mathbb{R}^3$.]

We begin by solving the simpler problem

$$\Delta\Psi = \Psi_{xx} + \Psi_{yy} + \Psi_{zz} = 0$$

(4.9) $\quad \Psi(0+, y, z) = \Psi(\pi-, y, z) = 0 \quad \Psi(x, 0+, z) = \Psi(x, \pi-, z) = 0$

$$\Psi(x, y, 0+) = 0 \quad \Psi(x, y, \pi-) = f_1(x, y)$$

Substituting $\Psi(x, y, z) = X(x)Y(y)Z(z)$ into Laplace's equation we obtain, upon dividing by $X(x)Y(y)Z(z)$,

$$\frac{X''(x)}{X(x)} + \frac{Y''(y)}{Y(y)} + \frac{Z''(z)}{Z(z)} = 0$$

and the homogeneous (zero) conditions in (4.9) yield

$$X(0) = X(\pi) = 0 \quad Y(0) = Y(\pi) = 0 \quad Z(0) = 0$$

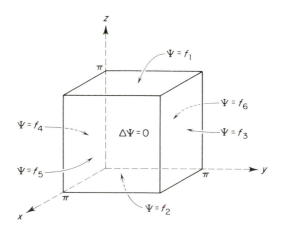

Figure 5.1 Dirichlet's problem for the cube C. The continuous function f on ∂C can be specified by six functions $f_1, f_2, \ldots, f_6$ on each of the six faces of ∂C.

These last two results lead us to

(4.10)

$$X''(x) = \lambda X(x) \qquad Y''(y) = \mu Y(y) \qquad Z''(z) = -(\lambda + \mu)Z(z)$$

$$X(0) = X(\pi) = 0 \qquad Y(0) = Y(\pi) = 0 \qquad Z(0) = 0$$

where λ and μ are constants. The first two problems in (4.10) yield the same eigenvalues and eigenfunctions as the wave equation (for $a = b = \pi$). Substituting those eigenvalues $\lambda_m = -m^2$ and $\mu_n = -n^2$ into the third problem in (4.10), we find the solution

$$Z_{mn}(z) = \sinh \delta_{mn} z$$

where δ_{mn} is, as for the wave equation, equal to $[m^2 + n^2]^{1/2}$. Any finite superposition of these separated solutions

(4.11)
$$\sum_{m,\,n=1}^{M,\,N} D_{mn} \sin mx \sin ny (\sinh \delta_{mn} z / \sinh \delta_{mn} \pi)$$

is found to be harmonic. Letting M and N tend to $+\infty$, we obtain a formal double series expression for Ψ

$$\Psi(x, y, z) = \sum_{m,\,n=1}^{\infty} D_{mn} \sin mx \sin ny (\sinh \delta_{mn} z / \sinh \delta_{mn} \pi)$$

where the coefficients D_{mn} are yet to be determined. If we set $z = \pi$, then we want $\Psi = f_1$ so we put

$$f_1 \sim \sum_{m,\,n=1}^{\infty} D_{mn} \sin mx \sin ny$$

Therefore, our solution to (4.9) is as follows.

(4.12)

$$\Psi(x, y, z) = \sum_{m,\,n=1}^{\infty} D_{mn} \sin mx \sin ny \, (\sinh \delta_{mn} z / \sinh \delta_{mn} \pi)$$

where

$$\delta_{mn} = [m^2 + n^2]^{1/2}$$

and

$$D_{mn} = (4/\pi^2) \int_0^\pi \int_0^\pi f_1(x, y) \sin mx \sin ny \, dx \, dy.$$

By adding solutions, as we did for the rectangle problems in §1 of Chapter 4, we obtain a solution Ψ for Dirichlet's problem as a sum of six series like the one shown in (4.12). In that connection, we should note that the problem

$$\Delta\Psi = 0$$

$$\Psi(0+, y, z) = \Psi(\pi -, y, z) = 0$$

$$\Psi(x, 0+, z) = f_4(x, z) \qquad \Psi(x, \pi -, z) = 0$$

$$\Psi(x, y, 0+) = \Psi(x, y, \pi -) = 0$$

has the series solution

$$\Psi(x, y, z) = \sum_{m,n=1}^{\infty} \left[\frac{4}{\pi^2} \int_0^\pi \int_0^\pi f_4(u, v) \sin mu \sin nv \, du \, dv \right]$$

$$\times \sin mx \sin nz \, \frac{\sinh \delta_{mn}(\pi - y)}{\sinh \delta_{mn}\pi}$$

The other four series solution that we need are obtained by permuting the variables x, y, and z in the two series above.

We will now briefly discuss the validity of (4.12) as a solution to (4.9). Similar arguments apply to the other series that occur in solving Dirichlet's problem for the cube $\mathscr{C}$.

Suppose that $|D_{mn}| \le A$, a constant, for all m and n. That is certainly true if $\int_0^\pi \int_0^\pi |f(x, y)| \, dx \, dy$ is finite. If $0 \le z \le \pi - \delta$ for $\delta > 0$, then

$$0 \le A \frac{\sinh \delta_{mn}z}{\sinh \delta_{mn}\pi} = A \frac{e^{\delta_{mn}z}}{e^{-\delta_{mn}\pi}} \frac{1 - e^{-2\delta_{mn}z}}{e^{2\delta_{mn}\pi} - 1} \le Ae^{-\delta_{mn}\delta} \frac{1}{e^{2\sqrt{2}\pi} - 1}$$

and $\sum_{m,n=1}^{\infty} Ae^{-\delta_{mn}\delta}[e^{2\sqrt{2}\pi} - 1]^{-1}$ converges by a comparison test. It follows that

(4.13)
$$\sum_{m,n=1}^{\infty} D_{mn} \sin mx \sin ny \, (\sinh \delta_{mn}z / \sinh \delta_{mn}\pi)$$

converges uniformly to Ψ if $0 \le z \le \pi - \delta$. Similar arguments show that (4.13) may be differentiated term by term any number of times provided that $0 \le z \le \pi - \delta$. Hence, letting δ pass to zero, we see that Ψ is harmonic in $\mathscr{C}$.

If f_1 is C^2 when extended as an odd periodic function, then Ψ satisfies the boundary conditions of Dirichlet's problem. We leave the details to the reader.

Finally, we note that the uniqueness of Ψ follows by the Maximum–Minimum Principle for harmonic functions [see Exercise (4.9), Chapter 4], or an argument like the one used above for the wave equation can also be used.

Heat Equation

By applying an argument like the one used in §1 of Chapter 3 to a rectangular box made of heat-conducting material, homogeneous in its

physical properties, we are led to the following equation for the temperature $u(x, y, z, t)$ at a point (x, y, z) at time t

(4.14) $$a^2[u_{xx} + u_{yy} + u_{zz}] = u_t \qquad (a^2 > 0)$$

Equation (4.14) is the *three-dimensional heat equation*. For the reader who
has difficulty deriving (4.14), we can do no better than to recommend
consulting Fourier's derivation of the heat equation [see Fourier (1955),
Chap. II, §IV, especially paragraphs 142–145]. An alternative derivation,
using the Divergence Theorem, can be found in Churchill and Brown
(1978), Chap. 1 §6.

We will limit ourselves for now to two types of boundary and initial
conditions for (4.14). First, suppose we have a box $B = [0, L] \times [0, M] \times [0, N]$ whose sides are held at zero temperature

(4.14')
$$u(0, y, z, t) = u(L, y, z, t) = 0 \qquad u(x, 0, z, t) = u(x, M, z, t) = 0$$
$$u(x, y, 0, t) = u(x, y, N, t) = 0$$

and whose initial temperature is given by f, that is

(4.14″) $$u(x, y, z, 0 +) = f(x, y, z)$$

We can solve (4.14) subject to (4.14′) and (4.14″) by employing a triple sine
series. Since the discussion proceeds in the same way as in the previous two
subsections, we leave the details to the reader (see Exercise (4.32)].

Second, suppose we have a thermally isolated (insulated) box $B = [-L, L] \times [-M, M] \times [-N, N]$ whose initial temperature is given by f, that
is

(4.15) $$u(x, y, z, 0 +) = f(x, y, z)$$

and whose subsequent $(t > 0)$ temperature u satisfies (4.14). We shall solve
(4.14) subject to (4.15); our method will employ the complex finite Fourier
transform. It is very convenient, though not at all necessary, to assume that
$a^2 = \frac{1}{2}$ and $L = M = N = \frac{1}{2}$. Expanding f in a triple Fourier series, we have

$$f \sim \sum_{k, m, n = -\infty}^{+\infty} c_{kmn} e^{i2\pi(kx + my + nz)}$$

$$c_{kmn} = \int_{-1/2}^{1/2} \int_{-1/2}^{1/2} \int_{-1/2}^{1/2} f(x, y, z) e^{-i2\pi(kx + my + nz)} \, dx \, dy \, dz$$

From now on we shall write $\iiint$ *in place of* $\int_{-1/2}^{1/2} \int_{-1/2}^{1/2} \int_{-1/2}^{1/2}$. If we express u
in a triple Fourier series, then

$$u \sim \sum_{k, m, n = -\infty}^{+\infty} C_{kmn}(t) e^{i2\pi(kx + my + nz)}$$

where

$$C_{kmn}(t) = \iiint u(x, y, z, t) e^{-i2\pi(kx + my + nz)} \, dx \, dy \, dz$$

It is implicit in the Fourier expansion of u above that we are considering u as periodic (period 1) in each variable x, y, and z. Differentiating under the integral sign gives

$$C'_{kmn}(t) = \iiint u_t(x, y, z, t)e^{-i2\pi(kx+my+nz)}\, dx\, dy\, dz$$

$$= \tfrac{1}{2}\iiint [u_{xx}(x, y, z, t) + u_{yy}(x, y, z, t)$$

$$+ u_{zz}(x, y, z, t)]e^{-i2\pi(kx+my+nz)}\, dx\, dy\, dz$$

since u satisfies (4.14). Integrating by parts twice with respect to x, y, and z separately, we obtain due to the periodicity of u

$$C'_{kmn}(t) = i\pi \iiint [ku_x + mu_y + nu_z]e^{-i2\pi(kx+my+nz)}\, dx\, dy\, dz$$

$$= -2\pi^2 \iiint [k^2 + m^2 + n^2]u(x, y, z, t)e^{-i2\pi(kx+my+nz)}\, dx\, dy\, dz$$

Thus

$$C'_{kmn}(t) = -2\pi^2[k^2 + m^2 + n^2]C_{kmn}(t)$$

morcover, for $t = 0$ we have $C_{kmn}(0) = c_{kmn}$. Solving these last two equations yields $C_{kmn}(t) = c_{kmn}e^{-2\pi^2[k^2+m^2+n^2]t}$. Thus, we propose the following triple series solution to (4.14) subject to (4.15).

(4.16)

$$u(x, y, z, t) = \sum_{k,\, m,\, n=-\infty}^{+\infty} c_{kmn}e^{-2\pi^2[k^2+m^2+n^2]t}e^{i2\pi(kx+my+nz)}$$

$$c_{kmn} = \int_{-1/2}^{1/2}\int_{-1/2}^{1/2}\int_{-1/2}^{1/2} f(x, y, z)e^{-i2\pi(kx+my+nz)}\, dx\, dy\, dz$$

We conclude this section with a short analysis of the validity of (4.16) as a solution. Suppose that $|c_{kmn}| \le A$, a constant, for all k, m, and n; that is certainly true if $\iiint |f(x, y, z)|\, dx\, dy\, dz$ is finite. Then for $t > 0$ and all integers k, m, and n we have

$$0 < Ae^{-2\pi^2[k^2+m^2+n^2]t} \le Ae^{-2\pi^2|k|t}e^{-2\pi^2|m|t}e^{-2\pi^2|n|t}$$

and

$$\sum_{k,\, m,\, n=-\infty}^{+\infty} Ae^{-2\pi^2|k|t}e^{-2\pi^2|m|t}e^{-2\pi^2|n|t}$$

converges. It follows that the series shown in (4.16) does converge, uniformly and absolutely, and that the function u defined by that series is continuous. Moreover, since $e^{-\theta}$ decreases to 0 faster than any power of θ

(when θ tends to $+\infty$), we conclude that the series in (4.16) may be partially differentiated term by term any number of times with respect to x, y, z, or t. In particular, u satisfies the heat equation (4.14) (with $a^2 = \frac{1}{2}$).

If f is C^2 when periodically extended, then it can be shown that the triple Fourier series for f converges uniformly and absolutely to f. In which case, u is continuous for all x, y, z, and $t \geq 0$, hence

(4.17) $$\lim_{t \to 0+} u(x, y, z, t) = f(x, y, z)$$

Rather than going into much detail here, we prefer to leave the verification of (4.17) to the next section where we shall prove that it holds under much more general conditions than f being C^2.

The uniqueness of u as a solution, when u is assumed to be C^2 for $t > 0$ and continuous for $t \geq 0$, and periodic (period 1) in x, y, and z, follows by an argument like the one used above for the wave equation.

Exercises

(4.18) Solve the following problem for z

$$c^2[z_{xx} + z_{yy}] = z_{tt} \qquad z(0, y, t) = z(a, y, t) = z(x, 0, t) = z(x, b, t) = 0$$

subject to

(a) $z(x, y, 0+) = 0.4 \sin x \, |\sin \pi y| \qquad a = \pi \qquad b = 1 \qquad c^2 = 1$
(b) $z(x, y, 0+) = 0.1(x - 1)x(2 - y)y \qquad a = 1 \qquad b = 2 \qquad c^2 = 4$
(c) $z(x, y, 0+) = 1 \qquad a = \pi \qquad b = \pi \qquad c^2 = 0.1$

(4.19) Complete the verification of solution (4.4).

(4.20) Assume that $a = b = \pi$. Show that every sum

$$a_{mn} \cos \delta_{mn} ct \sin mx \sin ny + b_{mn} \sin \delta_{mn} ct \sin mx \sin ny$$

can be expressed as $h_{mn} \sin(\delta_{mn} ct + \theta_{mn}) \sin mx \sin ny$, hence the series in (4.4) can be written as $\sum_{m, n=1}^{\infty} h_{mn} \sin(\delta_{mn} ct + \theta_{mn}) \sin mx \sin ny$.

(4.21) Show that each term $h_{mn} \sin(\delta_{mn} ct + \theta_{mn}) \sin mx \sin ny$ in the series in (4.20) has a time period of $T_{mn} = 2\pi / \delta_{mn} c = 2\pi / [m^2 + n^2]^{1/2} c$. Hence a time frequency of $w_{mn} = 2\pi / T_{mn} = [m^2 + n^2]^{1/2} c$.

(4.22) *Nodal lines.*

(a) Let $m = 1$ and $n = 2$ in problem (4.21). Show that the frequency $w_{12} = \sqrt{5} c$ has corresponding modes of vibration

$$z_{12} = h_{12} \sin(w_{12} t + \theta_{12}) \sin x \sin 2y$$

$$z_{21} = h_{21} \sin(w_{12} t + \theta_{21}) \sin 2x \sin y$$

which are zero when $y = \frac{1}{2}\pi$, $x = \frac{1}{2}\pi$, respectively. Graph those equations, obtaining Figure 5.2. The dotted lines in the graphs in Figure 5.2 are called *nodal lines* for the frequency $w_{12} = \sqrt{5} c$.

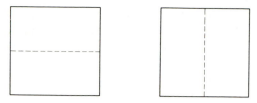

Figure 5.2 Nodal lines for the frequency w_{12}.

(b) Show that by varying the coefficients c and d in the linear combination $cz_{12} + dz_{21}$ we obtain an infinite number of nodal lines for the frequency w_{12}.

(c) Show that the frequency $w_{nn} = n\sqrt{2}c$ has only the nodal lines graphed in Figure 5.3.

(d) When will there be an infinite number of nodal lines for a frequency w_{mn}, as in (b), versus a finite number of nodal lines, as in (c)?

(4.23) Explain the following physical fact, using Fourier analysis: A rectangular drumhead does not exhibit tonal harmony, that is, you can keep a beat but you cannot play a tune. [*Remark*: The same thing occurs with circular drums, but the mathematics is more complicated; see Chapter 10, §4].

(4.24) Solve (4.9) subject to

(a) $f_1(x, y) = V$ a constant
(b) $f_1(x, y) = 0.2x(x - \pi)y(y - \pi)$
(c) $f_1(x, y) = \sin x + \frac{2}{3}\sin 2x$

(4.25) Solve the following three problems

(a) $$\Delta \Psi = 0 \qquad \text{on } (0, 2) \times (0, 3) \times (0, 5)$$

$$\Psi(0+, y, z) = 0 \qquad \Psi(2-, y, z) = A(y)B(z) \qquad \text{(see below)}$$
$$\Psi(x, 0+, z) = \Psi(x, 3-, z) = 0$$
$$\Psi(x, y, 0+) = \Psi(x, y, 5-) = 0$$

where

$$A(y) = \begin{cases} 0 & \text{for } |y - \frac{3}{2}| > \frac{3}{4} \\ 1 & \text{for } |y - \frac{3}{2}| < \frac{3}{4} \end{cases}$$

$$B(z) = \begin{cases} 0 & \text{for } |z - \frac{5}{2}| > \frac{5}{4} \\ 3 & \text{for } |z - \frac{5}{2}| < \frac{5}{4} \end{cases}$$

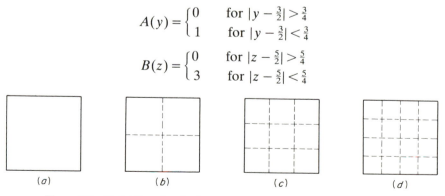

(a)   (b)   (c)   (d)

Figure 5.3 Nodal lines for the frequency w_{nn} [$n = 1(a)$, $2(b)$, $3(c)$, $4(d)$].

(b)
$$\Delta\Psi = 0 \quad \text{on } (0, 1) \times (0, \tfrac{1}{2}\pi) \times (0, \pi)$$

$$\Psi(0+, y, z) = 0.2y(y - \tfrac{1}{2}\pi)(z - \pi)z \qquad \Psi(1-, y, z) = 0$$

$$\Psi(x, 0+, z) = 0 \qquad \Psi(x, \tfrac{1}{2}\pi-, z) = 0.1x(x - 1)(z - \pi)z$$

$$\Psi(x, y, 0+) = \Psi(x, y, \pi-) = 0$$

(c)
$$\Delta\Psi = 0 \quad \text{on } (0, \pi) \times (0, 1) \times (0, 3)$$

$$\Psi(0+, y, z) = 0 \qquad \Psi(\pi-, y, z) = yz$$

$$\Psi(x, 0+, z) = \Psi(x, 1-, z) = 0$$

$$\Psi(x, y, 0+) = 4 \qquad \Psi(x, y, 3-) = 2$$

(4.26)   Prove that if f satisfies the hypotheses of Theorem (3.6), then Ψ as shown in (4.12) yields a solution to Dirichlet's problem (where $f = f_1$ on the top face of $\partial\mathscr{C}$ and 0 on the other faces). Verify the uniqueness of Ψ, either by Fourier analysis or by the maximum–minimum principle.

(4.27)   Solve (4.14) subject to (4.15) where a^2, f, K, M, and N are as follows.

(a) $f(x, y, z) = x^2y^2z^2 \qquad K = M = N = \tfrac{1}{2} \qquad a^2 = \tfrac{1}{2}$
(b) $f(x, y, z) = |x|^2 \, y^2(z^2 - N^2) \qquad K = M = N = \tfrac{1}{2} \qquad a^2 = \tfrac{1}{2}$
(c) f as in (a) $\quad K = M = N = \pi \qquad a^2 = 1$
(d) f as in (b) $\quad K = 1 \qquad M = 2 \qquad N = 3 \qquad a^2 = 0.1$

(4.28)   Find a series solution to

$$a^2[u_{xx} + u_{yy}] = u_t \qquad (t > 0)$$

$$u(x, y, 0+) = f(x, y) \qquad (-L \le x \le L \qquad -M \le y \le M)$$

Discuss the validity and uniqueness of this series solution.

(4.29)   Solve (4.28) explicitly, where a^2, L, M, and f are as follows.

(a) $f(x, y) = x^2y^2 \qquad L = \tfrac{1}{2} \qquad M = \tfrac{1}{2} \qquad a^2 = \tfrac{1}{2}$
(b) $f(x, y) = 5 \qquad L = M = \pi \qquad a^2 = 1$
(c) $f(x, y) = |xy| \qquad L = 2 \qquad M = 3 \qquad a^2 = 0.1$

(4.30)   Find a series solution to

$$a^2[u_{xx} + u_{yy}] = u_t \qquad (t \ge 0)$$

$$u(x, y, 0+) = f(x, y) \qquad (0 \le x \le L \qquad 0 \le y \le M)$$

$$u(x, 0) = u(x, M) = 0 \qquad u(0, y) = u(L, y) = 0.$$

Discuss the validity and uniqueness of this series solution.

(4.31)   Solve (4.30) explicitly, where a^2, M, L, and f are as follows.

(a) $f(x, y) = |x| \, y^2 \qquad L = M = \pi \qquad a^2 = 1$
(b) $f(x, y) = 8 \qquad L = M = 1 \qquad a^2 = \tfrac{1}{2}$

(4.32) Show that (4.14) through (4.14″) has a series solution based upon triple sine series.

(4.33) Prove that the temperature u shown in (4.16) satisfies

(4.33′) $$\lim_{t \to +\infty} u(x, y, z, t) = \int_{-1/2}^{1/2} \int_{-1/2}^{1/2} \int_{-1/2}^{1/2} f(x, y, z) \, dx \, dy \, dz$$

and interpret that limit physically.

(4.34) Discuss the existence of a limit such as (4.33′) for problem (4.30) and interpret that limit physically.

Poisson's Equation in a Box

Poisson's equation arises often in mathematical physics; for example, it is one consequence of Maxwell's equations for the electromagnetic field. The following problem involves Poisson's equation with boundary conditions on a rectangular box $B = (0, a) \times (0, b) \times (0, c)$. It is the subject of the following four exercises.

(4.35)
$$\Psi_{xx} + \Psi_{yy} + \Psi_{zz} = -F(x, y, z) \qquad \text{(Poisson's equation)}$$
$$\Psi(0, y, z) = \Psi(a, y, z) = 0$$
$$\Psi(x, 0, z) = \Psi(x, b, z) = 0 \qquad \text{(Boundary conditions)}$$
$$\Psi(x, y, 0) = \Psi(x, y, c) = 0$$

(4.36) Solve (4.35) using a triple series. [*Hint*: Set $\Psi(x, y, z) = \sum D_{kmn} \sin(k\pi x/a) \sin(m\pi y/b) \sin(n\pi z/c)$ and substitute into Poisson's equation.]

(4.37) Discuss the validity and uniqueness of the solution found in (4.36).

(4.38) Use the result of (4.36) to solve (4.35) for the following values of a, b, c, and F.

(a) $a = b = c = \pi$ $F(x, y, z) = |xyz|$
(b) $a = 1$ $b = 2$ $c = 3$ $F(x, y, z) = x(x - 1)y(y - 2)z(z - 3)$
(c) $a = \frac{1}{2}\pi$ $b = \frac{1}{4}\pi$ $c = \frac{1}{2}\pi$ $F(x, y, z) = 1$
(d) $a = \pi$ $b = \frac{1}{2}\pi$ $c = \pi$ $F(x, y, z) = |\sin x| \cdot |\sin 2\pi y| \cdot |\sin z|$

(4.39) Solve the following problem $[a = 1, \; b = 2, \; c = 3]$

$$\Delta \Psi = x(1 - x)y(y - 2)z(z - 3)) \qquad \text{[Take note of (4.38b).]}$$

$$\Psi(0, y, z) = 0 \qquad \Psi(1, y, z) = |\sin(\pi y/2)| \cdot |\sin(\pi z/3)|$$

$$\Psi(x, 0, z) = \Psi(x, 2, z) = 0 \qquad \Psi(x, y, 0) = \Psi(x, y, 3) = 0$$

*§5. Completeness of Complex Exponential Systems

In this section we shall prove the completeness of the complex exponential system $\{e^{i2\pi(kx+my+nz)}\}$. Although our approach is not the simplest, it has the advantage that we shall obtain some nontrivial results concerning the heat equation. The completeness of $\{e^{i(k\pi x/a+m\pi y/b+n\pi z/c)}\}$ follows from the completeness of the former system by a change of variables. Moreover, by a fairly obvious modification of the proof below we can prove that $\{e^{i2\pi(mx+ny)}\}$ is complete and hence so is $\{e^{i(m\pi x/a+n\pi y/b)}\}$ by a change of variables.

Let f be a continuous function on the cube $[-\frac{1}{2}, \frac{1}{2}] \times [-\frac{1}{2}, \frac{1}{2}] \times [-\frac{1}{2}, \frac{1}{2}]$. We shall continue the convention of writing $\int\int\int$ in place of $\int_{-1/2}^{1/2} \int_{-1/2}^{1/2} \int_{-1/2}^{1/2}$ that we adopted in the previous section. In that section we found for $t > 0$ [see (4.16)]

$$u(x, y, z, t) = \sum_{k,m,n=-\infty}^{+\infty} \left[\int\int\int f(s, v, w) e^{-i2\pi(ks+mv+nw)} \right.$$
$$\left. \times ds\, dv\, dw \right] e^{-2\pi^2(k^2+m^2+n^2)t} e^{i2\pi(kx+my+nz)}$$

solves the heat equation (4.14). The key part of our demonstration of completeness consists in showing that (4.17) also holds.

Due to uniformity of convergence for $t > 0$, we have

$$u(x, y, z, t) = \int\int\int f(s, v, w) \sum_{k,m,n=-\infty}^{\infty} e^{i2\pi[k(x-s)+m(y-v)+n(z-w)]}$$
$$\times e^{-2\pi^2(k^2+m^2+n^2)t} ds\, dv\, dw$$

$$= \int\int\int f(s, v, w)\theta(x-s, y-v, z-w) ds\, dv\, dw$$

where

$$\theta(s, v, w, t) = \sum_{k,m,n=-\infty}^{\infty} e^{i2\pi(ks+mv+nw)} e^{-2\pi^2(k^2+m^2+n^2)t} \qquad (t > 0).$$

The *three-dimensional theta function* θ defined above is periodic (period 1) in each variable s, v, and w. It follows, by changing variables and using periodicity, that

(5.1) $u(x, y, z, t) = \int\int\int f(x-s, y-v, z-w)\theta(s, v, w, t) ds\, dv\, dw$

The proof of (4.17) depends on the fact that θ is a (three-dimensional) summation kernel, parameterized by $t > 0$. To be precise, we have the following Lemma.

(5.2) Lemma. The three-dimensional theta function θ satisfies

(A_1) $\int\int\int \theta(s, v, w, t) ds\, dv\, dw = 1$ for all $t > 0$

(A$_2$) $\theta(s, v, w, t) \geq 0$ for all $t \geq 0$ and all s, v, and w

(A$_3$) Suppose that (s, v, w) belongs to $[-\frac{1}{2}, \frac{1}{2}] \times [-\frac{1}{2}, \frac{1}{2}] \times [-\frac{1}{2}, \frac{1}{2}]$. Given $\epsilon > 0$ and $\delta > 0$ we can have

$$0 \leq \theta(s, v, w, t) < \epsilon$$

provided that $[s^2 + v^2 + w^2]^{1/2} \geq \delta$ and t is taken close enough to 0.

Proof. By factoring the partial sums for θ, in an obvious way, we see that for each $t > 0$

(5.3)

$$\theta(s, v, w, t) = \sum_{k=-\infty}^{+\infty} e^{-2\pi^2 k^2 t} e^{i2\pi ks} \sum_{m=-\infty}^{+\infty} e^{-2\pi^2 m^2 t}$$

$$\times e^{i2\pi mv} \sum_{n=-\infty}^{+\infty} e^{-2\pi^2 n^2 t} e^{i2\pi nw}$$

$$= \theta_1(s, t) \theta_1(v, t) \theta_1(w, t)$$

where θ_1 is the theta function discussed in §7, Chapter 4. [See, in particular, formula (7.5) and Lemma (7.7) from that chapter.] That lemma implies, by a straightforward calculation, that (A$_1$) and (A$_2$) hold for θ because of (5.3). Moreover, Jacobi's identity (7.5) from Chapter 4 plus (5.3) above imply that

$$\theta(s, v, w, t) = (2\pi t)^{-3/2} \sum_{k,m,n=-\infty}^{+\infty} e^{-[(s-k)^2 + (v-m)^2 + (w-n)^2]/2t}$$

$$\leq (2\pi t)^{-3/2} e^{-\delta^2/2t}$$

$$+ (2\pi t)^{-3/2} \sum_{\substack{k,m,n=-\infty \\ (k,m,n)\neq(0,0,0)}}^{+\infty} e^{-[(s-k)^2 + (v-m)^2 + (w-n)^2]/2t}$$

Since $(s - k)^2 + (v - m)^2 + (w - n)^2$ equals the distance squared of (k, m, n) from (s, v, w), if $(k, m, n) \neq (0, 0, 0)$ then for some positive constant c

$$[(s - k)^2 + (v - m)^2 + (w - n)^2]/2t \geq c[k^2 + m^2 + n^2]/2t$$

Hence we can dominate the last series above by a convergent triple integral, and then change to spherical coordinates, obtaining

$$\theta(s, v, w, t) \leq 27(2\pi t)^{-3/2} e^{-\delta^2/2t} + (2\pi t)^{-3/2} \sum_{\substack{k,m,n=-\infty \\ (k^2+m^2+n^2>3)}}^{+\infty} e^{-c(k^2+m^2+n^2)/2t}$$

$$< 27(2\pi t)^{-3/2} e^{-\delta^2/2t} + (2\pi t)^{-3/2} \iiint\limits_{x^2+y^2+z^2\geq 1} e^{-c(x^2+y^2+z^2)/2t} \, dx \, dy \, dz$$

$$\leq 27(2\pi t)^{-3/2} e^{-\delta^2/2}$$

$$+ (2\pi t)^{-3/2} \int_0^\pi \int_0^{2\pi} \int_1^\infty e^{-cr^2/2t} r^2 \sin \Phi \, dr \, d\phi \, d\Phi$$

Thus, evaluating the first two integrals in the last line and making a change of variables, we have

$$0 \le \theta(s, v, w, t) < 27(2\pi t)^{-3/2} e^{-\delta^2/2t} + 4\pi(2\pi)^{-3/2} \int_{1/\sqrt{t}}^{\infty} e^{-cx^2/2} x^2 \, dx$$

As t tends to 0, both terms of the right side of the inequality above tend to 0. Therefore, (A_3) holds. ∎

Using Lemma (5.2) we get the following theorem.

(5.4) Theorem. If f is continuous with period 1 in x, y, and z, then

$$\lim_{t \to 0+} u(x, y, z, t) = f(x, y, z)$$

holds uniformly.

Proof. Because u is described by (5.1) and because Lemma (5.2) shows that θ is a summation kernel, we obtain the desired limit through an obvious generalization of Theorem (8.3), Chapter 2. ∎

Remark. Since θ is a summation kernel, Theorem (8.4), Chapter 2, also generalizes. Hence, if $\iiint |f(x, y, z)| \, dx \, dy \, dz$ is finite, and f has period 1 in each variable, then

$$\lim_{(x, y, z, t) \to (x_0, y_0, z_0, 0+)} \iiint f(x - s, y - v, z - w)\theta(s, v, w, t)$$

$$ds \, dv \, dw = f(x_0, y_0, z_0)$$

provided (x_0, y_0, z_0) is a point of continuity for f.

The completeness of $\{e^{i2\pi(kx+my+nz)}\}$ follows from the theorem just proved. First, suppose that f is continuous, real valued, and has period 1. Then Theorem (5.4) implies that for t_0 sufficiently small

$$\|u(x, y, z, t_0) - f(x, y, z)\|_2^2 = \iiint [u(x, y, z, t_0) - f(x, y, z)]^2 \, dx \, dy \, dz$$

$$< \iiint \epsilon^2 \, dx \, dy \, dx = \epsilon^2$$

where ϵ is some preassigned small positive constant. If we let $S_{KMN}^{t_0}$ stand for

$$\sum_{k=-K, m=-M, n=-N}^{K, M, N} c_{kmn} e^{-2\pi^2(k^2+m^2+n^2)t_0} e^{i2\pi(kx+my+nz)}$$

then for fixed $t_0 > 0$ we have

$$\lim_{K, M, N \to \infty} S_{KMN}^{t_0}(x, y, z) = u(x, y, z, t_0)$$

uniformly in (x, y, z) [due to the rapid decrease to zero of the factors $e^{-2\pi^2(k^2+m^2+n^2)t_0}$ as k, m, or n tends to infinity]. Therefore,

$$\|u(x, y, z, t_0) - S_{KMN}^{t_0}(x, y, z)\|_2 < \epsilon$$

provided *K*, *M*, and *N* are sufficiently large. Hence, by the triangle inequality

$$\|f - S^{t_0}_{KMN}\|_2 \le \|f - u\|_2 + \|u - S^{t_0}_{KMN}\|_2 < 2\epsilon$$

The Least Squares Theorem implies $\|f - S_{KMN}\|_2 < 2\epsilon$ for *K*, *M*, and *N* sufficiently large. Thus,

(5.5) $$\lim_{K, M, N \to \infty} \left[\iiint [f(x, y, z) - S_{KMN}(x, y, z)]^2 \, dx \, dy \, dz \right]^{1/2} = 0$$

provided *f* is continuous *and* periodic. We obtain (5.5) for every continuous function *f* on the cube $\mathscr{C} = [-\frac{1}{2}, \frac{1}{2}] \times [-\frac{1}{2}, \frac{1}{2}] \times [-\frac{1}{2}, \frac{1}{2}]$ by defining an auxiliary function g_ϵ which equals *f* up to within δ of the boundary $\partial \mathscr{C}$ then extends down (or up) to 0 in a continuous fashion.[9] We then complete our proof by taking δ small enough that $\|f - g_\epsilon\|_2 < \epsilon$ and using the triangle inequality and the Least Squares Theorem. The details are left to the reader as Exercise (5.7). [*Hint*: Mimic the completion of the proof of Theorem (6.8), Chapter 2.] Since completeness holds for *f* real valued, we obtain it for *f* complex valued by considering real and imaginary parts of *f*.

(5.6) Remark. If the reader completes the proof above, it will be seen that the key element is that we can obtain a continuous periodic function g_ϵ such that $\|f - g_\epsilon\|_2 < \epsilon$. this can be done, not only when *f* is continuous on the cube $\mathscr{C}$, but when *f* is square integrable over $\mathscr{C}$ (either in the sense of Riemann or Lebesgue). In that general case, we then also obtain (5.5).

Exercises

(5.7)   Finish the proof of completeness begun in the text, and verify that $\{e^{i2\pi(mx+ny)}\}$ is complete.

(5.8)   Prove Parseval's equality

$$\sum_{m, n = -\infty}^{+\infty} \lambda_{mn}[A_{mn}^2 + B_{mn}^2 + C_{mn}^2 + D_{mn}^2] = \frac{1}{\pi^2} \int_{-\pi}^{\pi} \int_{-\pi}^{\pi} f^2(x, y) \, dx \, dy$$

for a (continuous) real valued function *f* on $[-\pi, \pi] \times [-\pi, \pi]$.

(5.9)   Using (5.8), prove Parseval's equality (in complex form)

$$\sum_{m, n = -\infty}^{+\infty} |c_{mn}|^2 = \frac{1}{4\pi^2} \int_{-\pi}^{\pi} \int_{-\pi}^{\pi} |f(x, y)|^2 \, dx \, dy$$

(5.10)   Suppose that *f* is continuous on the cube $\mathscr{C} = [-\frac{1}{2}, \frac{1}{2}] \times [-\frac{1}{2}, \frac{1}{2}] \times [-\frac{1}{2}, \frac{1}{2}]$. Prove that there exists an infinitely differentiable periodic function *h* such that over $\mathscr{C}$ we have $\|f - h\|_2 < \epsilon$ where ϵ is any preassigned small positive constant.

---

9. That is, $g_\epsilon = fh_\epsilon$ where $h_\epsilon(x, y, z) = q_\epsilon(x)q_\epsilon(y)q_\epsilon(z)$ for $q_\epsilon(s) = 1$ if $|s| \le \frac{1}{2} - \delta$ and extended linearly down to 0 for $\frac{1}{2} - \delta \le |s| \le \frac{1}{2}$.

(5.11)   Generalize the results of Exercises (7.17) through (7.21) in Chapter 4 to the heat equation $u_t = a^2 \Delta u$ where Δ is the (two-) three-dimensional Laplacian.

(5.12)   Show that to prove Theorem (5.4) it would have sufficed for us to prove that θ satisfies (A_1), (A_2), *and*

$$(A_3') \quad \lim_{t \to 0+} \iiint_{(s^2+v^2+w^2 \geq \delta^2)} \theta(s, v, w, t) \, ds \, dv \, dw = 0, \qquad \text{for each } \delta > 0.$$

[Compare (8.31), Chapter 2.] Moreover, show that (A_3') follows easily from $\theta(s, v, w, t) = \theta_1(s, t)\theta_1(v, t)\theta_1(w, t)$.

References

For further discussion of multiple Fourier series, see Bochner (1959), Igari (1968), and Stein and Weiss (1971). Triple Fourier series play a major role in the theory of the structure of crystals and molecules; see Woolfson (1970) and Wheatley (1981). See also the remarks by H. Hauptman in the symposium "Mathematics: The Unifying Thread of Science," pp. 720–725 of Vol. 33, #5 (October 1986) issue of the *Notices of the Amer. Math. Soc.*

6

Basic Theory
of the Fourier Transform

In this chapter we shall examine the basic mathematical theory of the Fourier transform, which provides pure and applied mathematics with one of its most powerful tools. We shall compute many transforms and discuss the basic computational theorems that enable us to construct a whole calculus of transforms. Two forms of inversion for Fourier transforms will be discussed. One inversion theorem, based on Gauss–Weierstrass summation, generalizes easily to multiple transforms. Our second inversion theorem, due to Dirichlet, is presented for completeness of discussion of single variable transforms; it does not generalize easily to multiple transforms. Convolution and autocorrelation play an important role in transform theory; we shall discuss some aspects of these two concepts. We conclude the chapter with a discussion of multiple Fourier transforms, which are essential for the applications that we will treat in Chapter 7.

§1. Introduction

Fourier transforms are widely used in mathematics and science. Here is a list of some of the applications of Fourier transforms:

- Optics: Fraunhofer diffraction, interference theory, imaging with lenses
- Crystal structure: X-ray diffraction, spectroscopy
- Molecular structure: electron diffraction, spectroscopy
- Radio astronomy: aperture synthesis of radio waves received by antenna arrays
- Communications: theory of noise, sampling theory
- Probability and statistics: characteristic functions of distributions
- Quantum mechanics: equivalence of position and momentum representations

In addition to these, the Fourier transform is used in the fields of partial differential equations, electric networks, and geology. It is clear then that the Fourier transform is very useful. In the next chapter we shall examine a few of its applications.

We will now derive the Fourier transform. Suppose that f is a piecewise smooth function defined over $\mathbb{R}$. We cannot necessarily expand f in a Fourier series, *valid over all of* $\mathbb{R}$, because f may not be periodic. Let's begin then by considering f_L the periodic extension of f considered as a function on the interval $(-L, L)$. (See Figure 6.1).

Note that as L tends to $+\infty$ we recover f. If we expand f_L in a complex Fourier series, period $2L$, we get

(1.1)
$$f_L(x) = \sum_{n=-\infty}^{+\infty} c_n e^{i(n\pi x/L)}$$

$$c_n = \frac{1}{2L} \int_{-L}^{L} f(x) e^{-i(n\pi x/L)} \, dx$$

We define u_n to be the number $n/2L$ for each integer n. If we substitute the integral defining c_n into the series in (1.1) we obtain

(1.2)
$$f_L(x) = \sum_{n=-\infty}^{+\infty} \left[\frac{1}{2L} \int_{-L}^{L} f(x) e^{-i2\pi u_n x} \, dx \right] e^{i2\pi u_n x}$$

Since $\Delta u_n = u_{n+1} - u_n = 1/2L$ we can express (1.2) as

(1.3)
$$f_L(x) = \sum_{n=-\infty}^{+\infty} \left[\int_{-L}^{L} f(x) e^{-i2\pi u_n x} \, dx \right] e^{i2\pi u_n x} \Delta u_n$$

If we let L tend to $+\infty$, then f_L becomes f. Moreover, since the summation in (1.3) resembles a Riemann sum (using the variable u), we infer that (1.3) becomes

(1.4)
$$f(x) = \int_{-\infty}^{+\infty} \left[\int_{-\infty}^{+\infty} f(x) e^{-i2\pi u x} \, dx \right] e^{i2\pi u x} \, du$$

This transition from (1.3) to (1.4) was hardly a rigorous one; however, in

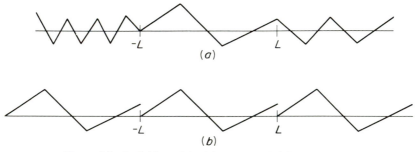

Figure 6.1 Definition of f_L. (a) Graph of f. (b) Graph of f_L.

§§4 and 5 we will prove theorems that describe conditions under which (1.4) is valid. The expression in brackets in (1.4) is called the Fourier transform of f. Here is a formal definition of this new concept.

(1.5) Definition. Let f be an absolutely integrable function over $\mathbb{R}$. The *Fourier transform* of f is denoted by $\hat{f}$ and is defined as a function of u by

$$\hat{f}(u) = \int_{-\infty}^{+\infty} f(x)e^{-i2\pi ux}\, dx \qquad \blacksquare$$

Thus formula (1.4) can be expressed as

(1.6)
$$f(x) = \int_{-\infty}^{+\infty} \hat{f}(u)e^{i2\pi ux}\, du$$

which is known as *Fourier inversion*.

(1.7) Remark. Sometimes in applications the transform $\hat{f}$ is called the (continuous) *spectrum* of f. Formula (1.6) expresses f as a continuous superposition (integral) of waves, a "wave bundle," having amplitudes given by $\hat{f}(u)$.

Here are a few examples of Fourier transforms.

(1.8) Examples

(a)   Let $f(x) = \begin{cases} 1 & \text{if } |x| < \frac{1}{2} \\ 0 & \text{if } |x| > \frac{1}{2} \end{cases}$

Then

$$\hat{f}(u) = \int_{-\infty}^{+\infty} f(x)e^{-i2\pi ux}\, dx = \int_{-1/2}^{1/2} e^{-i2\pi ux}\, dx$$

$$= \int_{-1/2}^{1/2} \cos 2\pi ux\, dx - i \int_{-1/2}^{1/2} \sin 2\pi ux\, dx$$

$$= \frac{\sin \pi u}{\pi u}$$

If we define Π by

$$\Pi(x) = \begin{cases} 1 & \text{if } |x| < \frac{1}{2} \\ 0 & \text{if } |x| > \frac{1}{2} \end{cases}$$

and sinc u by sinc $u = \sin \pi u / \pi u$, then $\hat{\Pi}(u) = \text{sinc } u$.

(b)   Let Λ be defined by

$$\Lambda(x) = \begin{cases} 1 - |x| & \text{if } |x| \le 1 \\ 0 & \text{if } |x| > 1 \end{cases}$$

Then

$$\hat{\Lambda}(u) = \int_{-1}^{1} (1 - |x|)e^{-i2\pi ux} \, dx = 2 \int_{0}^{1} (1 - x)\cos 2\pi ux \, dx$$

$$= \frac{1 - \cos 2\pi u}{2\pi^2 u^2} = \frac{\sin^2 \pi u}{\pi^2 u^2}$$

Thus $\hat{\Lambda}(u) = \text{sinc}^2 u$.

(c)   Let f be defined by $f(x) = e^{-2\pi|x|}$. Then

$$\hat{f}(u) = \int_{-\infty}^{+\infty} e^{-2\pi|x|} e^{-i2\pi ux} \, dx = 2 \int_{0}^{+\infty} e^{-2\pi x} \cos 2\pi ux \, dx$$

If we integrate by parts twice, we get

$$\hat{f}(u) = -\frac{1}{\pi} e^{-2\pi x} \cos 2\pi ux \,|_{x=0}^{+\infty} - 2u \int_{0}^{+\infty} e^{-2\pi x} \sin 2\pi ux \, dx$$

$$= \frac{1}{\pi} - 2u \int_{0}^{+\infty} e^{-2\pi x} \sin 2\pi ux \, dx$$

$$= \frac{1}{\pi} - u^2 \hat{f}(u)$$

Solving for $\hat{f}(u)$ yields $\hat{f}(u) = (1/\pi)[1/(1 + u^2)]$.

(d)   Let f be defined by $f(x) = e^{-2\pi|x|} \sin 2\pi x$. Since $\sin 2\pi x = (i/2)e^{-i2\pi x} - (i/2)e^{i2\pi x}$ we have

$$\hat{f}(u) = \frac{i}{2} \int_{-\infty}^{+\infty} e^{-2\pi|x|} e^{-i2\pi(u+1)x} \, dx - \frac{i}{2} \int_{-\infty}^{+\infty} e^{-2\pi|x|} e^{-i2\pi(u-1)x} \, dx$$

$$= \frac{1}{2\pi} \frac{i}{1 + (u + 1)^2} - \frac{1}{2\pi} \frac{i}{1 + (u - 1)^2}$$

Using the notation $f(x) \supset \hat{f}(u)$ the examples above can be summarized as follows

(1.9)

$$\Pi(x) \supset \text{sinc } u \qquad \Lambda(x) \supset \text{sinc}^2 u \qquad e^{-2\pi|x|} \supset \frac{1}{\pi} \frac{1}{1 + u^2}$$

$$e^{-2\pi|x|} \sin 2\pi x \supset \frac{1}{2\pi} \frac{i}{1 + (u + 1)^2} - \frac{1}{2\pi} \frac{i}{1 + (u - 1)^2}$$

Our derivation of the Fourier transform can be modified in various ways resulting in several alternative definitions. For example, the following definitions of Fourier transform are commonly used:

$$\int_{-\infty}^{+\infty} f(x)e^{-iux} \, dx \qquad \frac{1}{\sqrt{2\pi}} \int_{-\infty}^{+\infty} f(x)e^{-iux} \, dx \qquad \int_{-\infty}^{+\infty} f(x)e^{iux} \, dx$$

For each of the alternate definitions above, there is a different version of Fourier inversion. Our choice of definition of Fourier transform was based on mathematical aesthetics; the most elegant forms of the various formulas that we shall use are obtained through using our definition. Our form for $\hat{f}$ is also very popular among engineers.

Exercises

(1.10) Find the Fourier transforms of the following functions.

(a) $f(x) = \begin{cases} 1 & \text{if } |x| < 1 \\ 0 & \text{if } |x| > 1 \end{cases}$

(b) $f(x) = \begin{cases} 1 & \text{if } 0 < x < 1 \\ -1 & \text{if } -1 < x < 0 \\ 0 & \text{if } |x| > 1 \end{cases}$

(c) $f(x) = e^{-|x|}$

(d) $f(x) = e^{-|x|} \cos 2\pi x$

(e) $f(x) = e^{-|x|} + i\Pi(x)$

(f) $f(x) = \Lambda(x) + ie^{-3|x|} \cos 2\pi x$

(1.11) Modify the details of the derivation of the Fourier transform so as to obtain the first two alternate definitions listed above and obtain the corresponding inversion formulas.

(1.12) Use real Fourier series to obtain (in a nonrigorous way) the following result:

$$f(x) = 2 \int_0^{+\infty} \left[\int_{-\infty}^{+\infty} f(s) \cos 2\pi u(s - x)\, ds \right] du$$

which is known as *Fourier's integral theorem*.

(1.13) *Cosine Transforms.* (a) Prove that if f is an even, absolutely integrable function, then $\hat{f}(u) = 2 \int_0^{+\infty} f(x) \cos 2\pi ux\, dx$. (b) If f is absolutely integrable over $(0, +\infty)$, then we define the *cosine transform* $\hat{f}_C$ by $\hat{f}_C(u) = 2 \int_0^{+\infty} f(x) \cos 2\pi ux\, dx$. Prove that the cosine transform of f equals the Fourier transform of the even extension of f and find the cosine transforms of the following functions:

$$f(x) = e^{-x} \qquad h(x) = \begin{cases} 1 & \text{if } 0 < x < 2 \\ 0 & \text{if } 2 < x \end{cases}$$

$$k(x) = \begin{cases} 1 - 2x & \text{if } 0 < x < \frac{1}{2} \\ 0 & \text{if } x > \frac{1}{2} \end{cases}$$

(1.14) *Sine transforms.* (a) Prove that if f is an odd and absolutely integrable function, then $\hat{f}(u) = -2i \int_0^{+\infty} f(x) \sin 2\pi ux\, dx$. (b) If f is absolutely integrable over $(0, +\infty)$, then we define the *sine transform* $\hat{f}_S$ by $\hat{f}_S(u) = 2 \int_0^{+\infty} f(x) \sin 2\pi ux\, dx$. Prove that the sine transform of f equals i times the Fourier transform of the odd extension of f and find the sine transforms of the functions given in (1.13b).

§2. Improper Integrals Dependent on a Parameter

A Fourier transform $\int_{-\infty}^{+\infty} f(s)e^{-i2\pi us}\, ds$ is a special case of an improper integral that depends upon a parameter u. We shall now discuss some of the principal theorems concerning these types of integrals and apply our results to Fourier transforms. Many more applications will be given in the rest of this chapter and Chapter 7.

Let $F(s, u)$ denote a function of s for $-\infty < s < +\infty$ and u for $a \le u \le b$; sometimes we will allow $a = -\infty$ or $b = +\infty$. We shall assume that $F(s, u) = f(s)G(s, u)$ where f is piecewise continuous[1] and G is continuous. This special form for F is necessary for treating the Fourier transform, where $F(s, u) = f(s)e^{-i2\pi us}$, as well as several other applications. An improper integral dependent on a parameter u is an infinite improper integral having one of the following forms

$$(2.1) \qquad \int_{p}^{+\infty} F(s, u)\, ds \qquad \int_{-\infty}^{p} F(s, u)\, ds \qquad \int_{-\infty}^{+\infty} F(s, u)\, ds$$

where p is a fixed real number. Since

$$\int_{-\infty}^{p} F(s, u)\, ds = \int_{-p}^{+\infty} F(-s, u)\, ds$$

and

$$\int_{-\infty}^{+\infty} F(s, u)\, ds = \int_{0}^{+\infty} F(s, u)\, ds + \int_{0}^{+\infty} F(-s, u)\, ds$$

we will restrict our attention to integrals of the form $\int_{p}^{+\infty} F(s, u)\, ds$ *with the understanding that all of our results apply to the other two integrals in* (2.1).

We now come to the main definition of this section.

(2.2) Definition.   The integral $\int_{p}^{+\infty} F(s, u)\, ds$ is *uniformly convergent* for $a \le u \le b$ if

(a) $\int_{p}^{+\infty} F(s, u)\, ds = \lim_{R \to +\infty} \int_{p}^{R} F(s, u)\, ds$ exists for $a \le u \le b$.
(b)   Given $\delta > 0$, no matter how small δ may be, there exists a constant $A \ge p$ such that $|\int_{R}^{+\infty} F(s, u)\, ds| < \delta$ for all $R \ge A$ and $a \le u \le b$. ∎

A uniformly convergent integral is the continuous analogue of a uniformly convergent infinite series. Compare Definition (5.1), Chapter 1, with the one above; the variable s replaces the index n and the process of integration replaces the process of summation. The following theorem is analogous to Weierstrass' M-test for infinite series.

(2.3) Theorem.   If for all $s \ge q \ge p$ and all $a \le u \le b$, we have $|F(s, u)| \le g(s)$ and $\int_{q}^{+\infty} g(s)\, ds$ converges, then the integral $\int_{p}^{+\infty} F(s, u)\, ds$ is uniformly convergent.

---

1. Our results remain true when f is merely Riemann (or Lebesgue) integrable over every closed interval.

Proof. For each u value $\int_p^{+\infty} F(s, u)\,ds$ converges (absolutely) by the Comparison Test for infinite integrals. Moreover, we have for $R \geq q$

$$\left| \int_R^{+\infty} F(s, u)\,ds \right| \leq \int_R^{+\infty} |F(s, u)|\,ds \leq \int_R^{+\infty} g(s)\,ds$$

Since $\int_q^{+\infty} g(s)\,ds$ converges, the last term of the inequality above can be made smaller than any given $\delta > 0$ if $R \geq A$ for A sufficiently large. Thus, for $R \geq A$ and all $a \leq u \leq b$ we have $|\int_R^{+\infty} F(s, u)\,ds| < \delta$. ∎

(2.4) Example. Suppose that f is absolutely integrable. Then, since

$$|f(s)e^{-i2\pi us}| = |f(s)|$$

we see from Theorem (2.3) that the Fourier transform $\int_{-\infty}^{+\infty} f(s)e^{-i2\pi us}\,ds$ is uniformly convergent for all real u values. Moreover, we have

$$|\hat{f}(u)| = \left| \int_{-\infty}^{+\infty} f(s)e^{-i2\pi us}\,ds \right|$$

$$\leq \int_{-\infty}^{+\infty} |f(s)e^{-i2\pi us}|\,ds = \int_{-\infty}^{+\infty} |f(s)|\,ds$$

which proves that $\hat{f}$ is a bounded function, bounded by the constant $\int_{-\infty}^{+\infty} |f(s)|\,ds$.

Just as Theorem (2.3) generalizes Weierstrass's M-test, there are theorems for uniformly convergent integrals that generalize the theorems on continuity, integration term by term, and differentiation term by term of infinite series. In these next two theorems we shall make explicit use of the form $F(s, u) = f(s)G(s, u)$ that we defined for F above.

(2.5) Theorem. Let $a \leq u \leq b$ and let $F(u) = \int_p^{+\infty} f(s)G(s, u)\,ds$ where f is piecewise continuous and G is continuous. If $\int_p^{+\infty} f(s)G(s, u)\,ds$ converges uniformly, then we have

(a) $\int_p^{+\infty} f(s)G(s, u)\,ds = F(u)$ is a continuous function of u.
(b) For a and b finite, and $a \leq a' \leq b' \leq b$

$$\int_{a'}^{b'} \left[\int_p^{+\infty} f(s)G(s, u)\,ds \right] du = \int_p^{+\infty} f(s)\left[\int_{a'}^{b'} G(s, u)\,du \right] ds$$

(c) If $b = +\infty$, and f is absolutely integrable, and $\int_a^{+\infty} G(s, u)\,du$ converges uniformly for $p \leq s < +\infty$, then

$$\int_a^{+\infty} \left[\int_p^{+\infty} f(s)G(s, u)\,ds \right] du = \int_p^{+\infty} f(s)\left[\int_a^{+\infty} G(s, u)\,du \right] ds$$

provided that the iterated integral on the right side is defined (finite).[2]

---

2. That is $\lim_{R \to +\infty} \int_p^R f(s)[\int_a^{+\infty} G(s, u)\,du]\,ds$ exists.

Proof.   Let u_0 be in the interval $[a, b]$, then for $R > 0$

$$|F(u) - F(u_0)| = \left| \int_p^{+\infty} f(s)G(s, u)\, ds - \int_p^{+\infty} f(s)G(s, u_0)\, ds \right|$$

$$\leq \left| \int_p^R f(s)[G(s, u) - G(s, u_0)]\, ds \right| + \left| \int_R^{+\infty} f(s)G(s, u)\, ds \right|$$

$$+ \left| \int_R^{+\infty} f(s)G(s, u_0)\, ds \right|$$

Given $\epsilon > 0$, we can choose $A \geq p$ so that each of the last two terms above is less than ϵ for all $R \geq A$. Having fixed such a number R we have

$$|F(u) - F(u_0)| < \left| \int_p^R f(s)[G(s, u) - G(s, u_0)]\, ds \right| + 2\epsilon$$

For $p \leq s \leq R$ and $a \leq u, u_0 \leq b$, we can invoke the uniform continuity of G to ensure that if u is sufficiently close to u_0, then $|G(s, u) - G(s, u_0)| < \epsilon/M$ for $p \leq s \leq R$ and $M = \int_p^R |f(s)|\, ds + 1$. Then

$$\left| \int_p^R f(s)[G(s, u) - G(s, u_0)]\, ds \right| < (\epsilon/M) \int_p^R |f(s)|\, ds < \epsilon$$

Thus, for u sufficiently close to u_0 we have $|F(u) - F(u_0)| < 3\epsilon$. In other words, $\int_p^{+\infty} f(s)G(s, u)\, ds = F(u)$ is continuous at u_0 for each u_0 in $[a, b]$. That proves (a).

We now turn to (b). First, assume that F is continuous. For $R > p$ we know from advanced calculus that the following equality of iterated integrals is valid

(2.6) $$\int_p^R \left[\int_{a'}^{b'} F(s, u)\, du \right] ds = \int_{a'}^{b'} \left[\int_p^R F(s, u)\, ds \right] du$$

Moreover, if $F(s, u) = f(s)G(s, u)$ where f is piecewise continuous and G is continuous, then (2.6) still holds if we express $\int_p^R$ as a sum of integrals over subintervals where f (and hence F) is continuous. From (2.6) we have

$$\left| \int_{a'}^{b'} \left[\int_p^{+\infty} f(s)G(s, u)\, ds \right] du - \int_p^R \left[\int_{a'}^{b'} f(s)G(s, u)\, du \right] ds \right|$$

$$= \left| \int_{a'}^{b'} \left[\int_R^{+\infty} f(s)G(s, u)\, ds \right] du \right| \leq \int_{a'}^{b'} \left| \int_R^{+\infty} f(s)G(s, u)\, ds \right| du$$

If $A \geq p$ is chosen large enough, then the integrand in the final integral above can be made less than δ in magnitude for all $R \geq A$ and all u values. Therefore, noting that

$$\int_p^R \left[\int_{a'}^{b'} f(s)G(s, u)\, du \right] ds = \int_p^R f(s) \left[\int_{a'}^{b'} G(s, u)\, du \right] ds,$$

we have for all $R \geq A$

$$\left| \int_{a'}^{b'} \left[\int_p^{+\infty} f(s)G(s, u)\, ds \right] du - \int_p^R f(s) \left[\int_{a'}^{b'} G(s, u)\, du \right] ds \right| < \delta(b' - a')$$

Because δ may be taken arbitrarily small, this last result proves that

$$\int_p^{+\infty} f(s) \left[\int_{a'}^{b'} G(s, u)\, du \right] ds$$

converges to

$$\int_{a'}^{b'} \left[\int_p^{+\infty} f(s)G(s, u)\, ds \right] du.$$

Thus, (b) is proved.

From (b) for each $R \geq a$ we have

$$\int_a^R \left[\int_p^{+\infty} f(s)G(s, u)\, ds \right] du = \int_p^{+\infty} f(s) \left[\int_a^R G(s, u)\, du \right] ds$$

Hence, we have

$$\left| \int_p^{+\infty} f(s) \left[\int_a^{+\infty} G(s, u)\, du \right] ds - \int_a^R \left[\int_p^{+\infty} f(s)G(s, u)\, ds \right] du \right|$$

$$\leq \int_p^{+\infty} |f(s)| \cdot \left| \int_R^{+\infty} G(s, u)\, du \right| ds < \delta \int_p^{+\infty} |f(s)|\, ds$$

provided R is taken sufficiently large. Thus, we have

$$\lim_{R \to +\infty} \int_a^R \left[\int_p^{+\infty} f(s)G(s, u)\, ds \right] du = \int_p^{+\infty} f(s) \left[\int_a^{+\infty} G(s, u)\, du \right] ds$$

which proves (c). ∎

Combining Example (2.4) and Theorem (2.5), we obtain the following result.

(2.7) Theorem. If f is piecewise continuous and absolutely integrable, then its Fourier transform is a bounded continuous function, bounded by $\int_{-\infty}^{+\infty} |f(s)|\, ds$.

More applications of Theorem (2.5) will be made in §§4 and 5. Our next theorem is frequently used in problems in differential equations.

(2.8) Theorem. Suppose that $F(s, u) = f(s)G(s, u)$ where f is piecewise continuous and G is continuous. Further, suppose that $\int_p^{+\infty} f(s)G(s, u)\, ds$ converges for $a \leq u \leq b$, that $\partial G / \partial u$ is continuous, and that $\int_p^{+\infty} f(s)(\partial G / \partial u)(s, u)\, ds$ converges uniformly for $a \leq u \leq b$. Then $\int_p^{+\infty} f(s)G(s, u)\, ds$ is a continuously differentiable function of u with

$$\int_p^{+\infty} f(s)(\partial G / \partial u)(s, u)\, ds = (d/du) \int_p^{+\infty} f(s)G(s, u)\, ds$$

Proof. By Theorem (2.5b) we have for each u

$$\int_a^u \left[\int_p^{+\infty} f(s)(\partial G/\partial v)(s, v)\, ds\right] dv = \int_p^{+\infty} f(s)\left[\int_a^u (\partial G/\partial v)(s, v)\, dv\right] ds$$

$$= \int_p^{+\infty} f(s)[G(s, u) - G(s, a)]\, ds$$

$$= \int_p^{+\infty} f(s)G(s, u)\, ds - \int_p^{+\infty} f(s)G(s, a)\, ds$$

Differentiating the above relations with respect to u we obtain[3]

$$\int_p^{+\infty} f(s)(\partial G/\partial u)(s, u)\, ds = (d/du)\int_p^{+\infty} f(s)G(s, u)\, ds \qquad \blacksquare$$

Here is an application of Theorem (2.8) to the calculation of a Fourier transform.

(2.9) Example.   Let f be defined by $f(x) = e^{-\pi x^2}$. Then

$$\hat{f}(u) = \int_{-\infty}^{+\infty} e^{-\pi s^2} e^{-i2\pi us}\, ds = 2\int_0^{+\infty} e^{-\pi s^2} \cos 2\pi us\, ds$$

Since $2\int_0^{+\infty} e^{-\pi s^2}\, ds$ and $2\int_0^{+\infty} 2\pi s e^{-\pi s^2}\, ds$ both converge, we see that Theorem (2.8) may be applied to $\hat{f}$. Thus

$$d\hat{f}/du = 2\int_0^{+\infty} e^{-\pi s^2}(\partial/\partial u)(\cos 2\pi us)\, ds = 2\int_0^{+\infty} (-2\pi se^{-\pi s^2}) \sin 2\pi us\, ds$$

Integrating by parts yields

$$d\hat{f}/du = 2e^{-\pi s^2} \sin 2\pi us\, |_{s=0}^{s=+\infty} - 4\pi u \int_0^{+\infty} e^{-\pi s^2} \cos 2\pi\, us\, ds$$

$$= 0 - 2\pi u\hat{f}(u)$$

Therefore $\hat{f}$ satisfies the differential equation $d\hat{f}/du = -2\pi u\hat{f}$ that has the solution $\hat{f}(u) = \hat{f}(0)e^{-\pi u^2}$. We need only find the value of $\hat{f}(0) = \int_{-\infty}^{+\infty} e^{-\pi x^2}\, dx$ to finish our calculation. To do that we employ an ingenious trick of Liouville's.

Since the limit of a product is the product of the limits, we have

$$\int_{-\infty}^{+\infty} e^{-\pi x^2}\, dx \cdot \int_{-\infty}^{+\infty} e^{-\pi y^2}\, dy = \lim_{n \to +\infty} \left[\left[\int_{-n}^n e^{-\pi x^2}\, dx\right]\left[\int_{-n}^n e^{-\pi y^2}\, dy\right]\right]$$

$$= \lim_{n \to +\infty} \int_{-n}^n \int_{-n}^n e^{-\pi(x^2+y^2)}\, dx\, dy$$

This last limit equals $\int_{-\infty}^{+\infty} \int_{-\infty}^{+\infty} e^{-\pi(x^2+y^2)}\, dx\, dy$. Therefore, since x and y are

3. By the fundamental theorem of calculus, and the fact that $\int_p^{+\infty} f(s)G(s, a)\, ds$ is a constant.

just dummy variables,

$$\hat{f}(0)^2 = \left[\int_{-\infty}^{+\infty} e^{-\pi x^2} \, dx\right]\left[\int_{-\infty}^{+\infty} e^{-\pi y^2} \, dy\right] = \int_{-\infty}^{+\infty}\int_{-\infty}^{+\infty} e^{-\pi(x^2+y^2)} \, dx \, dy$$

Converting to polar coordinates yields

$$\hat{f}(0)^2 = \int_0^{2\pi}\int_0^{+\infty} e^{-\pi r^2} r \, dr \, d\phi = 1$$

Thus, since $\hat{f}(0) > 0$ we know that $\hat{f}(0) = 1$, hence $\hat{f}(u) = e^{-\pi u^2}$.

This last example is quite important. The function $f(x) = e^{-\pi x^2}$ satisfies $\hat{f}(u) = f(u)$; that fact will play a key role in the inversion theory discussed in §4. If we use the symbol $\mathscr{F}$ to denote the taking of the Fourier transform, then this last equation can be expressed as $\mathscr{F}(f) = f$. For future reference, we record the result of Example (2.9) below

(2.10)
$$e^{-\pi x^2} \supset e^{-\pi u^2}$$

We close this section with an example whose result we shall need in §5.

(2.11) Example. Consider the integral $I(a) = \int_0^{+\infty} e^{-ax}(\sin x/x) \, dx$ for $a > 0$. Since $|e^{-ax}(\sin x/x)| \le e^{-ax}$ and $|e^{-ax}\sin x| \le e^{-ax}$ and $\int_0^{+\infty} e^{-ax} \, dx = 1/a$, we conclude from Theorems (2.3) and (2.8) that

$$I'(a) = -\int_0^{+\infty} e^{-ax}\sin x \, dx = -1/(1+a^2)$$

where the last equality was obtained by two integration by parts. Performing antidifferentiation yields $I(a) = C - \tan^{-1} a$ for $a > 0$. To find the value of the constant C we let a tend to $+\infty$ observing that

$$|I(a)| \le \int_0^{+\infty} |e^{-ax}(\sin x/x)| \, dx \le \int_0^{+\infty} e^{-ax} \, dx = 1/a$$

and $1/a$ tends to 0. Therefore, since $\lim_{a \to +\infty} \tan^{-1} a = \frac{1}{2}\pi$, we conclude that

(2.12) $$I(a) = \int_0^{+\infty} e^{-ax}(\sin x/x) \, dx = \frac{1}{2}\pi - \tan^{-1} a \qquad (a > 0).$$

We now prove that $\lim_{a \to 0+} I(a) = \int_0^{+\infty}(\sin x/x) \, dx = I(0)$, that is, that $I(a)$ is continuous at $a = 0$. This proof is complicated because $\int_0^{+\infty} |\sin x/x| \, dx$ diverges [see Exercise (2.20)] so we cannot use Theorem (2.3) to infer uniform convergence of $I(a)$.

Two integration by parts yields for $a > 0$

$$\int e^{-ax}\sin x \, dx = -\frac{e^{-ax}\cos x + a e^{-ax}\sin x}{1+a^2} + C$$

and the equality also holds for $a = 0$. Hence, again by integration by parts, we have for $a \geq 0$ and $R \geq 0$

$$\int_R^{+\infty} e^{-ax} \frac{\sin x}{x} dx = \frac{e^{-aR}}{1+a^2} \left(\frac{\cos R + a \sin R}{R} \right) - \int_R^{+\infty} \frac{e^{-ax}}{1+a^2} \frac{\cos x + a \sin x}{x^2} dx$$

If we assume that $0 \leq a \leq 1$, then for every $x \geq 0$

$$\left| \frac{e^{-ax}}{1+a^2} \right| \leq 1 \qquad |\cos x + a \sin x| \leq 1 + a \leq 2$$

hence

$$\left| \int_R^{+\infty} e^{-ax} \frac{\sin x}{x} dx \right| \leq \frac{2}{R} + \int_R^{+\infty} \frac{2}{x^2} dx = \frac{4}{R}$$

Since $4/R$ can be made less than $\delta > 0$ if R taken sufficiently large, we have proved that $I(a) = \int_0^{+\infty} e^{-ax}(\sin x/x) dx$ converges uniformly for $0 \leq a \leq 1$. Thus, by Theorem (2.5a), $I(a)$ is continuous for $0 \leq a \leq 1$. In particular,

$$\int_0^{+\infty} (\sin x/x) dx = I(0+) = \tfrac{1}{2}\pi$$

From that result we have $\int_{-\infty}^{+\infty}(\sin x/x) dx = \pi$ and by a simple change of variables

(2.13) $$\int_{-\infty}^{+\infty} \frac{\sin 2\pi cs}{\pi s} ds = 1 \qquad \text{for each } c > 0$$

Equality (2.13) will be needed in §5.

Exercises

(2.14)   Suppose that f and xf are absolutely integrable functions. Prove that the Fourier transform $\hat{f}$ is continuously differentiable.

(2.15)   Prove that the sine and cosine transforms $\hat{f}_C$ and $\hat{f}_S$ are continuous, bounded functions.

(2.16)   Find the Fourier transforms of the following functions.

(a) $f(x) = xe^{-\pi x^2}$ [*Hint*: Use the result of (2.14).]
(b) $f(x) = e^{-2\pi c|x|}$ for $c > 0$
(c) $x^2 e^{-\pi x^2}$

(2.17)   Generalize the theorems of this section, so as to apply to integrals

$$\int_0^{+\infty} \int_0^{+\infty} f(r, s)G(r, s, u, v) \, dr \, ds$$

where f is absolutely integrable (continuous) and G is continuous.

(2.18)   *Alternate Definition of the Fourier Transform.* The following problems focus on the alternate definition of the Fourier transform of f

which we denote by $\bar{f}$ and define by

$$\bar{f}(u) = \frac{1}{\sqrt{2\pi}} \int_{-\infty}^{+\infty} f(x)e^{-iux}\, dx$$

(a)   Find the alternate Fourier transform $\bar{f}$ of the following functions.

$$f(x) = e^{-x^2/2} \qquad g(x) = \begin{cases} \sqrt{2\pi} & \text{if } |x| < 1 \\ 0 & \text{if } |x| > 1 \end{cases}$$

$$\Lambda(x) = \begin{cases} 1 - |x| & \text{if } |x| < 1 \\ 0 & \text{if } |x| > 1 \end{cases} \qquad k(x) = e^{-|x|}$$

(b)   Prove that the alternate Fourier transform $\bar{f}$ is always a bounded continuous function, and that $f(x) \supset \hat{f}(u)$ if and only if $f(x/\sqrt{2\pi}) \supset \bar{f}$ $(u/\sqrt{2\pi})$.

The alternate Fourier transform $\bar{f}$ is commonly used in quantum mechanics.

(2.19)   For $a > 0$ show that

$$\int_0^{+\infty} \frac{\tan^{-1}ax - \tan^{-1}x}{x}\, dx = \tfrac{1}{2}\pi \ln a.$$

(2.20)   Prove   that $\int_0^{+\infty} |\sin x/x|\, dx$ diverges.   [*Hint:*   Consider $\int_0^{n\pi} |\sin x/x|\, dx = \sum_{k=1}^n \int_{(k-1)\pi}^{k\pi} |\sin x/x|\, dx.$]

(2.21)   Show that $\int_0^{+\infty} (\sin^3 x/x)\, dx = \tfrac{1}{4}\pi.$ [*Hint:* Find a trigonometric identity for $\sin^3 x.$]

§3. Basic Computational Theorems

As the reader has probably discovered, finding a Fourier transform can often involve evaluating a difficult integral. Just as there are extensive tables of indefinite integrals, there are also extensive tables of Fourier transforms. A few of those tables are listed in Appendix B. Just as for definite integrals in calculus, however, it is worthwhile to be able to quickly calculate simple Fourier transforms without having to consult a table. In this section, we shall discuss several basic properties that are useful for computing transforms. Our first theorem summarizes the four most basic properties.

(3.1) Theorem.   The   Fourier   transform   enjoys   the   following   four properties.

(a)   *Linearity.* If $f(x) \supset \hat{f}(u)$ and $g(x) \supset \hat{g}(u)$, then for all complex numbers a and b

$$af(x) + bg(x) \supset a\hat{f}(u) + b\hat{g}(u)$$

(b)   *Shifting.* If $f(x) \supset \hat{f}(u)$, then for each real number c

$$f(x - c) \supset \hat{f}(u)e^{-i2\pi cu}$$

(c) *Change of Scale.* If $f(x) \supset \hat{f}(u)$ and $c > 0$, then

$$f(cx) \supset (1/c)\hat{f}(u/c)$$

(d) *Modulation.* If $f(x) \supset \hat{f}(u)$ and c is a real number, then

$$f(x)e^{i2\pi cx} \supset \hat{f}(u - c)$$

Moreover,

$$f(x) \cos(2\pi cx) \supset \tfrac{1}{2}\hat{f}(u - c) + \tfrac{1}{2}\hat{f}(u + c)$$

$$f(x) \sin(2\pi cx) \supset (i/2)\hat{f}(u + c) - (i/2)\hat{f}(u - c)$$

Proof.

(a) By the linearity of integration, we have

$$af(x) + bg(x) \supset \int_{-\infty}^{+\infty} [af(x) + bg(x)]e^{-i2\pi ux} \, dx$$

$$= a \int_{-\infty}^{+\infty} f(x)e^{-i2\pi ux} \, dx + b \int_{-\infty}^{+\infty} g(x)e^{-i2\pi ux} \, dx$$

$$= a\hat{f}(u) + b\hat{g}(u)$$

(b) Substituting $v = x - c$ and factoring the relative constant $e^{-i2\pi cu}$ outside the integral yields

$$f(x - c) \supset \int_{-\infty}^{+\infty} f(x - c)e^{-i2\pi ux} \, dx$$

$$= \int_{-\infty}^{+\infty} f(v)e^{-i2\pi uv} \, dv \, e^{-i2\pi uc} = \hat{f}(u)e^{-i2\pi cu}$$

(c) Substituting $v = cx$ yields

$$f(cx) \supset \int_{-\infty}^{+\infty} f(cx)e^{-i2\pi ux} \, dx$$

$$= (1/c) \int_{-\infty}^{+\infty} f(v)e^{-i2\pi(u/c)v} \, dv = (1/c)\hat{f}(u/c)$$

(d) Since $e^{i2\pi cx}e^{-i2\pi ux} = e^{-i2\pi(u-c)x}$ we have

$$f(x)e^{i2\pi cx} \supset \int_{-\infty}^{+\infty} f(x)e^{-i2\pi(u-c)x} \, dx = \hat{f}(u - c)$$

Moreover, because $\cos(2\pi cx) = \tfrac{1}{2}e^{i2\pi cx} + \tfrac{1}{2}e^{-i2\pi cx}$ and $\sin(2\pi cx) = (i/2)e^{-i2\pi cx} - (i/2)e^{i2\pi cx}$, the rest of the theorem follows by linearity.

∎

Here are some examples of the power of the four simple properties in Theorem (3.1).

(3.2) Examples

(a) For $c > 0$ we find, by applying change of scale to the third formula in (1.9) that

$$e^{-2\pi c|x|} \supset \frac{1}{\pi} \frac{1/c}{1 + (u/c)^2} = \frac{1}{\pi} \frac{c}{c^2 + u^2}$$

Therefore

$$e^{-c|x|} \supset \frac{2c}{c^2 + (2\pi u)^2}$$

(b) Let $c > 0$. Using a change of scale on (2.11), we have

$$e^{-cx^2} \supset (\pi/c)^{1/2} e^{-\pi^2 u^2/c}$$

(c) Using modulation and the second result of (a), we find that for $c > 0$ and b a real constant

$$e^{-c|x|} \cos bx \supset \frac{c}{c^2 + (2\pi u - b)^2} + \frac{c}{c^2 + (2\pi u + b)^2}$$

(d) By linearity and modulation applied to the result in (b), we have for A a complex constant and b a real constant

$$A e^{ibx} e^{-cx^2} \supset A(\pi/c)^{1/2} e^{-(2\pi u - b)^2/4c}$$

(e) By linearity and the shifting property we have

$$\frac{i}{2} \Pi(x - \tfrac{1}{2}) - \frac{i}{2} \Pi(x + \tfrac{1}{2}) \supset \frac{\sin^2 \pi u}{\pi u}$$

From Example (3.2b) we have the following result, which was used in §7 of Chapter 4.

(3.3) $$\int_{-\infty}^{+\infty} e^{-cx^2} e^{-i2\pi ux}\, dx = (\pi/c)^{1/2} e^{-\pi^2 u^2/c} \qquad (c > 0).$$

Our next two theorems connect the operation of differentiation with the Fourier transform.

(3.4) Theorem. Let f be piecewise continuous and absolutely integrable. If the function $xf(x)$ is absolutely integrable, then the Fourier transform $\hat{f}$ is continuously differentiable. Moreover,

$$xf(x) \supset \frac{i}{2\pi} \hat{f}'(u)$$

Proof. From the relation $|sf(s)e^{-i2\pi us}| = |sf(s)|$, we infer from Theorem

(2.8) that $\hat{f}(u)$ is continuously differentiable with

$$\hat{f}'(u) = \int_{-\infty}^{+\infty} f(s)(\partial/\partial u)[e^{-i2\pi us}]\,ds = -i2\pi \int_{-\infty}^{+\infty} sf(e)^{-i2\pi us}\,ds$$

The result $xf(x) \supset (i/2\pi)\hat{f}'(u)$ follows immediately. ∎

(3.5) Theorem.   Let f be continuous and absolutely integrable. If f' is piecewise continuous and absolutely integrable, then

$$f'(x) \supset i2\pi u\hat{f}(u)$$

Proof.   Since $f(x) = \int_0^x f'(s)\,ds + f(0)$ and $\int_0^{+\infty} f'(s)\,ds$, $\int_0^{-\infty} f'(s)\,ds$ both converge, it follows that

$$\lim_{x \to +\infty} f(x) = f(+\infty) \qquad \lim_{x \to -\infty} f(x) = f(-\infty)$$

both exist. Since f is absolutely integrable, it must be that $f(-\infty) = f(+\infty) = 0$. The theorem follows by integration by parts

$$f'(x) \supset \int_{-\infty}^{+\infty} f'(x)e^{-i2\pi ux}\,dx$$

$$= 0 + i2\pi u \int_{-\infty}^{+\infty} f(x)e^{-i2\pi ux}\,dx \qquad ∎$$

The following corollary is an immediate consequence of the previous two theorems.

(3.6) Corollary.   If $f^{(m)}$ is piecewise continuous and $x^k f^{(m)}(x)$ is absolutely integrable, then

$$x^k f^{(m)}(x) \supset i^{m+k}(2\pi)^{m-k} u^m \hat{f}^{(k)}(u)$$

(3.7) Examples

(a)   Since $xe^{-|x|}$ is absolutely integrable, we have

$$xe^{-|x|} \supset \frac{i}{2\pi}\frac{d}{du}\left[\frac{2}{1+4\pi^2 u^2}\right] = \frac{-i8\pi u}{(1+4\pi^2 u^2)^2}$$

(b)   Consider the function $i\pi e^{-2\pi|x|}\,\mathrm{sgn}\,x$ where

$$\mathrm{sgn}\,x = \begin{cases} 1 & \text{if } x > 0 \\ -1 & \text{if } x < 0 \end{cases}$$

Then $i\pi e^{-2\pi|x|}\,\mathrm{sgn}\,x = f'(x)$ where $f(x) = -(i/2)e^{-2\pi|x|}$. Therefore

$$i\pi e^{-2\pi|x|}\,\mathrm{sgn}\,x \supset i2\pi u\left(-\frac{i}{2}\right)\frac{1}{\pi}\frac{1}{1+u^2} = \frac{u}{1+u^2}$$

(c)   By Theorem (3.4) and formula (2.10), we have

$$xe^{-\pi x^2} \supset \frac{i}{2\pi}\frac{d}{du}[e^{-\pi u^2}] = -iue^{-\pi u^2}$$

Thus, for $f(x) = xe^{-\pi x^2}$, we have $\mathscr{F}(f) = -if$.

Exercises

(3.8)   Find the Fourier transforms of the following functions.
  (a) $\Pi(x - \frac{1}{2})$
  (b) $\Pi\!\left(\dfrac{x - \frac{1}{2}a}{a}\right)$ for $a > 0$
  (c) $\Pi(x)\,\text{sgn}\,x$
  (d) $e^{-c|x-b|}$ for $c > 0$ b real
  (e) $e^{-(x-b)^2/c}$ for $c > 0$ b real
  (f) $e^{-c|x|}\sin bx$ for $c > 0$ b real

(3.9)   Show that $\int_0^{+\infty} e^{-ax}\sin cx\,dx = c/(a^2 + c^2)$ and $\int_0^{+\infty} e^{-ax}\cos cx\,dx = a/(a^2 + c^2)$ where $a > 0$ and c is a real number.

(3.10)   Find the Fourier transforms for the following functions.

  (a) $f(x) = \begin{cases} 1 & \text{if } |x| < \frac{1}{2}c \\ 0 & \text{if } |x| > \frac{1}{2}c \end{cases}$

  (b) $f(x) = \begin{cases} \cos(\pi x/c) & \text{if } |x| < \frac{1}{2}c \\ 0 & \text{if } |x| > \frac{1}{2}c \end{cases}$

  (c) $f(x) = \begin{cases} \cos^2(\pi x/c) & \text{if } |x| < \frac{1}{2}c \\ 0 & \text{if } |x| > \frac{1}{2}c \end{cases}$

  (d) $f(x) = \dfrac{1}{c}\Lambda(x/c)$ for $c > 0$

  (e) $f(x) = x^2 e^{-\pi x^2}$
  (f) $f(x) = (4\pi x^2 - 1)e^{-\pi x^2}$

(3.11)   Find the cosine and sine transforms $\hat{f}_C$ and $\hat{f}_S$ for the following functions. [*Hint*: Exercises (1.13(b)) and (1.14b) are helpful.]

  (a) $f(x) = \begin{cases} \cos bx & \text{if } |x| < c \\ 0 & \text{if } |x| > c \end{cases}$

  (b) $f(x) = e^{-3x}\cos x$

(3.12)   Prove that for real numbers $a \neq 0$ and b we have

$$f(x) \supset \hat{f}(u) \qquad \text{implies} \qquad f(ax + b) \supset \frac{1}{|a|}e^{i2\pi bu/a}\hat{f}\!\left(\frac{u}{a}\right)$$

(3.13) Prove Theorems (3.1), (3.4), and (3.5), mutatis mutandis, for the alternate Fourier transform $\tilde{f}$. [See Exercise (2.18).] Find the alternate Fourier transform $\tilde{f}$ of the following functions.

(a) $f(x) = (1/c)e^{-x^2/2c}$ for $c > 0$
(b) $f(x) = e^{-c|x|}$ for $c > 0$
(c) $(1/c)\Lambda(x/c)$ for $c > 0$
(d) $f(x) = (1/c)\Pi(x/c)$ for $c > 0$
(e) $f(x) = xe^{-c|x|}\cos bx$ for $c > 0$ and b real

§4. Fourier Inversion, Gauss–Weierstrass Summation

In this section we shall prove a Fourier inversion theorem by a method known as *Gauss–Weierstrass summation*. This method of summation has its basis in the solution of the heat equation (which we shall discuss in the next chapter). It is the continuous analogue of the method involving the theta function that we treated in §7 of Chapter 4.

The key element of Gauss–Weierstrass summation is the existence of the following reciprocal Fourier transforms (where c is a positive constant)

(4.1) $e^{-\pi c^2 x^2} \supset (1/c)e^{-\pi u^2/c^2}$ $(1/c)e^{-\pi x^2/c^2} \supset e^{-\pi c^2 u^2}$ $(c > 0)$

Consider the expressions

$$\int_{-\infty}^{+\infty} \hat{f}(u)e^{i2\pi ux}e^{-\pi c^2 u^2}\, du = \int_{-\infty}^{+\infty}\left[\int_{-\infty}^{+\infty} f(s)e^{-i2\pi us}\, ds\right]e^{i2\pi ux}e^{-\pi c^2 u^2}\, du$$

$$= \int_{-\infty}^{+\infty}\left[\int_{-\infty}^{+\infty} f(s)e^{-\pi c^2 u^2}e^{-i2\pi u(s-x)}\, ds\right]du$$

where f is piecewise continuous and absolutely integrable. For each fixed x value, because of Theorem (2.5c),[4] we can interchange the last two integrals above obtaining by (4.1)

$$\int_{-\infty}^{+\infty} \hat{f}(u)e^{i2\pi ux}e^{-\pi c^2 u^2}\, du = \int_{-\infty}^{+\infty} f(s)\left[\int_{-\infty}^{+\infty} e^{-\pi c^2 u^2}e^{-i2\pi(s-x)u}\, du\right]ds$$

$$= \int_{-\infty}^{+\infty} f(s)(1/c)e^{-\pi(s-x)^2/c^2}\, ds$$

Replacing $(s-x)^2$ by $(x-s)^2$ and changing variables, we obtain

(4.2) $$\int_{-\infty}^{+\infty} \hat{f}(u)e^{i2\pi ux}e^{-\pi c^2 u^2}\, du = \int_{-\infty}^{+\infty} f(x-s)W(s;c)\, ds$$

where

(4.3) $$W(s;c) = (1/c)e^{-\pi s^2/c^2}   (c > 0)$$

---

4. Where $G(s, u) = e^{-\pi c^2 u^2}e^{-i2\pi u(s-x)}$ for each x value and positive constant c.

We now prove that if f is a continuous, absolutely integrable function, then

(4.4)
$$\lim_{c \to 0+} \int_{-\infty}^{+\infty} \hat{f}(u) e^{i2\pi ux} e^{-\pi c^2 u^2} \, du = f(x)$$

for all x values. Once we prove (4.4), then we can easily show that

(4.4')
$$\int_{-\infty}^{+\infty} \hat{f}(u) e^{i2\pi ux} \, du = f(x)$$

provided that $\hat{f}$ is absolutely integrable. Formula (4.4') is our desired Fourier inversion theorem.

To prove (4.4) we need the following Lemma, which states that $W(s;c)$ is a *summation kernel over* $\mathbb{R}$, parameterized by $c > 0$. In probabilistic terms, it says that we have a family of normal distributions of mean 0 and variances $c^2/2\pi \to 0$ as $c \to 0$.

(4.5) Lemma. The functions $W(s;c)$ satisfy

(A_1) $\displaystyle\int_{-\infty}^{+\infty} W(s;c) \, ds = 1$ for each $c > 0$

(A_2) $W(s;c) \geq 0$ for all s values and each $c > 0$

(A_3) For every $\epsilon > 0$ and $\delta > 0$ we have

$$0 \leq W(s;c) < \epsilon \qquad 0 \leq \int_{|s| \geq \delta} W(s;c) \, ds < \epsilon$$

provided $|s| \geq \delta$ and c is taken close enough to 0.

Proof. Using (4.1) we have

$$\int_{-\infty}^{+\infty} W(s;c) \, ds = \int_{-\infty}^{+\infty} (1/c) e^{-\pi s^2/c^2} e^{-i2\pi 0 s} \, ds$$
$$= e^{-\pi c^2 0^2} = 1$$

which proves (A_1). Property (A_2) is clear because of the positivity of the exponential function. If $|s| \geq \delta$, then

(4.6)
$$0 \leq W(s;c) \leq (1/c) e^{-\pi \delta^2/c^2}$$

Since $e^{-\theta}$ decreases to 0, as θ tends to $+\infty$, faster than any power of θ (i.e., $\lim_{\theta \to +\infty} \theta^p/e^\theta = 0$ for all positive powers p) we see that if c is taken close enough to 0 then $(1/c) e^{-\pi \delta^2/c^2} < \epsilon$. Therefore, from (4.6), we have $0 \leq W(s;c) < \epsilon$. Moreover,

$$0 \leq \int_{|s| \geq \delta} W(s;c) \, ds = 2 \int_{\delta}^{+\infty} (1/c) e^{-\pi s^2/c^2} \, ds$$

Substituting $s = cr$ we have

$$2 \int_{\delta}^{+\infty} (1/c) e^{-\pi s^2/c^2} \, ds = 2 \int_{\delta/c}^{+\infty} e^{-\pi r^2} \, dr$$

Due to the convergence of $2 \int_0^{+\infty} e^{-\pi r^2} \, dr$, if c is taken close enough to 0 we obtain $0 \le \int_{|s| \ge \delta} W(s; c) \, ds < \epsilon$. ∎

Using Lemma (4.5) we can easily establish (4.4).

(4.7) Theorem. Let f be a continuous, absolutely integrable function. Then for all x values

$$\lim_{c \to 0+} \int_{-\infty}^{+\infty} \hat{f}(u) e^{i2\pi ux} e^{-\pi c^2 u^2} \, du = f(x)$$

Proof.   Because of (4.2) it suffices for us to show that

(4.8)
$$\lim_{c \to 0+} \int_{-\infty}^{+\infty} f(x - s) W(s; c) \, ds = f(x)$$

Since Lemma (4.5) shows that $W(s; c)$ is a summation kernel over $\mathbb{R}$, we know from Theorem (8.10), Chapter 2, that

(4.9)
$$\lim_{(x, c) \to (x_0, 0+)} \int_{-\infty}^{+\infty} f(x - s) W(s; c) \, ds = f(x_0)$$

holds for *all* points x_0 in $\mathbb{R}$. Since (4.8) is a simplified form of (4.9), we are done. ∎

(4.10) Remark.   Because $W(s; c)$ is a summation kernel over $\mathbb{R}$, formula (4.9) actually holds whenever f is absolutely integrable over $\mathbb{R}$ and x_0 is a point of continuity of f. Moreover, it also holds, *uniformly for all x_0 in $\mathbb{R}$*, when f is continuous and $\lim_{|x| \to +\infty} f(x) = 0$. [By Theorems (8.10) and (8.11) in Chapter 2.]

Now that (4.4) is proved we can easily verify (4.4').

(4.11) Theorem: Fourier Inversion.   Suppose that f is continuous and absolutely integrable. If $\hat{f}$ is absolutely integrable, then

$$\int_{-\infty}^{-\infty} \hat{f}(u) e^{i2\pi ux} \, du = f(x)$$

Proof.   Since $|\hat{f}(u) e^{i2\pi ux} e^{-\pi c^2 u^2}| \le |\hat{f}(u)|$ for all $c \ge 0$, it follows that

$$F(x, c) = \int_{-\infty}^{+\infty} \hat{f}(u) e^{i2\pi ux} e^{-\pi c^2 u^2} \, du$$

is a continuous function of x and c $(c \ge 0)$. Since F is continuous at $c = 0$, we have

$$\lim_{c \to 0+} F(x, c) = F(x, 0) = \int_{-\infty}^{+\infty} \hat{f}(u) e^{i2\pi ux} \, du$$

However, Theorem (4.7) tells us that $\lim_{c \to 0+} F(x, c) = f(x)$. Comparing these last two limits yields the desired conclusion. ∎

One defect of Theorem (4.11) is that it requires that $\hat{f}$ be absolutely integrable. Unfortunately, no necessary and sufficient conditions are known concerning f which will ensure that $\hat{f}$ is absolutely integrable. See Exercise (4.16) for a sufficient condition.

Here is an application of Theorem (4.11). If $f(x) = e^{-2\pi|x|}$ then $\hat{f}(u) = (1/\pi)[1/(1 + u^2)]$ which is absolutely integrable. Therefore,

$$(4.12) \qquad \frac{1}{\pi} \int_{-\infty}^{+\infty} \frac{e^{i2\pi ux}}{1 + u^2} \, du = e^{-2\pi|x|}$$

Similarly, we have for all $c > 0$

$$(4.13) \qquad \int_{-\infty}^{+\infty} \frac{1}{c} \frac{\sin^2 c\pi u}{(\pi u)^2} e^{i2\pi ux} \, du = \begin{cases} 1 - |x|/c & \text{if } |x| \le c \\ 0 & \text{if } |x| > c \end{cases}$$

Formula (4.13) allows us to define a summation kernel over $\mathbb{R}$ known as *Cesàro's kernel* [see Exercise (4.19)].

Another defect of Theorem (4.11) is that it requires that f be continuous. In the next section we shall discuss a method of Fourier inversion that allows f to be discontinuous, and also overcomes the other defect of Theorem (4.11) mentioned above.

Exercises

*(4.14)   Prove that if f and $\hat{f}$ are continuous and absolutely integrable, then $\hat{\hat{f}}(x) \supset f(-u)$. Use this result to check that $1/(1 + x^2) \supset \pi e^{-2\pi|u|}$ and $\text{sinc}^2 x \supset \Lambda(u)$.

(4.15)   Show that $\int_{-\infty}^{+\infty} \sin^2 x / x^2 \, dx = \pi$.

(4.16)   Prove that if the second derivative f'' is piecewise continuous and absolutely integrable, then $\hat{f}$ is absolutely integrable. [*Hint*: Use (3.6).]

(4.17)   Prove the following uniqueness theorem.

Theorem: Uniqueness of Fourier Transforms.   Suppose that f and g are continuous and absolutely integrable. If $\hat{f} = \hat{g}$ then $f = g$.

(4.18)   Using Theorem (4.11) prove the following theorem.

Theorem: Cosine and Sine Inversion.   Suppose that f is continuous and absolutely integrable over $[0, +\infty)$. If the cosine transform $\hat{f}_C$ is absolutely integrable, then

$$f(x) = 2 \int_0^{+\infty} \hat{f}_C(u) \cos 2\pi ux \, du$$

Moreover, if $f(0) = 0$ and the sine transform $\hat{f}_S$ is absolutely integrable, then

$$f(x) = 2i \int_0^{+\infty} \hat{f}_S(u) \sin 2\pi ux \, du$$

(4.19)   *Cesàro Summation.*   For $c > 0$ define the *Cesàro kernel* $C(s;c)$ by

$$C(s;c) = \frac{1}{c} \frac{\sin^2(c\pi s)}{(\pi s)^2}$$

Solve the following two problems.

(a)   Using (4.13), prove that $C(s;c)$ is a summation kernel over $\mathbb{R}$, parameterized by $c \to +\infty$ instead of $\tau \to 0+$.
(b)   Prove the following theorem.

Theorem.   Let f be continuous over $\mathbb{R}$. If either (1) f is absolutely integrable, or (2) f is bounded, then for all x values

$$\lim_{c \to +\infty} \int_{-c}^{c} \hat{f}(u) \left[1 - \frac{|u|}{c} \right] e^{i2\pi ux} \, du = f(x)$$

Moreover, in either case, if $\lim_{|x| \to +\infty} f(x) = 0$, then the limit above holds *uniformly* for all x values.

(4.20)   Suppose that f is piecewise continuous and absolutely integrable. Prove that for all x values

$$\lim_{c \to 0+} \int_{-\infty}^{+\infty} f(x - s) W(s;c) \, ds = \tfrac{1}{2} f(x +) + \tfrac{1}{2} f(x -)$$

(4.21)   Suppose that f is piecewise continuous and absolutely integrable. Prove that for all x values

$$\lim_{c \to 0+} \int_{-\infty}^{+\infty} \hat{f}(u) e^{-2\pi c|u|} e^{i2\pi ux} \, dx = \tfrac{1}{2} f(x +) + \tfrac{1}{2} f(x -)$$

Use this result to give a second proof of Theorem (4.11).

(4.22)   Suppose that f and g are both piecewise continuous and absolutely integrable. Prove that if $\hat{f} = \hat{g}$ then $f = g$ except possibly for a finite number of points in every finite interval.

(4.23)   Give an example of a continuous, bounded function that is not absolutely integrable. Conversely, give an example of a continuous, absolutely integrable function that is not bounded.

(4.24)   Let f be a *discontinuous*, piecewise continuous function for which $0 < \int_{-\infty}^{+\infty} |f(x)| \, dx < +\infty$. Prove that $\hat{f}$ cannot be absolutely integrable.

(4.25)   Suppose that f is continuous and absolutely integrable. Prove that

$$v(x, t) = \int_{-\infty}^{+\infty} \hat{f}(u) e^{-(2\pi)^2 u^2 t} e^{i2\pi ux} \, du$$

satisfies $v_t = v_{xx}$ and $v(x, 0 +) = f(x)$ for $t > 0$ and all x values.

§5. Fourier Inversion, Dirichlet's Method

In the previous section we saw that when f and $\hat{f}$ are continuous and absolutely integrable, then

(5.1)
$$\int_{-\infty}^{+\infty} \hat{f}(u)e^{i2\pi ux}\, du = f(x)$$

Often, however, the integral in (5.1) is not defined. For instance, by Example (3.7b), we have for $f(x) = i\pi e^{-2\pi|x|}\, \mathrm{sgn}\, x$ that

(5.2)
$$\hat{f}(u) = \frac{u}{1+u^2}$$

Hence, for $x = 0$, the integral in (5.1) is not defined (even in an improper sense). The same thing happens with $f(x) = (i/2)\Pi(x - \tfrac{1}{2}) - (i/2)\Pi(x + \tfrac{1}{2})$ for which $\hat{f}(u) = (\sin^2 \pi u)/\pi u$ [see Exercise (5.18)].

Dirichlet discovered that in such cases the integral in (5.1) can be replaced by

(5.3)
$$\lim_{c \to +\infty} \int_{-c}^{c} \hat{f}(u)e^{i2\pi ux}\, du$$

which we call the *principal value* of $\int_{-\infty}^{+\infty}\hat{f}(u)e^{i2\pi ux}\, du$ and denote by

$$\mathrm{P.V.} \int_{-\infty}^{+\infty} \hat{f}(u)e^{i2\pi ux}\, du$$

In this section, we shall investigate the validity of the relation

(5.4)
$$\mathrm{P.V.} \int_{-\infty}^{+\infty} \hat{f}(u)e^{i2\pi ux}\, du = f(x)$$

Sometimes $f(x)$ must be replaced in (5.4) by $\tfrac{1}{2}[f(x+) + f(x-)]$, as in Theorem (5.17). Our results are very similar to those for Fourier series that we discussed in §§2 and 3 of Chapter 2.

Observe that for $c > 0$

$$\int_{-c}^{c} \hat{f}(u)e^{i2\pi ux}\, du = \int_{-\infty}^{+\infty} \hat{f}(u)e^{i2\pi ux}\, \Pi\!\left(\frac{u}{2c}\right) du$$

Since $\Pi(x/2c) \supset 2c \, \mathrm{sinc}\, 2cu = (\sin 2\pi cu)/\pi u$ we obtain, by the same steps as we used to derive (4.2),

$$\int_{-c}^{c} \hat{f}(u)e^{i2\pi ux}\, du = \int_{-\infty}^{+\infty} f(x-s)\frac{\sin 2\pi cs}{\pi s}\, ds$$

Since $(\sin 2\pi cs)/\pi s$ is an even function of s we obtain

(5.5)
$$\int_{-c}^{c} \hat{f}(u)e^{i2\pi ux}\, du = \int_{-\infty}^{+\infty} f(x+s)\frac{\sin 2\pi cs}{\pi s}\, ds$$

The function $(\sin 2\pi cs)/\pi s$ is sometimes called *Dirichlet's kernel* and, in all important respects, behaves in the same way as the Dirichlet kernel $\sin(n+\frac{1}{2})u/\sin(\frac{1}{2}u)$ for Fourier series. We shall use $D(s;c)$ to denote $(\sin 2\pi cs)/\pi s$.

From (2.13) we have

(5.6)
$$\int_{-\infty}^{+\infty} D(s;c)\,ds = 1 \qquad (c>0)$$

and, since $(\sin 2\pi cs)/\pi s$ is an even function of s,

(5.6')
$$\int_{0}^{+\infty} D(s;c)\,ds = \int_{-\infty}^{0} D(s;c)\,ds = \tfrac{1}{2}$$

Using (5.5) and (5.6) we have

(5.7)
$$\left| \int_{-c}^{c} \hat{f}(u)e^{i2\pi ux}\,du - f(x) \right| = \left| \int_{-\infty}^{+\infty} [f(x+s)-f(x)]D(s;c)\,ds \right|$$

If we split $\int_{-\infty}^{+\infty}$ into $\int_{|s|\geq\delta} + \int_{-\delta}^{+\delta}$, for $\delta>0$, and express

$$\int_{|s|\geq\delta} [f(x+s)-f(x)]D(s;c)\,ds$$

as

$$\int_{|s|\geq\delta} f(x+s)D(s;c)\,ds - f(x)\int_{|s|\geq\delta} D(s;c)\,ds,$$

then we obtain upon applying the triangle inequality for absolute values to the right side of (5.7)

(5.8)
$$\left| \int_{-c}^{c} \hat{f}(u)e^{i2\pi ux}\,du - f(x) \right|$$
$$\leq \left| \int_{|s|\geq\delta} f(x+s)D(s;c)\,ds \right| + |f(x)| \cdot \left| \int_{|s|\geq\delta} D(s;c)\,ds \right|$$
$$+ \left| \int_{-\delta}^{\delta} [f(x+s)-f(x)]D(s;c)\,ds \right|$$

We can show that the first two terms on the right side of (5.8) will have magnitudes smaller than ϵ when c is taken sufficiently large (ϵ being some preassigned small number). To ensure that the last term in (5.8) is negligibly small, we will have to assume something about the behavior of f at or near x (besides its piecewise continuity). Our analysis of (5.8) is more delicate than our work in the previous section because $D(s;c)$ is not a summation kernel.

First, we note that

$$\lim_{c\to+\infty} \int_{|s|\geq\delta} D(s;c)\,ds = \lim_{c\to+\infty} 2\int_{\delta}^{+\infty} \frac{\sin 2\pi cs}{\pi s}\,ds$$
$$= \lim_{c\to+\infty} 2\int_{2c\delta}^{+\infty} \frac{\sin \pi s}{\pi s}\,ds = 0$$

Therefore, for every $\delta > 0$, we have

(5.9)
$$\lim_{c \to +\infty} |f(x)| \cdot \left| \int_{|s| \geq \delta} D(s; c) \, ds \right| = 0$$

for each fixed x value.

Second, we need the following lemma.

(5.10) Lemma. If g is piecewise continuous and absolutely integrable, then

$$\lim_{c \to +\infty} \int_{-\infty}^{+\infty} g(s) \sin 2\pi cs \, ds = 0$$

Proof. Let $\epsilon > 0$ be given. Because g is absolutely integrable, if N is sufficiently large, then

$$\left| \int_{|s| \geq N} g(s) \sin 2\pi cs \, ds \right| \leq \int_{|s| \geq N} |g(s)| \, ds < \epsilon$$

Now, we restrict our attention to the closed interval $[-N, N]$. Let's suppose that we are dealing with a *step function* $G(s) = \sum_{j=1}^{n} a_j h_j(s)$ where

$$h_j(s) = \begin{cases} 1 & \text{if } c_j \leq s < d_j \\ 0 & \text{otherwise} \end{cases}$$

and the a_j's are constants. Then

$$\left| \int_{-N}^{N} G(s) \sin 2\pi cs \, ds \right| = \left| \sum_{j=1}^{n} a_j \int_{c_j}^{d_j} \sin 2\pi cs \, ds \right|$$

$$\leq \frac{1}{2\pi c} \left[\sum_{j=1}^{n} 2 |a_j| \right]$$

$$< \epsilon$$

provided c is taken sufficiently large.

Since g is piecewise continuous, we can prove that there exists a step function G such that

(5.11)
$$\int_{-N}^{N} |g(s) - G(s)| \, ds < \epsilon$$

From (5.11) we obtain, using the inequalities above,

$$\left| \int_{-\infty}^{+\infty} g(s) \sin 2\pi cs \, ds \right| = \left| \int_{|s| \geq N} g(s) \sin 2\pi cs \, ds \right.$$

$$+ \int_{-N}^{N} [g(s) - G(s)] \sin 2\pi cs \, ds$$

$$+ \left. \int_{-N}^{N} G(s) \sin 2\pi cs \, ds \right|$$

$$\leq \int_{|s| \geq N} |g(s)| \, ds + \int_{-N}^{N} |g(s) - G(s)| \, ds$$

$$+ \left| \int_{-N}^{N} G(s) \sin 2\pi cs \, ds \right|$$

$$< \epsilon + \epsilon + \epsilon = 3\epsilon$$

provided c is sufficiently large. Thus our lemma is proved as soon as we establish (5.11).

The piecewise continuity of g implies that g is continuous on each of a finite number (say M) of subintervals of $[-N, N]$ defined by

$$x_0 = -N < x_1 < x_2 < \cdots < x_{M-1} < N = x_M$$

(provided g is defined by its limits at the endpoints). *Let $[a, b]$ stand for any one of those M subintervals $[x_{j-1}, x_j]$.* Since g is continuous on $[a, b]$ it is uniformly continuous, hence for $\delta > 0$ sufficiently small we will have for $a \leq s, s' \leq b$ that

$$|g(s) - g(s')| < \frac{\epsilon}{M(b-a)}$$

if $|s - s'| < \delta$. If we partition $[a, b]$ into K subintervals of equal length $[(b-a)/K] < \delta$ by the points

$$s_0 = a < s_1 < s_2 < \cdots < s_{K-1} < b = s_K$$

then we can define a step function G by $G(s) = \sum_{j=1}^{K} g(s_j) h_j(s)$ where

$$h_j(s) = \begin{cases} 1 & \text{if } s_{j-1} \leq s < s_j \\ 0 & \text{otherwise} \end{cases}$$

We then have for each s in $[a, b]$ and some j for which $s_{j-1} \leq s \leq s_j$

$$|g(s) - G(s)| = |g(s) - g(s_j)| < \frac{\epsilon}{M(b-a)}$$

Therefore,

$$\int_a^b |g(s) - G(s)| \, ds < \int_a^b \frac{\epsilon}{M(b-a)} \, ds = \frac{\epsilon}{M}$$

Hence, because $[a, b]$ stands for an arbitrary subinterval $[x_{j-1}, x_j]$, we have defined the step function G on $[-N, N]$ and

$$\int_{-N}^{N} |g(s) - G(s)| \, ds = \sum_{j=1}^{M} \int_{x_{j-1}}^{x_j} |g(s) - G(s)| \, ds < M \frac{\epsilon}{M} = \epsilon$$

Thus, (5.11) is established so our lemma is proved. ∎

(5.12) Corollary. Suppose that f is piecewise continuous and absolutely integrable. For $\delta > 0$, we have for each x value

$$\lim_{c \to +\infty} \int_{|s| \geq \delta} f(x + s) D(s; c) \, ds = 0$$

Proof. Let g be defined by

$$g(s) = \begin{cases} f(x + s)/\pi s & \text{if } |s| \geq \delta \\ 0 & \text{if } |s| < \delta \end{cases}$$

and apply Lemma (5.10). ∎

We can now prove our first inversion theorem by Dirichlet's method.

(5.13) Theorem. If f is piecewise continuous and absolutely integrable, then P.V. $\int_{-\infty}^{+\infty} \hat{f}(u) e^{i2\pi ux} \, du = f(x)$ provided f has a derivative at x.

Proof. Because

$$\lim_{s \to 0} \frac{f(x + s) - f(x)}{s} = f'(x)$$

we certainly have for $\delta > 0$ sufficiently small

$$\left| \frac{f(x + s) - f(x)}{\pi s} \right| \leq \frac{1}{\pi} |f'(x)| + 1$$

if $|s| < \delta$. Denoting $(1/\pi) |f'(x)| + 1$ by A, we then have for all c

$$\left| \int_{-\delta}^{\delta} [f(x + s) - f(x)] D(s; c) \, ds \right| \leq \int_{-\delta}^{\delta} \left| \frac{f(x + s) - f(x)}{\pi s} \right| \, ds$$

$$\leq 2\delta A < \epsilon$$

when $\delta < \epsilon/2A$. Therefore, choosing such a value of δ and applying (5.9) and Corollary (5.12) to (5.8) we have

$$\left| \int_{-c}^{c} \hat{f}(u) e^{i2\pi ux} \, du - f(x) \right| < \epsilon + \epsilon + \epsilon = 3\epsilon$$

provided c is taken sufficiently large. That proves that

$$\lim_{c \to +\infty} \int_{-c}^{c} \hat{f}(u) e^{i2\pi ux} \, du = f(x) \qquad ∎$$

(5.14) Remark. The crucial point in the proof above was that for $\delta > 0$

sufficiently small

$$\left|\frac{f(x+s)-f(x)}{\pi s}\right| \leq A \qquad (0<|s|<\delta)$$

In words, $[f(x+s)-f(x)]/\pi s$ *is bounded in magnitude for all s near zero.* For example, we have the following result whose proof is left to the reader as Exercise (5.19). [Compare Theorem (3.13), Chapter 2.]

(5.15) Theorem.   Suppose that f is piecewise continuous and absolutely integrable. If f satisfies a Lipschitz condition, order one, at x then P.V. $\int_{-\infty}^{+\infty} \hat{f}(u)e^{i2\pi ux}\, du = f(x)$.

(5.16) Examples

(a)   Let $f(x) = \Pi(x)$. By Theorem (5.13), we have for all $x \neq \pm\frac{1}{2}$, P.V. $\int_{-\infty}^{+\infty} (\sin \pi u/\pi u)e^{i2\pi ux}\, du = \Pi(x)$

(b)   Suppose that f is piecewise smooth and absolutely integrable. For each point x where f is differentiable, by Theorem (5.13), we have

$$\text{P.V.} \int_{-\infty}^{+\infty} \hat{f}(u)e^{i2\pi ux}\, du = f(x)$$

and we know that this result holds at all but a finite number of x values in every finite interval.

(c)   The equality in (b) will actually hold for all x values where f is continuous. For, even if $f'(x)$ does not exist, the one sided limits $f'(x+)$ and $f'(x-)$ both exist. Therefore, by l'Hôpital's rule

$$\lim_{s\to 0+} \frac{f(x+s)-f(x)}{\pi s} = \lim_{s\to 0+} \frac{1}{\pi}f'(x+s) = \frac{1}{\pi}f'(x+)$$

$$\lim_{s\to 0-} \frac{f(x+s)-f(x)}{\pi s} = \frac{1}{\pi}f'(x-)$$

Since $[f(x+s)-f(x)]/\pi s$ tends to finite limits as s tends to zero from either side, it follows that $[f(x+s)-f(x)]/\pi s$ is bounded in magnitude for all s near enough to zero. Therefore, as we noted in Remark (5.14), we must have

$$\text{P.V.} \int_{-\infty}^{+\infty} \hat{f}(u)e^{i2\pi ux}\, du = f(x)$$

Our last theorem for this section is due to Dirichlet; it extends his theorem on convergence of Fourier series to the case of Fourier transforms.

(5.17) Theorem.   Let f be piecewise smooth and absolutely integrable. Then for every x value

$$\text{P.V.} \int_{-\infty}^{+\infty} \hat{f}(u)e^{i2\pi ux}\, du = \tfrac{1}{2}[f(x+)+f(x-)]$$

In particular, at each point x of continuity of f, we have

$$\text{P.V.} \int_{-\infty}^{+\infty} \hat{f}(u) e^{i2\pi u x} \, du = f(x)$$

Proof. Due to Examples (5.16b and c), we only need to discuss the case of x a point of discontinuity of f. We shall prove that

$$\lim_{c \to +\infty} \int_{0}^{+\infty} f(x+s)D(s;c) \, ds = \tfrac{1}{2}f(x+)$$

$$\lim_{c \to +\infty} \int_{-\infty}^{0} f(x+s)D(s;c) \, ds = \tfrac{1}{2}f(x-)$$

Because of (5.5) these last two limits are sufficient for establishing our result. Let's first consider $\int_{0}^{+\infty} f(x+s)D(s;c) \, ds$. Using (5.6′) we have for $\delta > 0$

$$\left| \int_{0}^{+\infty} f(x+s)D(s;c) \, ds - \tfrac{1}{2}f(x+) \right| = \left| \int_{0}^{+\infty} [f(x+s) - f(x+)]D(s;c) \, ds \right|$$

$$\leq \left| \int_{\delta}^{+\infty} f(x+s)D(s;c) \, ds \right|$$

$$+ |f(x+)| \cdot \left| \int_{\delta}^{+\infty} D(s;c) \, ds \right|$$

$$+ \int_{0}^{\delta} \left| \frac{f(x+s) - f(x+)}{\pi s} \right| \, ds$$

l'Hôpital's Rule implies that $[f(x+s) - f(x+)]/(\pi s)$ is bounded in magnitude for all s sufficiently near to zero. Hence, using the same type of arguments as we used to prove Theorem (5.13), we conclude that

$$\lim_{c \to +\infty} \int_{0}^{+\infty} f(x+s)D(s;c) \, ds = \tfrac{1}{2}f(x+)$$

A similar argument establishes that

$$\lim_{c \to +\infty} \int_{-\infty}^{0} f(x+s)D(s;c) \, ds = \tfrac{1}{2}f(x-)$$

Hence,

$$\lim_{c \to +\infty} \int_{-c}^{c} \hat{f}(u) e^{i2\pi u x} \, du = \lim_{c \to +\infty} \int_{-\infty}^{+\infty} f(x+s)D(s;c) \, ds$$

$$= \tfrac{1}{2}f(x+) + \tfrac{1}{2}f(x-) \quad \blacksquare$$

Remark. The only place in this section where we made use of piecewise continuity was in the proof of (5.11). A careful examination of the arguments given above will show that if a function f on $\mathbb{R}$ is arbitrarily closely approximable by step functions in the sense of (5.11), then Theorems (5.13), (5.15), and (5.17) will apply to f. For example, if f is Riemann integrable on every closed interval of $\mathbb{R}$ and $|f|$ is improperly integrable over $\mathbb{R}$, then

$$\text{P.V.} \int_{-\infty}^{+\infty} \hat{f}(u) e^{i2\pi ux} \, du = f(x)$$

provided f satisfies a Lipschitz condition at x. [Similar remarks apply to Lebesgue integrable functions.]

Exercises

(5.18)   Show that $\hat{f}(u) = u/(1 + u^2)$ [see (5.2)] is not integrable, even in an improper sense, over $\mathbb{R}$. Show the same thing for $\hat{f}(u) = \sin^2 \pi u / \pi u$.

(5.19)   Prove Theorem (5.15).

(5.20)   Let f be piecewise continuous and absolutely integrable. Suppose that f has left- and right-hand derivatives at x_0 [see (3.24), Chapter 2]. Prove that

$$\text{P.V.} \int_{-\infty}^{+\infty} \hat{f}(u) e^{i2\pi ux_0} \, du = \tfrac{1}{2}[f(x_0 +) + f(x_0 -)].$$

(5.21)   Suppose that f is piecewise continuous over (a, b), prove that

$$\lim_{c \to \infty} \int_a^b f(x) \cos cx \, dx = \lim_{c \to \infty} \int_a^b f(x) \sin cx \, dx = 0.$$

Prove also that $\hat{f}(u) \to 0$ as $|u| \to +\infty$ under the same hypotheses as in (5.10).

Remark.   The results in (5.21) are called the *Riemann–Lebesgue lemma*.

(5.22)   Let f be piecewise continuous and absolutely integrable. Suppose that f satisfies a Lipschitz condition, order $\alpha > 0$, at x_0 [see (3.26), Chapter 2]. Prove that P.V. $\int_{-\infty}^{+\infty} \hat{f}(u) e^{i2\pi ux_0} \, du = f(x_0)$

(5.23)   We say that f is *Lipschitz from both sides at* x if

$$|f(x + s) - f(x +)| \le A s^\alpha \qquad \text{for } \delta_1 > s > 0$$
$$|f(x + s) - f(x -)| \le B |s|^\beta \qquad \text{for } -\delta_2 < s < 0$$

where A, B, α, β, δ_1, and δ_2 are all positive constants. Prove the following Theorem, which generalizes (5.20) and (5.22).

Theorem.   Let f be piecewise continuous and absolutely integrable. If f is Lipschitz from both sides at x then

$$\text{P.V.} \int_{-\infty}^{+\infty} \hat{f}(u) e^{i2\pi ux} \, du = \tfrac{1}{2}[f(x +) + f(x -)]$$

(5.24)   Prove Fourier's integral theorem, which is stated in (1.12).

(5.25)   Generalize Theorem (5.17) to cover the inversion of cosine and sine transforms [cf. Exercise (4.18)].

(5.26)   Prove that P.V. $\int_{-\infty}^{+\infty} \hat{f}(u)e^{i2\pi ux} \, du = L$ if and only if

$$\lim_{c \to +\infty} \int_{-\delta}^{\delta} f(x+s)D(s;c) \, ds = L$$

for each $\delta > 0$ (no matter how small δ may be).

(5.27) Remark.   The result in (5.26) is known as the *localization principle*; it says that the ability to perform Fourier inversion at x depends only upon the nature of f on an arbitrarily small interval about x.

*(5.28)   Prove that all hypotheses that ensure convergence of Fourier series will also ensure that Fourier inversion by Dirichlet's method is possible for Fourier transforms (provided the function being transformed is absolutely integrable). [*Hint*: Show that a localization principle, see (5.27), also holds for Fourier series, then compare the Dirichlet kernels for Fourier series and Fourier transforms.]

*(5.29)   Let g be a continuous function of period 2π whose Fourier series diverges at $x = 0$. Prove that for the function f defined by $f(x) = g(x)\Pi(x/2\pi)$ the principal value P.V. $\int_{-\infty}^{+\infty} \hat{f}(u)e^{i2\pi ux} \, du$ *fails to exist* at $x = 0$. [*Hint*: See the hint for (5.28).]

§6. Convolution

The operation of convolution and the closely related concept of autocorrelation are important tools in Fourier transform theory. Each of the areas listed in §1 finds an interpretation and use for convolution. For example, in optics the theories of interference and image formation can be treated using convolution, whereas in electric circuits, the circuit response can be treated using convolution. In probability theory, if X and Y are two independent random variables with probability density functions f and g, then $X + Y$ is a random variable with probability density function equal to the convolution of f and g.

If f and g are piecewise continuous (and absolutely integrable), then the *convolution* of g and f is defined as [see (8.13), Chapter II]

(6.1) $$g * f(x) = \int_{-\infty}^{+\infty} g(s)f(x-s) \, ds$$

By a simple change of variables we see that

(6.2) $$g * f(x) = f * g(x) = \int_{-\infty}^{+\infty} f(s)g(x-s) \, ds$$

We have already seen examples of convolution in §§4 and 5. Using (6.2), the integral on the right side of (4.2) can be written as

(6.3)
$$_cW * f(x) = \int_{-\infty}^{+\infty} f(s)W(x - s; c)\,ds$$

where $_cW$ stands for the Gauss–Weierstrass kernel $_cW(s) = W(s; c)$. In §5 we found that $\int_{-c}^{c} \hat{f}(u)e^{i2\pi ux}\,du$ could be expressed as

(6.4)
$$_cD * f(x) = \int_{-\infty}^{+\infty} f(x - s)D(s; c)\,ds$$

where $_cD$ stands for the Dirichlet kernel $_cD(s) = D(s; c)$.

The interpretation of convolution is greatly helped by examining (6.3) and (6.4) in the light of the results of the previous two sections. Let's begin with formula (6.3), the Gauss–Weierstrass case. In Figure 6.2, we illustrate the geometric meaning of (6.3). The kernel $_cW$ is said to "scan" across the function f. As pointed out in Remark (4.10), for very small values of c we

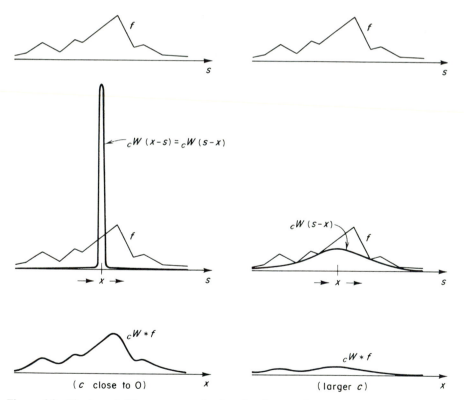

Figure 6.2 The kernel $_cW$ scans across the function f as x varies, to generate the convolution $_cW * f$. The figures on the left side above illustrate that for c close to 0, $_cW * f$ provides a smooth approximation to f. On the right side, we illustrate the fact that for c too large, some of the variation (fine detail) in f is lost in $_cW * f$.

have that $_cW * f$ is a uniformly close approximation to f, provided that $\lim_{|x| \to +\infty} f(x) = 0$. Moreover, because $(d^n/dx^n)[_cW(x)] = p_n(x)e^{-\pi x^2/c^2}$ for some polynomial p_n, we can differentiate $_cW * f$ under the integral sign as many times as we wish. Thus

(6.5) $f \mapsto {}_cW * f$

describes a mapping (linear transformation) from the set of piecewise continuous functions (that are bounded or absolutely integrable) to the set of infinitely continuously differentiable functions. Convolution in this case is said to "smooth", or "regularize," the given function f.

We now turn to formula (6.4), the Dirichlet case. In Figure 6.3 we illustrate the geometric content of (6.4). Notice the oscillation of $_cD * f$ near the jump discontinuities of f in Figure 6.3. This is Gibb's phenomenon for Fourier transforms, it is called "ringing" in optics and we shall encounter it again in §10 of Chapter 8. The proof that Gibb's phenomenon always occurs at discontinuities of piecewise smooth, absolutely integrable functions is not greatly different from the discussion for Fourier series in §5, Chapter 2, so we will not go into it.

We now turn to an important theoretical result: The Fourier transform of the convolution of two functions yields the product of their transforms.

(6.6) Theorem: Convolution Theorem. If f and g are absolutely integrable, then

$$f * g(x) \supset \hat{f}(u)\hat{g}(u)$$

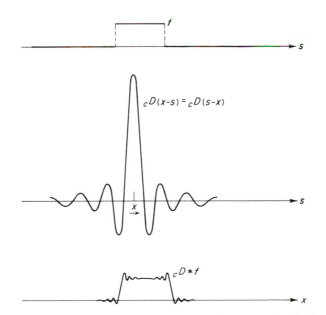

Figure 6.3   Convolution of $_cD$ with f. Notice the Gibb's phenomenon in $_cD * f$ near the jump discontinuities of f.

Proof.   First, we observe that

(6.7) $$\int_{-\infty}^{+\infty}\left[\int_{-\infty}^{+\infty}|f(s)g(x-s)|\,dx\right]ds = \int_{-\infty}^{+\infty}|f(s)|\,ds\int_{-\infty}^{+\infty}|g(v)|\,dv$$

is finite. It follows from the theorems of Fubini and Tonelli[5] that

$$\int_{-\infty}^{+\infty}\left[\int_{-\infty}^{+\infty}f(s)g(x-s)\,ds\right]e^{-2i\pi ux}\,dx = \int_{-\infty}^{+\infty}\left[\int_{-\infty}^{+\infty}g(x-s)e^{-i2\pi ux}\,dx\right]f(s)\,ds$$

$$= \int_{-\infty}^{+\infty}\left[\int_{-\infty}^{+\infty}g(v)e^{-i2\pi uv}\,dv\right]f(s)e^{-i2\pi us}\,ds$$

$$= \hat{g}(u)\int_{-\infty}^{+\infty}f(s)e^{-i2\pi us}\,ds = \hat{f}(u)\hat{g}(u)$$

Thus $f*g(x)\supset\hat{f}(u)\hat{g}(u)$.   ∎

(6.8) Remark.   It follows from (6.7) and Fubini's and Tonelli's theorems that $f*g$ is absolutely integrable and

(6.9) $$\int_{-\infty}^{+\infty}|f*g(x)|\,dx \le \int_{-\infty}^{+\infty}|f(x)|\,dx\int_{-\infty}^{+\infty}|g(x)|\,dx$$

Here are a couple of simple examples of the convolution theorem.

(6.10) Examples

(a)   We have $\Lambda(x)\supset\operatorname{sinc}^2 u = \hat{\Pi}(u)\hat{\Pi}(u)$. Hence, by Fourier inversion, we conclude that $\Pi*\Pi(x)=\Lambda(x)$ which can also be obtained by a direct calculation.

(b)   Let a and b be positive real numbers. Consider the convolution

$$f*g = \left[\frac{1}{\sqrt{2\pi}a}e^{-x^2/2a^2}\right]*\left[\frac{1}{\sqrt{2\pi}b}e^{-x^2/2b^2}\right]$$

(which is the convolution of two normal probability distributions of mean 0 and variances a^2 and b^2). Using Example (3.2b) and the convolution theorem, we have

$$f*g\supset e^{-2a^2\pi^2u^2}e^{-2b^2\pi^2u^2}=e^{-2(a^2+b^2)\pi^2u^2}$$

Therefore, applying Fourier inversion, we conclude that

$$\left[\frac{1}{\sqrt{2\pi}a}e^{-x^2/2a^2}\right]*\left[\frac{1}{\sqrt{2\pi}b}e^{-x^2/2b^2}\right]=\frac{1}{\sqrt{2\pi}(a^2+b^2)^{1/2}}e^{-x^2/2(a^2+b^2)}$$

(This last results shows that the probability distribution of the sum of two independent normally distributed random variables of mean

5. The theorems of Fubini and Tonelli state that if the iterated integral $\int[\int|G(s,x)|\,dx]\,ds$ exists, then the two iterated integrals $\int[\int G(s,x)\,ds]\,dx$ and $\int[\int G(s,x)\,dx]\,ds$ both exist and are equal. See Wheeden and Zygmund (1977) for further discussion.

zero is also normally distributed with mean zero, and the variances add.)

The examples above illustrate how we can use the convolution theorem to find the convolution of two functions without performing the integration in (6.1) directly.

We now turn to the concept of autocorrelation, which is closely related to convolution. If f is a *complex-valued* absolutely integrable function, then the *autocorrelation* of f is the function denoted by $f \star f$ and defined by (recall that the values of $\bar{f}$ are the complex conjugates to the values of f)

(6.11)
$$f \star f(x) = \int_{-\infty}^{+\infty} f(s)\bar{f}(s-x)\,ds$$

The following theorem is used frequently in applications.

(6.12) Theorem: Autocorrelation Theorem. If $f(x) \supset \hat{f}(u)$, then $f \star f(x) \supset |\hat{f}(u)|^2$.

Proof. Noting that $\bar{f}(s-x) = \bar{f}[-(x-s)]$, the convolution theorem implies that

$$f \star f(x) \supset \int_{-\infty}^{+\infty} f(x)e^{-i2\pi ux}\,dx \cdot \int_{-\infty}^{+\infty} \bar{f}(-x)e^{-i2\pi ux}\,dx$$

Since x is real, we have

$$\int_{-\infty}^{+\infty} \overline{f(-x)}e^{-i2\pi ux}\,dx = \overline{\int_{-\infty}^{+\infty} f(-x)e^{i2\pi ux}\,dx}$$

where the bar denotes taking complex conjugates. Substituting $x' = -x$ in the integral on the right above we get

$$\overline{\int_{-\infty}^{+\infty} f(-x)e^{i2\pi ux}\,dx} = \overline{\int_{-\infty}^{+\infty} f(x')e^{-i2\pi ux'}\,dx'} = \overline{\hat{f}(u)}$$

Thus, $f \star f(x) \supset \hat{f}(u)\overline{\hat{f}(u)} = |\hat{f}(u)|^2$. ∎

Based on the autocorrelation theorem, some autocorrelations can be computed by going through the transform domain.

(6.13) Examples

(a) We have $\Pi \star \Pi(x) \supset |\hat{\Pi}(u)|^2 = \text{sinc}^2 u$. Thus, $\Pi \star \Pi$ has the same transform as $\Pi * \Pi$, which leads us to conclude that $\Pi \star \Pi = \Pi * \Pi$. This last result is easily confirmed since Π is a real valued even function [see Exercise (6.29)].

(b) Let

$$f(x) = \begin{cases} e^{-cx} & \text{for } x \geq 0 \\ 0 & \text{for } x < 0 \end{cases} \qquad \text{where } c > 0$$

Then

$$\hat{f}(u) = \int_0^{+\infty} e^{-cx} e^{-i2\pi ux} \, dx$$

$$= \int_0^{+\infty} e^{(-i2\pi u - c)x} \, dx = \frac{1}{c + i2\pi u}$$

Therefore $f \star f(x) \supset [1/(c^2 + 4\pi^2 u^2)]$. From Example (3.2a), we conclude by Fourier inversion that $f \star f(x) = (1/2c)e^{-c|x|}$. This last result could be confirmed by direct evaluation of $f \star f$, but the transform argument is much easier.

One of the important corollaries of the autocorrelation theorem is the following result, known as *Parseval's Equality*. Recall that a function f is *square integrable* if $\int_{-\infty}^{+\infty} |f(x)|^2 \, dx$ is finite.

(6.14) Theorem: Parseval's Equality. Suppose that the piecewise continuous function f is absolutely integrable and square integrable, then $\hat{f}$ is square integrable and

$$\int_{-\infty}^{+\infty} |f(x)|^2 \, dx = \int_{-\infty}^{+\infty} |\hat{f}(u)|^2 \, du$$

Proof. We will not be making any applications of Parseval's equality beyond this section, therefore we shall prove the theorem using the following simplifying assumptions

(6.15) $f \star f$ is continuous

(6.16) $\int_{-\infty}^{+\infty} |\hat{f}(u)|^2 \, du$ is finite

In the miscellaneous Exercises for this chapter we shall outline the proof that those assumptions follow from the hypotheses of our theorem.

Applying the Fourier inversion theorem (4.11) we have

$$f \star f(x) = \int_{-\infty}^{+\infty} |\hat{f}(u)|^2 \, e^{i2\pi ux} \, du$$

Setting $x = 0$ and noting that $f \star f(0) = \int_{-\infty}^{+\infty} |f(s)|^2 \, ds$, we obtain

$$\int_{-\infty}^{+\infty} |f(s)|^2 \, ds = \int_{-\infty}^{+\infty} |\hat{f}(u)|^2 \, du \quad \blacksquare$$

Parseval's equality asserts the equality of two integrals; one mathematical application of this theorem is to the evaluation of certain integrals. For example, consider the integral $\int_{-\infty}^{+\infty} (\sin^4 x / x^4) \, dx$. Since $\Lambda(x) \supset \text{sinc}^2 u$, we find by Parseval's equality that

(6.17) $$\int_{-\infty}^{+\infty} \Lambda^2(x) \, dx = \int_{-\infty}^{+\infty} \text{sinc}^4 u \, du$$

Moreover, because $\int_{-\infty}^{+\infty} \Lambda^2(x)\, dx = 2/3$, we obtain the result

(6.18)
$$\int_{-\infty}^{+\infty} \frac{\sin^4 x}{x^4}\, dx = \frac{2\pi}{3}$$

Proving the following corollary to Parseval's equality is left as an exercise for the reader.

(6.19) Corollary. If f and g satisfy the hypotheses of Parseval's equality, then

$$\int_{-\infty}^{+\infty} f(x)\bar{g}(x)\, dx = \int_{-\infty}^{+\infty} \hat{f}(u)\bar{\hat{g}}(u)\, du$$

Exercises

(6.20) Suppose f is piecewise continuous and absolutely integrable (bounded). Prove that $\lim_{c\to+\infty} {}_c W * f(x) = 0$ uniformly over $\mathbb{R}$.

(6.21) Suppose f and g are both 0 outside a closed interval $[-a, a]$. If f is k times continuously differentiable and g is m times continuously differentiable, then prove that $f * g$ is 0 outside the interval $[-2a, 2a]$ and is $k + m$ times continuously differentiable.

(6.22) Prove that if f is piecewise continuous and absolutely integrable, then ${}_c D * f$ is infinitely differentiable.

(6.23) Prove that $\int_{-\infty}^{+\infty} |{}_c W * f(x)|\, dx \le \int_{-\infty}^{+\infty} |f(x)|\, dx$ provided f is absolutely integrable.

(6.24) Verify (6.2) and check that $f * (g * h) = (f * g) * h$.

(6.25) Prove that $\lim_{c\to0+} \int_{-\infty}^{+\infty} |{}_c W * f(x) - f(x)|\, dx = 0$ if f is piecewise continuous and absolutely integrable.

(6.26) If f is continuous and $m \le f(x) \le M$, then prove that $m \le {}_c W * f(x) \le M$.

(6.27) Show by a direct application of (6.1) that $\Pi * \Pi(x) = \Lambda(x)$.

(6.28) (a) Find the autocorrelation function for $f(x) = 1/(1 + x^2)$. (b) Find the convolution $[1/(a^2 + x^2)] * [1/(b^2 + x^2)]$.

(6.29) Prove that if f is real valued and even, then $f \star f = f * f$.

(6.30) Show that for $a > 0$, $b > 0$, and c, d real constants

$$\frac{1}{\sqrt{2\pi a}} e^{-(x-c)^2/2a^2} * \frac{1}{\sqrt{2\pi b}} e^{-(x-d)^2/2b^2} = \frac{1}{\sqrt{2\pi\delta}} e^{-[x-(c+d)]^2/2\delta^2}$$

where $\delta = [a^2 + b^2]^{1/2}$.

Schwartz's Space

Let $\mathscr{S}$ denote the set of functions f over $\mathbb{R}$ for which $|x^k f^{(j)}(x)|$ is bounded (by different constants) for all nonnegative integers k and j. The set $\mathscr{S}$ is called the space of *infinitely continuously differentiable functions of rapid decrease*, or *Schwartz's space*. The following exercises concern Schwartz's space, which has extensive applications in the theory of partial differential equations. [See, e.g., Vladimirov (1971).]

(6.31)   Prove that $p(x)e^{-bx^2}$ belongs to $\mathscr{S}$ for each polynomial $p(x)$ and positive constant b.

(6.32)   Prove that if f belongs to $\mathscr{S}$, then so does $\hat{f}$. [*Hint*: Use (3.6).] Then show that the mapping

$$\mathscr{F}: \mathscr{S} \to \mathscr{S}$$

$$f \mapsto \hat{f}$$

is one to one and onto.

(6.33)   Prove that if f is continuous and is 0 outside a closed interval, then $_cW * f$ belongs to $\mathscr{S}$. Moreover, prove that if g is continuous and $\lim_{|x| \to +\infty} g(x) = 0$ then g can be uniformly closely approximated by functions from $\mathscr{S}$.

(6.34)   Suppose that f, g belong to $\mathscr{S}$ and a, b are complex constants. Prove that $af + bg$, fg, and $f * g$ all belong to $\mathscr{S}$. Moreover, if p is a polynomial in x, then pf belongs to $\mathscr{S}$.

§7. Multiple Fourier Transforms

In this section, we define double Fourier transforms and discuss the generalizations of the theorems above for single variable transforms. Triple order transforms will be treated briefly in the exercises.

The *double Fourier transform* of an absolutely integrable function f on $\mathbb{R}^2$ is defined by

(7.1)
$$\hat{f}(u, v) = \int\limits_{-\infty}^{+\infty}\!\!\int f(x, y)e^{-i2\pi(ux+vy)}\,dx\,dy$$

We shall use the notation $f(x, y) \supset \hat{f}(u, v)$ to denote this transform operation.

The fundamental theorems discussed in §2 hold for double integrals dependent upon parameters, provided that the obvious changes in notation are made. For instance, the double Fourier transform $\hat{f}$ is always a continuous, bounded function on $\mathbb{R}^2$. We shall, therefore, devote most of our discussion in this section to the new phenomena that arise in considering double Fourier transforms.

Theorem (3.1) generalizes easily to double Fourier transforms.

(7.2) Theorem.   The double Fourier transform enjoys the following four properties.

(a)   *Linearity.*   If $f(x, y) \supset \hat{f}(u, v)$ and $g(x, y) \supset \hat{g}(u, v)$, then for all complex numbers a and b

$$af(x, y) + bg(x, y) \supset a\hat{f}(u, v) + b\hat{g}(u, v)$$

(b)   *Shifting.*   If $f(x, y) \supset \hat{f}(u, v)$, then for each element (c, d) of $\mathbb{R}^2$

$$f(x - c, y - d) \supset \hat{f}(u, v)e^{-i2\pi(cu + dv)}$$

(c)   *Change of Scales.*   If $f(x, y) \supset \hat{f}(u, v)$ and $c > 0$, $d > 0$ then

$$f(cx, dy) \supset \frac{1}{cd}\hat{f}\left(\frac{u}{c}, \frac{v}{d}\right)$$

(d)   *Modulation.*   If $f(x, y) \supset \hat{f}(u, v)$ and (c, d) is in $\mathbb{R}^2$, then

$$f(x, y)e^{i2\pi(cx + dy)} \supset \hat{f}(u - c, v - d)$$

Moreover,

$$f(x, y) \cos 2\pi(cx + dy) \supset (i/2)\hat{f}(u - c, v - d) + (i/2)\hat{f}(u + c, v + d)$$
$$f(x, y) \sin 2\pi(cx + dy) \supset (i/2)\hat{f}(u + c, v + d) - (i/2)\hat{f}(u - c, v - d)$$

Proof.   We will prove property (c); the remaining proofs are left to the reader. If we make the substitutions $x' = cx$ and $y' = dy$ then

$$f(cx, dy) \supset \int\!\!\!\int_{-\infty}^{+\infty} f(cx, dy)e^{-i2\pi(ux + vy)} \, dx \, dy$$

$$= \frac{1}{cd}\int\!\!\!\int_{-\infty}^{+\infty} f(x', y')e^{-i2\pi[(u/c)x' + (v/d)y']} \, dx' \, dy' = \frac{1}{cd}\hat{f}\left(\frac{u}{c}, \frac{v}{d}\right) \qquad \blacksquare$$

Furthermore, the following two theorems, which generalize Theorems (3.4) and (3.5), are easily proved, so we leave them as exercises.

(7.3) Theorem.   Let f be absolutely integrable over $\mathbb{R}^2$. If $xf(x, y)$ [or $yf(x, y)$] is absolutely integrable, then $\partial\hat{f}/\partial u$ (or $\partial\hat{f}/\partial v$) exists and is continuous on $\mathbb{R}^2$. Moreover,

$$xf(x, y) \supset \left(\frac{i}{2\pi}\right)\frac{\partial\hat{f}}{\partial u}(u, v) \qquad \left[\text{or } yf(x, y) \supset \left(\frac{i}{2\pi}\right)\frac{\partial\hat{f}}{\partial v}(u, v)\right]$$

(7.4) Theorem.   Let f and $\partial f/\partial x$ (or $\partial f/\partial y$) be continuous and absolutely integrable over $\mathbb{R}^2$, then

$$\frac{\partial f}{\partial x} \supset i2\pi u \hat{f}(u, v) \qquad \left[\frac{\partial f}{\partial y} \supset i2\pi v \hat{f}(u, v)\right]$$

Our final basic theorem allows us to calculate many two-dimensional transforms by applying our knowledge of one-dimensional transforms. Its proof is quite simple so we leave it to the reader as an exercise.

(7.5) Theorem: Separation Property. If $f(x, y) = g(x)h(y)$ and $g(x) \supset \hat{g}(u)$, $h(y) \supset \hat{h}(v)$ then $f(x, y) \supset \hat{g}(u)\hat{h}(v)$.

Here are some applications of these theorems.

(7.6) Examples

(a)   Let $f(x, y) = e^{-\pi(x^2+y^2)}$. Since $e^{-\pi(x^2+y^2)} = e^{-\pi x^2}e^{-\pi y^2}$ we obtain by the separation property

$$e^{-\pi(x^2+y^2)} \supset e^{-\pi u^2}e^{-\pi v^2} = e^{-\pi(u^2+v^2)}$$

(b)   Let $f(x, y) = e^{-|x|-y^2}$, then by the separation property along with

$$e^{|x|} \supset \frac{2}{1 + 4\pi^2 u^2} \qquad e^{-y^2} \supset \sqrt{\pi}e^{-\pi^2 v^2}$$

we obtain

$$e^{-|x|-y^2} \supset \frac{2\sqrt{\pi}e^{-\pi^2 v^2}}{1 + 4\pi^2 u^2}$$

(c)   Let

$$A(x, y) = \begin{cases} 1 & \text{for } |x| < \tfrac{1}{2}a \text{ and } |y| < \tfrac{1}{2}b \\ 0 & \text{for } |x| > \tfrac{1}{2}a \text{ or } |y| > \tfrac{1}{2}b \end{cases} \qquad \text{then } A(x, y) = \Pi\left(\frac{x}{a}\right)\Pi\left(\frac{y}{b}\right)$$

Hence by the separation property and change of scales

$$\hat{A}(u, v) = ab \text{ sinc } au \text{ sinc } bv = \frac{\sin \pi au}{\pi u} \frac{\sin \pi bv}{\pi v}$$

(d)   Suppose f is defined by $f(x, y) = A(x - \tfrac{1}{2}c, y + \tfrac{1}{2}d) + A(x + \tfrac{1}{2}c, y - \tfrac{1}{2}d)$ where A is the function defined in (c) and c and d are real constants. Then by the shifting property and linearity

$$\hat{f}(u, v) = \hat{A}(u, v)[e^{-i\pi(cu-dv)} + e^{i\pi(cu-dv)}]$$

$$= 2\frac{\sin \pi au}{\pi u} \frac{\sin \pi bv}{\pi v} \cos \pi(cu - dv)$$

Our next example is a nontrivial calculation of a transform that is important in the theory of solutions to Laplace's equation.

(7.7) Example.   Show that

$$e^{-2\pi(x^2+y^2)^{1/2}} \supset \frac{1}{2\pi} \frac{1}{[1 + u^2 + v^2]^{3/2}}$$

Solution. The following ingenious calculation is due to Bochner and Stein.[6] The interchanges of integrals that we shall perform are all justified by Fubini's theorem [see the footnote following formula (6.7)].

For $s \geq 0$, Fourier inversion yields

$$e^{-2\pi s} = \frac{1}{\pi} \int_{-\infty}^{+\infty} \frac{e^{i2\pi xs}}{1+x^2} dx$$

Also,

$$\frac{1}{1+x^2} = \int_0^{+\infty} e^{-(1+x^2)w} dw$$

Hence,

$$e^{-2\pi s} = \frac{1}{\pi} \int_{-\infty}^{+\infty} \left[\int_0^{+\infty} e^{-(1+x^2)w} dw \right] e^{i2\pi xs} dx$$

Interchanging integrals and applying Example (3.2b) we obtain

$$e^{-2\pi s} = \frac{1}{\pi} \int_0^{+\infty} e^{-w} \left[\int_{-\infty}^{+\infty} e^{-wx^2} e^{i2\pi xs} dx \right] dw$$

$$= \pi^{-1/2} \int_0^{+\infty} w^{-1/2} e^{-w} e^{-\pi^2 s^2/w} dw$$

Therefore, letting $s = (x^2 + y^2)^{1/2}$ we have

$$e^{-2\pi(x^2+y^2)^{1/2}} = \pi^{-1/2} \int_0^{+\infty} w^{-1/2} e^{-w} e^{-\pi^2(x^2+y^2)/w} dw$$

Hence,

$$e^{-2\pi(x^2+y^2)^{1/2}} \supset \pi^{-1/2} \int\int_{-\infty}^{+\infty} \left[\int_0^{+\infty} w^{-1/2} e^{-w} e^{-\pi^2(x^2+y^2)/w} dw \right] e^{-i2\pi(ux+vy)} dx\, dy$$

Interchanging integrals, and applying Example (7.6a) along with a change of scales to the double Fourier transform in the inner integral, we have

$$e^{-2\pi(x^2+y^2)^{1/2}} \supset \pi^{-1/2} \int_0^{+\infty} (w/\pi) e^{-w(u^2+v^2)} e^{-w} w^{-1/2} dw$$

$$= \pi^{-3/2} \int_0^{+\infty} w^{1/2} e^{-(1+u^2+v^2)w} dw$$

If we let $r = (1 + u^2 + v^2)w$, then substitution into the last integral above, followed by a further substitution of $r = s^2$ yields

$$e^{-2\pi(x^2+y^2)^{1/2}} \supset \frac{\pi^{-3/2}}{[1+u^2+v^2]^{3/2}} \int_{-\infty}^{+\infty} s^2 e^{-s^2} ds$$

---

6. See Stein (1970), p. 61.

Performing an integration by parts ($u = s/2$, $dv = 2se^{-s^2} ds$), we obtain

$$e^{-2\pi(x^2+y^2)^{1/2}} \supset \frac{\pi^{-3/2}}{[1+u^2+v^2]^{3/2}} \frac{1}{2} \int_{-\infty}^{+\infty} e^{-s^2} ds$$

$$= \frac{1}{2\pi} \frac{1}{[1+u^2+v^2]^{3/2}}$$

which completes our verification.

We now turn to the question of Fourier inversion. The following lemma shows that $W(r, s; c)$ is a summation kernel over $\mathbb{R}^2$, parameterized by $c > 0$. It provides the basis for Fourier inversion of double Fourier transforms.

(7.8) Lemma.   The function $W(r, s; c) = (1/c^2)e^{-\pi(r^2+s^2)/c^2}$ satisfies

(A_1) $\iint_{-\infty}^{+\infty} W(r, s; c) \, dr \, ds = 1$ for each $c > 0$
(A_2) $W(r, s; c) \geq 0$ for all values of r and s, and each $c > 0$
(A_3)   For every $\epsilon > 0$ and $\delta > 0$ we have

$$0 \leq W(r, s; c) < \epsilon \qquad 0 \leq \iint_{(r^2+s^2)^{1/2} \geq \delta} W(r, s; c) \, dr \, dr < \epsilon$$

provided $(r^2 + s^2)^{1/2} \geq \delta$ and c is taken close enough to 0.

Proof.   Since $W(r, s; c) = W_1(r; c)W_1(s; c)$ where W_1 is the function considered in Lemma (4.5), properties (A_1) and (A_2) follow immediately. If $(r^2 + s^2)^{1/2} \geq \delta$, then

$$0 \leq W(r, s; c) \leq (1/c^2)e^{-\pi\delta^2/c^2} < \epsilon$$

when c is sufficiently close to 0, for the same reasons that we gave in the proof of Lemma (4.5). Moreover, changing to polar coordinates ρ, ϕ we see that

$$0 \leq \iint_{(r^2+s^2)^{1/2} \geq \delta} W(r, s; c) \, dr \, ds = \int_0^{2\pi} \int_\delta^{+\infty} (1/c^2)e^{-\pi\rho^2/c^2} \rho \, d\rho \, d\phi$$

$$= e^{-\pi\delta^2/c^2} < \epsilon$$

for the same small values of c.   ∎

From Lemma (7.8) we obtain the following Fourier inversion theorem. We leave its proof, which is almost identical to that of Theorem (4.11), to the reader.

(7.9) Theorem: Fourier Inversion.   Suppose that f and $\hat{f}$ are continuous and absolutely integrable over $\mathbb{R}^2$. Then

$$\iint_{-\infty}^{+\infty} \hat{f}(u, v)e^{i2\pi(ux+vy)} \, du \, dv = f(x, y)$$

Sometimes we do not need to have $\hat{f}$ absolutely integrable in order to perform inversion. For example, if $f(x, y) = g(x)h(y)$ where g and h are piecewise smooth, then by Theorem (5.17) and the separation property we obtain

$$\tfrac{1}{4}[g(x+)+g(x-)][h(y+)+h(y-)] = \lim_{a,b \to +\infty} \int_{-a}^{a} \int_{-b}^{b} \hat{f}(u, v)e^{i2\pi(ux+vy)} \, du \, dv$$

for each point (x, y). In particular, at each point (x, y) where x is a point of continuity for g and y is a point of continuity for h, the left side of the equality above is simply $g(x)h(y) = f(x, y)$. We obtain the same kind of results when f is a sum of constant multiples of such functions. A complete generalization of Theorem (5.17) to double Fourier transforms is much harder to obtain; see Bochner (1959) for some results.

The *convolution* of two absolutely integrable functions f and g over $\mathbb{R}^2$ is defined by

(7.10) $$f * g(x, y) = \int\!\!\!\int_{-\infty}^{+\infty} f(w, z)g(x - w, y - z) \, dw \, dz$$

while the *autocorrelation* of f is defined by

(7.11) $$f \star f(x, y) = \int\!\!\!\int_{-\infty}^{+\infty} f(w, z)\bar{f}(w - x, z - y) \, dw \, dz$$

The convolution and autocorrelation theorems generalize in the obvious ways:

(7.12) $f * g(x, y) \supset \hat{f}(u, v)\hat{g}(u, v)$ $f \star f(x, y) \supset |\hat{f}(u, v)|^2$

and so does Parseval's equality.

We close this section by discussing one aspect of double Fourier transforms that differs radically from the single variable case. Suppose we have a function f on $\mathbb{R}^2$ that depends only on the radial coordinate $r = (x^2 + y^2)^{1/2}$, and not upon the angular coordinate ϕ in the polar coordinate system. Such a function f is called a *radial function*. Some examples of radial functions are $e^{-\pi(x^2+y^2)} = e^{-\pi r^2}$ and $e^{-2\pi(x^2+y^2)^{1/2}} = e^{-2\pi r}$. For a radial function $f(r)$, the Fourier transform is also a radial function. If we introduce polar coordinates to the u–v plane: $u = \rho \cos \Phi$ and $v = \rho \sin \Phi$, then $f(r) \supset \hat{f}(u, v) = \hat{f}(\rho)$. For example, the results of Examples (7.6a) and (7.7) are expressed in polar coordinates as

$$e^{-\pi r^2} \supset e^{-\pi \rho^2} \qquad e^{-2\pi r} \supset (2\pi)^{-1}[1 + \rho^2]^{-3/2}$$

That these last two results do indeed reflect the general phenomenon is the content of our next Theorem.

(7.13) Theorem.   Suppose that $f(x, y) = f(r)$ is a radial function. If $\hat{f}$ is defined, then $\hat{f}(u, v) = \hat{f}(\rho)$ is a radial function.

Proof.   Expressing the integral that defines $\hat{f}$ in polar coordinates yields

$$\hat{f}(u, v) = \int_0^{+\infty} \int_0^{2\pi} f(r) e^{-i2\pi\rho r(\cos\phi\cos\Phi + \sin\phi\sin\Phi)} \, r \, d\phi \, dr$$

$$= \int_0^{+\infty} \left[\int_0^{2\pi} e^{-i2\pi\rho r\cos(\phi-\Phi)} \, d\phi \right] f(r) r \, dr$$

If we substitute $\theta = \phi - \Phi$ into the inner integral above we obtain (using periodicity)

$$\int_0^{2\pi} e^{-i2\pi\rho r\cos(\phi-\Phi)} \, d\phi = \int_0^{2\pi} e^{-i2\pi\rho r\cos\theta} \, d\theta$$

Thus,

$$\hat{f}(u, v) = \hat{f}(\rho) = \int_0^{+\infty} f(r) G(2\pi\rho r) r \, dr$$

(7.14)

$$G(2\pi\rho r) = \int_0^{2\pi} e^{-2\pi\rho r\cos\theta} \, d\theta$$

Formula (7.14) shows that $\hat{f}$ is a radial function.   ∎

This last result suffers from the fact that $G(2\pi\rho r)$ is expressed as an integral. Nevertheless, if we substitute $v = 2\pi\rho r$, and $\theta = \phi + \frac{1}{2}\pi$, we obtain (using periodicity of $e^{iv\sin\phi}$ in ϕ)

$$G(v) = \int_0^{2\pi} e^{iv\sin\phi} \, d\phi$$

which is a well-known function in mathematical physics. As we shall see in Chapter 10 [see Theorem (2.2)], G equals $2\pi J_0$ where J_0 is the *Bessel function of the first kind, order* 0. Therefore, (7.14) can be stated as follows.

(7.15) Theorem.   If $f(x, y) = f(r)$ is the radial function that possesses a double Fourier transform, then that transform $\hat{f}$ is a radial function and satisfies

$$\hat{f}(\rho) = 2\pi \int_0^{+\infty} f(r) J_0(2\pi\rho r) r \, dr   \blacksquare$$

Theorem (7.15) gives us a couple of *Fourier–Bessel transforms* (also known as *Hankel transforms*) using previously calculated double Fourier transforms

$$\int_0^{+\infty} e^{-\pi r^2} J_0(2\pi\rho r) r \, dr = (2\pi)^{-1} e^{-\pi\rho^2}$$

$$\int_0^{+\infty} e^{-2\pi r} J_0(2\pi\rho) r \, dr = (2\pi)^{-2} [1 + \rho^2]^{-3/2}$$

Exercises

(7.16)   Prove the rest of Theorem (7.2) and prove Theorems (7.3)–(7.5).

(7.17)   Find double Fourier transforms for the following functions.

(a) $f(x, y) = e^{-\pi(x^2+y^2)/c^2}$ for $c > 0$

(b) $f(x, y) = e^{-c(x^2+y^2)^{1/2}}$ for $c > 0$

(c) $f(x, y) = \begin{cases} \sin ax & \text{for } -c \leq x \leq c, \; -d < y < d \\ 0 & \text{otherwise} \end{cases}$

(d) $f(x, y) = e^{-2\pi(x^2+y^2)^{1/2}} \sin ax \cos by$

(e) $f(x, y) = xye^{-\pi(x^2+y^2)}$

(f) $f(x, y) = x^2ye^{-\pi(x^2+y^2)}$

(g) $f(x, y) = xe^{-2\pi(x^2+y^2)^{1/2}}$

(7.18) Prove the following *uniqueness theorem* for double Fourier transforms.

Theorem: Uniqueness.   If f and g are continuous and absolutely integrable over $\mathbb{R}^2$ and $\hat{f} = \hat{g}$ identically, then $f = g$ identically.

(7.19)   Find double Fourier transforms for the following functions

(a) $f(x, y) = [4 + x^2 + y^2]^{-3/2}$

(b) $f(x, y) = [6 + x^2 - 2x + y^2 + 4y]^{-3/2}$

(c) $f(x, y) = \dfrac{x}{[1 + x^2 + y^2]^{5/2}}$

*(7.20)   *Poisson Kernel for Upper Half Space.*   Define $P = P(x, y; z)$ as follows for all $z > 0$ and all x, y.

$$P(x, y; z) = \int\int_{-\infty}^{+\infty} e^{-2\pi(\alpha^2+\beta^2)^{1/2}z} e^{-i2\pi(\alpha x+\beta y)} \, d\alpha \, d\beta$$

Prove the following results.

(a) $P(x, y; z) = \dfrac{1}{2\pi} \dfrac{z}{[x^2 + y^2 + z^2]^{3/2}}$

(b) $\Delta P = P_{xx} + P_{yy} + P_{zz} = 0$ for all $z > 0$

(c) P is a summation kernel over $\mathbb{R}^2$, parameterized by $z > 0$ [see (7.8)].

(7.21)   Evaluate

$$\int_0^{+\infty} e^{-ar^2} J_0(br) r \, dr \qquad \text{for } a > 0 \text{ and } b > 0$$

The next two exercises deal with the *triple Fourier transform*, which we define as follows

$$\hat{f}(u, v, w) = \int\int\int_{-\infty}^{+\infty} f(x, y, z) e^{-i2\pi(ux+vy+wz)} \, dx \, dy \, dz$$

(7.22)   Using the definition above, generalize Theorems (7.2) through (7.5) to triple Fourier transforms; also, prove an inversion theorem.

(7.23)   Find the triple Fourier transforms of the following functions.

(a) $f(x, y, z) = e^{-(2x^2 + 3y^2 + 4z^2)}$

(b) $f(x, y, z) = xyze^{-\pi(x^2 + y^2 + z^2)}$

(c) $f(x, y, z) = \begin{cases} 1 & \text{if } |x| < a \quad |y| < b \quad |z| < c \\ 0 & \text{otherwise} \end{cases}$

(d) $f(x, y, z) = e^{-\pi(x^2 + y^2 + z^2)} \cos x \sin y \cos z$

(e) $f(x, y, z) = e^{-2\pi(x^2 + y^2 + z^2)^{1/2}}$

(f) $f(x, y, z) = \begin{cases} \cos \pi x & \text{if } |x| < \frac{1}{2} \quad |y| < \frac{1}{2} \quad |z| < \frac{1}{2} \\ 0 & \text{otherwise} \end{cases}$

§8. Miscellaneous Exercises

Hermite Polynomials

In the following exercises, we shall mostly consider the alternate Fourier transform $\bar{f}$ defined in (2.18). The Hermite polynomials play a major role in quantum mechanics, especially in the consideration of the harmonic oscillator.

(8.1)   Define the *nth Hermite polynomial* H_n by

$$H_n(x) = (-1)^n e^{x^2} \frac{d^n}{dx^n}(e^{-x^2}) \qquad (n = 0, 1, 2, \ldots)$$

Prove the following statements about these polynomials.

(a)   The degree of H_n is n.

(b) $H_0(x) = 1 \qquad H_1(x) = 2x \qquad H_2(x) = 4x^2 - 2 \qquad H_3(x) = 8x^3 - 12x$

(c)   For $n = 0, 1, 2, \ldots, \ H_{n+1}(x) = 2xH_n(x) - H_n'(x)$.

(8.2)   Let f be a piecewise continuous, absolutely integrable function for which $\int_{-\infty}^{+\infty} |f(x)| \, dx > 0$. Prove that if $\bar{f}(u) = cf(u)$ where c is a constant, then c equals 1, -1, i, or $-i$. [*Hint*: What does four applications of the Fourier transform do to f?]

(8.3)   Prove that using the alternate Fourier transform $\bar{f}$ defined in (2.18), we have $H_n(x)e^{-x^2/2} \supset (-i)^n H_n(u)e^{-u^2/2}$ for every Hermite polynomial H_n. [*Hint*: First prove that $H_{n+1}(x)e^{-x^2/2} = xH_n(x)e^{-x^2/2} - (d/dx)[H_n(x)e^{-x^2/2}]$, then apply mathematical induction.]

(8.4)   Prove that

$$\int_{-\infty}^{+\infty} H_m(x)e^{-x^2/2}H_n(x)e^{-x^2/2} \, dx = \begin{cases} 0 & \text{if } m \neq n \\ 2^n n! \sqrt{2\pi} & \text{if } m = n \end{cases}$$

[*Hint*: Use the formula for H_n in (8.1) and integrate by parts repeatedly.]

(8.5) Prove that for the Fourier transform $\hat{f}$ considered in this chapter,
$H_n(\sqrt{2\pi}x)e^{-\pi x^2} \supset (-i)^n H_n(\sqrt{2\pi}u)e^{-\pi u^2}$.

Parseval's Equality

Throughout the following discussion we will suppose that *f and g are piecewise continuous, absolutely integrable, and square integrable over* $\mathbb{R}$.

(8.6) Prove that $f * g$ is a continuous function. [*Hint*: Use the Schwarz inequality for $\int_{-\infty}^{+\infty}$ and the fact that over a finite interval $[-R, R]$ we may replace a piecewise continuous function g by a continuous function.] From this result conclude that $f \star f$ is a continuous function. Thus, (6.15) is satisfied.

(8.7) *If* $\lim_{n \to +\infty} \int_{-n}^{n} |\hat{f}(u)|^2 \, du = +\infty$, *then prove that*

$$\lim_{n \to +\infty} \int_{-n}^{n} |\hat{f}(u)|^2 [1 - |u|/n] \, du = +\infty$$

[*Hint*: Consider $\int_{-\frac{1}{2}n}^{\frac{1}{2}n} |\hat{f}(u)|^2 [1 - |u|/n] \, du$.]

(8.8) Combining (8.7) and (4.19b), prove that if f satisfies the hypotheses of Theorem (6.14), then $\int_{-\infty}^{+\infty} |\hat{f}(u)|^2 \, du$ is finite. Thus, (6.16) is satisfied. Since we have shown that the two assumptions, (6.15) and (6.16), used in the proof of Theorem (6.14) are satisfied, Parseval's equality is now completely proved.

References

For further reading, see Bochner and Chandrasekharan (1949), Titchmarsh (1937), Stein and Weiss (1971), Bochner (1959), or Weiner (1958). Modern generalizations of Fourier transforms are discussed in Bachman (1964), Gel'fand and Shilov (1964), or Vladimirov (1971).

7

Applications
of Fourier Transforms

This chapter is divided into three parts; these parts cover topics in *Partial Differential Equations, Fourier Optics,* and *Signal Processing.* The first part treats Laplace's equation and the heat equation in various infinite regions. Then we turn to the modern form of optics, Fourier optics, where we discuss Fraunhofer diffraction and imaging with lenses. Diffraction of light by rectangular and circular apertures, interference effects, diffraction gratings and spectra, and basic imaging theory will all be examined. The last part treats some topics in signal processing: sampling theory, Poisson summation, and fast Fourier transforms. The first two topics are major tools in modern communication and information theory, whereas the last topic has revolutionized all areas of applied Fourier analysis.

These applications by no means exhaust the uses of Fourier transforms. In the References for this chapter we list some sources that discuss other applications.

Part A. Partial Differential Equations

§1. Laplace's Equation

In this section we consider Laplace's equation in various infinite domains. *Let f denote a given continuous function over* $\mathbb{R}$. We shall solve the following problem for Ψ.

(1.1)
$$\begin{cases} \Delta\Psi = \Psi_{xx} + \Psi_{yy} = 0 & (-\infty < x < +\infty \quad y > 0) \\ \Psi(x, 0) = f(x) & (-\infty < x < +\infty) \\ \Psi \text{ is continuous for } -\infty < x < +\infty \text{ and } y \geq 0 \end{cases}$$

Problem (1.1) is Dirichlet's problem for the upper half plane $\mathbb{R}^2_+$. [See (3.1), Chapter 4.]

To solve (1.1) we first write

(1.2)
$$\hat{\Psi}(v, y) = \int_{-\infty}^{+\infty} \Psi(x, y)e^{-i2\pi vx}\, dx$$

and proceed under the temporary assumption that all necessary integrals are defined. Using Theorem (3.5), Chapter 6, we have

(1.3)
$$\Psi_x \supset i2\pi v\hat{\Psi} \qquad \Psi_{xx} \supset -(2\pi)^2 v^2\hat{\Psi}$$

Differentiating under the integral sign in (1.2) yields

(1.4)
$$\Psi_y \supset \frac{\partial\hat{\Psi}}{\partial y} \qquad \Psi_{yy} \supset \frac{\partial^2\hat{\Psi}}{\partial y^2}$$

Formulas such as (1.3) and (1.4) are the basis for the application of Fourier transforms to (1.1) as well as other problems in partial differential equations. Using these formulas, problem (1.1) becomes

(1.5)
$$-(2\pi)^2 v^2\hat{\Psi}(v, y) = -\partial^2\hat{\Psi}/\partial y^2 \qquad \hat{\Psi}(v, 0) = \hat{f}(v)$$

From (1.5) we obtain $\hat{\Psi}(v, y) = \hat{f}(v)e^{-2\pi y(v^2)^{1/2}} = \hat{f}(v)e^{-2\pi y|v|}$. Performing Fourier inversion yields

(1.6)
$$\Psi(x, y) = \int_{-\infty}^{+\infty} \hat{f}(v)e^{-2\pi y|v|}e^{i2\pi vx}\, dv$$

In Eq. (1.6) we are performing Fourier inversion on the *product of two transforms*. Therefore, by the convolution theorem, we expect that Ψ is a convolution of two functions. Since $f(x) \supset \hat{f}(v)$ and $(1/\pi)[y/(x^2 + y^2)] \supset e^{-2\pi y|v|}$ (for $y > 0$) we propose that Ψ has the following form.

For $y > 0$ and all x values

(1.7)
$$\Psi(x, y) = \int_{-\infty}^{+\infty} f(s)P(x - s; y)\, ds = \int_{-\infty}^{+\infty} f(x - s)P(s; y)\, dy$$

where $P(s; y) = (1/\pi)[y/(s^2 + y^2)]$. For $y = 0$ we define $\Psi(x, 0) = f(x)$.

The function P in (1.7) is called *Poisson's kernel* for the upper half plane. Sometimes we might express (1.7) as

(1.7′)
$$\Psi(x, y) = f *_y P(x) = {}_y P * f(x) \qquad [{}_y P(x) = P(x; y)]$$

Poisson's kernel P has two main properties. First, it is a *summation kernel over* $\mathbb{R}$ [see Example (8.9b) in Chapter 2]. Second, as the reader may easily check, *P is harmonic for $y > 0$ and all x values*. [See Exercise (1.20).]

We will now show that the function Ψ in (1.7) provides a solution to (1.1).

(1.8) Theorem. Let f be continuous and absolutely integrable over $\mathbb{R}$. Then Ψ as given in (1.7) solves (1.1).

Proof. To see that Ψ is harmonic, we observe that by differentiation under the integral sign

(1.9)
$$\Delta\Psi = \int_{-\infty}^{+\infty} f(s)\, \Delta P(x-s;y)\, ds = \int_{-\infty}^{+\infty} f(s)\, 0\, ds = 0$$

Of course, we must justify bringing the Laplacian Δ inside the integral sign in (1.9). To do this, suppose that $y \geq \delta > 0$. Then

$$|f(s)P(x-s;y)| \leq |f(s)|\frac{1}{\pi}\frac{1}{y} \leq |f(s)|\frac{1}{\pi}\frac{1}{\delta}$$

which proves, by Theorems (2.3) and (2.5) of Chapter 6, that Ψ is continuous for $y \geq \delta$ and all x values. Moreover,

$$\left|f(s)\frac{\partial P}{\partial x}(x-s;y)\right| = |f(s)| \cdot \left|-\frac{2}{\pi}\frac{(x-s)y}{[(x-s)^2+y^2]^2}\right|$$

$$\leq |f(s)|\, y\, \frac{2}{\pi}\frac{|x-s|}{[(x-s)^2+\delta^2]^2}$$

$$\leq |f(s)|\, yA$$

where

$$A = \underset{0\leq v<+\infty}{\text{maximum}}\; \frac{2}{\pi}\, v/(v^2+\delta^2)^2$$

It follows from Theorems (2.3) and (2.8), Chapter 6, that for $y \geq \delta$

(1.10)
$$\frac{\partial\Psi}{\partial x} = \int_{-\infty}^{+\infty} f(s)\frac{\partial P}{\partial x}(x-s;y)\, ds$$

Also, for $y \geq \delta$

$$\left|f(s)\frac{\partial P}{\partial y}(x-s;y)\right| = |f(s)|\frac{2}{\pi}\frac{|(x-s)^2-y^2|}{[(x-s)^2+y^2]^2} \leq |f(s)|\frac{2}{\pi}\frac{1}{(x-s)^2+y^2}$$

$$\leq |f(s)|\, B$$

where

$$B = \underset{0\leq v<+\infty}{\text{maximum}}\; \frac{2}{\pi}[v^2+\delta^2]^{-1}$$

Hence, again for $y \geq \delta$

(1.11)
$$\frac{\partial\Psi}{\partial y} = \int_{-\infty}^{+\infty} f(s)\frac{\partial P}{\partial y}(x-s;y)\, ds$$

Formulas (1.10) and (1.11) show [when $\delta \to 0$] that $\Psi = f *_y P$ may be differentiated under the integral sign with respect to x or y. Similar arguments show that we can do this twice and then (1.9) is justified. Thus $\Psi = f *_y P$ is harmonic for $y > 0$ and all x values.

Since $\Psi(x, 0)$ is defined as equal to $f(x)$ in (1.7), all that remains for us to show is that Ψ is continuous for $-\infty < x < +\infty$ and $y \geq 0$. Because we have already shown that Ψ is continuous when $y > 0$ and we are given that f is continuous, our task reduces to proving that

$$\textbf{(1.12)} \qquad \lim_{(x, y) \to (x_0, 0+)} \Psi(x, y) = f(x_0)$$

for each x_0 in $\mathbb{R}$. But, since Poisson's kernel is a summation kernel, (1.12) must hold because of Theorem (8.10) in Chapter 2. ∎

(1.13) Remark.   The structure of the proof of Theorem (1.8) can be summarized as follows. The harmonic function P induces a harmonic function Ψ through convolution $\Psi(x, y) = f *_y P(x)$ (which can be interpreted as a kind of continuous superposition of harmonic functions). The continuity of Ψ at the boundary of the upper half plane is due to Ψ being a convolution $\Psi(x, y) = {}_y P * f$ and the summation kernel P forcing ${}_y P * f \to f$ as $y \to 0+$.

The boundary function f need not be absolutely integrable for (1.7) to solve (1.1).

(1.14) Theorem.   Let f be continuous and bounded over $\mathbb{R}$. Then Ψ as given in (1.7) solves (1.1).

Proof.   We must again justify (1.9). Suppose that $|f(s)| \leq M$, a constant, for $-\infty < s < +\infty$. Then

$$|f(s)P(x - s; y)| \leq MP(x - s; y)$$

so $\Psi(x, y)$ is defined as an absolutely convergent integral for each $y > 0$ and all x values. Furthermore, if $|x| \leq R$, a positive number, then we have for $|s| \geq 3R$

$$\left| f(s) \frac{\partial P}{\partial x}(x - s; y) \right| \leq \frac{M}{\pi} \frac{2|x - s|y}{[y^2 + (x - s)^2]^2} \leq \frac{2My}{\pi} \frac{|s| + R}{[y^2 + s^2 - 2R|s|]^2}$$

For each $y > 0$ the last quantity above is an absolutely integrable function of s for $|s| \geq 3R$. Therefore, because of Theorems (2.3) and (2.8) in Chapter 6, *formula* (1.10) *holds for* $|x| \leq R$. Similarly, for $y \geq \delta$, $|x| \leq R$, and $|s| \geq 3R$

$$\left| f(s) \frac{\partial P}{\partial y}(x - s; y) \right| \leq \frac{2M}{\pi} \frac{1}{(x - s)^2 + y^2} \leq \frac{2M}{\pi} \frac{1}{s^2 - 2R|s| + \delta^2}$$

and the last function is absolutely integrable for $|s| \geq 3R$. Hence, for $y \geq \delta$ and $|x| \leq R$, *formula* (1.11) *holds*.

Using similar arguments we establish that Ψ can be differentiated under the integral sign twice with respect to x or y for $|x| \leq R$ and $y \geq \delta$. In particular, (1.9) is valid. Hence Ψ is harmonic for $y > 0$ and all x values (letting $\delta \to 0$ and $R \to +\infty$).

Formula (1.12) also holds again. Except that now we apply Theorem (8.11), Chapter 2. ∎

Unlike Dirichlet's problem for *bounded* open sets, uniqueness of solutions to (1.1) is not guaranteed. In fact, both $\Psi = 0$ and $\Psi(x, y) = y$ solve (1.1) when $f = 0$. It follows that (1.7) does not provide a unique solution to (1.1); we may always add a nonzero multiple of y.

The following theorem shows that we can still have uniqueness if we require bounded solutions to (1.1).

(1.15) Theorem: Uniqueness. If f is continuous and bounded, then there exists only one *bounded* solution to problem (1.1) and it is given in (1.7).

Proof. That there can be only one bounded solution to (1.1) follows from Corollary (5.10), Chapter 4.

If $|f(s)| \leq M$ for all s values, then

$$-M \leq f(s) \leq M$$

Hence, using the fact that $P \geq 0$ and $\int_{-\infty}^{+\infty} P(x - s; y)\, ds = 1$, we have

$$-M \leq \int_{-\infty}^{+\infty} f(s) P(x - s; y)\, ds \leq M$$

Thus, $\Psi = f *_y P$ is bounded just like f. ∎

For uniqueness in the case of f absolutely integrable, see Remark (1.32).

The results described above generalize nicely to the upper half space $\mathbb{R}_+^3 = \{(x, y, z): (x, y) \text{ in } \mathbb{R}^2 \text{ and } z > 0\}$. Dirichlet's problem consists in finding a function Ψ that solves

(1.16)
$$\begin{cases} \Delta \Psi = \Psi_{xx} + \Psi_{yy} + \Psi_{zz} = 0 & [(x, y, z) \text{ in } \mathbb{R}_+^3] \\ \Psi(x, y, 0) = f(x, y) \\ \Psi \text{ continuous for all } (x, y) \text{ in } \mathbb{R}^2 \text{ and } z \geq 0 \end{cases}$$

where f is a given continuous function on $\mathbb{R}^2$.

Using double Fourier transforms we are led to defining Ψ as follows.

For $z > 0$ and all (x, y) in $\mathbb{R}^2$

(1.17)
$$\Psi(x, y, z) = f *_z P(x, y) = {_z}P * f(x, y)$$

where

$$_zP(x, y) = \frac{1}{2\pi} \frac{z}{(z^2 + x^2 + y^2)^{3/2}}$$

For $z = 0$ we define $\Psi(x, y, 0) = f(x, y)$.

Proving the following two theorems is left to the reader [see Exercise (1.24)].

(1.18) Theorem. If f is continuous and absolutely integrable over $\mathbb{R}^2$, then the function Ψ described in (1.17) solves (1.16).

(1.19) Theorem. If f is continuous and bounded over $\mathbb{R}^2$, then the function Ψ given in (1.17) is a bounded solution to (1.16).

Remark. In connection with Theorem (1.19) it should be noted that Corollary (5.10), Chapter 4, generalizes to the upper half space $\mathbb{R}^3_+$ (see Exercise (9.11), Chapter 8). Hence the function Ψ described in (1.17) is the only bounded solution to (1.16) when f is continuous and bounded over $\mathbb{R}^2$.

Exercises

(1.20) Prove that Poisson's kernel $P(x;y) = (1/\pi)[y/(y^2 + x^2)]$ is harmonic, $P_{xx} + P_{yy} = 0$, for $y > 0$ and all x values.

(1.21) Solve (1.1) given the following functions f

(a) $f(x) = e^{-3|x|}$
(b) $f(x) = e^{-2x^2}$

(c) $f(x) = \dfrac{1}{1+x}$

(d) $f(x) = \dfrac{1}{1+x^2}$

(1.22) Prove that (1.12) holds uniformly for $-\infty < x_0 < +\infty$ if $\lim_{x \to \pm\infty} f(x) = 0$.

(1.23) Show that (1.7) describes a solution to

$$(*) \quad \Delta\Psi = 0 \quad \Psi(x, 0+) = f(x)$$

when f is piecewise continuous and absolutely integrable, or bounded, over $\mathbb{R}$. [*Note:* f may require redefinition at some points.] Show that (1.12) may not hold.

(1.24) Derive (1.17) and prove Theorems (1.18) and (1.19).

(1.25) Let f be continuous and bounded for all $x \geq 0$ and suppose that $f(0) = 0$. Solve the following problem for Ψ

$$(*) \quad \begin{cases} \Delta\Psi = 0 \quad (x > 0 \quad y > 0) \\ \Psi(x, 0) = f(x) \quad \Psi(y, 0) = 0 \\ \Psi \text{ continuous and bounded for } x \geq 0 \text{ and } y \geq 0 \end{cases}$$

Prove that your solution is unique, given f.

(1.26) Write specific solutions to (1.25 $*$) given the following functions for f.

(a) $f(x) = xe^{-x}$
(b) $f(x) = [\sin x - x]/x$
(c) $f(x) = x^2 e^{-3x}$

(1.27) Generalize problem (1.25) and its solution to the case of three variables x, y, and z.

*(1.28)** Using Fourier transforms, derive the expression

$$(*) \quad y(x, t) = \int_{-\infty}^{+\infty} [\hat{f}(u) \cos(2\pi cut) + \hat{g}(u)t \cdot \text{sinc}(2cut)] e^{i2\pi ux} \, du$$

as a solution to

$$(**) \qquad y_{tt} = c^2 y_{xx} \qquad (-\infty < x < +\infty \qquad t > 0)$$
$$y(x, 0) = f(x) \qquad y_t(x, 0) = g(x)$$

Show that the function y in ($*$) can be expressed in *D'Alembert's form*

$$y(x, t) = \tfrac{1}{2}[f(x + ct) + f(x - ct)] + (2c)^{-1} \int_{x-ct}^{x+ct} g(s) \, ds$$

and discuss the validity of this form for y as a solution to ($**$).

Mean Convergence of Poisson Integrals

In this set of 3 exercises we investigate mean convergence of the Poisson integral defined by (1.7′).

(1.29) Suppose that g is continuous over $\mathbb{R}$ and $g(x) = 0$ whenever $|x|$ is sufficiently large. Prove that

$$\lim_{y \to 0+} \int_{-\infty}^{+\infty} |_y P * g(x) - g(x)| \, dx = 0$$

(1.30) Suppose that f is piecewise continuous and absolutely integrable over $\mathbb{R}$. Prove that

$$\lim_{y \to 0+} \int_{-\infty}^{+\infty} |_y P * f(x) - f(x)| \, dx = 0$$

[*Hint*: First find a continuous function g, as in (1.29), such that

$$\int_{-\infty}^{+\infty} |f(x) - g(x)| \, dx < \epsilon.]$$

(1.31) Prove that for f piecewise continuous and absolutely integrable over $\mathbb{R}$

$$\int_{-\infty}^{+\infty} |_y P * f(x)| \, dx \le \int_{-\infty}^{+\infty} |f(x)| \, dx$$

for every $y > 0$.

(1.32) Remark. Problems (1.30) and (1.31) characterize the function $\Psi = {}_y P * f = f * {}_y P$ defined in (1.7′) [or (1.7)] as a solution to (1.23$*$) when f is piecewise continuous and absolutely integrable. The solution $\Psi = {}_y P * f$

is unique among all harmonic functions Ψ over $\mathbb{R}^2$ that satisfy

$$\lim_{y\to0+} \int_{-\infty}^{+\infty} |\Psi(x, y) - f(x)|\, dx = 0 \qquad \int_{-\infty}^{+\infty} |\Psi(x, y)|\, dx < C$$

for some positive constant C. The second condition above says that Ψ belongs to the set of harmonic functions known as the *Hardy space* $H^1(\mathbb{R}_+^2)$. This uniqueness result also applies to Theorem (1.8) since (1.23∗) is a generalization of that problem. The proof of this uniqueness result is beyond the scope of this text; see Stein and Weiss (1971), Theorems 2.3 and 2.5b in Chapter 2.

§2. The Heat Equation

In this section we shall study the following heat conduction problem

(2.1)
$$\begin{cases} u_t = u_{xx} & (t > 0 \qquad -\infty < x < +\infty) \\ u(x, 0) = f(x) & (-\infty < x < +\infty) \\ u \text{ continuous for } -\infty < x < +\infty \text{ and } t \geq 0 \end{cases}$$

Problem (2.1) is the analogue for heat conduction of Dirichlet's problem (1.1). Our method of solving (2.1) is quite similar to the method used in §1 to solve (1.1). We begin by writing

(2.2)
$$\hat{u}(v, t) = \int_{-\infty}^{+\infty} u(x, t) e^{-i2\pi vx}\, dx$$

and proceed under the temporary assumption that all necessary integrals are defined. Using Theorem (3.5), Chapter 6, we have

(2.3)
$$u_x \supset i2\pi v\hat{u} \qquad u_{xx} \supset -(2\pi)^2 v^2 \hat{u}$$

Differentiating under the integral sign in (2.2) yields

(2.4)
$$u_t \supset \partial\hat{u}/\partial t$$

Formulas (2.3) and (2.4) allow us to transform problem (2.1) into

(2.5)
$$\partial\hat{u}/\partial t = -(2\pi)^2 v^2 \hat{u} \qquad \hat{u}(v, 0) = \hat{f}(v)$$

Hence we put $\hat{u}(v, t) = \hat{f}(v) e^{-(2\pi)^2 v^2 t}$. Performing Fourier inversion on this last expression for $\hat{u}$, we obtain (for $t > 0$)

(2.6)
$$u(x, t) = \int_{-\infty}^{+\infty} \hat{f}(v) e^{-(2\pi)^2 v^2 t} e^{i2\pi vx}\, dv$$

We now put (2.6) into a convolution integral form. If we let c equal $\sqrt{4\pi t}$, then by formula (4.2) in Chapter 6, we have

---

For $t > 0$ and all x values

$$(2.7) \quad u(x, t) = \int_{-\infty}^{+\infty} f(x - s)H(s; t)\, ds = \int_{-\infty}^{+\infty} f(s)H(x - s; t)\, ds$$

where

$$H(s; t) = \frac{1}{\sqrt{4\pi t}}\, e^{-s^2/4t}$$

For $t = 0$ we define $u(x, 0) = f(x)$.

The function H in (2.7) is called the *heat kernel* (in one dimension). *It satisfies the heat equation*

$$(2.8) \qquad\qquad H_t = H_{ss} \qquad (t > 0 \qquad \text{all } s \text{ values})$$

as an easy computation will show. Moreover, the heat kernel is a *summation kernel*. To see that, we note that $H(s; t)$ equals the Gauss–Weierstrass kernel $W(s; c)$ for $c = \sqrt{4\pi t}$ and so H is a summation kernel over $\mathbb{R}$, parameterized by $t > 0$.

Writing $_tH(s)$ in place of $H(s; t)$ we can express the convolution integrals in (2.7) in the form $u(x, t) = {_tH} * f(x) = f * {_tH}(x)$. Because H solves the heat equation the convolution $u = f * {_tH}$ will, as a general rule, also solve the heat equation. And, the convolution $u = {_tH} * f$ will tend to f as t tends to 0 because H is a summation kernel.

As with the Poisson kernel, the general remarks just made are valid when f is continuous and absolutely integrable.

(2.9) Theorem. If f is continuous and absolutely integrable over $\mathbb{R}$, then the function u given in (2.7) solves (2.1).

Proof. The proof is quite similar to the proof of Theorem (1.8). To see that u solves the heat equation, we must justify the following differentiation under the integral sign

$$(2.10) \qquad u_{xx} - u_t = \int_{-\infty}^{+\infty} f(s)\left[\frac{\partial^2 H}{\partial x^2}(x - s; t) - \frac{\partial H}{\partial t}(x - s; t) \right] ds$$

$$= \int_{-\infty}^{+\infty} f(s)\, 0\, ds = 0$$

Suppose $t \geq \delta > 0$. Then

$$|f(s)H(x - s; t)| = |f(s)| \frac{1}{\sqrt{4\pi t}}\, e^{-(x-s)^2/4t} \leq |f(s)| \frac{1}{\sqrt{4\pi \delta}}$$

and, similarly,

$$\left| f(s) \frac{\partial H}{\partial x} (x - s; t) \right| = |f(s)| \frac{|x - s|}{2\sqrt{4\pi} \, t^{3/2}} e^{-(x-s)^2/4t} \le B(t) \, |f(s)|$$

where

$$B(t) = \underset{0 \le v < +\infty}{\text{maximum}} \frac{v}{2\sqrt{4\pi} \, t^{3/2}} e^{-v^2/4t}$$

is a constant for each fixed t value. Therefore, by Theorems (2.3), (2.5), and (2.8) in Chapter 6, we know that

(2.11)
$$u_x = \int_{-\infty}^{+\infty} f(s) \frac{\partial H}{\partial x} (x - s; t) \, ds$$

for all x values and for $t > 0$ (letting $\delta \to 0$). *Similar arguments show that u may be differentiated, under the integral sign, infinitely often with respect to either x or t.* In particular, (2.10) is verified for $t > 0$ and all x values.

The rest of the proof is exactly like the proof of (1.8), using the summation kernel H in place of the summation kernel P. ∎

As with (1.7), formula (2.7) has the value of allowing us to extend our class of solutions to (2.1) beyond those described in (2.9). However, *unlike Poisson's kernel, the heat kernel decreases very rapidly to 0 as $|x| \to +\infty$.* [For example, $\lim_{|x| \to +\infty} e^{|x|} H(x; t) = 0$.] To see how this affects things, let's consider the following example.

(2.12) Example. Let f be the function defined by $f(x) = Be^{cx^2}$ where B and c are positive constants. Then

$$u(x, t) = \frac{B}{\sqrt{1 - 4ct}} e^{cx^2/(1-4ct)}$$

solves

$$u_t = u_{xx} \qquad (-\infty < t < 1/4c \qquad -\infty < x < +\infty)$$
$$u(x, 0) = Be^{cx^2} \qquad (-\infty < x < +\infty)$$

In particular, u is continuous for $0 \le t < 1/4c$ and all x values, u solves the heat equation for $0 < t < 1/4c$ and all x values, and $u(x, 0) = Be^{cx^2}$.

Remarks. We will show how we obtained the function u in (2.12) in a moment. First, however, observe that u satisfies

$$u(x, t) = B\sqrt{4\pi} \, H(i\sqrt{4c} \, x; 1 - 4ct)$$

where H is the heat kernel. Since $1 - 4ct > 0$ when $-\infty < t < 1/4c$, it follows from (2.8) and the chain rule that $u_t = u_{xx}$ when $-\infty < t < 1/4c$ and $-\infty < x < +\infty$. Thus, u is continuous for all such values of x and t (hence for $0 \le t < 1/4c$ and $-\infty < x < +\infty$ as well). That $u(x, 0) = Be^{cx^2}$ is obvious.

Although it may appear that u only partially solves (2.1) for $f(x) = Be^{cx^2}$, we will see that it provides a basis, through (2.7), of a vast extension of the *class of functions for which* (2.1) *is solvable. This class will include bounded continuous functions (as in* §1) *as well as all polynomials and exponential functions (unlike* §1).

Before we discuss this extension, however, we will briefly describe how we obtained the function u in Example (2.12). Substituting $f(s) = Be^{cs^2}$ into the second convolution integral in (2.7) we get

$$u(x, t) = \int_{-\infty}^{+\infty} Be^{cs^2} \frac{1}{\sqrt{4\pi t}} e^{-(x-s)^2/4t} \, ds$$

Factoring out $B/\sqrt{4\pi t}$ and $e^{-x^2/4t}$ from the integral and combining exponents in the remaining integrand, we obtain

(2.13) $$u(x, t) = \frac{B}{\sqrt{4\pi t}} e^{-x^2/4t} \int_{-\infty}^{+\infty} e^{\{(xs/2t) + [c - (1/4t)]s^2\}} \, ds.$$

For every real number a and *negative number* b we have

(2.14) $$\int_{-\infty}^{+\infty} e^{(as + bs^2)} \, ds = (\pi/-b)^{1/2} e^{-a^2/4b} \qquad (b < 0)$$

[Formula (2.14) is verified by completing the square in the exponent of the integrand.] Applying (2.14), for $a = x/2t$ and $b = c - (1/4t)(0 < t < 1/4c)$, we obtain from (2.13)

$$u(x, t) = \frac{B}{\sqrt{1 - 4ct}} e^{cx^2/(1 - 4ct)}$$

which is the function u in Example (2.12).

The following definition will allow us to describe those functions for which (2.7) gives a solution to (2.1).

(2.15) Definition. If two functions f and g satisfy, for some positive constant B,

$$|f(x)| \le B |g(x)| \qquad (-\infty < x < +\infty)$$

then we write $f(x) = 0[g(x)]$. ∎

For example, a polynomial of degree N is $0(x^N)$. Every polynomial is $0(e^{cx^2})$ for all constants $c > 0$. Each exponential function $f(x) = e^{ax}$, where a is a real constant, is $0(e^{cx^2})$ for every constant $c > 0$.

(2.16) Theorem. Suppose that f is continuous and $f(x) = 0(e^{cx^2})$ for every constant $c > 0$. Then the function u given in (2.7) solves (2.1).

Proof. Suppose that $|x| \le R$ and $\delta \le t \le 1/4c - \delta$ where R and δ are positive constants. Then, for $c > 0$

$$|f(s)H(x - s; t)| \le Be^{cs^2} \frac{1}{\sqrt{4\pi t}} e^{-(x-s)^2/4t}$$

$$\le \frac{B}{\sqrt{4\pi \delta}} e^{-x^2/4t} e^{xs/2t} e^{[4ct - 1]s^2/4t}$$

$$\le \frac{B}{\sqrt{4\pi \delta}} e^{R|s|/2\delta} e^{-[4c^2\delta/(1 - 4c\delta)]s^2} = g(s)$$

The function g has a finite integral over $\mathbb{R}$. By Theorems (2.3) and (2.5a), Chapter 6, it follows that

$$u(x, t) = \int_{-\infty}^{+\infty} f(s)H(x - s; t)\, ds$$

is continuous for $\delta \le t \le 1/4c - \delta$ and $|x| \le R$. Moreover,

$$\left| f(s) \frac{\partial H}{\partial x}(x - s; t) \right| \le Be^{cs^2} \frac{|x - s|}{4\sqrt{\pi}\, t^{3/2}} e^{-(x-s)^2/4t}$$

$$\le B \frac{R + |s|}{4\sqrt{\pi}\, \delta^{3/2}} e^{-x^2/4t} e^{xs/2t} e^{[4ct - 1]s^2/4t}$$

$$\le B \frac{R + |s|}{4\sqrt{\pi}\, \delta^{3/2}} e^{R|s|/2\delta} e^{-[4c^2\delta/(1 - 4c\delta)]s^2} = h(s)$$

The function h has a finite integral over $\mathbb{R}$. Therefore, Theorems (2.3) and (2.8) in Chapter 6 tell us that

(2.17) $$u_x = \int_{-\infty}^{+\infty} f(s) \frac{\partial H}{\partial x}(x - s; t)\, ds$$

Similar arguments show that *u may be differentiated under the integral sign any number of times with respect to x or t.* Combining that fact with (2.8) yields $u_t = u_{xx}$ for all x values and $0 < t < 1/4c$ (upon letting R tend to $+\infty$ and δ tend to 0). Since c is an arbitrary positive constant, when we let $c \to 0$ we see that u solves the heat equation in (2.1).

It only remains for us to show that

(2.18) $$\lim_{(x, t) \to (x_0, 0+)} u(x, t) = f(x_0) \qquad (-\infty < x_0 < +\infty)$$

Let $\epsilon > 0$ be given. Using the fact that the heat kernel H is a summation kernel we have [see (8.5), Chapter 2]

(2.19)
$$|u(x, t) - f(x_0)| \le \int_{|s| \ge \delta} |f(x - s)|\, H(s; t)\, ds + |f(x_0)| \int_{|s| \ge \delta} H(s; t)\, ds$$
$$+ \int_{|s| < \delta} |f(x - s) - f(x_0)|\, H(s; t)\, ds$$

Since f is continuous at x_0 we have

(2.20) $$\int_{|s| < \delta} |f(x - s) - f(x_0)|\, H(s; t)\, ds < \epsilon$$

after choosing δ sufficiently small and x close enough to x_0. [Compare the proof of (8.6), Chapter 2.]

Moreover, for t sufficiently close to 0

(2.21) $$|f(x_0)| \int_{|s| \ge \delta} H(s; t)\, ds < \epsilon\, |f(x_0)|$$

Using the fact that $f(x) = 0(e^{cx^2})$ we have

(2.22) $$\int_{|s| \geq \delta} |f(x-s)| H(s;t)\, ds \leq \int_{|s| \geq \delta} B e^{c(x-s)^2} H(s;t)\, ds$$

If $|x| \leq R$ for R a positive constant, then

$$B e^{c(x-s)^2} = B e^{cx^2} e^{-2cxs} e^{cs^2} \leq B e^{cR^2} e^{2cR|s|} e^{cs^2}$$

Since $e^{2cR|s|} = 0(e^{cs^2})$ we have for some positive constant C

$$B e^{c(x-s)^2} \leq C e^{2cs^2}$$

Therefore

$$\int_{|s| \geq \delta} B e^{c(x-s)^2} H(s;t)\, ds \leq \int_{|s| \geq \delta} \frac{C}{\sqrt{4\pi t}} e^{-[(1/4t)-2c]s^2}\, ds$$

$$= \frac{C}{\sqrt{\pi - 8c\pi t}} \int_{|r| \geq \delta[(1/4t)-2c]^{1/2}} e^{-r^2}\, dr$$

where $r = [(1/4t) - 2c]^{1/2} s$. Letting t tend to 0 we see that the last integral above tends to 0 because $\delta[(1/4t) - 2c]^{1/2}$ tends to $+\infty$.

Hence for $|x| \leq R$ and t close enough to 0 we have

(2.23) $$\int_{|s| \geq \delta} |f(x-s)| H(s;t)\, ds < \epsilon$$

Combining (2.20), (2.21), and (2.23), we see from (2.19) that

(2.24) $$|u(x,t) - f(x_0)| < \epsilon[1 + |f(x_0)| + 1]$$

provided $|x| \leq R$, x is close to x_0 (so we must choose x_0 so that $|x_0| < R$), and t is close to 0. Since ϵ can be chosen arbitrarily small in (2.24) we see that (2.18) holds for $|x_0| < R$. Letting $R \to +\infty$ we have proved (2.18) for every real number x_0. ∎

(2.25) Remarks. (a) Because of Theorem (2.16) we have that (2.7) solves (2.1) whenever $f(x)$ is continuous and $0(x^N)$ for a positive integer N, or $0(e^{b|x|})$ for a positive constant b. In particular, (2.7) solves (2.1) when f is a polynomial, or exponential. (b) In order for (2.18) to hold it is not necessary that f be continuous. For example, (2.18) will hold at every point of continuity x_0 of f provided f is piecewise continuous[1] and $0(e^{cx^2})$ for *some* positive constant c. The proof is unchanged. (c) Likewise, the function u in (2.7) will satisfy the heat equation in (2.1) when f is piecewise continuous[1] and $0(e^{cx^2})$ for *every* constant $c > 0$.

We will prove a uniqueness result for Theorem (2.16) in the next section. We close this section by noting that our results generalize to problems involving more than one space variable. The results are summarized below; we leave the details to the reader as exercises.

---

1. Or Riemann (Lebesgue) integrable in every finite interval.

To solve

(2.26)
$$\begin{cases} u_t = u_{xx} + u_{xx} & [(x, y) \text{ in } \mathbb{R}^2 \quad t > 0] \\ u(x, y, 0) = f(x, y) & (f \text{ continuous on } \mathbb{R}^2) \\ u \text{ continuous for } (x, y) \text{ in } \mathbb{R}^2 \text{ and } t \geq 0 \end{cases}$$

for u, we use double Fourier transforms. If f is continuous and absolutely integrable over $\mathbb{R}^2$, then we derive (for $t > 0$)

(2.27)
$$u(x, y, t) = \int\!\!\int_{-\infty}^{+\infty} \hat{f}(v, w) e^{-(2\pi)^2(v^2 + w^2)t} e^{i2\pi(vx + wy)} \, dv \, dw$$

We then put (2.27) into the form

For $t > 0$ and (x, y) in $\mathbb{R}^2$

(2.28)
$$u(x, y, t) = \int\!\!\int_{-\infty}^{+\infty} f(x - r, y - s) H(r, s; t) \, dr \, ds$$
$$= \int\!\!\int_{-\infty}^{+\infty} f(r, s) H(x - r, y - s; t) \, dr \, ds$$

where

$$H(r, s; t) = \frac{1}{4\pi t} e^{-(r^2 + s^2)/4t}$$

For $t = 0$ we define $u(x, y, 0) = f(x, y)$.

The function H in (2.28) is called the *heat kernel*, or *fundamental solution to the heat equation*, in two dimensions; it satisfies the heat equation and is a summation kernel.

(2.29) Theorem. If f is continuous and absolutely integrable over $\mathbb{R}^2$, then the function u in (2.28) solves (2.26).

(2.30) Theorem. If f is continuous and $f(x, y) = 0[e^{c(x^2 + y^2)}]$ for every constant $c > 0$, then the function u in (2.28) solves (2.26).

Exercises

(2.31) Prove that for $t > 0$ and $t' > 0$ and all x values: $_t H * _{t'} H(x) = _{(t+t')} H(x)$.

(2.32) Prove that for f continuous and absolutely integrable [or $0(e^{cx^2})$ for some constant $c > 0$], (2.18) holds uniformly for all x_0 in each closed interval.

(2.33)   Solve the following problem for u

$$(*) \quad \begin{cases} u_t = a^2 u_{xx} & (-\infty < x < +\infty \quad t > 0) \\ u(x, 0) = f(x) & (f \text{ continuous}) \\ u \text{ continuous for } -\infty < x < +\infty & t \geq 0. \end{cases}$$

(2.34)   Solve (2.33∗) given the following functions f and constants a^2. For (a), (c), and (e), you should find *closed form* solutions.

(a) $f(x) = e^{-x^2} \qquad a^2 = \frac{1}{2}$
(b) $f(x) = (1 + x^2)^{-1} \qquad a^2 = 3$
(c) $f(x) = x^n \qquad (n = 0, 1, 2, 3, \ldots) \qquad a^2 = 1$
(d) $f(x) = (1 + x)^{-1} \qquad a^2 = 2$
(e) $f(x) = e^{-4x^2} \qquad a^2 = 3.$

(2.35)   Show that (2.18) cannot hold if x_0 is a point of discontinuity for a piecewise continuous function f that is $0(e^{cx^2})$ for some $c > 0$. Prove that instead we can only have $\lim_{t \to 0+} u(x_0, t) = \frac{1}{2}[f(x_0+) + f(x_0-)]$.

(2.36)   Show that there are continuous, absolutely integrable functions that are not $0(e^{cx^2})$, no matter what positive constant c is used.

(2.37)   Suppose that f is continuous over $\mathbb{R}$ and $f(x) = 0$ whenever $|x|$ is sufficiently large. Prove that

$$\lim_{t \to 0+} \int_{-\infty}^{+\infty} |_t H * f(x) - f(x)| \, dx = 0$$

Then prove that this same limit holds if f is piecewise continuous and absolutely integrable over $\mathbb{R}$.

(2.38)   Show that for f piecewise continuous and absolutely integrable over $\mathbb{R}$

$$\int_{-\infty}^{+\infty} {}_t H * f(x) \, dx = \int_{-\infty}^{+\infty} f(x) \, dx$$

Show that a similar result holds for ${}_y P * f(x)$.

(2.39)   Prove that if f is piecewise continuous and absolutely integrable over $\mathbb{R}$ then

$$\int_{-\infty}^{+\infty} |_t H * f(x)| \, dx \leq \int_{-\infty}^{+\infty} |f(x)| \, dx$$

for every $t > 0$.

(2.40)   Prove that if f is piecewise continuous and absolutely integrable over $\mathbb{R}$, then for $t > 0$

$$|_t H * f(x)| \leq (4\pi t)^{-1/2} \int_{-\infty}^{+\infty} |f(s)| \, ds$$

Thus $u(x, t) = {}_t H * f(x)$ is bounded for $-\infty < x < +\infty$, as soon as $t > 0$.

(2.41)   Prove that (2.7) solves (2.1) if f is continuous and $|f(x)|\,e^{-cx^2}$ is absolutely integrable for every constant $c>0$.

★(2.42)   Show that if $f(x)=0(e^{cx^2})$ then the function u in (2.7) satisfies for $\delta>0$

$$|u(x,\,t)|\le Ae^{bx^2}\qquad (0\le t\le(1/4c)-\delta)$$

for some positive constants A and b. [*Hint*: Consider (2.12).]

(2.43)   Suppose $f(x)=e^{bx}$ for b a real constant. Find a closed form solution to problem (2.1).

(2.44)   Solve the following problem for u, given that f is continuous and absolutely integrable (or bounded) over $\mathbb{R}$ and $f(0)=0$.

$$(*)\quad\begin{cases}u_t=u_{xx}&(0<x<\infty\qquad t>0)\\u(x,\,0)=f(x)\qquad u(0,\,t)=0\\u\text{ continuous for }0\le x<+\infty\text{ and }t\ge0\end{cases}$$

(2.45)   Solve (2.44★) given the following functions for f

(a) $f(x)=xe^{-x}$
(b) $f(x)=1-e^{-x^2}$
(c) $f(x)=\Lambda(x-1)$

(2.46)   Show that the solution found for (2.44★) can be expressed (for $t>0$) as

$$u(x,\,t)=\int_0^{+\infty}f(s)[H(x-s;t)-H(x+s;t)]\,ds$$

Prove that u solves (2.44★) if f is continuous and $0(e^{cx^2})$ for each $c>0$.

(2.47)   If $f=1$ prove that the function u in (2.46) can be expressed by

$$u(x,\,t)=erf(x/\sqrt{2t})\quad\text{where}\quad erf(s)=2\pi^{-1/2}\int_0^s e^{-r^2}\,dr.$$

(2.48)   Derive (2.27) and (2.28), and prove Theorems (2.29) and (2.30).

(2.49)   Solve (2.26), in closed form, given the following functions for f.

(a) $f(x,\,y)=e^{-(x^2+y^2)}$
(b) $f(x,\,y)=x+y^2$
(c) $f(x,\,y)=e^{-2(x^2+y^2)}$

(2.50)   Generalize (2.26) through (2.30) to the case of three $(n>3)$ space variables x, y, z $(x_1,\,x_2,\,\ldots,\,x_n)$.

*§3. Uniqueness of Solutions to the Heat Equation

In this section we shall discuss the uniqueness of solutions to problem (2.1) as well as some other similar problems. Uniqueness for problem (2.1) is not possible without added restrictions on the type of solution allowed. For, if we let

$$u_0(x, t) = \sum_{k=0}^{\infty} g^{(k)}(t) \frac{x^{2k}}{(2k)!}$$

where

$$g(t) = \begin{cases} e^{-t^{-2}} & \text{for } t > 0 \\ 0 & \text{for } t \le 0 \end{cases}$$

then u_0 solves (2.1) with $f = 0$. But, since $u_0(0, t) = g(t)$, we see that u_0 is not identically zero. [For details of the proof that u_0 solves (2.1), see Widder (1975), pp. 49, 50, and 52; or Copson (1975), Chapter 12, §5.] Moreover, the function $u_0(x, t - c)$ solves (2.1) with $f = 0$ provided c is a positive constant. These nontrivial solutions to (2.1), when $f = 0$, are called *null solutions*. By adding any null solution to any solution of (2.1), for some function f, we obtain distinct solutions of (2.1).

We will see below that uniqueness can be obtained for solutions to (2.1) if we require that solutions do not grow too quickly as $x \to \infty$. The following definition describes this growth restriction.

(3.1) Definition.   Let u be a function of x and t. We say that u is $0(e^{ax^2})$ for $0 \le t < b$ if there are positive constants A and a such that

$$|u(x, t)| \le A e^{ax^2} \qquad (-\infty < x < +\infty)$$

for $0 \le t < b$. We do allow b to be $+\infty$ as well as finite. Furthermore, we say that u is of class $\mathscr{E}$ for $0 \le t < b$ if u is $0(e^{ax^2})$ for some $a > 0$ on every subinterval $0 \le t < d$ where $d < b$. ∎

For instance $u(x, t) = t^2 e^{x^2}$ is $0(e^{x^2})$ for $0 \le t < d$ when $d < +\infty$. Therefore, that function u is of class $\mathscr{E}$ for $0 \le t < +\infty$. The function

$$u(x, t) = (1 - 4ct)^{-1/2} e^{cx^2/(1-4ct)} \qquad (-\infty < t < 1/4c)$$

from Example (2.12) is of class $\mathscr{E}$ for $0 \le t < 1/4c$ [see Exercise (2.42)]. But the function $u(x, t) = (t + 1)e^{x^3}$ is not $0(e^{ax^2})$ on any interval $0 \le t < b$, no matter how large a is chosen.

The following Theorem is due to Tikhonov; it is our fundamental uniqueness result.

(3.2) Theorem.   Suppose that u solves

$$(*) \quad \begin{cases} u_t = u_{xx} & (-\infty < x < +\infty \qquad 0 < t < b) \\ u(x, 0) = 0 \\ u \text{ continuous for } 0 \le t \le b \text{ and } -\infty < x < +\infty \end{cases}$$

If u is $0(e^{ax^2})$ for $0 \le t < b$ then u is *identically zero* for $0 \le t < b$ and all x values.

Proof.   Our proof is based on the one given in Widder (1975), pp.

28–29. Since u is $0(e^{ax^2})$ for $0 \le t < b$, we have

(3.3) $\qquad\qquad |u(x, t)| \le Ae^{ax^2} \qquad (-\infty < x < +\infty \qquad 0 \le t < b)$

for some positive constants A and a. Let C be a positive constant satisfying $C > a$ and $C > 1/4b$. Consider the function v defined by [compare (2.12)]

$$v(x, t) = (1 - 4Ct)^{-1/2} e^{Cx^2/(1-4Ct)}$$

The function v satisfies the heat equation $v_t = v_{xx}$, and $v(x, t) > 0$ for $0 \le t \le 1/8C$ and all x values. Moreover, for each fixed x value $v(x, t)$ increases as t increases. In particular,

(3.4) $\qquad\qquad e^{Cx^2} = v(x, 0) \le v(x, t) \qquad (0 \le t \le 1/8C)$

Let (x_0, t_0) be an arbitrarily chosen point in the infinite strip described by $0 < t \le 1/8C$. Choose a constant L such that $L > |x_0|$. We now compare the functions $|u(x, t)|$ and

$$w(x, t) = Ae^{(a-C)L^2} v(x, t)$$

on the boundary ∂R of the rectangle

$$R = \{(x, t): \qquad -L \le x \le L \qquad 0 \le t \le 1/8C\}$$

On the base $(t = 0)$ of ∂R we have

(3.5) $\qquad\qquad |u(x, 0)| = 0 < Ae^{(a-C)L^2} e^{Cx^2} = w(x, 0)$

While on the vertical sides $(x = \pm L, \, 0 < t \le 1/8C)$

$$|u(\pm L, t)| \le Ae^{aL^2}$$

and, because of (3.4),

(3.6) $\qquad \begin{aligned} w(\pm L, t) &= Ae^{(a-C)L^2} v(\pm L, t) \\ &\ge Ae^{(a-C)L^2} e^{CL^2} = Ae^{aL^2} \ge |u(\pm L, t)| \end{aligned}$

Combining (3.5) and (3.6), we see that on $\partial_1 R$ (using the notation of Chapter 4, §7)

(3.7) $\qquad\qquad -w(x, t) \le u(x, t) \le w(x, t)$

Hence, because maxima and minima for solutions to the heat equation must occur on $\partial_1 R$ [see Theorem (7.1), Chapter 4] we conclude that (3.7) holds throughout all of R.[2] In particular, (3.7) holds at the point (x_0, t_0) in R. Therefore,

(3.8) $\qquad \begin{aligned} |u(x_0, t_0)| &\le Ae^{(a-C)L^2} v(x_0, t_0) \\ &= Ke^{-(C-a)L^2} \end{aligned}$

where K denotes the constant $Av(x_0, t_0)$. Letting L tend to $+\infty$, we see that $Ke^{-(C-a)L^2}$ tends to 0 (because $C > a$). Therefore, (3.8) implies that $u(x_0, t_0)$

---

2. Note: w is just a constant multiple of v so it satisfies the heat equation, and (3.7) is equivalent to $0 \le u + w$ and $0 \ge u - w$ on $\partial_1 R$, hence on R.

must be 0. Since (x_0, t_0) was chosen arbitrarily, we have proved that u is identically zero for $0 \le t \le 1/8C$ and all x values.

If we repeat the proof above for the function $u_1(x, t) = u[x, t + (1/8C)]$ we find that u is identically zero for $0 \le t \le 2/8C$ and all x values. Repeating this procedure n times if necessary, we find that u is identically zero for $0 \le t \le n/8C$ and all x values. Clearly, every point in the infinite strip described by $0 \le t < b$ is eventually contained in one of these strips. Therefore, u is identically zero for $0 \le t < b$ and all x values. ∎

(3.9) Corollary.   If u solves $(3.2*)$ and u is of class $\mathscr{E}$ for $0 \le t < b$ then u is *identically zero* for $0 \le t < b$ and all x values.

Proof.   If $0 < d < b$ then u is $0(e^{ax^2})$ for some $a > 0$ on $0 \le t < d$. Hence, by the previous theorem, u is identically zero for $0 \le t < d$ and all x values. Letting $d \to b$ yields the desired result. ∎

Another corollary to Theorem (3.2) is the following result.

(3.10) Theorem.   Let f be a continuous function on $\mathbb{R}$ and $0(e^{cx^2})$ for some constant $c > 0$. Then the function u defined in (2.7) is the only solution of

$$(*) \quad \begin{cases} u_t = u_{xx} & (0 < t < 1/4c \quad -\infty < x < +\infty \\ u(x, 0) = f(x) \\ u \text{ continuous for } 0 \le t < 1/4c \text{ and } -\infty < x < +\infty \end{cases}$$

that belongs to the class $\mathscr{E}$.

Proof.   It was shown in the proof of Theorem (2.16) that the function u, defined in (2.7), solves $(3.10*)$. Since $f = 0(e^{cx^2})$ we see that for $0 < t \le (1/4c) - \delta \ (0 < \delta < 1/4c)$

$$|u(x, t)| \le \int_{-\infty}^{+\infty} |f(s)| \, H(x - s; t) \, ds$$

$$\le \int_{-\infty}^{+\infty} Be^{cs^2} H(x - s; t) \, ds$$

for some positive constants B and c. Using Example (2.12), we have

$$|u(x, t)| \le B(1 - 4ct)^{-1/2} e^{cx^2/(1-4ct)}$$

$$\le B(4c\delta)^{-1/2} e^{x^2/4\delta} \qquad (0 \le t \le (1/4c) - \delta)$$

Thus u is $0(e^{x^2/4\delta})$ for $0 \le t < (1/4c) - \delta$. Since δ can be taken as close to 0 as we wish, it follows that u is of class $\mathscr{E}$ for $0 \le t < 1/4c$.

If $u^{\#}$ is a function of class $\mathscr{E}$ for $0 \le t < 1/4c$ then $\bar{u} = u - u^{\#}$ is also of class $\mathscr{E}$ for $0 \le t < 1/4c$. Moreover, if $u^{\#}$ also solves $(3.10*)$, then $\bar{u}$ solves $(3.2*)$ for $b = 1/4c$. Hence, by Corollary (3.9), we would have u identically zero and then $u = u^{\#}$. Thus u is the only function of class $\mathscr{E}$ that solves $(3.10*)$. ∎

We can now prove a uniqueness theorem for problem (2.1).

(3.11) Theorem. Let f be a continuous function on $\mathbb{R}$ and $0(e^{cx^2})$ for every constant $c > 0$. Then the function u defined in (2.7) is the only function of class $\mathscr{E}$ that solves (2.1).

Proof. We know from Theorem (2.16) that the function u given in (2.7) does solve (2.1). Moreover, from the previous proof we know that for each $c > 0$ the function u is $0(e^{x^2/4\delta})$ for $0 \le t < (1/4c) - \delta$. Since c can be taken as close to 0 as we wish, it follows that u is in class $\mathscr{E}$ for $0 \le t < +\infty$. Moreover, if $u^{\#}$ is in class $\mathscr{E}$ for $0 \le t < +\infty$ and $u^{\#}$ solves (2.1), then the previous theorem implies that $u = u^{\#}$ for $0 \le t < 1/4c$ and all x values. Again, letting $c \to 0+$ yields $u = u^{\#}$ for $0 \le t < +\infty$ and all x values. ∎

Exercises

(3.12) Prove uniqueness theorems for Theorem (2.30) and exercises (2.33 ∗) and (2.44 ∗).

(3.13) Prove that if u and v are of class $\mathscr{E}$, then $au + bv$ are of class $\mathscr{E}$ for all real constants a and b. Moreover, uv is of class $\mathscr{E}$.

(3.14) Prove that every polynomial in x and t is of class $\mathscr{E}$.

(3.15) Prove the following Theorem.

Theorem. Suppose that f is continuous and bounded over $\mathbb{R}$. Prove that the function u given in (2.7) is the only bounded solution to (2.1).

(3.16) Suppose that u is a bounded (class $\mathscr{E}$) solution to the heat equation. Prove that for $t_0 > 0$ and $t > 0$

$$u(x, t + t_0) = \int_{-\infty}^{+\infty} u(x - s, t_0)H(s; t)\, ds = \int_{-\infty}^{+\infty} u(s, t_0)H(x - s; t)\, ds$$

Moreover, prove that u is infinitely continuously differentiable in x and t for $t > 0$ and $-\infty < x < +\infty$.

(3.17) Suppose that $\{u_n\}$ is a sequence of bounded solutions to the heat equation that are all bounded by the same constant for $t > 0$ and $-\infty < x < +\infty$. Prove that if $\lim_{n \to \infty} u_n(x, t) = u(x, t)$ uniformly for $t > 0$ and all x values, then u is a solution to the heat equation.

Heat Polynomials and Exponentials

The next four exercises explore polynomial, and exponential, solutions to the heat equation $u_t = u_{xx}$ for $-\infty < t < +\infty$ and $-\infty < x < +\infty$.

(3.18) Prove that the function $u(x, t) = e^{ax + bt}$ solves the heat equation if and only if $b = a^2$. Prove that $u(x, t) = e^{ax + a^2 t}$ is the only solution of class $\mathscr{E}$ that satisfies $u(x, 0) = e^{ax}$.

(3.19) Let $p(x)$ be a polynomial in x. Prove that there exists a *unique* polynomial in x and t, say $q(x, t)$, that satisfies the heat equation for all x and all t and for which $q(x, 0) = p(x)$.

(3.20)   Prove that for $p(x) = x^n$ $(n = 0, 1, 2, 3, \ldots)$ the polynomial

$$v_n(x, t) = n! \sum_{k=0}^{[n/2]} \frac{t^k}{k!} \frac{x^{n-2k}}{(n-2k)!}$$

is the unique polynomial discussed in (3.19). [The expression $[n/2]$ stands for the largest integer $\leq n/2$.] In particular, $v_0(x, t) = 1$, $v_1(x, t) = x$, $v_2(x, t) = x^2 + 2t$, $v_3(x, t) = x^3 + 6xt$.

(3.21)   Define w_n as $n!$ times the nth coefficient in the power series expansion in z of e^{zx+z^2t}, that is, $e^{zx+z^2t} = \sum_{n=0}^{\infty} w_n(x, t)(z^n/n!)$. Prove that (a) w_n is a polynomial in x and t, (b) $w_n(x, 0) = x^n$, (c) w_n solves the heat equation, and (d) $w_n(x, t) = v_n(x, t)$ from (3.20).

Remark.   The *heat polynomial* w_n described in (3.21) is closely related to the Hermite polynomial H_n discussed in the Miscellaneous Exercises for Chapter 6. Putting t equal to -1 and replacing x by $2x$ in the power series expansion given in (3.21), we obtain

(3.22) $$e^{2xz-z^2} = \sum_{n=0}^{\infty} w_n(2x, -1) \frac{z^n}{n!}$$

and from (3.21d) we have

(3.23) $$w_n(2x, -1) = n! \sum_{k=0}^{[n/2]} \frac{(-1)^k}{k!} \frac{(2x)^{n-2k}}{(n-2k)!}$$

The left side of (3.22) is the *generating function* for the Hermite polynomials, while the right side of (3.23) is the explicit expression for the nth Hermite polynomial $H_n(x)$. [See Davis (1975), p. 368.] Thus $H_n(x) = w_n(2x, -1)$.

Part B. Fourier Optics

§4. Fraunhofer Diffraction

Fraunhofer diffraction of light is fundamental in optics. It plays a key role in imaging with lenses, two important cases being telescopes and microscopes. Fraunhofer diffraction also occurs when a crystal diffracts an X-ray beam; this diffraction opens the way to identifying the underlying crystal structure. *In terms of Fourier analysis, Fraunhofer diffraction is a physical Fourier transform operation.*

Since this is not an optics text, we do not have sufficient space for a rigorous discussion of diffraction. Instead, we will only give an outline [similar to the one given in Lipson and Lipson (1981), pp. 163–165] of how the Fourier transform comes into play. *In §§5–9, we shall see that this relationship is amply verified by the quantitative and qualitative description it gives of actual diffraction patterns.* The reader who desires a complete discussion of the underlying theory of diffraction should consult Goodman (1968), Iizuka (1985), or Lipson and Lipson (1981).

Suppose that a plane parallel wave of light, of wavelength λ, is cast upon an opaque screen in which there is a tiny aperture. Let a lens collect all the light emerging from the aperture and project it to its focal plane. As shown in Figure 7.1, the lens takes each bundle of parallel light rays and brings it to a focus.

Let the optic axis of the lens lie along the z axis and the aperture plane be the x–y plane located at $z = 0$. Suppose that $\mathcal{P}$ is a point in the aperture with coordinates $(x, y, 0)$. We will now argue that, because the wave front emerging from the point $\mathcal{P}$ has an amplitude $A(x, y)$, *the physical effect of the shuffling of light rays,* shown in Figure 1(b), *is to Fourier transform this amplitude function $A(x, y)$*. In Figure 7.2 we show light rays from $\mathcal{O}$ and $\mathcal{P}$, each parallel to the unit vector $\mathbf{u} = (\ell, m, n)$, coming to a focus $\mathcal{Q}$ after passing through points $\mathcal{O}_1$ and $\mathcal{P}_1$ in the lens. By Hamilton's version of Fermat's principle, light rays always lie normal to wave fronts of equal phase. Therefore, we shall imagine that the light ray from $\mathcal{P}$ has instead emanated from $\mathcal{P}_0$, the projection of $\mathcal{P}$ onto the plane normal to $\mathbf{u}$ and passing through $\mathcal{O}$ (as shown in Figure 7.2). Then the wave fronts along the rays $\mathcal{P}_0\mathcal{P}_1\mathcal{Q}$ and $\mathcal{O}\mathcal{O}_1\mathcal{Q}$ will arrive at $\mathcal{Q}$ in phase. Thus (because the wavelength is λ)

(4.1) $$e^{i(2\pi/\lambda)|\mathcal{P}_0\mathcal{P}_1\mathcal{Q}|} = e^{i(2\pi/\lambda)|\mathcal{O}\mathcal{O}_1\mathcal{Q}|}$$

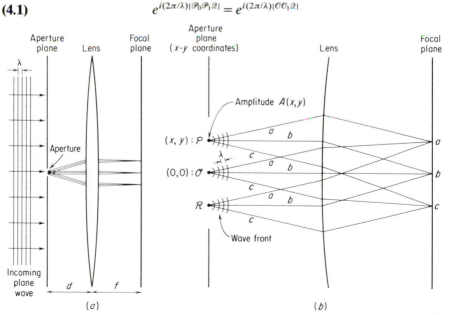

Figure 7.1 Focusing of bundles of parallel light rays by a lens. (*b*) An enlarged view of (*a*).

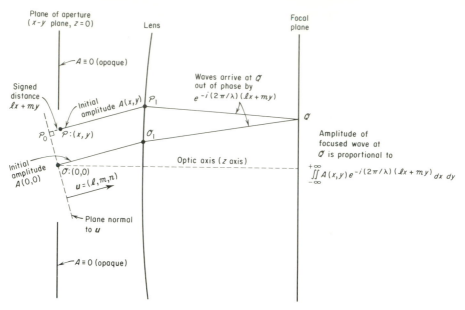

Figure 7.2   Focusing of two rays parallel to $\mathbf{u} = (\ell, m, n)$.

where $|\mathcal{P}_0\mathcal{P}_1\mathcal{Q}|$ and $|\mathcal{O}\mathcal{O}_1\mathcal{Q}|$ are the (optical) lengths of the two rays. Since $|\mathcal{P}_0\mathcal{P}_1| = \ell x + my$ we have the following phase relations

$$e^{i(2\pi/\lambda)|\mathcal{P}_0\mathcal{P}_1\mathcal{Q}|} = e^{i(2\pi/\lambda)|\mathcal{P}_0\mathcal{P}|}e^{i(2\pi/\lambda)|\mathcal{P}\mathcal{P}_1\mathcal{Q}|}$$

(4.2)

$$= e^{i(2\pi/\lambda)(\ell x + my)}e^{i(2\pi/\lambda)|\mathcal{P}\mathcal{P}_1\mathcal{Q}|}$$

Comparing (4.2) with (4.1) we obtain

(4.3) $$e^{i(2\pi/\lambda)|\mathcal{P}\mathcal{P}_1\mathcal{Q}|} = e^{i(2\pi/\lambda)|\mathcal{O}\mathcal{O}_1\mathcal{Q}|}e^{-i(2\pi/\lambda)(\ell x + my)}$$

Thus the wave fronts from $\mathcal{O}$ and $\mathcal{P}$ arrive at $\mathcal{Q}$ with amplitudes $A(0,0) \cdot e^{i(2\pi/\lambda)|\mathcal{O}\mathcal{O}_1\mathcal{Q}|}$ and $A(x,y)e^{i(2\pi/\lambda)|\mathcal{O}\mathcal{O}_1\mathcal{Q}|}e^{-i(2\pi/\lambda)(\ell x + my)}$, respectively. Therefore, the total amplitude Ψ at $\mathcal{Q}$ is given by the superposition integral

(4.4) $$\Psi = e^{i(2\pi/\lambda)|\mathcal{O}\mathcal{O}_1\mathcal{Q}|}\int\!\!\!\int\limits_{-\infty}^{+\infty} A(x,y)e^{-i(2\pi/\lambda)(\ell x + my)}\,dx\,dy$$

Since the screen is opaque outside the aperture, we may assume that $A(x,y) = 0$ for (x,y) outside the aperture. Hence the integral in (4.4) is actually taken only over a finite region of the x–y plane.

Introducing the variables

(4.5) $$u = \ell/\lambda \qquad v = m/\lambda$$

we can write (4.4) in the form

(4.6) $$\Psi = e^{i(2\pi/\lambda)|\mathcal{O}\mathcal{O}_1\mathcal{Q}|}\hat{A}(u,v)$$

Experimentally, it is found that when the wave amplitude Ψ is recorded, either photographically or by striking an observation screen, we record only the intensity $I = |\Psi|^2$. Hence, from (4.6) we have

(4.7) $$I = |\Psi|^2 = |\hat{A}(u, v)|^2$$

Because of (4.6) and (4.7) we say that the focal plane of the lens is *coincident* with the Fourier transform $(u-v)$ plane. The variables u and v are called the *spatial frequencies* of the light amplitude emanent from the aperture.

If we introduce Cartesian coordinates X, Y into the focal plane (with origin on the optic axis), then we wish to know the relationship between those coordinates and the spatial frequencies. In the *paraxial approximation* $[(\ell, m, n) \approx (\ell, m, 1)]$ the relationship is approximately *linear*

(4.8) $$X \approx f\ell = (f\lambda)u \qquad Y \approx fm = (f\lambda)v$$

where f is the distance from the lens to the focal plane.[3] Thus, *the greater the wavelength λ the greater the $X - Y$ dimensions of the diffraction pattern formed by Ψ in the focal plane.*

In (4.6) there is a phase factor $e^{i(2\pi/\lambda)|\mathcal{O}\mathcal{O}_1\mathcal{Q}|}$ that depends upon the point $\mathcal{Q}$ in the focal plane. *This dependency can be removed by placing the aperture plane exactly f units from the lens.* In this case, a plane wave emanating in parallel from the focal plane (rays parallel to the optic axis) is focused by the lens onto the point $\mathcal{O}$. Conversely, all waves emanating from $\mathcal{O}$ arrive at the focal plane in phase (Fermat's principle). Thus $e^{i(2\pi/\lambda)|\mathcal{O}\mathcal{O}_1\mathcal{Q}|} = e^{i(2\pi/\lambda)2f}$ where $e^{i(2\pi/\lambda)2f}$ is the phase change along the ray from $\mathcal{O}$ to the focal plane that lies along the optic axis. Thus (4.6) becomes

(4.9) $$\Psi = e^{i4\pi f/\lambda}\hat{A}(u, v)$$

Formula (4.9) shows that the light amplitude Ψ at the focal plane is, except for a constant factor, *equal to the Fourier transform $\hat{A}$ of the amplitude function A of the light emanating from the aperture.*

The simplest form for the amplitude function A is

(4.10) $$A(x, y) = \begin{cases} 1 & \text{if } (x, y) \text{ is in the aperture} \\ 0 & \text{if } (x, y) \text{ is not in the aperture} \end{cases}$$

This function A corresponds to perfect transmission of the incoming plane wave (assumed to have constant amplitude 1 and phase e^{i0} when it strikes the aperture screen) through the aperture. We will call the function in (4.10) an *aperture function*. Other types of amplitude functions will correspond to partial absorption and/or partial phase variations introduced by apertures of varying material compositions.

---

3. Formula (4.8) is obtained by considering the *straight* ray $\mathcal{R}\mathcal{R}_1\mathcal{Q}$ parallel to $(\ell, m, 1)$ that passes through the point $\mathcal{R}_1$ on the lens that lies on the optic axis.

(4.11) Remark. Fraunhofer diffraction also occurs without a lens. This happens when X-rays are diffracted by crystals (or when a narrow laser beam shines through a small aperture to a screen on the other side of an optics lab). Roughly speaking, the analysis above will still apply if the light rays arriving at each point in the observation screen can be considered as approximately parallel. This is the case with both of the examples mentioned above (if only a small portion of the observation screen is viewed). For more details, see the references above, or Hecht and Zajec (1974).

Exercises

(4.12) Explain why light in the red spectrum is diffracted more than light in the blue spectrum. (How does this compare with refraction of light?) Why is white light split into a spectrum of colors by diffraction?

(4.13) Suppose that (4.8) applies throughout the focal plane. Show that the result of the lens system shown in Figure 7.3 is to produce an inverted and magnified (reduced) image of the aperture screen. The magnification (reduction) factor being $M = f_2/f_1$.

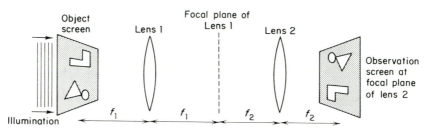

Figure 7.3 Double lens imaging system.

(4.14) In this section we assumed that the lens collects all the light from the aperture. A more realistic situation is depicted in Figure 7.4, where $P(u, v) = 0$ for sufficiently large values of u or v. Show that the result of

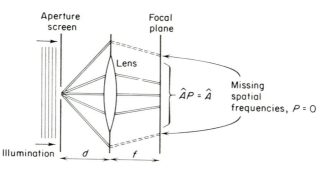

Figure 7.4 Diffraction limitation of lens. [Note that this figure requires that we observe only a portion of the focal plane near the optic axis to see the transform $|\hat{A}|^2$.]

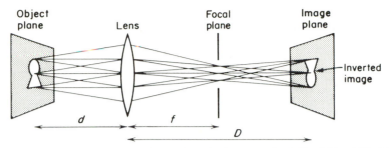

Figure 7.5 Imaging in geometrical optics. The lens equation $1/d + 1/D = 1/f$ holds.

(4.13) must be modified; the image of the aperture has the form $P_2[U_g * h_1]$ where h_1, P_2, and U_g are defined in an appropriate way.

(4.15) The classic imaging diagram of geometrical optics is shown in Figure 7.5. Interpret this figure in terms of Fourier analysis. How does the more realistic situation in Figure 7.4 affect what is depicted in Figure 7.5?

§5. Rectangular Apertures

The simplest apertures for Fourier analysis are the rectangular ones. Suppose we have a rectangular aperture of width $a > 0$ in the x direction and height $b > 0$ in the y direction, and the aperture is centered at the origin. [See Figure 7.8(a)]. If we let

$$A(x, y) = \begin{cases} 1 & \text{if } |x| < \tfrac{1}{2}a \text{ and } |y| < \tfrac{1}{2}b \\ 0 & \text{otherwise} \end{cases}$$

then by Example (7.6c), Chapter 6, we have

$$\hat{A}(u, v) = \frac{\sin \pi a u}{\pi u} \frac{\sin \pi b v}{\pi v}$$

Therefore, the intensity $I = |\hat{A}|^2$ is given by

(5.1) $$I(u, v) = \frac{\sin^2 \pi a u}{(\pi u)^2} \frac{\sin^2 \pi b v}{(\pi v)^2} = a^2 b^2 \operatorname{sinc}^2 au \operatorname{sinc}^2 bv$$

The function I has the graph shown in Figure 7.6. Based on Figure 7.6 we can predict a diffraction pattern like the one shown in Figure 7.7. There are regions of brightness, marked off by fringes of zero intensity along the lines $u = \pm 1/a, \pm 2/a, \pm 3/a, \ldots$, and $v = \pm 1/b, \pm 2/b, \pm 3/b, \ldots$, that are the zeroes of $\operatorname{sinc}^2 au$ and $\operatorname{sinc}^2 bv$. The zone of highest brightness is a lozenge shaped region centered at the origin with u width $2/a$ and v height $2/b$. Notice the reciprocal relationship between the dimensions of the rectangular aperture and the dimensions of the central zone of brightness in the diffraction pattern, as well as the distances $\Delta u = 1/a$ and $\Delta v = 1/b$ between the fringes of zero intensity.

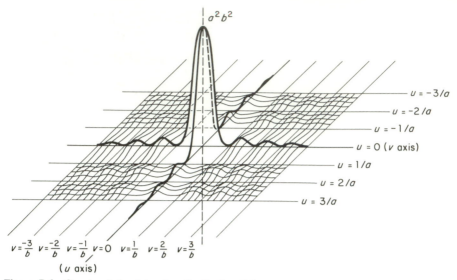

Figure 7.6 Graph of the intensity distribution I for a rectangular aperture. The graph is obtained by treating $1/(\pi u)^2[1/(\pi v)^2]$ as an envelope for $\sin^2 \pi au$ [$\sin^2 \pi bv$] and noting that $\lim_{u \to 0} \sin^2 \pi au/(\pi u)^2 = a^2$ [$\lim_{v \to 0} \sin^2 \pi bv/(\pi v)^2 = b^2$].

An actual diffraction pattern is shown in Figure 7.8(b). The reciprocal relationship between aperture detail and diffraction pattern detail, which we noted above, will be manifested in all of our examples. Diffraction patterns are often said to exist in *reciprocal space* (relative to apertures in real space).

If b is considerably larger than a then we obtain a *vertical slit*. Because the details of the pattern in Figure 7.7 *in the v direction* are proportional to $1/b$, for large b the pattern in Figure 7.7 will be squashed in the v direction. In Figure 7.9 we have a photograph of the diffraction pattern of a vertical slit.

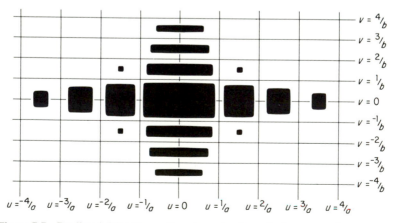

Figure 7.7 Predicted diffraction pattern (negative image) of a rectangular aperture.

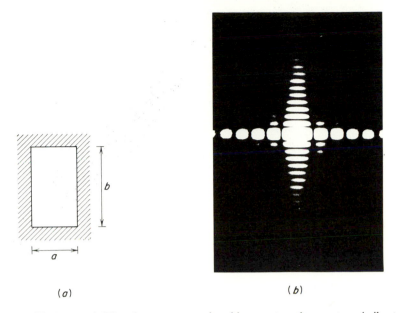

(a)   (b)

Figure 7.8 (b) An actual diffraction pattern produced by a rectangular aperture similar to the one shown in (a). (Photograph courtesy of S. G. Lipson.)

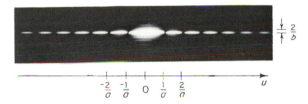

Figure 7.9 Diffraction pattern for a vertical slit. (Photograph courtesy of S. G. Lipson.)

Exercises

(5.2) Show that the maximum intensity of the diffraction pattern of a rectangular aperture is proportional to the square of the area of the aperture.

(5.3) Suppose two identical rectangular apertures are separated by a distance d along the x axis, as shown in Figure 7.10. Describe the diffraction pattern from such an aperture. What happens as d is increased?

(5.4) Suppose two identical rectangular apertures have their centers located at $(\frac{1}{2}c, -\frac{1}{2}d)$ and $(-\frac{1}{2}c, \frac{1}{2}d)$ as shown in Figure 7.11. Describe the diffraction pattern from such an aperture.

(5.5) Suppose the amplitude function A has the form shown in Figure 7.12. Describe the diffraction pattern from such an aperture and compare it

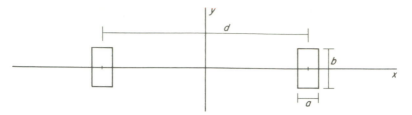

Figure 7.10   Aperture for Exercise (5.3).

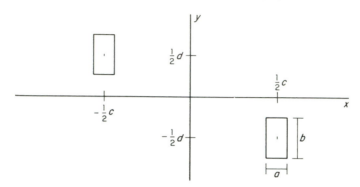

Figure 7.11   Aperture for Exercise (5.4).

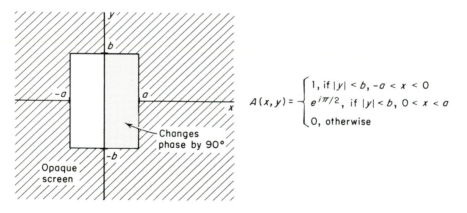

$$A(x,y) = \begin{cases} 1, & \text{if } |y| < b, -a < x < 0 \\ e^{i\pi/2}, & \text{if } |y| < b, 0 < x < a \\ 0, & \text{otherwise} \end{cases}$$

Figure 7.12   Aperture and aperture function for Exercise (5.5).

with Figure 7.8(b). **Remark.** Such an aperture can be made by placing a sheet of mica over half of a simple rectangular aperture; the mica induces a 90° phase shift of the incoming light wave.

§6. Circular Apertures

Consider a circular aperture of radius a. The aperture function in this case is a radial function (§7, Chapter 6)

$$A(r) = \begin{cases} 1 & \text{if } 0 \leq r < a \\ 0 & \text{if } r > a \end{cases}$$

where $r = (x^2 + y^2)^{1/2}$. Using Theorem (7.15), Chapter 6, we have

(6.1)
$$\hat{A}(\rho) = 2\pi \int_0^a J_0(2\pi\rho r) r \, dr$$

where $\rho = (u^2 + v^2)^{1/2}$. If we let $s = 2\pi\rho r$, then

(6.2)
$$\hat{A}(\rho) = \frac{1}{2\pi\rho^2} \int_0^{2\pi\rho a} s J_0(s) \, ds$$

Let's consider the function $H(x) = \int_0^x s J_0(s) \, ds$ for $x \geq 0$. We have

(6.3)
$$dH/dx = x J_0(x) \qquad (x \geq 0)$$

But, by a recurrence relation for Bessel functions [(2.1b), Chapter 10] we have

(6.3′)
$$(d/dx)[x J_1(x)] = x J_0(x) \qquad (x \geq 0)$$

where [see (2.2), Chapter 10]

$$J_1(x) = \frac{1}{2\pi} \int_{-\pi}^{\pi} e^{ix \sin \phi} e^{-i\phi} \, d\phi$$

Noting that J_1 is continuous at $x = 0$, and has the value 0 there, we conclude from (6.3) and (6.3′) that $H(x) = x J_1(x)$ for $x \geq 0$. Therefore, we can rewrite (6.2) as

(6.4)
$$\hat{A}(\rho) = \frac{a}{\rho} J_1(2\pi\rho a) = \pi a^2 \left[\frac{2 J_1(2\pi\rho a)}{2\pi\rho a} \right]$$

Therefore, the intensity I is given by

(6.5)
$$I(\rho) = (\pi a^2)^2 \left[\frac{2 J_1(2\pi\rho a)}{2\pi\rho a} \right]^2 \qquad [\rho = (u^2 + v^2)^{1/2}]$$

The function J_1 has been extensively tabulated[4] and so the graph of I is well known. (See Figure 7.13.) The graph of I can also be sketched using asymptotic formulas for J_1 that we shall discuss in Chapter 10.

Figure 7.13(b) gives a nice prediction for the diffraction pattern, as we can see from Figure 7.14. The rings in the pattern are called *Airy's rings* and the whole pattern is called *Airy's pattern*.

Notice that the radius of the central disk, bounded by the zero intensity (dark) ring at $\rho \doteq 3.83/2\pi a$, grows larger with decreasing radius a. Furthermore, as this radius decreases the location of Airy's rings (bright and dark) increases away from the origin. These results illustrate again the reciprocal relationship between aperture detail and diffraction pattern (transform) detail.

Airy's rings are observed when a telescope with too small an aperture attempts to resolve the image of a distant star. *The resulting image is the*

---

4. See Watson (1944), pp. 667–697.

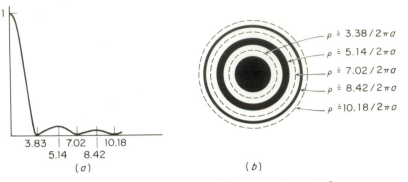

Figure 7.13 Diffraction from a circular aperture. (*a*) Graph of $[2J_1(x)/x]^2$. (*b*) Prediction for the diffraction pattern.

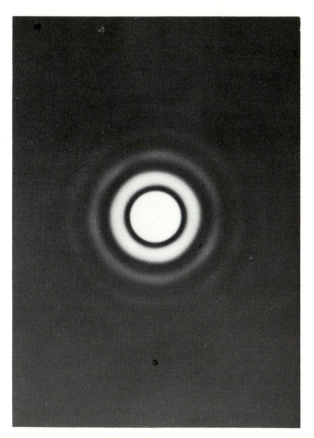

Figure 7.14 Diffraction pattern from a circular aperture. (Photograph courtesy of S. G. Lipson.)

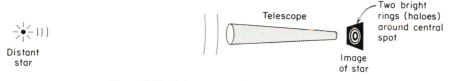

Figure 7.15   Airy pattern observed in telescopes.

diffraction pattern of the aperture. (See Figure 7.15.) The astronomer Airy first predicted this ring pattern. Because of the large wavelengths used, this diffraction is a serious problem encountered with radio telescopes (where the aperture is the collecting dish).

Exercises

(6.6)   Show that the intensity of Airy's pattern increases with the square of the area of the circular aperture, and that the position of Airy's rings increases inversely to the circumference of the circular aperture.

(6.7)   Explain why a circular aperture with a small enough radius will have a Fraunhofer diffraction pattern consisting only of a disk. [*Hint*: See (4.11) and Figure 7.4.] Such an aperture is called a *pinhole*.

(6.8)   How would the Airy patterns differ between light of wavelength $\lambda_1 = 750 \times 10^{-9}$ m (red light) versus light of $\lambda_2 = 500 \times 10^{-9}$ m (bluish green light)? Describe the effect of shining white light through circular apertures of various radii. (This phenomenon occurs in color movies occasionally.)

(6.9)   Let the aperture consist of an annular ring with aperture function

$$A(x, y) = \begin{cases} 1 & \text{if } a < (x^2 + y^2)^{1/2} < b \\ 0 & \text{otherwise} \end{cases}$$

Describe the resulting diffraction patterns for $a \ll b$ and for a close to b.

(6.10)   *Babinet's Principle.* One aperture is called *complementary* to a second aperture if their aperture functions A_1, A_2 satisfy $A_1 + A_2 = A_0$ where A_0 is a large circular (or rectangular) aperture function. Show that *the two complementary apertures have identical diffraction patterns*, except for a small region about the origin in the u–v plane.

§7.   Interference

The study of interference is an important area of application of Fourier analysis in optics. In this section we will treat some simple examples of interference. The more important cases of diffraction gratings and the array theorem will be discussed in the next two sections.

Suppose that two circular apertures of radius a are centered at $x = \pm\frac{1}{2}d$ on the x axis. In this case our aperture function will be (assuming $d > 2a$)

$$A(x, y) = A_1(x + \tfrac{1}{2}d, y) + A_1(x - \tfrac{1}{2}d, y)$$

where A_1 is the aperture function for a circular aperture that we considered in the previous section. Using the shift property and linearity we obtain

$$\hat{A}(u, v) = \hat{A}_1(u, v)[e^{i\pi\, du} + e^{-i\pi\, du}] = 2 \cos \pi\, du \hat{A}_1(u, v)$$

hence for the intensity I we have

(7.1) $$I(u, v) = 4 \cos^2 \pi\, du \, |\hat{A}_1(u, v)|^2 = 4 \cos^2 \pi\, du \, I_1(u, v)$$

where I_1 is the intensity for a single circular aperture.

Since $4 \cos^2 \pi\, du$ equals 0 when $u = \pm 1/(2d)$, $\pm 3/(2d)$, $\pm 5/(2d)$, ... , we expect zero intensity along vertical lines in the u–v plane defined by those u values. In view of (7.1) we expect dark vertical strips, *interference fringes*, to overlap the Airy pattern shown in Figure 7.14. (Also, we expect amplification along the lines $u = 0$, $\pm 1/d$, $\pm 2/d$, ... , where $4 \cos^2 \pi\, du$ has its maximum value of 4.) In Figure 7.16 we show the diffraction pattern from two such circular apertures.

For a second example, consider two identical rectangular apertures positioned as indicated in Figure 7.11. The aperture function in this case is

$$A(x, y) = A_1(x - \tfrac{1}{2}c, y + \tfrac{1}{2}d) + A_1(x + \tfrac{1}{2}c, y - \tfrac{1}{2}d)$$

which was considered in Example (7.6d), Chapter 6. Using the transform found in that example, we obtain

$$I(u, v) = 4 \cos^2 \pi(cu - dv)I_1(u, v)$$

Figure 7.16 A diffraction pattern from two horizontally separated circular apertures. (Photograph courtesy of S. G. Lipson.)

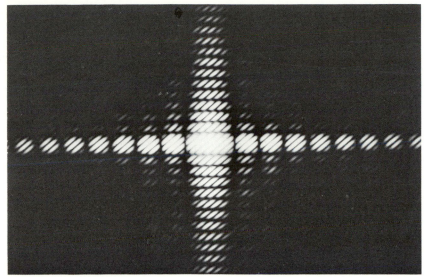

Figure 7.17   Diffraction pattern from two rectangular apertures (here $c/d \doteq 1$). (Photograph courtesy of S. G. Lipson.)

where I_1 is the intensity distribution for a single rectangular aperture. The interference fringes lie along lines given by $cu - dv = \frac{1}{2}k$ for $k = \pm 1, \pm 3, \pm 5, \ldots$. Those dark fringes lie over the diffraction pattern for a single rectangular aperture at parallel directions with slope c/d, hence perpendicular to the line segment that connects the centers of the two apertures. Figure 7.17 shows the actual diffraction pattern that results in this case.

Exercises

(7.2)   Show that interference fringes always occur when an aperture consists of two identical apertures.

(7.3)   Show that the number of interference fringes per unit length in both the examples for this section is proportional to the distance between the centers of the apertures.

(7.4)   Describe the diffraction pattern of three equally spaced similar vertical slits. [*Hint*: Add the transform of the center slit to that of the two outer ones.] Now do four equally spaced vertical slits.

(7.5)   Suppose we have four pinholes [see Exercise (6.7)] arranged in a parallelogram, that is, the pinholes are placed at (d_1, d_2), $(c_1, -c_2)$, $(-d_1, -d_2)$, and $(-c_1, c_2)$ where c_1, c_2, d_1, and d_2 are positive. Describe the resulting diffraction pattern.

§8. Diffraction Gratings

Diffraction gratings were first constructed by Fraunhofer in the 1830's. They are an essential tool of modern science.

Suppose that an aperture consists of a large number, say N, of vertical slits spaced equally far apart and very close together. Such an aperture is called a *diffraction grating*. Diffraction gratings with 100,000 slits have been constructed. If we assume that the first slit on the left side of the grating is centered at the origin, then $A(x, y) = \sum_{n=0}^{N-1} A_1(x - nd, y)$ where A_1 is the aperture function for a single vertical slit (see the end of §5), and d is the distance between the central axes of the slits. By the shifting and linearity properties we have

$$\hat{A}(u, v) = \sum_{n=0}^{N-1} \hat{A}_1(u, v)e^{-i2\pi n\, du} = \hat{A}_1(u, v) \sum_{n=0}^{N-1} [e^{-i2\pi\, du}]^n$$

$$= \hat{A}_1(u, v) \frac{1 - e^{-i2\pi N\, du}}{1 - e^{-i2\pi\, du}}$$

where we summed a finite geometric series to obtain the last quantity above. Factoring out $e^{-i\pi N\, du}/e^{-i\pi\, du}$ from the last fraction above we get

$$\hat{A}(u, v) = \hat{A}_1(u, v) \frac{e^{-i\pi N\, du}}{e^{-i\pi\, du}} \frac{\sin N\pi\, du}{\sin \pi\, du}$$

Hence, the intensity distribution I is given by

(8.1)
$$I = I_1 \frac{\sin^2 N\pi\, du}{\sin^2 \pi\, du} = I_1 S$$

where I_1 is the intensity for a single vertical slit. In Figure 7.18 we have graphed the function S, called the *structure factor* for the grating, and its intensity distribution in the u–v plane. [Note: $S(u) = N \cdot F_N(2\pi\, du)$ where F_N is *Fejér's kernel*.] As an intensity distribution in the u–v plane, S is a sequence of vertical strips centered along the vertical lines $u = 0$, $\pm 1/d$, $\pm 2/d, \ldots$, each strip having a width no greater than $2/Nd$ and intensity N^2 along their central lines. When this sequence of strips is multiplied by I_1 (see Figure 7.9) it acts as a mask of *amplifiers* of power N^2 and we obtain a sequence of bright dots centered at $u = 0$, $\pm 1/d$, $\pm 2/d, \ldots$. (See Figure 7.19.) In Figure 7.20 we have an actual diffraction pattern resulting from a grating with a small number of lines ($N = 55$). As the reader can see, *away from the central vertical axis*, the prediction of a sequence of bright dots is confirmed in Figure 7.20. The slight discrepancy between Figures 7.19 and 7.20 along that central vertical axis is easily explained [see Exercise (8.4)].

Diffraction gratings are primarily used for the production of spectra. If we shine white light through the diffraction grating considered above, then because of formula (4.5) we will obtain a *spectral decomposition* of that light. From Figure 7.19, we see that the spacing between dots is $\Delta u = 1/d$ hence from (4.5) we have[5]

(8.2)
$$\Delta \ell = \lambda \Delta u = \lambda/d$$

---

5. We are assuming here that white light is a linear superposition of all its wavelengths and that passing the light through the grating is a linear process, assumptions that are confirmed in practice.

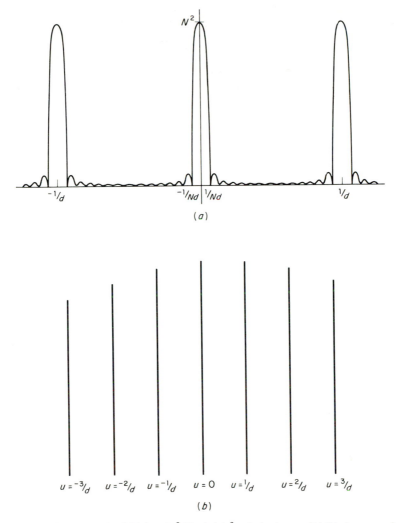

Figure 7.18 (*a*) The graph of $S(u) = \sin^2 N\pi\,du/\sin^2 \pi\,du$ is shown. (*b*) We have graphed the intensity distribution in the u–v plane that is generated by $S(u)$, *ignoring secondary peaks*.

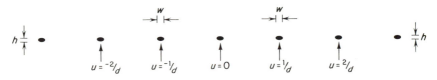

Figure 7.19 The intensity distribution resulting from a diffraction grating (negative image). The width w of each dot is no more than $2/Nd$, their height h is no more than $2/b$ where b is the height in the y direction of each vertical slit in the grating.

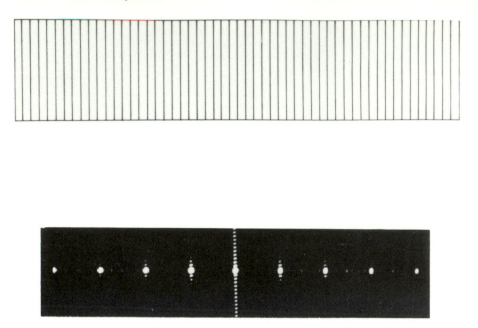

Figure 7.20 A grating of 55 narrow lines (the negative of the drawing on top), and its diffraction pattern (the photograph on the bottom). [Reproduced with permission from Lipson (1972), p. 358.]

is the *angular spacing* between the dots. Here we may take $\ell = \sin\theta$ where θ is the angle of diffracted rays to the optic axis (since $m \doteq 0$ because we have vertical slits). Comparing (8.2) with the ratios of the wavelengths of visible light in Table 7.1, we see that (8.2) describes a spreading out (dispersion) of white light into its spectrum of colors. (See Figure 7.21.) From Table 7.1 and Figure 7.21 we can see that the second-order spectrum is longer (greater dispersion) than the first-order spectrum. The third-order spectrum, which begins at $3\lambda_{red}/2d$ and ends at $3\lambda_{red}/d$, actually overlaps half of the second-order spectrum, which begins at λ_{red}/d and ends at $2\lambda_{red}/d$. The third-order spectrum is even more dispersed than the first two spectra.

Table 7.1 Approximate Wavelengths λ for the Visible Spectrum and Their Ratios to $\lambda_{red} = 780 \times 10^{-9}$ m

Color	λ (1×10^{-9} m)	λ/λ_{red}
Red	780–622	1.–0.797
Orange	622–597	0.797–0.765
Yellow	597–577	0.765–0.739
Green	577–492	0.739–0.631
Blue	492–455	0.631–0.583
Violet	455–390	0.583–0.5

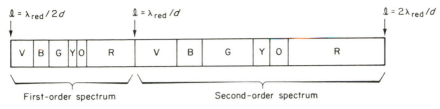

Figure 7.21   Spectra of white light from a diffraction grating.

Since each dot in Figure 7.19 has a width w, no greater than $2/Nd$, there will be a smearing of the colors in the spectra obtained by diffraction gratings. However, if Nd is very large, then $2/Nd$ will be very small and there will be less smearing (better *resolution*). In any case, the resolution is inversely proportional to Nd, the *length of the grating*. [See Exercise (8.7).]

We assumed above that white light was shined through the grating. The grating can be used, of course, to disperse the spectra of other kinds of light. By passing white light through a gas and then observing that light through a diffraction grating, we obtain spectra characterized by dark fringes (absorption lines) in the normal spectrum of white light. Depending on the element(s) composing the gas, certain characteristic absorption lines appear. Gas absorption spectroscopy is of some use in chemistry. Moreover, by observing the spectra of starlight, the composition of stars can be analyzed. The classic example is the discovery of helium in the sun. The absorption lines in solar light are sometimes called *Fraunhofer lines*. Another application is the analysis of the spectra of the light emitted from chemical reactions (emission spectra). For all of these applications, the problem of resolution of spectra is a vital one. [See Bell (1972).]

Exercises

(8.3)   Sketch the graph of $y = I_1 S$ as a function of u when $v = 0$.

(8.4)   Explain why there is a vertical strip of dots in the center of the diffraction pattern in Figure 7.20. Show that if Nd is large enough, then the dots in Figure 7.19 will actually be thin lines.

(8.5)   Show that if the distance d between the vertical slits in a diffraction grating is equal to $2a$ (twice the width of each slit) then the second-order spectrum will be eliminated. Show also that the first-order spectrum is not overlapped by any of the higher order spectra. What value should d have so that the third (nth)-order spectra is eliminated?

(8.6)   Let $d \doteq 8a$ and let h_0, h_1, h_2, h_3 stand for the first four highest maxima of the intensity function $I = I_1 S$. Show that

$$h_0 \doteq N^2 a^2 b^2 \qquad h_1 \doteq 0.95 N^2 a^2 b^2 \qquad h_2 \doteq 0.81 N^2 a^2 b^2 \qquad h_3 \doteq 0.62 N^2 a^2 b^2$$

(8.7)   Suppose that light containing two wavelengths $\lambda_1 < \lambda_2$ is transmitted through a diffraction grating. If λ_1 and λ_2 are close together, then their

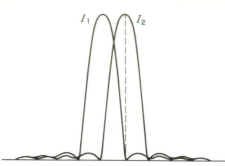

Figure 7.22   A central peak for I_2 lying over a first zero for I_1. This is known as *Rayleigh's criterion* for the resolution of two spectral lines.

spectral lines [see (8.4)] might smear together. How large does N have to be in order for a central peak (highest maxima) of I_2, the intensity function for λ_2, to lie over a first zero for I_1, the intensity function for λ_1? [See Figure 7.22] *Note*: your answer should depend on which spectrum (first, second, third, etc.) for λ_1 and λ_2 is being looked at.

(8.8)   Show that resolution by Rayleigh's criterion [see problem (8.7)] is easier to obtain the higher the order of the spectrum.

(8.9)   In spectroscopy, the second-order spectrum is usually preferred among all the orders. Can you think of any reasons for this?

§9. The Array Theorem

We now turn to a beautiful result in optics known as the *Array Theorem*. Let c and d be positive constants. Suppose that we have a collection of $M \cdot N$ identical apertures positioned in a rectangular array at points (mc, nd) for $m = 0, 1, \ldots, M - 1$ and $n = 0, 1, \ldots, N - 1$. (See Figure 7.23.) We assume that c and d are large enough so that none of the apertures overlaps.

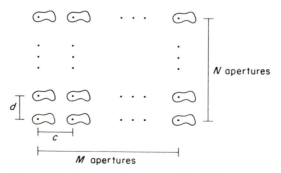

Figure 7.23   A rectangular array of identical apertures.

Letting A_1 denote the aperture function for a single aperture, and supposing that the origin in the x-y plane is located in the aperture at the lower left hand corner of the array, we have

$$A(x, y) = \sum_{m,n=0}^{M-1,N-1} A_1(x - mc, y - nd)$$

is the aperture function for the whole array. By linearity and shifting, we get

$$\hat{A}(u, v) = \sum_{m, n=0}^{M-1, N-1} \hat{A}_1(u, v)e^{-i2\pi(mcu + n\,dv)}$$

$$= \hat{A}_1(u, v)\left[\sum_{m=0}^{M-1} e^{-i2\pi mcu}\right]\left[\sum_{n=0}^{N-1} e^{-i2\pi n\,dv}\right]$$

From the result of the discussion of diffraction gratings in the previous section, we obtain

(9.1) $$I(u, v) = I_1(u, v)\frac{\sin^2 M\pi cu}{\sin^2 \pi cu}\frac{\sin^2 N\pi\,dv}{\sin^2 \pi\,dv}$$

Thus the intensity distribution I for the whole array is the product of the intensity distribution I_1 for a single aperture with the function S defined by

$$S(u, v) = \frac{\sin^2 M\pi cu}{\sin^2 \pi cu}\frac{\sin^2 N\pi\,dv}{\sin^2 \pi\,dv}$$

The function I_1 is called the *form factor* and the function S is called the *structure factor*.

We see that S is the product of the type of function treated in the previous section. The factor $\sin^2 N\pi\,dv/\sin^2 \pi\,dv$ has an intensity distribution consisting of horizontal strips lying along the lines $v = 0$, $\pm 1/d$, $\pm 2/d, \dots$, and having intensity N^2. Hence, noting the results of the last section (especially Figure 7.18), we conclude that the structure factor S has an intensity distribution in the u-v plane consisting of a *rectangular array of dots* of intensity M^2N^2 centered at the intersections of vertical lines at $u = 0$, $\pm 1/c$, $\pm 2/c, \dots$ and horizontal lines at $v = 0$, $\pm 1/d$, $\pm 2/d, \dots$. These dots have $u - v$ dimensions no larger than $(2/Mc) \times (2/Nd)$. Between these dots are spaces of essentially zero intensity (if we ignore secondary maxima in S).

Thus the intensity distribution I in (9.1) is obtained by overlaying the intensity distribution for a single aperture by an array of dot amplifiers of internsity M^2N^2 located at the points $(m/c, n/d)$ for integers m and n. The larger the values of c and d the closer the dot amplifiers. The array of amplifiers is called the *reciprocal lattice* to the array of apertures.

(9.2) **Example.** Suppose that a square array of 11×11 circular apertures each having radius 0.5 mm is formed by setting $c = d = 3$ mm. The amplification of each dot amplifier will be 11^4. These dots will be positioned on a square array with $\Delta u = \Delta v = 0.333 \dots$. An Airy pattern for one circular

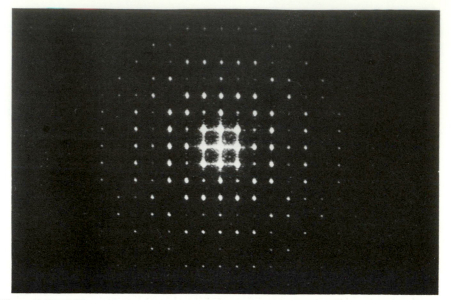

Figure 7.24 Diffraction pattern of 11×11 circular apertures. We have a grainy Airy pattern. The fine structure between the 9 dots nearest the center is due to secondary maxima in the structure factor. (Photograph courtesy of S. G. Lipson.)

aperture is overlayed by this square array of dot amplifiers. (See Figure 7.24.)

One important application of the array theorem is in radio astronomy. A large dish aerial will bring a radio wave to focus on its receiving device; thus, such an aperture acts like a *diffracting circular aperture*. By setting up an array of such dish aerials, the amplification of the array theorem can be brought into play. According to S. Lipson [see p. 384 of Lipson (1972)], "the sensitivity of such an array is almost as good as a filled aerial of the same dimensions." In other words, *by building a large array of dish aerials we can achieve the equivalent of a gigantic dish aerial covering the area of the array* (and the resolution improves when the dishes in the array are farther apart). [See Lipson and Lipson (1981), §11.6.]

The array theorem has also been applied in the study of protein molecules. The Fourier transform of an electron photomicrograph of a protein might contain high intensity dots lying along a (reciprocal) lattice, a sure sign that there is an *underlying periodic structure in the original micrograph* (that might not be evident due to random interference, called *noise*). For an excellent nontechnical discussion, see Unwin and Henderson (1984). See also Lipson (1972), pp. 401–413.

Exercises

(9.3) Describe the diffraction pattern of an array of 15×15 square apertures of dimensions $0.1 \text{ mm} \times 0.1 \text{ mm}$ formed by letting $c = d = 0.2 \text{ mm}$.

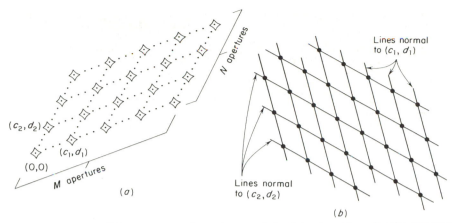

Figure 7.25 A parallelogram array of apertures and the intensity pattern for the structure factor. (*a*) Parallelogram array. (*b*) Reciprocal lattice.

(9.4) Show that the fine structure about the 9 central dots in Figure 7.24 consists of tiny diffraction patterns resembling those for a square aperture of $x - y$ dimensions 33×33 mm^2. [*Hint*: Compare the intensity patterns for $S(u, v)$ and the function $I(u, v)$ in Figure 7.6.] *Can you show the same result by applying the convolution theorem? *Note*: Beautiful pictures illustrating these ideas can be found in plate 13 of Harburn, Taylor, and Welberry (1975).

★(9.5) *Parallelogram Array.* Suppose that $M \cdot N$ identical apertures are placed in a parallelogram array as shown in Figure 7.25(*a*). Show that the intensity distribution I for the diffraction pattern is equal to $I_1 S$ where I_1 is the intensity distribution for a single aperture and S is a structure factor. Show that S consists of an array of dot amplifiers, of intensity $M^2 N^2$, situated at points of intersection of lines in a parallelogram configuration, where those lines are perpendicular to those of the array of apertures. [See Figure 7.25(*b*).] Describe the total diffraction pattern resulting from $I = I_1 S$. [*Hint*: Proceed as in the rectangular array case, but at an appropriate point substitute $\theta = c_1 u + d_1 v$ and $\phi = c_2 u + d_2 v$.]

Remark. The array of dots in Figure 7.25(*b*) is called the *reciprocal lattice* for the array of apertures.

(9.6) Same problem as (9.3), but the square apertures are situated on a parallelogram array formed by $\mathbf{v}_1 = (0.4 \text{ mm}, 0.4 \text{ mm})$ and $\mathbf{v}_2 = (0.1 \text{ mm}, 0.2 \text{ mm})$.

§10. Imaging Theory

In this section we shall briefly describe the theory of lens imaging via Fourier analysis. This theory is due to Abbe and Zernike.

In §4 we showed that the amplitude Ψ at each point $\mathcal{Q}$ in the focal plane of a lens is given by [see (4.6)]

(10.1) $$\Psi(u, v) = e^{i(2\pi/\lambda)|\mathcal{O}_1\mathcal{Q}|}\hat{A}(u, v)$$

provided the object (aperture) is illuminated by monochromatic coherent light (i.e., a plane parallel wave with wavelength λ). This type of illumination is very common in science, due to the invention of the laser; it is also approximately fulfilled in microscopy (the illumination of the slide of a microscope is approximately coherent).

To see how an image is formed by a lens we must allow the rays, that have converged at the focal plane in Figure 7.1, to diverge and proceed to a further plane as shown in Figure 7.26. If we reason as in §4, then we can calculate the amplitude $\varphi(x, y)$ at a point $\mathcal{R}_2$ with coordinates $(x, y, d + D)$ in the image plane. We find that

(10.2) $$\varphi(x, y) = \int\!\!\!\int_{-\infty}^{+\infty} \Psi(u, v)e^{i(2\pi/\lambda)|\mathcal{Q}\mathcal{R}_2|} \, du \, dv$$

Hence we must determine the (optical) distance $|\mathcal{Q}\mathcal{R}_2|$ of each ray $\mathcal{Q}\mathcal{R}_2$. By the law of cosines (see Figure 7.27)

(10.3)
$$|\mathcal{Q}\mathcal{R}_2| = [|\mathcal{Q}\mathcal{O}_2|^2 + (x^2 + y^2) - 2(x^2 + y^2)^{1/2}|\mathcal{Q}\mathcal{O}_2| \cos \tilde{\gamma}]^{1/2}$$
$$\approx |\mathcal{Q}\mathcal{O}_2| - (x^2 + y^2)^{1/2} \cos \tilde{\gamma}$$

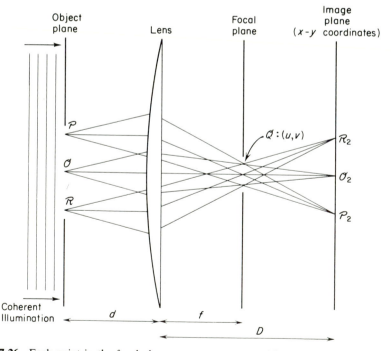

Figure 7.26 Each point in the focal plane emanates a wave, with amplitude Ψ. These waves converge at points in the image plane, a second Fourier transform operation.

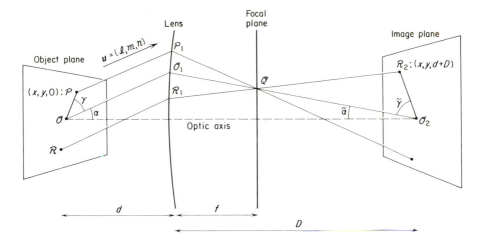

Figure 7.27   Ray diagram for the image–object derivation.

To justify the approximation in (10.3) we must assume that $(x^2 + y^2)^{1/2} \ll |2O_2|$ which will be the case if the image is assumed to be a small one.

Letting $\bar{\alpha}$ be the complementary angle (to $\bar{\gamma}$) that $2O_2$ makes with the z axis (optic axis) we can rewrite (10.3) as

(10.4) $$|2R_2| = |2O_2| - (x^2 + y^2)^{1/2} \sin \bar{\alpha}$$

We now invoke the *Abbe sine condition* which states the angles α and $\bar{\alpha}$ should satisfy

(10.5) $$\sin \alpha = M \sin \bar{\alpha} \qquad \text{(Abbe sine condition)}$$

where M is a positive constant. Using (10.5), formula (10.4) becomes

(10.6)
$$|2R_2| = |2O_2| - \frac{(x^2 + y^2)^{1/2} \sin \alpha}{M}$$
$$= |2O_2| - \frac{(x^2 + y^2)^{1/2}(\ell^2 + m^2 + n^2)^{1/2} \cos \gamma}{M}$$

Here γ is the complementary angle (to α) that OO_1 makes with OP, and we have replaced 1 by the length of the unit vector $\mathbf{u} = (\ell, m, n)$. Since OO_1 is parallel to $\mathbf{u}$ we obtain from (10.6), using a well known formula from vector calculus,

$$|2R_2| = |2O_2| - \frac{(x, y, 0) \cdot (\ell, m, n)}{M} = |2O_2| - (\ell x + m y)/M$$

This last result allows us to rewrite (10.2), using (10.1) and (4.5),

$$\varphi(x, y) = \int\!\!\int_{-\infty}^{+\infty} \Psi(u, v)e^{i(2\pi/\lambda)|\mathcal{Q}\mathcal{O}_2|}e^{-i(2\pi/\lambda)[\ell x+my]/M}\, du\, dv$$

(10.7)

$$= \int\!\!\int_{-\infty}^{+\infty} e^{i(2\pi/\lambda)[|\mathcal{O}\mathcal{O}_1\mathcal{Q}|+|\mathcal{Q}\mathcal{O}_2|]}\hat{A}(u, v)e^{-i2\pi[u(x/M)+v(y/M)]}\, du\, dv$$

The phase factor $e^{i(2\pi/\lambda)[|\mathcal{O}\mathcal{O}_1\mathcal{Q}|+|\mathcal{Q}\mathcal{O}_2|]}$ is a *constant, provided* the object plane and the image plane are *conjugate planes.*[6] In that case, as depicted in the classic imaging diagram shown in Figure 7.26, all the rays from $\mathcal{O}$, passing through various points $\mathcal{Q}$ in the focal plane, converge on a single point $\mathcal{O}_2$. Therefore, invoking Hamilton's form of Fermat's principle as we did in §4, we conclude that the wave fronts along those rays converge on $\mathcal{O}_2$ *in phase*. In particular, $e^{i(2\pi/\lambda)[|\mathcal{O}\mathcal{O}_1\mathcal{Q}|+|\mathcal{Q}\mathcal{O}_2|]}$ is a *constant* for all points $\mathcal{Q}$, equal to $e^{i(2\pi/\lambda)(d+D)}$ (taking $\mathcal{Q}$ on the optic axis). Thus, denoting this constant by e^{ic} for convenience, Eq. (10.7) becomes by Fourier inversion

$$\varphi(x, y) = e^{ic}\int\!\!\int_{-\infty}^{+\infty} \hat{A}(u, v)e^{-i2\pi[u(x/M)+v(y/M)]}\, du\, dv$$

(10.8)

$$= e^{ic}A\left(-\frac{x}{M}, -\frac{y}{M}\right)$$

Equation (10.8) expresses the well known fact that the image is an inverted copy of the object (represented by its amplitude function A) *magnified by the factor $M > 0$.* (If $0 < M < 1$ then we might say *reduced* by the factor M.) If we observe φ by photographic film or by a viewing screen, then we record $|\varphi|^2 = |A(-x/M, -y/M)|^2$.

The key element of the discussion above was the Abbe sine condition (10.5), which is *much less restrictive* than the *small angle condition* $\alpha = M\bar{\alpha}$ of Gaussian optics (paraxial approximation).[7] Note also that our derivation reveals that imaging results from two diffraction (Fourier transform) processes. The second diffraction was treated by endowing the focal plane with *spatial frequency* coordinates (u and v) rather than spatial coordinates.

Not all optical systems obey the Abbe sine condition; high quality microscopes do but ordinary lenses do not. It is shown in optics texts [e.g., Goodman (1968) or Lipson and Lipson (1981)] that the upper limit to the angular field of view is provided by Abbe's sine condition; the lower limit is not really zero but rather the small angle condition mentioned above.

Until now we have pretended that the lens captures all the light from the object plane. We will now consider what happens when the lens is of finite size. As shown in Figure 7.4, we must now modify the transform $\hat{A}$ by allowing for missing spatial frequencies. We replace $\hat{A}$ in (10.8) by $\hat{A}P$ where P is a function for which $P(u, v) = 0$ when u or v is large enough.

---

6. In geometrical optics, this occurs when the *lens equation* $(1/d) + (1/D) = 1/f$ holds.

7. For this case, M is found to be D/d.

Then, by the convolution theorem

(10.9)
$$\varphi(x, y) = e^{ic} \int\!\!\int_{-\infty}^{+\infty} \hat{A}(u, v)P(u, v)e^{-i2\pi[u(x/M)+v(y/M)]} \, du \, dv$$

$$= e^{ic}(A * \hat{P})\left(-\frac{x}{M}, -\frac{y}{M}\right)$$

Formula (10.9) says that the image is the magnified and inverted image of the convolution $A * \hat{P}$. The function P is called a *pupil function*. A typical example is

(10.10)
$$P(u, v) = \begin{cases} 1 & \text{if } (u^2 + v^2)^{1/2} < R \\ 0 & \text{if } (u^2 + v^2)^{1/2} > R \end{cases}$$

which corresponds to a spherical lens (well corrected for aberrations). In Figure 7.28 we show how this pupil function and its transform are related to the imaging of an object. In this example, the sharp cutoff (discontinuity) of P in the higher spatial frequencies results in the image $A * \hat{P}$ exhibiting a Gibb's phenomenon called *ringing*.

In many situations, such as natural light illumination, the object screen is not illuminated coherently. We do not have space to discuss the case of incoherent illumination. We shall only briefly describe the principal results. In this case the different phases in the illuminating wave combine, when

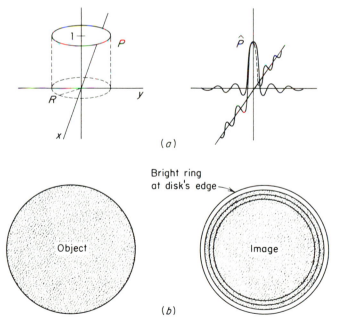

Figure 7.28 Ringing under coherent illumination. The bright ring on the image disk is a type of Gibb's phenomenon (for transforms). (*a*) A pupil function P and its transform $\hat{P}$. (*b*) Imaging of a disk. [See also Figures 6-12 and 6-13 in Goodman (1968).]

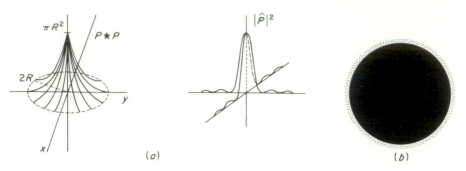

Figure 7.29 Loss of ringing, but blurring of image, under incoherent illumination. (*a*) Graphs of $P \star P$ and $|\hat{P}|^2$. Same P as in Figure 7.28. (*b*) Image of a black disk. [See also Figures 6-12 and 6-13 in Goodman (1968)].

focused onto the image plane, to yield only an *average* (over time) *intensity*. Formula (10.9) is replaced by

(10.11)
$$|\varphi(x, y)|^2 = |A|^2 * |\hat{P}|^2 \left(-\frac{x}{M}, -\frac{y}{M} \right)$$

Note that $|\hat{P}|^2$ is the transform of $P \star P$, the autocorrelation of the pupil function P. The details of deriving (10.11) are given in Iizuka (1985), §10.3. In Figure 7.29 we show that the ringing observed in coherent illumination is decreased to a blurring under incoherent illumination.

For further discussion of imaging, see Chapter 7 of Goodman (1968), Chapter 11 of Iizuka (1985), and Chapter 9 of Lipson and Lipson (1981).

Exercises

(10.12) Find $\hat{P}$ when $P(u, v) = \Pi(u/a)\Pi(v/b)$. Is there ringing, as shown in Figure 7.28, for this pupil function?

(10.13) Justify the graphs made of the pupil function P, its autocorrelation $P \star P$, and their transforms, shown in Figures 7.28 and 7.29.

(10.14) For what objects would rings (haloes, fringes) appear in the images, even under incoherent illumination, using the lens system described in Figure 7.29.

(10.15) Justify, using Fourier analysis, the following well known principle of optics: *The larger the lens opening (aperture) the better the image.*

(10.16) Show that (10.9) can be expressed in the form (ignoring the constant e^{ic})

$$(*) \quad \varphi(x, y) = h * \varphi_g(x, y) = \int\int_{-\infty}^{+\infty} h(x - r, y - s)\varphi_g(r, s)\, dr\, ds$$

where

$$h(r, s) = \frac{1}{M} \hat{P}\left(-\frac{r}{M}, -\frac{s}{M}\right) \qquad \varphi_g(r, s) = A\left(-\frac{r}{M}, -\frac{s}{M}\right)$$

(φ_g stands for the image predicted by geometrical optics).

(10.17) Suppose that a sequence of m lenses is arranged along a single optic axis in such a way that the image plane of each lens coincides with the object plane of the next lens in the sequence. Justify the following optical principle: *The image quality from a system of lenses is limited by the smallest lens opening (aperture) in the system.*

(10.18) Show that, using coherent illumination and a circular lens, the smallest resolvable spatial detail in an object is approximately $1.22(\lambda d/a)$, where λ is the wavelength of the illumination, a is the radius of the circular lens aperture, and d is the distance from the object to the lens. [*Hint:* Use (4.8), (10.9), (10.10), and $M = D/d$.]

Part C. Signal Processing

§11. A Brief Introduction to Sampling Theory

The basic idea of sampling theory is the reconstruction of a function from an isolated collection, usually a periodic array, of data points. Sampling theory has major applications in several areas such as communications (telephone signal and optical processing), radio astronomy, and sound reproduction (digital recording).

Let's begin with a typical one-dimensional sampling theorem. A function g defined over $\mathbb{R}$ is said to be *limited* if for some positive constant c we have $g(x) = 0$ when $|x| \geq c$. The following theorem has been proved by several people working in different areas. It is called the sampling theorem by Nyquist, and by Shannon, and it is called the cardinal theorem of interpolation theory by Whittaker.

(11.1) Theorem. Suppose that f is continuous and absolutely integrable over $\mathbb{R}$ and $\hat{f}$ is limited. Then

$$f(x) = \sum_{n=-\infty}^{+\infty} f\left(\frac{n}{2c}\right) \text{sinc}(2cx - n)$$

for all x values, provided $\hat{f}(u) = 0$ for $|u| \geq c$.

Proof. Because $\hat{f}(u) = 0$ for $|u| \geq c$, Fourier inversion yields

(11.2)
$$f(x) = \int_{-c}^{c} \hat{f}(u) e^{i2\pi ux} \, du$$

In particular, for each integer n

(11.2′)
$$f\left(\frac{n}{2c}\right) = \int_{-c}^{c} \hat{f}(u)e^{i(n\pi/c)u}\,du$$

This last integral is just a constant multiple of the $-n$th Fourier coefficient in a complex Fourier series for $\hat{f}$. Computing such a Fourier series, we obtain

(11.3)
$$\hat{f} \sim \sum_{n=-\infty}^{+\infty} c_n e^{i(n\pi/c)u} \qquad c_n = \left(\frac{1}{2c}\right)\int_{-c}^{c} \hat{f}(u)e^{-i(n\pi/c)u}\,du$$

$$= \left(\frac{1}{2c}\right)f\left(\frac{-n}{2c}\right).$$

Performing Fourier inversion on the left side of the correspondence in (11.3), and applying the same integration term by term to the right side, we get

(11.4)
$$\int_{-c}^{c} \hat{f}(u)e^{i2\pi ux}\,du = \sum_{n=-\infty}^{+\infty} c_n \int_{-c}^{c} e^{i(n\pi/c)u}e^{i2\pi ux}\,du$$

The fact that we obtain equality in (11.4), *even though* (11.3) *may not be an equality, is one of the remarkable aspects of Fourier analysis.* We will prove that (11.4) is valid in a moment; first we note that by a straightforward integration

$$\int_{-c}^{c} e^{i(n\pi/c)u}e^{i2\pi ux}\,du = \frac{2c\,\sin[2\pi cx + n\pi]}{2\pi cx + n\pi} = 2c\,\mathrm{sinc}(2cx + n)$$

Since $c_n = (1/2c)f(-n/2c)$ we obtain, from combining this last equality with (11.4) and (11.2),

$$f(x) = \sum_{n=-\infty}^{+\infty} f\left(\frac{-n}{2c}\right)\mathrm{sinc}(2cx + n)$$

Substituting $-n$ for n in the series above yields the desired series expansion for f.

We will now prove (11.4). For each positive integer N we have

$$\left| \int_{-c}^{c} \hat{f}(u)e^{i2\pi ux}\,du - \sum_{n=-N}^{+N} c_n \int_{-c}^{c} e^{i(n\pi/c)u}e^{i2\pi ux}\,du \right|$$

$$= \left| \int_{-c}^{c} \left[\hat{f}(u) - \sum_{n=-N}^{+N} c_n e^{i(n\pi/c)u} \right]e^{i2\pi ux}\,du \right|$$

$$\leq \int_{-c}^{c} \left| \hat{f}(u) - \sum_{n=-N}^{+N} c_n e^{i(n\pi/c)u} \right|\,du$$

Applying Schwarz's inequality to the last integral above, we get

$$\left| \int_{-c}^{c} \hat{f}(u) e^{i2\pi ux}\, du - \sum_{n=-N}^{+N} c_n \int_{-c}^{c} e^{i(n\pi/c)u} e^{i2\pi ux}\, du \right|$$

$$\leq \left[\int_{-c}^{c} \left| \hat{f}(u) - \sum_{n=-N}^{+N} c_n e^{i(n\pi/c)u} \right|^2 du \right]^{1/2} \left[\int_{-c}^{c} 1^2\, du \right]^{1/2}$$

$$= (2c)^{1/2} \left[\int_{-c}^{c} \left| \hat{f}(u) - \sum_{n=-N}^{+N} c_n e^{i(n\pi/c)u} \right|^2 du \right]^{1/2}$$

The completeness of $\{e^{i(n\pi/c)u}\}$ over $[-c, c]$ implies that the last quantity above tends to 0 as N tends to $+\infty$ [Theorem (7.21), Chapter 2]. Thus

(11.4′) $$\lim_{N\to+\infty} \sum_{n=-N}^{+N} c_n \int_{-c}^{c} e^{i(n\pi/c)u} e^{i2\pi ux}\, du = \int_{-c}^{c} \hat{f}(u) e^{i2\pi ux}\, du$$

and that verifies (11.4). ∎

(11.5) Remark.   The series in Theorem (11.1) is called an *interpolation* (or *sampling*) *series* for f.

By orthogonality of the system $\{e^{i(n\pi/c)u}\}$ we have [see Exercise (11.11)]

(11.6)
$$\int_{-c}^{c} \left| \hat{f}(u) - \sum_{n=-N}^{+N} c_n e^{i(n\pi/c)u} \right|^2 du = \sum_{|n|>N} |c_n|^2 \, \|e^{i(n\pi/c)u}\|_2^2$$

$$= 2c \sum_{|n|>N} |c_n|^2$$

Using $c_n = (1/2c)f(-n/2c)$ along with (11.2) and (11.6), we obtain from the last inequality in the proof of Theorem (11.1)

(11.7) $$\left| f(x) - \sum_{n=-N}^{+N} f\left(\frac{n}{2c}\right) \text{sinc}(2cx - n) \right| \leq \left[\sum_{|n|>N} \left| f\left(\frac{n}{2c}\right) \right|^2 \right]^{1/2}$$

Formula (11.7) is of some use in estimating the error in taking a partial sum as an approximation for the infinite interpolation series of f. Moreover, since the right side of (11.7) is independent of all x values, it follows that *the interpolation series for f converges uniformly to $f(x)$ for all x values.*

Consider the positive constant c referred to in Theorem (11.1). Suppose $c' > c$. Then $\hat{f}(u) = 0$ for all $|u| \geq c'$ and we can write

$$f(x) = \sum_{n=-\infty}^{+\infty} f\left(\frac{n}{2c'}\right) \text{sinc}(2c'x - n)$$

But, the series above is at a disadvantage in comparison with the one in (11.1) since the function f needs to be evaluated at more *sample points* ($2c'$ vs. $2c$) per unit length. The least number of sample points per unit length, $2C$, is obtained when C is the minimum positive constant such that $\hat{f}(u) = 0$ for $|u| \geq C$. That minimum number of sample points per unit length is called the *Nyquist sampling rate*.

The sampling theorem (11.1) has many applications. For example, it allows for the reconstruction of a complete telephone signal when only periodic portions (pulses) of the signal are transmitted. In the lapses between the pulses of one signal, more signals can be sent. Using fiber optics up to 8000 messages have been transmitted along a single cable. Furthermore, the pulsing of a signal permits the use of *error correcting codes* (and noise reduction). The modern science of information theory can be briefly defined as the study of efficient means of replacing continuous (analog) signals by discrete (pulsed) signals. See Raisbeck (1963), Hamming (1980), and also p. 397 of Lipson and Lipson (1981). A major recent achievement is the digital reproduction of sound (compact disc players). See Monforte (1984) and Mathews and Pierce (1987).

In all applications it is sometimes necessary to use an interpolation series for a function f even when $\hat{f}$ is not limited. Or, for either practical reasons or because of ignorance of $\hat{f}$, a sampling rate of less than the Nyquist rate is used. In such instances, the interpolation series (also called the sampling series) will only approximate f. The error involved is called the *aliasing error*. This whole procedure is called *aliasing* in the jargon of sampling theory.

Theorem (11.1) is a one-dimensional sampling theorem. The following sampling theorem is two-dimensional; its proof is basically the same as the proof of (11.1) so we leave the details to the reader.

(11.8) Theorem.   Suppose that f is a continuous function that is absolutely integrable over $\mathbb{R}^2$. If $\hat{f}$ is limited, then

$$f(x, y) = \sum_{m,n=-\infty}^{+\infty} f\left(\frac{m}{2c}, \frac{n}{2d}\right) \operatorname{sinc}(2cx - m) \operatorname{sinc}(2dy - n)$$

provided that $\hat{f}(u, v) = 0$ for $|u| \geq c$ *or* $|v| \geq d$.

Remark.   The final statement in (11.8) simply means that $\hat{f}(u, v) = 0$ for all points (u, v) which lie outside the rectangle $(-c, c) \times (-d, d)$.

Theorem (11.8) has many applications. In optics, parts of an image, or picture, may be used to reconstruct the total image. Only small portions (pixals) of a picture need be transmitted in order for the complete image to be reconstructed [see Goodman (1968), Allan (1973), and Kapany (1967)]. In radio astronomy, antennas can be positioned in arrays and the signals received from these arrays can then be used to reconstruct the larger radiograph [see Hjellming and Bignell (1982)].

Exercises

(11.9)   Verify (11.6), (11.7), and prove Theorem (11.8).

(11.10)   Prove that if f is piecewise continuous and limited as a function on $\mathbb{R}$, then $\hat{f}(u) = \sum_{n=-\infty}^{+\infty} \hat{f}(n/2c) \operatorname{sinc}(2cu - n)$ provided $f(x) = 0$ for $|x| \geq c$.

*(11.11) Show that if f satisfies the hypotheses of Theorem (11.1), then

$$\int_{-c}^{c} \left| \hat{f}(u) - \sum_{n=-M}^{+N} c_n e^{i(n\pi/c)u} \right|^2 du = 2c \left[\sum_{n>N} |c_n|^2 + \sum_{n<-M} |c_n|^2 \right]$$

then show that

$$\lim_{M, N \to +\infty} \sum_{n=-M}^{N} f\left(\frac{n}{2c}\right) \operatorname{sinc}(2cx - n) = f(x)$$

uniformly over $\mathbb{R}$.

(11.12) Show that for a finite aperture with aperture function A (see §4), the autocorrelation $A \star A$ is a limited function on $\mathbb{R}^2$. Describe how the array theorem (§9) could be used to reconstruct the diffraction pattern of a single aperture (too weak to observe directly) from a periodic array of such apertures.

(11.13) Using the result of (11.12), which of the separation constants $d = 0.8, 0.4, 0.2$ mm for forming a square array is sufficient for the validity of the equation

$$I_1(x, y) = \sum_{m,n=-\infty}^{+\infty} I_1\left(\frac{m}{d}, \frac{n}{d}\right) \operatorname{sinc}(2dx - m) \operatorname{sinc}(2dy - n)$$

where I_1 is the intensity function for the diffraction pattern from one square aperture of side length 0.1 mm? What if the array consists of equilateral triangles of side length 0.1 mm? For the case of the squares, draw diagrams of the array diffraction patterns when $d = 0.4$ and 0.2 mm.

(11.14) Suppose that $f(x) = \sin^2(0.1x)/(\pi x)^2$. Find the Nyquist sampling rate, C, for f. Find a value of N so that $\sum_{n=-N}^{N} f(n/2C) \operatorname{sinc}(2Cx - n)$ approximates $f(x)$ to within $\pm 10^{-6}$ for all x values.

(11.15) Let f be continuous and $\hat{f}(x) = 0$ for $|x| > \frac{1}{2}$. Prove that $\operatorname{sinc} * f = f$.

*(11.16) *Parallelogram Sampling.* Find as general conditions as you can that ensure that a function f can be reconstructed by an interpolation (sampling) series where the data points lie on a parallelogram lattice [see (9.5)] formed by the points $(mc_1 + nc_2, md_1 + nd_2)$. [*Hint:* Change coordinates.][8]

§12. The Poisson Summation Formula

The theories of Fourier series and Fourier transforms are closely related. For example, Fourier series and transforms were used together to prove the sampling theorem(s) in the previous section. Our next theorem connects these two operations in a very explicit way.

---

8. For the solution, see Marks (1986).

(12.1) Theorem: Poisson Summation. Suppose that f is a continuous and absolutely integrable function over $\mathbb{R}$. If $\sum_{n=-\infty}^{+\infty} f(x-n)$ converges uniformly for $|x| \leq \frac{1}{2}$ and $\sum_{n=-\infty}^{+\infty} |\hat{f}(n)|$ converges, then

$$\sum_{n=-\infty}^{+\infty} f(x-n) = \sum_{n=-\infty}^{+\infty} \hat{f}(n) e^{i2\pi nx}$$

for all x values.

Proof. Let $g(x) = \sum_{n=-\infty}^{+\infty} f(x-n)$. The function g is continuous and has period 1. Since g is periodic we expand it in a Fourier series $g \sim \sum c_k e^{i2\pi kx}$ where

$$c_k = \int_{-1/2}^{+1/2} \sum_{n=-\infty}^{+\infty} f(x-n) e^{-i2\pi kx} \, dx$$

$$= \sum_{n=-\infty}^{+\infty} \int_{-1/2}^{+1/2} f(x-n) e^{-i2\pi kx} \, dx$$

The interchange of summation and integration is allowed by the uniform convergence of $\sum f(x-n)$. Substituting $s = x - n$ and using periodicity, we obtain

$$c_k = \sum_{n=-\infty}^{+\infty} \int_{-1/2-n}^{+1/2-n} f(s) e^{-i2\pi ks} \, ds = \int_{-\infty}^{+\infty} f(s) e^{-i2\pi ks} \, ds = \hat{f}(k)$$

Since $\sum |\hat{f}(n)|$ converges, it follows by the M-test that $\sum \hat{f}(n) e^{i2\pi nx}$ converges uniformly. The completeness of $\{e^{i2\pi nx}\}$ implies that this last series converges to g (since it has the same Fourier coefficients). Thus, for all x values

$$\sum_{n=-\infty}^{+\infty} \hat{f}(n) e^{i2\pi nx} = g(x) = \sum_{n=-\infty}^{+\infty} f(x-n) \qquad \blacksquare$$

The following corollary is also sometimes called the Poisson summation formula.

(12.2) Corollary. Under the same hypotheses as (12.1) we have

$$\sum_{n=-\infty}^{+\infty} \hat{f}(n) = \sum_{n=-\infty}^{+\infty} f(n)$$

Proof. Substitute $x = 0$ into the Poisson summation formula in (12.1) and then replace $-n$ by n in $\sum f(-n)$. $\blacksquare$

The following theorem shows that the Poisson summation formula can be used for linking Fourier series of all periods with the Fourier transform. Its proof involves only obvious modifications of the proof of (12.1) so we leave the details to the reader.

(12.3) Theorem: Poisson Summation. Suppose that f is continuous and absolutely integrable over $\mathbb{R}$ and p is a positive constant. If $\sum_{n=-\infty}^{+\infty} f(x-np)$

converges uniformly for $|x| \le \frac{1}{2}p$ and $\sum_{n=-\infty}^{+\infty} |\hat{f}(n/p)|$ converges, then

$$\sum_{n=-\infty}^{+\infty} f(x - np) = \sum_{n=-\infty}^{+\infty} \frac{1}{p} \hat{f}\left(\frac{n}{p}\right) e^{i(2n\pi/p)x}$$

for all x values. In particular, $\sum_{n=-\infty}^{+\infty} f(np) = (1/p)\sum_{n=-\infty}^{+\infty} \hat{f}(n/p)$.

Here are some applications of Poisson summation.

(12.4) Example.   Let

$$f(x) = (2\pi t)^{-1/2} e^{-x^2/2t} \quad \text{for} \quad t > 0$$

Using (3.2b) from Chapter 6, for $c = 1/2t$, we have $\hat{f}(u) = e^{-2\pi^2 u^2 t}$. Poisson summation yields

(a) $$(2\pi t)^{-1/2} \sum_{n=-\infty}^{+\infty} e^{-(x-n)^2/2t} = \sum_{n=-\infty}^{+\infty} e^{-2\pi^2 n^2 t} e^{i2\pi n x} \qquad (t > 0)$$

which is Jacobi's Identity [see (7.4″) and (7.5), Chapter 4]. For $x = 0$, we obtain a more specific identity which is often called Jacobi's identity

(b) $$(2\pi t)^{-1/2} \sum_{n=-\infty}^{+\infty} e^{-n^2/2t} = \sum_{n=-\infty}^{+\infty} e^{-2\pi^2 n^2 t} \qquad (t > 0)$$

(12.5) Example.   Let $f(x) = (1/\pi)[y/(y^2 + x^2)]$ for $y > 0$. Then $\hat{f}(u) = e^{-2\pi y |u|}$. Hence, by Poisson summation

(a) $$\frac{1}{\pi} \sum_{n=-\infty}^{+\infty} \frac{y}{y^2 + (x - 2n\pi)^2} = \frac{1}{2\pi} \sum_{n=-\infty}^{+\infty} e^{-y|n|} e^{inx} \qquad (y > 0)$$

The right side of (a) can be expressed in closed form by treating it as a sum of two geometric series, that is

$$\sum_{n=-\infty}^{+\infty} e^{-y|n|} e^{inx} = \sum_{n=0}^{\infty} [e^{-y}e^{ix}]^n + \sum_{n=1}^{\infty} [e^{-y}e^{-ix}]^n$$

$$= (1 - e^{-y}e^{ix})^{-1} + e^{-y}e^{-ix}(1 - e^{-y}e^{-ix})^{-1}$$

$$= \frac{1 - e^{-2y}}{1 - 2e^{-y}\cos x + e^{-2y}}.$$

Thus, (a) becomes

(a′) $$\frac{1}{\pi} \sum_{n=-\infty}^{+\infty} \frac{y}{y^2 + (x - 2n\pi)^2} = \frac{1}{2\pi} \frac{1 - e^{-2y}}{1 - 2e^{-y}\cos x + e^{-2y}} \qquad (y > 0)$$

(12.6) Remark.   If we let r equal e^{-y} for $y > 0$ and replace x by α, then the right side of (a′) equals $1/2\pi$ times $P_r(\alpha)$ the Poisson kernel for the *unit disk* $B_1(0, 0)$ in $\mathbb{R}^2$ [see (2.17), Chapter 4]. The last example has thus established an analogy between the Poisson kernels for the unit disk and the upper half plane:

(a) $\dfrac{1}{2\pi}\dfrac{1-r^2}{1+r^2-2r\cos\alpha}$ $\leftrightarrow$ $\dfrac{1}{\pi}\dfrac{y}{y^2+x^2}$

    (unit disk)   (upper half plane)

Poisson summation generalizes to multiple Fourier series and transforms. For instance, the following theorem is proved in essentially the same way as (12.3).

(12.7) Theorem. Suppose that f is continuous and absolutely integrable over $\mathbb{R}^2$. If $\sum_{m,n=-\infty}^{+\infty} f(x-mp, y-nq)$ converges uniformly for $|x|\le\frac{1}{2}p$ and $|y|\le\frac{1}{2}q$ and $\sum_{m,n=-\infty}^{+\infty} |\hat{f}[(m/p),(n/q)]|$ converges, then

$$\sum_{m,n=-\infty}^{+\infty} f(x-mp, y-nq) = \sum_{m,\,n=-\infty}^{+\infty} \frac{1}{pq}\hat{f}\left(\frac{m}{p},\frac{n}{q}\right)e^{i2\pi(mx/p+ny/q)}$$

for all points (x, y) in $\mathbb{R}^2$. In particular,

$$\sum_{m,\,n=-\infty}^{+\infty} f(mp, nq) = \frac{1}{pq}\sum_{m,\,n=-\infty}^{+\infty} \hat{f}\left(\frac{m}{p},\frac{n}{q}\right)$$

(12.8) Example. Let

$$f(x, y) = \frac{1}{2\pi}\frac{z}{(z^2+x^2+y^2)^{3/2}}\quad\text{for}\quad z>0$$

Then $\hat{f}(u, v) = e^{-2\pi z(u^2+v^2)^{1/2}}$. Hence by Poisson summation

(a) $\dfrac{1}{2\pi}\displaystyle\sum_{m,\,n=-\infty}^{+\infty}\dfrac{z}{[z^2+(x-2m\pi)^2+(y-2n\pi)^2]^{3/2}}$

$$=\frac{1}{4\pi^2}\sum_{m,\,n=-\infty}^{+\infty} e^{-z(m^2+n^2)^{1/2}}e^{i(mx+ny)}$$

See Exercise (12.14) for an application of (a).

Exercises

(12.9) Prove Theorems (12.3) and (12.7).

(12.10) Find the sums of the following series

(a) $\displaystyle\sum_{n=1}^{\infty}\frac{1}{1+n^2}$

(b) $\displaystyle\sum_{n=1}^{\infty}\frac{1}{n^2-2n+5}$

(12.11) Assuming the value of π is known exactly, find the value of $\sum_{n=-\infty}^{+\infty} e^{-2\pi^2 n^2(0.01)}$ to within 36 significant digits. [*Hint*: Use (12.4b).]

(12.12) Show that if f and $\hat{f}$ are continuous and $|f(x)|\le A[1+|x|]^{-1-\alpha}$,

$|\hat{f}(u)| \le B[1 + |u|]^{-1-\beta}$ where A, α, B, β are positive constants, then Poisson summation as in (12.3) can be performed. Generalize this result to two (three) dimensions.

(12.13) Describe how the array theorem (§9) is related to Poisson summation (when M, $N \gg 1$).

(12.14) (a) Using separation of variables, derive a series solution $\Psi = \sum c_{mn} e^{imx} e^{iny} e^{-(m^2+n^2)^{1/2}z}$ to the following problem

$$(*) \quad \begin{cases} \Psi_{xx} + \Psi_{yy} + \Psi_{zz} = 0 \quad (0 < x < \pi \quad 0 < y < \pi \quad 0 < z < +\infty) \\ \Psi(0, y, z) = \Psi(\pi, y, z) = \Psi(x, 0, z) = \Psi(x, \pi, z) = 0 \\ \Psi(x, y, 0+) = f(x, y) \quad \lim_{z \to +\infty} \Psi(x, y, z) = 0 \text{ [uniformly in } (x, y)] \end{cases}$$

(b) Prove that this series can be expressed in the form

$$\Psi(x, y, z) = \frac{1}{4\pi^2} \int_{-\pi}^{\pi} \int_{-\pi}^{\pi} F(x - u, y - v) E(u, v; z) \, du \, dv$$

for some function E, where F is a suitably defined extension of f.

(c) Using (12.8a) show that the function E in (b) is a two-dimensional summation kernel over $[-\pi, \pi] \times [-\pi, \pi]$. [*Hint*: Modify the proof of (7.7), Chapter 4.]

(d) Verify that the function Ψ in (b) rigorously solves (*).

*Summation Kernels Resulting from Poisson Summation

In the following five exercises, we outline how Poisson summation yields summation kernels (see Chapter 2, §8). Throughout these exercises assume that f is continuous and absolutely integrable over $\mathbb{R}$ and

$$\int_{-\infty}^{+\infty} f(x) \, dx = 1 \qquad |f(x)| \le A[1 + |x|]^{-1-\alpha} \qquad |\hat{f}(u)| \le B[1 + |u|]^{-1-\beta}$$

where A, α, B, β are positive constants [see (12.12)].

(12.15) For each positive number c, verify the Poisson summation formula

$$\sum_{k=-\infty}^{+\infty} \hat{f}(ck) e^{i2\pi ks} = \sum_{k=-\infty}^{+\infty} \frac{1}{c} f\left[\frac{1}{c}(s - m)\right]$$

holds for all s values.

(12.16) Let K be defined by $K(u; c) = \sum_{k=-\infty}^{+\infty} \hat{f}(ck) e^{i2\pi ku}$. Prove that K is a summation kernel, with period 1 in u, over $[-\frac{1}{2}, \frac{1}{2}]$ in the generalized sense of (8.23), Chapter 2. [*Hint*: Modify the proof of (7.7), Chapter 4.]

(12.17) Let K be the summation kernel defined in (12.16). If $g \sim \sum c_k e^{i2\pi kx}$ then prove that $K * g(x) = \sum_{k=-\infty}^{+\infty} c_k \hat{f}(ck) e^{i2\pi kx}$ holds for all x values, provided g is piecewise continuous and has period 1. [*Hint*: Use (8.15c), Chapter 2.]

(12.18)   Generalize the results of (12.15)–(12.17) to period $p = 2a$, and find the functions f and $\hat{f}$ which give rise to the Fejér kernel F_n and de la Vallée Poussin's kernel V_{2m} (for appropriate constants c).

(12.19)   Show that Hann's kernel H_n [see (8.19), Chapter 2] is a summation kernel over $[-\pi, \pi]$ in the generalized sense of (8.23), Chapter 2. Also, do (8.28c), Chapter 2, using (12.18) above.

§13. Computerized Fourier Analysis (Fast Fourier Transforms)

During the last two decades there has been a revolution in applications of Fourier analysis. This has happened mainly because of the computer. In this section we shall briefly describe how a computer can be used to calculate Fourier transforms and Fourier series. First, we describe the basic tool for those calculations, the *discrete Fourier transform*. Second, we shall examine the computer algorithm for the discrete Fourier transform, the *fast Fourier transform*.

Discrete Fourier Transforms

We begin by showing how the discrete Fourier transform is related to an approximation of an ordinary, continuous, Fourier transform. Suppose that f is a function on $\mathbb{R}$ and that f is limited to $[0, \Omega]$, that is, $f(x) = 0$ if $x < 0$ or $x > \Omega$. Such functions occur quite often in applications; for example, f might describe the intensity, or amplitude, of an electrical (acoustic, electromagnetic) signal that is only transmitted for a finite amount of time (then t would usually be written in place of x above).

Since f is limited to $[0, \Omega]$ its Fourier transform $\hat{f}$ is given by

(13.1)
$$\hat{f}(u) = \int_0^\Omega f(x)e^{-i2\pi ux}\, dx$$

By the sampling theorem described in Exercise (13.13), we can recover $\hat{f}$ from its samples $\{\hat{f}(k/\Omega)\}_{k=-\infty}^{+\infty}$. From (13.1) we have

(13.2)
$$\hat{f}\left(\frac{k}{\Omega}\right) = \int_0^\Omega f(x)e^{-i2\pi(k/\Omega)x}\, dx$$

The integral in (13.2) can be approximated by a Riemann sum

(13.3)
$$\int_0^\Omega f(x)e^{-i2\pi(k/\Omega)x}\, dx \doteq \sum_{j=0}^{N-1} f\left(\frac{j\Omega}{N}\right)e^{-i2\pi jk/N}\frac{\Omega}{N}$$

Moreover, if f is piecewise continuous (or Riemann integrable over $[0, \Omega]$), then the approximation in (13.3) can be made *uniformly* close for any *finite* collection of points $\{k/\Omega\}_{k=-M}^{+M}$ by taking N sufficiently large. Combining

(13.3) and (13.2) we have

(13.4)
$$\hat{f}\left(\frac{k}{\Omega}\right) \doteq \frac{\Omega}{N} \sum_{j=0}^{N-1} f\left(\frac{j\Omega}{N}\right) e^{-i2\pi jk/N}$$

Formula (13.4) shows the importance of the following definition.

(13.5) Definition. Let $\{f_j\}_{j=0}^{N-1}$ be a finite sequence of N complex numbers. The (*N-point*) *discrete Fourier transform* (DFT) of such a sequence is defined, for all integers k, by

$$F_k = \sum_{j=0}^{N-1} f_j e^{-i2\pi jk/N} \qquad \blacksquare$$

We see that the right side of (13.4) is Ω/N times the DFT of $\{f(j\Omega/N)\}_{j=0}^{N-1}$. Furthermore, *the sampled transform* $\{\hat{f}(k/\Omega)\}$ *is approximated* (at least for a finite number of samples) *by this constant multiple of a DFT.*

Besides Fourier transforms, we can also approximate Fourier series by using DFTs. The basis for this is the Poisson summation formula (12.3). Given a function f on $[0, \Omega]$, its Fourier series of period Ω is

(13.6)
$$f \sim \sum_{k=-\infty}^{+\infty} \frac{1}{\Omega} \hat{f}\left(\frac{k}{\Omega}\right) e^{i2\pi(k/\Omega)x}$$

where $\hat{f}$ is given in (13.1). *Hence* (13.4) *can be used to approximate* (at least for a finite number of terms) *the Fourier series for f in* (13.6).

We conclude this subsection with the following fundamental theorem.

(13.7) Theorem. Let $\{F_k\}$ be the N-point DFT of $\{f_j\}$. Then $\{F_k\}$ has period N, that is
$$F_{k+N} = F_k$$

for all integers k. Moreover, we have the following *inversion formula*

$$f_j = \frac{1}{N} \sum_{k=0}^{N-1} F_k e^{i2\pi jk/N} \qquad (j = 0, 1, \ldots, N-1)$$

Proof. Let $W = e^{-i2\pi/N}$. Then

(13.8)
$$F_k = \sum_{j=0}^{N-1} f_j W^{jk}$$

and, since $W^N = 1$,

$$F_{k+N} = \sum_{j=0}^{N-1} f_j W^{j(k+N)} = \sum_{j=0}^{N-1} f_j W^{jk} (W^N)^j$$
$$= \sum_{j=0}^{N-1} f_j W^{jk} = F_k$$

and periodicity is proved.

Just like periodicity, the key aspect of inversion is the fact that $W^N = 1$.

Moreover, for $m = 1, 2, \ldots, N - 1$ we have $(W^m)^N = 1$ *and* $W^m \neq 1$. Thus $x = W^m$ is a nontrivial root of $x^N - 1 = (x - 1)(x^{N-1} + x^{N-2} + \cdots + x + 1)$. By nontrivial we mean that, for $m = 1, 2, \ldots, N - 1$,

(13.9) $$(W^m)^{N-1} + (W^m)^{N-2} + \cdots + W^m + 1 = 0$$

Now, for $j = 0, 1, \ldots, N - 1$, we get

$$\frac{1}{N} \sum_{k=0}^{N-1} F_k e^{i2\pi jk/N} = \frac{1}{N} \sum_{k=0}^{N-1} \left(\sum_{n=0}^{N-1} f_n W^{nk} \right) W^{-jk}$$

$$= \frac{1}{N} \sum_{n=0}^{N-1} \left(\sum_{k=0}^{N-1} W^{(n-j)k} \right) f_n$$

Using (13.9) for $m = n - j$ we obtain

$$\frac{1}{N} \sum_{k=0}^{N-1} F_k e^{i2\pi jk/N} = \frac{1}{N} \left(\sum_{k=0}^{N-1} 1 \right) f_j = f_j$$

and inversion is proved. ∎

(13.10) Remarks.   Because of inversion, each distinct sequence $\{f_j\}_{j=0}^{N-1}$ has a different DFT. Also $\{f_j\}_{j=0}^{N-1}$ is a subsequence of an infinite sequence, *having period N,* defined by

$$f_j = \frac{1}{N} \sum_{j=0}^{N-1} F_k e^{i2\pi jk/N}$$

for each integer j. From now on, we shall stick to the notation

$$F_k = \sum_{j=0}^{N-1} f_j W^{jk}$$

where $W = e^{-i2\pi/N}$.

Fast Fourier Transforms

A direct calculation of an N-point DFT requires $(N - 1)^2$ multiplications [and $N(N - 1)$ additions]. For large N (say $N > 1000$) this would require too much computer time. We shall now describe an algorithm that greatly reduces the number of operations needed to calculate a DFT. A DFT calculated by such an algorithm is called a *fast Fourier transform* (FFT). For N on the order of 1000, a computer can perform an FFT in seconds.

   Throughout this discussion we shall assume that N is a power of 2. (This is adequate for most applications, for FFTs when N is not a power of 2 see the references.)

   The FFT is begun by noting that $W^{\frac{1}{2}N} = -1$ and then splitting the DFT into two sums in the following way

$$F_k = \sum_{j=0}^{N-1} f_j W^{jk} = \sum_{j=0}^{\frac{1}{2}N-1} f_j W^{jk} + \sum_{j=0}^{\frac{1}{2}N-1} f_{\frac{1}{2}N+j} W^{(\frac{1}{2}N+j)k}$$

$$= \sum_{j=0}^{\frac{1}{2}N-1} f_j W^{jk} + \sum_{j=0}^{\frac{1}{2}N-1} f_{\frac{1}{2}N+j}(-1)^k W^{jk} \qquad (W^{\frac{1}{2}N} = -1)$$

$$= \sum_{j=0}^{\frac{1}{2}N-1} [f_j + (-1)^k f_{\frac{1}{2}N+j}] W^{jk}$$

Now we halve $\{F_k\}$ into two subsequences, according to whether k is even or odd

$$F_{2k} = \sum_{j=0}^{\frac{1}{2}N-1} [f_j + f_{\frac{1}{2}N+j}](W^2)^{jk}$$

(13.11) $\qquad\qquad\qquad\qquad (k = 0, 1, \ldots, \tfrac{1}{2}N - 1)$

$$F_{2k+1} = \sum_{j=0}^{\frac{1}{2}N-1} [f_j - f_{\frac{1}{2}N+j}] W^j (W^2)^{jk}$$

In (13.11), the even and odd indexed outputs are expressed as $\frac{1}{2}N$-point DFTs of the transformed inputs $\{f_0 + f_{\frac{1}{2}N}, f_1 + f_{\frac{1}{2}N+1}, \ldots, f_{\frac{1}{2}N-1} + f_{N-1}\}$ and $\{(f_0 - f_{\frac{1}{2}N})W^0, (f_1 - f_{\frac{1}{2}N+1})W, \ldots, (f_{\frac{1}{2}N-1} - f_{N-1})W^{\frac{1}{2}N-1}\}$, respectively. In Figure 7.30 we have illustrated this halving operation. Notice that the basic transformation of the input $\{f_j\}$ is the *butterfly*

(13.12)
$$\begin{array}{ccc} f_j & \longrightarrow & f_j + f_{\frac{1}{2}N+j} \\ & \times & \\ f_{\frac{1}{2}N+j} & \longrightarrow & [f_j - f_{\frac{1}{2}N+j}]W^j \end{array}$$

performed for $j = 0, 1, \ldots, \frac{1}{2}N - 1$. The top portion of each butterfly is

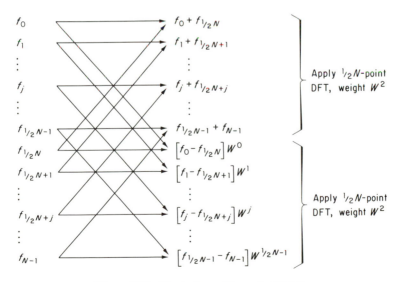

Figure 7.30   Illustration of Formula (13.11).

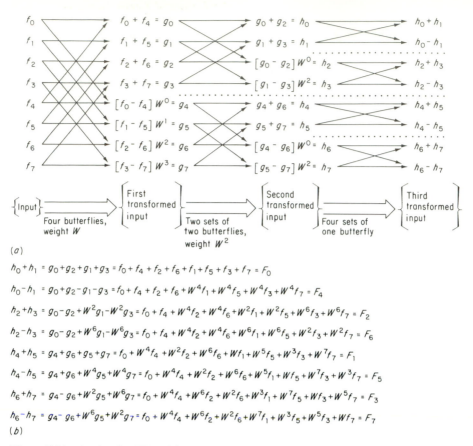

(a)

$$h_0+h_1 = g_0+g_2+g_1+g_3 = f_0+f_4+f_2+f_6+f_1+f_5+f_3+f_7 = F_0$$

$$h_0-h_1 = g_0+g_2-g_1-g_3 = f_0+f_4+f_2+f_6+W^4f_1+W^4f_5+W^4f_3+W^4f_7 = F_4$$

$$h_2+h_3 = g_0-g_2+W^2g_1-W^2g_3 = f_0+f_4+W^4f_2+W^4f_6+W^2f_1+W^2f_5+W^6f_3+W^6f_7 = F_2$$

$$h_2-h_3 = g_0-g_2+W^6g_1-W^6g_3 = f_0+f_4+W^4f_2+W^4f_6+W^6f_1+W^6f_5+W^2f_3+W^2f_7 = F_6$$

$$h_4+h_5 = g_4+g_6+g_5+g_7 = f_0+W^4f_4+W^2f_2+W^6f_6+Wf_1+W^5f_5+W^3f_3+W^7f_7 = F_1$$

$$h_4-h_5 = g_4+g_6+W^4g_5+W^4g_7 = f_0+W^4f_4+W^2f_2+W^6f_6+W^5f_1+Wf_5+W^7f_3+W^3f_7 = F_5$$

$$h_6+h_7 = g_4-g_6+W^2g_5+W^6g_7 = f_0+W^4f_4+W^6f_2+W^2f_6+W^3f_1+W^7f_5+Wf_3+W^5f_7 = F_3$$

$$h_6-h_7 = g_4-g_6+W^6g_5+W^2g_7 = f_0+W^4f_4+W^6f_2+W^2f_6+W^7f_1+W^3f_5+W^5f_3+Wf_7 = F_7$$

(b)

Figure 7.31 An 8-point FFT. (a) Structure of 8-point FFT. (There are $3 = \log_2 8$ stages.) (b) Output of calculation described in (a). [Note: $W^4 = -1$ and $W^8 = 1$.]

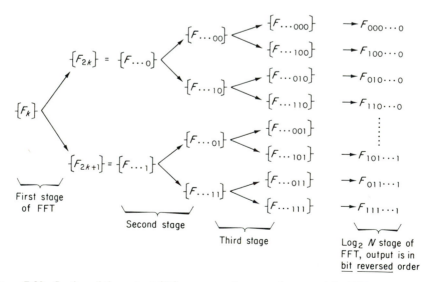

Figure 7.32 Sorting of the output $\{F_k\}$ corresponding to each stage of the FFT. Each index k is written in binary (base 2) notation (e.g., $13 = 1101 = 00\cdots01101$).

simply an addition $f_j + f_{\frac{1}{2}N+j}$, while the bottom portion is a subtraction $f_j - f_{\frac{1}{2}N+j}$ *multiplied* by W^j (called the "twiddle factor").

We now repeat the halving operation, to divide each of the two $\frac{1}{2}N$-point DFTs into two $\frac{1}{4}N$-point DFTs. Repeating this halving we eventually reach the stage of performing $\frac{1}{2}N$, 2-point DFTs. The results for $N = 8$ are shown in Figure 7.31. Notice that the final output in Figure 7.31 is $\{F_k\}$, *but in a scrambled order.* The final step in the FFT is to unscramble this output. To see how to do this, examine Figure 7.32. We see from that figure that the successive stages of the FFT correspond to *sorting the output* $\{F_k\}$ *into successive halves according to the binary expansions of the indices k,* **read in reverse order.** Therefore, to unscramble the output of the FFT we have to perform a *bit reversal* as shown in Figure 7.33 for $N = 8$. This bit reversal of the binary expansions of the indices of the output of the FFT completes the algorithm.

As previously stated, the advantage of the FFT consists of the tremendous savings that it gives in the number of multiplications (and additions) needed to compute a DFT. Each stage of the algorithm requires $\frac{1}{2}N$ multiplications [by various powers of W, see Figures 7.30 and 7.31(a)] and N additions. Since there are $\log_2 N$ stages, this results in $\frac{1}{2}N \log_2 N$ multiplications and $N \log_2 N$ additions. This is an enormous savings over the direct calculation of an N-point DFT, which requires $(N - 1)^2$ multiplications and $N(N - 1)$ additions. For instance, if $N = 1024 = 2^{10}$ then the FFT requires 5120 multiplications as opposed to 1,046,529 multiplications for a direct DFT.

This savings, and consequent speed in computer processing, has made "real time" Fourier analysis possible. Using a moderately powerful minicomputer, data streaming in at around 1024 data points/sec can be Fourier transformed every second (i.e., in the same time frame as it is being collected). For example, some university chemistry departments now have FFT analyzers that perform Fourier analysis on optical spectra from chemical reactions *as they are occurring* [for more details, see Bell (1972)]. FFTs have also helped to revolutionize optics. With their help, the basic theory of Fourier optics, described above in §§4 and 10, can be implemented on a practical basis. A very clear, detailed, description of this kind

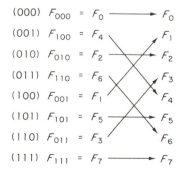

Figure 7.33   Bit reversal reordering of FFT output for $N = 8$.

of work can be found in the article: "Effect of aberrations and apodizations on the performance of coherent optical systems (I and II)," by J. P. Mills and B. J. Thompson, *J. Opt. Soc. Am. A* **3**(5): 694 (1986).

Computer programs for FFTs can be found in Brigham (1974) and Press et al. (1986). In addition to those references, see Elliot and Rao (1982) and Nussbaumer (1982) for more discussion of FFTs.

Exercises

(13.13)   Suppose that f is piecewise continuous and $f(x) = 0$ for $x < 0$ and $x > \Omega$. Prove that $\hat{f}(u) = \sum_{k=-\infty}^{+\infty} \hat{f}(k/\Omega)g_k(u)$ for appropriately defined functions $\{g_k\}$. [*Hint*: Shift f to the left and apply (11.10).]

(13.14)   Using the FFT described above, compute the DFTs of the following sequences.

(a)   1, 1, 0, 0, 1, 1, 0, 0
(b)   1, 1, 1, 1, 1, 1, 1, 1
(c)   1, −1, 1, −1, 1, −1, 1, −1
(d)   1, 1, 1, 1, 0, 0, 0, 0

(13.15)   Make a table of the bit reversal unscrambling, like the one shown in Figure 7.33, for the output of a 16-point FFT.

(13.16)   Make a table, like the one shown in Figure 7.31(*a*), showing the calculational scheme for a 16-point FFT. Use this FFT to calculate the DFTs of the following sequences.

(a)   0, 0, 0, 0, 1, 1, 1, 1, 0, 0, 0, 0, 1, 1, 1, 1
(b)   1, −1, 1, −1, 1, −1, 1, −1, 1, −1, 1, −1, 1, −1, 1, −1

(13.17)   *Zero Padding.* Suppose that we wish to use (13.4) to approximate $\hat{f}$, but we have only 1000 samples of f on $[0, \Omega]$. Show that by adjoining 24 zeroes to the end of the sequence of samples of f and then performing a 1024-point FFT we can obtain an approximation of $\hat{f}$.

Remark.   Exercise (13.17) illustrates why it is often sufficient to take N as a power of 2 for FFT algorithms. Either data collection can be arbitrarily required to obtain a power of 2 number of data points, *or* the data can be "padded" with zeroes as described in (13.17).

(13.18)   Describe how the Fourier transform (or Fourier series) of a function limited to $[-c, c]$ can be approximated using an FFT.

(13.19)   Derive the double DFT:

$$F_{kn} = \sum_{j,\, m=0}^{N-1,\, N-1} f_{jm} W^{(jk+mn)} = \sum_{j=0}^{N-1} \left[\sum_{m=0}^{N-1} f_{jm} W^{mn} \right] W^{jk}$$

where $W = e^{-i2\pi/N}$, by way of approximating the double Fourier transform of a (Riemann integrable) function limited to the square $[0, \Omega] \times [0, \Omega]$.

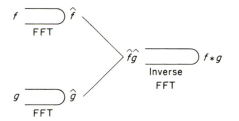

Figure 7.34   Scheme for approximating convolution using FFT.

Show how the double DFT can be calculated by iterating single FFTs in the iterated sum above, and that this involves $N^2 \log_2 N$ multiplications.

(13.20)   Describe how the double Fourier transform (or double Fourier series) of a (Riemann integrable) function limited to a square $[-c, c] \times [-c, c]$ can be approximated using the iterated FFT method described in (13.19).

(13.21)   Show that the *inverse* DFT

$$f_j = \frac{1}{N} \sum_{j=0}^{N-1} F_k \bar{W}^{jk}$$

can also be computed by an FFT algorithm.

$\star$**(13.22)**   Explain how convolution of two functions could be approximated using the scheme shown in Figure 7.34.

(13.23)   Prove *Parseval's theorem* for DFTs:

$$\sum_{j=0}^{N-1} |f_j|^2 = \frac{1}{N} \sum_{k=0}^{N-1} |F_k|^2$$

References

For those readers who wish to read further in applications here is a partial list.

Partial Differential Equations.   In addition to the references for Chapter 3, see Garabedian (1964), Courant and Hilbert (1953), John (1982), Sneddon (1951), Vladimirov (1971), Shilov (1968), and Gel'fand and Shilov (1964). A good set of problems is in Lebedev, Skalskaya, and Uflyand (1979).

Optics.   See Goodman (1968), Iizuka (1985), Lipson and Lipson (1981), Papoulis (1968), Lipson (1972), Sommerfeld (1949), Pauli (1973), Allan (1973), and Kapany (1967). The grand treatise on the subject is by Born and Wolf (1965). Beautiful diffraction patterns can be found in Harburn, Taylor, and Welberry (1975).

Communications and Information Theory.   See Higgins (1985) and the references therein, particularly the articles by Shannon, Jerri, Peterson and Middleton, and Weston. See also Blackman and Tukey (1959), Hamming (1980), and Raisbeck (1963).

Crystal and Molecular Structure.   See Woolfson (1970), Lipson and Cochran (1966), Wheatley (1981), and the references therein.

Probability and Statistics.   See Gnedenko (1978), Cramér (1974), and Bochner (1960).

Electric Circuits.   See Papoulis (1962) and Papoulis (1968), and the references therein.

Quantum Mechanics.   See White (1968), Feynman and Hibbs (1965) (especially Chapter 1), Messiah (1958), Schiff (1968), Kemble (1937), and Landau and Lifshitz (1977).

Miscellaneous.   See Bracewell (1978), Terras (1985), and Dym and McKean (1972) for further applications. There are a number of accessible articles in the *Journal of the Optical Society of America, Applied Optics*, and *Transactions and Proceedings of the Institute of Electrical and Electronic Engineers* (*I.E.E.E.*). For example, the following articles from the *Journal of the Optical Society of America* have been used by the author in seminars

Harris, J. L., Diffraction and Resolving Power, *J. Opt. Soc. Am.* **54**:931 (1964).
Jansson, P. A., Hunt, R. H., and Player, E. K., Resolution Enhancement of
    Spectra, *J. Opt. Soc. Am.* **60**:596 (1970).
Marks, R. J., Multi-Dimensional Signal Sampling Dependency at Nyquist Densities,
    J. Opt. Soc. Am **3**:268 (1986).
Mills, J. P., and Thompson, B. J., Effect of Aberrations and Apodizations on the
    Performance of Coherent Optical Systems (I and II), *J. Opt. Soc. Am. A*
    3:694 (1986).
Morris, C. E., Richards, M. A., and Hayes, M. H., Iterative Deconvolution
    Algorithm with Quadratic Convergence, *J. Opt. Soc. Am. A* **4**:200 (1987).
Smith, R. C., and Marsh, J. S., Diffraction Patterns of Simple Apertures, *J. Opt.
    Soc. Am.* **64**:798 (1974).
Som, S. C., Multiple Reproduction by Sampling, *J. Opt. Soc. Am.* **60**:1628 (1970).

8

Legendre Polynomials and Spherical Harmonics

Legendre polynomials and spherical harmonics play a crucial role in the study of harmonic functions of three variables. Spherical harmonics find many uses in electrostatics and quantum mechanics. The fact that spherical harmonics are the eigenfunctions for angular momentum in quantum mechanics illustrates their importance for that theory. To understand spherical harmonics requires some background in the theory of Legendre polynomials. Legendre polynomials are the most important orthogonal system of functions besides the trigonometric systems. The quickest way to treat Legendre polynomials is by using their generating function, which is how Legendre first discovered them. From the generating function we obtain a recursion relation that allows us to prove Rodrigues' formula and Laplace's integral form. Those results lead to Legendre's differential equation. The orthogonality of Legendre's polynomials, which allows us to construct generalized Fourier series, is a consequence of Legendre's differential equation. Once we can form series expansions, we can then solve Dirichlet's problem for spheres under special symmetry conditions.

The topics just described occupy the first four sections of the present chapter. In the remaining five sections we discuss the considerably more difficult theory of spherical harmonics. Spherical harmonics are a generalization of Legendre polynomials. We use them to show that, without any assumed symmetry, Dirichlet's problem for spheres can be solved by series of spherical harmonics. the rigorous verification of our series solutions depends upon Poisson's integral over spheres. Using Poisson's integral we can derive many properties of three-dimensional harmonic functions analogous to those found in Chapter 4 for two dimensions. The reader who merely wishes to learn about the general properties of harmonic functions can do so by reading §§7 and 9 only.

§1. A Problem in Mechanics, Legendre Polynomials

Let's begin by considering a slightly simplified version of a problem first studied by Laplace and Legendre. Suppose that three masses m_1, m_2, m_3 are located at the points O (for origin), P and Q, respectively. (See Figure 8.1.) As shown in Figure 8.1 we have set the distance between O and P

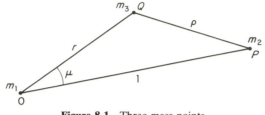

Figure 8.1   Three mass points.

equal to 1, while r equals the distance between O and Q and ρ equals the distance between P and Q. We know that the gravitational force acting on m_3 equals the sum of the forces exerted by m_1 and m_2 along the line segments connecting O and P with Q. The magnitude F_{13} of the force of m_1 on m_3 is given by *Newton's Law of Gravitation*[1] to be

$$F_{13} = -m_3 \frac{\partial}{\partial r}\left(\frac{Gm_1}{r}\right) = \frac{Gm_1 m_3}{r^2} \qquad (G = \text{gravitation constant})$$

The function $\Psi_1 = Gm_1/r$ is called the *potential function* for the mass m_1 at the point O. Likewise, the magnitude F_{23} of the force of m_2 on m_3 is given by

$$F_{23} = -m_3 \frac{\partial}{\partial \rho}(\Psi_2)$$

where $\Psi_2 = Gm_2/\rho$ is the *potential function* for the mass m_2 at the point Q.

The second potential Ψ_2 can be related to the first potential Ψ_1 as follows. Let μ equal the angle $\measuredangle POQ$. By the law of cosines $\rho^2 = 1 + r^2 - 2r\cos\mu$. Hence

$$\Psi_2 = \frac{Gm_2}{[1 + r^2 - 2r\cos\mu]^{1/2}}$$

so Ψ_2 depends on r in a more complicated fashion than Ψ_1. The constant Gm_2 is not important here; we will consider instead the factor $[1 + r^2 - 2r\cos\mu]^{-1/2}$.

Let

(1.1) $z = \cos\mu$

---

1. If static electric charges are used instead of masses, then Coulomb's Law is used instead of Newton's Law.

and define the function F by

(1.2) $F(r, z) = [1 + r^2 - 2rz]^{-1/2}$

In considering (1.2) we shall allow r to assume negative values.

   Legendre saw that the function F could be manipulated to yield what we now call Legendre polynomials. Expanding F in a Taylor series with respect to r at $r = 0$ we obtain

(1.3) $$F(r, z) = \sum_{n=0}^{\infty} P_n(z)r^n$$

where

$$P_n(z) = \frac{1}{n!} \frac{\partial^n F}{\partial r^n}\bigg|_{r=0}$$

For example,

(1.4) $P_0(z) = 1$ $P_1(z) = z$ $P_2(z) = \frac{1}{2}(3z^2 - 1)$

Let's assume that the series in (1.3) converges for $|r| < 1$ and $|z| \leq 1$; we will verify that assumption in the next section. Since a power series may be differentiated term by term we have

$$\frac{\partial F}{\partial r} = (z - r)(1 + r^2 - 2rz)^{-3/2} = \sum_{n=1}^{\infty} nP_n(z)r^{n-1}$$

Hence

(1.5)
$$(1 + r^2 - 2rz)\frac{\partial F}{\partial r} = \sum_{n=1}^{\infty} nP_n(z)r^{n-1}$$
$$+ \sum_{n=1}^{\infty} nP_n(z)r^{n+1} - \sum_{n=1}^{\infty} 2nzP_n(z)r^n$$

On the other hand

(1.5')
$$(1 + r^2 - 2rz)\frac{\partial F}{\partial r} = (z - r)(1 + r^2 - 2rz)^{-1/2}$$
$$= (z - r)F(r, z)$$
$$= \sum_{n=0}^{\infty} zP_n(z)r^n - \sum_{n=0}^{\infty} P_n(z)r^{n+1}$$

Equating the coefficients of r^n in the power series in (1.5) and (1.5') we obtain (for $n \geq 1$)

(1.6) $(n + 1)P_{n+1}(z) + (n - 1)P_{n-1}(z) - 2nzP_n(z) = zP_n(z) - P_{n-1}(z)$

Table 8.1   The First Eight Legendre Polynomials

$$P_0(z) = 1 \qquad P_1(z) = z \qquad P_2(z) = \tfrac{1}{2}(3z^2 - 1) \qquad P_3(z) = \tfrac{1}{2}(5z^3 - 3z)$$

$$P_4(z) = \tfrac{1}{8}(35z^4 - 30z^2 + 3) \qquad P_5(z) = \tfrac{1}{8}(63z^5 - 70z^3 + 15z)$$

$$P_6(z) = \tfrac{1}{16}(231z^6 - 315z^4 + 105z^2 - 5)$$

$$P_7(z) = \tfrac{1}{16}(429z^7 - 693z^5 + 315z^3 - 35z)$$

From (1.6) we get

(1.7) $(n + 1)P_{n+1}(z) - (2n + 1)zP_n(z) + nP_{n-1}(z) = 0 \qquad (n \geq 1)$

Starting with P_0 and P_1 from (1.4) we can generate, by successive applications of (1.7), all of the functions P_n, which we see are polynomials in z. The first eight *Legendre polynomials* P_n are listed in Table 8.1. Formula (1.7) is called a *recursion formula*. It provides the quickest way of generating the sequence of Legendre polynomials.

The function $F(r, z) = (1 + r^2 - 2rz)^{-1/2}$ is called the *generating function* for the Legendre polynomials. The generating function provides the most effective means for studying Legendre polynomials. Here is an interesting property of the Legendre polynomials that is easily proved using the generating function

(1.8) $P_n(1) = 1 \qquad (n = 0, 1, 2, \ldots)$

To prove (1.8) substitute $z = 1$ into F; obtaining from (1.2) and (1.3)

$$(1 - r)^{-1} = \sum_{n=0}^{\infty} P_n(1)r^n$$

Hence, using the familiar geometric series expansion of $(1 - r)^{-1}$ we have

$$\sum_{n=0}^{\infty} 1 \cdot r^n = \sum_{n=0}^{\infty} P_n(1)r^n$$

from which (1.8) follows immediately by comparing the coefficients of r^n.

Exercises

(1.9)   Verify Table 8.1 and construct a table of the first nine derivatives of the Legendre polynomials, that is, $P_n^{(1)}(z)$ for $n = 1, 2, 3, \ldots, 9$.

(1.10)   Prove that the degree of P_n is n and that P_n is an even (odd) function if n is even (odd).

(1.11) Show that $P_n(-1) = (-1)^n$ for each n, that $P_{2n}(0) = (-1)^n \cdot (2n)![2^{2n}(n!)^2]^{-1}$, and that $P_{2n-1}(0) = 0$.

(1.12) Solve $z^2 = \sum_{j=0}^{2} c_j P_j(z)$ and $z^3 = \sum_{j=0}^{3} d_j P_j(z)$ for c_0, c_1, c_2 and d_0, d_1, d_2, d_3.

(1.13) Prove that $z^n = \sum_{j=0}^{n} c_j P_j(z)$ is uniquely solvable for $c_0, c_1, \ldots, c_n$. Hence every polynomial is uniquely expressible as a linear combination of $P_0, \ldots, P_n$ where n equals the degree of that polynomial.

(1.14) Prove that $\int_{-1}^{1} P_n^2(z) \, dz = 2/(2n+1)$ for $n = 0, 1, 2$.

(1.15) Show that the potential $\Psi_1 = Gm_1/r$ is harmonic as a function of x, y, z except at the origin $O = (0, 0, 0)$. Using Exercise (1.5), Chapter 4, show that $\Psi_2 = Gm_2/\rho$ is harmonic as a function of x, y, z except at the point $P = (x_0, y_0, z_0)$ where $\rho = 0$.

§2. Rodrigues' Formula and Laplace's Integral

In this section we shall express the Legendre polynomial P_n in two forms. The first form is a differential form due to Rodrigues, whereas the second form is an integral form due to Laplace. These two complementary forms will prove quite useful in the sequel.

(2.1) Theorem: Rodrigues' Formula.   For each n the Legendre polynomial P_n satisfies

$$P_n(z) = \frac{1}{2^n n!} \frac{d^n}{dz^n} (z^2 - 1)^n$$

Proof.   Define $R_n(z)$ to be the right side of the equality stated above. We see immediately that $R_0 = P_0$ and $R_1 = P_1$. Our proof consists in showing that R_n satisfies the same recursion formula as P_n.

We have

(2.2) $R_{n+1}(z) = \dfrac{1}{2^{n+1}(n+1)!} \dfrac{d^{n+1}}{dz^{n+1}} (z^2 - 1)^{n+1} = \dfrac{1}{2^n n!} \dfrac{d^n}{dz^n} [z(z^2 - 1)^n]$

By *Leibniz's Rule* [see Exercise (2.17)]

(2.3) $\dfrac{d^n}{dz^n} [f(z)g(z)] = \sum_{k=0}^{n} \binom{n}{k} \dfrac{d^k f}{dz^k} \dfrac{d^{n-k} g}{dz^{n-k}}$

applied to $f(z) = z$ and $g(z) = (z^2 - 1)^n$, we obtain from (2.2)

$$R_{n+1}(z) = \frac{1}{2^n n!} \left[z \frac{d^n}{dz^n} (z^2 - 1)^n + n \frac{d^{n-1}}{dz^{n-1}} (z^2 - 1)^n \right]$$

(2.4)

$$= zR_n(z) + \frac{n}{2^n n!} \frac{d^{n-1}}{dz^{n-1}} (z^2 - 1)^n$$

On the other hand, if we take another derivative in (2.2) we obtain

$$R^{n+1}(z) = \frac{1}{2^n n!} \frac{d^{n-1}}{dz^{n-1}} [(z^2 - 1)^n + 2nz^2(z^2 - 1)^{n-1}]$$

$$= \frac{1}{2^n n!} \frac{d^{n-1}}{dz^{n-1}} [(2n + 1)(z^2 - 1)^n + 2n(z^2 - 1)^{n-1}]$$

$$= \frac{2n + 1}{2^n n!} \frac{d^{n-1}}{dz^{n-1}} (z^2 - 1)^n + R_{n-1}(z)$$

Therefore,

(2.5) $$R_{n+1}(z) - R_{n-1}(z) = \frac{2n + 1}{2^n n!} \frac{d^{n-1}}{dz^{n-1}} (z^2 - 1)^n$$

Eliminating $(d^{n-1}/dz^{n-1})(z^2 - 1)^n$ from (2.4) and (2.5) yields

$$(n + 1)R_{n+1}(z) - (2n + 1)zR_n(z) + nR_{n-1}(z) = 0 \qquad (n \geq 1)$$

which is identical in form to (1.7). Therefore $R_n = P_n$ for every n.   ∎

If we differentiate (2.5) we obtain another useful recursion formula

(2.6) $$P'_{n+1}(z) - P'_{n-1}(z) = (2n + 1)P_n(z) \qquad (n \geq 1)$$

We now turn to Laplace's form for P_n.

(2.7) Theorem.   For each n the Legendre polynomial P_n satisfies for $|z| \leq 1$

$$P_n(z) = \frac{1}{\pi} \int_0^\pi [z + i(1 - z^2)^{1/2} \cos \phi]^n \, d\phi$$

Proof.   Our proof is adapted from Churchill and Brown (1978). We use the fact that

$$\frac{1}{2\pi} \int_{-\pi}^\pi e^{im\phi} \, d\phi = \begin{cases} 0 & \text{if } m = 1, 2, \ldots \\ 1 & \text{if } m = 0 \end{cases}$$

If k is a nonnegative integer the binomial expansion of $(z + we^{i\phi})^k$ will have factors $e^{im\phi}$, for $m = 0, 1, 2, \ldots, k$, in each term and hence

$$\frac{1}{2\pi} \int_{-\pi}^\pi (z + we^{i\phi})^k \, d\phi = z^k$$

provided z and w are independent of ϕ. It follows by linearity of integration that

$$q(z) = \frac{1}{2\pi} \int_{-\pi}^\pi q(z + we^{i\phi}) \, d\phi$$

whenever q is a polynomial in z. For instance, if q is the Legendre polynomial P_n, then

$$P_n(z) = \frac{1}{2\pi} \int_{-\pi}^{\pi} P_n(z + we^{i\phi}) \, d\phi$$

Rodrigues' formula for P_n in the integrand above yields

(2.8)
$$P_n(z) = \frac{1}{2\pi} \int_{-\pi}^{\pi} \frac{1}{2^n n!} \frac{d^n}{du^n} (u^2 - 1)^n |_{u = z + we^{i\phi}} \, d\phi$$

$$= \frac{1}{2\pi} \int_{-\pi}^{\pi} p^{(n)}(z + we^{i\phi}) \, d\phi$$

for $p(u) = (u^2 - 1)^n/(2^n n!)$. Performing an integration by parts we obtain for each integer $k = 0, 1, \dots, n$

(2.9)
$$\int_{-\pi}^{\pi} e^{-ik\phi} p^{(n-k)}(z + we^{i\phi}) \, d\phi$$

$$= \frac{k+1}{w} \int_{-\pi}^{\pi} e^{-i(k+1)\phi} p^{(n-k-1)}(z + we^{i\phi}) \, d\phi$$

[See Exercise (2.18).] Applying (2.9) to (2.8) successively for each k yields

(2.10)
$$P_n(z) = \frac{n!}{2\pi w^n} \int_{-\pi}^{\pi} e^{-in\phi} p(z + we^{i\phi}) \, d\phi$$

If $|z| \le 1$ and we set $w = i(1 - z^2)^{1/2}$ then algebra reveals that [see Exercise (2.19)]

(2.10′)
$$\frac{n!}{w^n} e^{-in\phi} p(z + we^{i\phi}) = \frac{e^{-in\phi}}{2^n w^n} [(z + we^{i\phi})^2 - 1]^n$$

$$= [z + i(1 - z^2)^{1/2} \cos \phi]^n$$

Hence we may express (2.10) as

$$P_n(z) = \frac{1}{2\pi} \int_{-\pi}^{\pi} [z + i(1 - z^2)^{1/2} \cos \phi]^n \, d\phi$$

Since $\cos \phi$ is even, we obtain Laplace's integral form for P_n. ■

A common alternative way of expressing Laplace's integral consists in writing $(z^2 - 1)^{1/2}$ in place of $i(1 - z^2)^{1/2}$, which yields

(2.11) $\qquad P_n(z) = \frac{1}{\pi} \int_0^{\pi} [z + (z^2 - 1)^{1/2} \cos \phi]^n \, d\phi \qquad (|z| \le 1)$

We conclude this section by describing some important consequences of Laplace's integral. If we recall that $z = \cos \mu$ [see (1.1)], then

$$P_n(\cos \mu) = \frac{1}{\pi} \int_0^{\pi} [\cos \mu + i \sin \mu \cos \phi]^n \, d\phi$$

Because $|\cos \mu + i \sin \mu \cos \phi| = [1 - \sin^2 \mu \sin^2 \phi]^{1/2}$ we obtain

$$|P_n(\cos \mu)| \leq \frac{1}{\pi} \int_0^\pi [1 - \sin^2 \mu \sin^2 \phi]^{n/2} \, d\phi$$

$$\leq \frac{1}{\pi} \int_0^\pi 1 \, d\phi = 1$$

It follows that for each n

(2.12) $|P_n(z)| \leq 1$ $(|z| \leq 1)$

From (2.12) we conclude by the M-test that $\sum_{n=0}^\infty P_n(z)r^n$ converges for $|z| \leq 1$ and $|r| < 1$, and converges uniformly when $|r| \leq R < 1$. Moreover, using (2.12) and (2.6), it is a straightforward exercise to verify that

(2.13) $|P_n'(z)| \leq \dfrac{n(n+1)}{2}$ $(|z| \leq 1)$

From (2.13) we get [using Theorem (5.5), Chapter 1, for fixed r, $|r| < 1$]

(2.14) $\dfrac{\partial}{\partial z}\left[\sum_{n=0}^\infty P_n(z)r^n\right] = \sum_{n=0}^\infty P_n'(z)r^n$

for $|z| \leq 1$ and $|r| < 1$.

One consequence of these results is that formula (1.3) is valid for $|z| \leq 1$ and $|r| < 1$. Fix a value of z, say z_0. The function $f(r) = F(r, z_0)$ solves the ordinary differential equation

(2.15) $[1 + r^2 - 2rz_0]\left(\dfrac{df}{dr}\right) - (z_0 - r)f = 0$

for $|r| < 1$. The method of integrating factors proves that f solves (2.15) uniquely, up to multiplication by a constant. Using the recursion formula (1.7), however, we see that

$$g(r) = \sum_{n=0}^\infty P_n(z_0)r^n$$

solves (2.15) also. Since $g(0) = 1 = f(0)$ we conclude that $g = f$ identically in r. Thus, since z_0 was an arbitrary choice, we have

---

(2.16) $F(r, z) = [1 + r^2 - 2rz]^{-1/2} = \sum_{n=0}^\infty P_n(z)r^n$ $(|z| \leq 1, |r| < 1)$

Exercises

(2.17)   Prove Leibniz's rule (2.3). [*Hint*: Use mathematical induction.]

(2.18)   Verify (2.9). [*Hint*: Write the integral on the left side as

$$\int_{-\pi}^\pi e^{-i(k+1)\phi} \frac{d}{d\phi}\left[\frac{-i}{w}p^{(n-k-1)}(z + we^{i\phi})\right] d\phi]$$

(2.19)   Check (2.10′). [*Hint*: Bring $e^{-in\phi}/(2^n w^n)$ inside $[(z + we^{i\phi})^2 - 1]^n$ and use $z^2 - 1 = w^2$.]

(2.20)   Verify (2.13).

(2.21)   Show that $\text{maximum}_{-1 \leq z \leq 1} |P_n(z)| = 1$.

(2.22)   Show that $|P_n''(z)| \leq \frac{1}{2}n^4$ for all $|z| \leq 1$.

(2.23)   Use (2.22) to prove that $\partial^2 F/\partial z^2 = \sum_{n=2}^{\infty} P_n''(z)r^n$ for $|z| \leq 1$ and $|r| < 1$.

(2.24)   Expand $(5 - 4z)^{-1/2}$ in a series of Legendre polynomials.

(2.25)   Prove that
$$P_n(z) = \sum_{k=0}^{N} \frac{(-1)^k (2n - 2k)!}{2^n k! (n - k)! (n - 2k)!} z^{n-2k}$$
where N is the greatest integer less than or equal to $n/2$. [*Hint*: Expand $(z^2 - 1)^n$ in Rodrigues' formula using the binomial theorem.]

§3. Legendre's Equation, Orthogonality of the Legendre Polynomials

The Legendre polynomials form an orthogonal set of continuous functions over the interval $[-1, 1]$. We will show that this fact follows from a differential equation known as Legendre's equation. The following ingenious argument is due to Legendre.

A simple calculation shows that

(3.1)
$$(1 + r^2 - 2rz)\frac{\partial F}{\partial z} - rF = 0$$

From (1.5′) we obtain

(3.2)
$$(1 + r^2 - 2rz)\frac{\partial F}{\partial r} - (z - r)F = 0$$

Eliminating F from (3.1) and (3.2) yields, after canceling $(1 + r^2 - 2rz)$,

(3.3)
$$r\frac{\partial F}{\partial r} - (z - r)\frac{\partial F}{\partial z} = 0$$

Substituting the power series form for F into (3.3) and equating coefficients of r^n we obtain for $n \geq 1$

(3.4)
$$nP_n(z) - zP_n'(z) + P_{n-1}'(z) = 0$$

Since $r(\partial/\partial r)(rF) = r^2(\partial F/\partial r) + rF$, when we use $r^2(\partial F/\partial r) = r(z - r)\frac{\partial F}{\partial z}$ from (3.3) and $rF = (1 + r^2 - 2rz)\frac{\partial F}{\partial z}$ from (3.1), we obtain

$$r\left(\frac{\partial}{\partial r}\right)(rF) - (1 - rz)\left(\frac{\partial F}{\partial z}\right) = 0$$

Substituting the power series form for F into the equation above yields for $n \geq 1$

(3.5) $$nP_{n-1}(z) - P'_n(z) + zP'_{n-1}(z) = 0$$

From (3.4) it follows that

(3.6) $$P'_{n-1}(z) = zP'_n(z) - nP_n(z)$$

Substituting this form for P'_{n-1} into (3.5) we get for $n \geq 1$

$$nP_{n-1}(z) + (z^2 - 1)P'_n(z) - nzP_n(z) = 0$$

Differentiating the expression above and applying (3.6) we obtain

---

(3.7) $(1 - z^2)P''_n(z) - 2zP'_n(z) + n(n + 1)P_n(z) = 0$ $(n \geq 0)$

Equation (3.7) is called *Legendre's equation*. We will now show that it implies the orthogonality of the Legendre polynomials over the interval $[-1, 1]$.

(3.8) Theorem. The Legendre polynomials satisfy the orthogonality relations

$$\int_{-1}^{1} P_m(x)P_n(x)\, dx = \begin{cases} 0 & \text{if } m \neq n \\ \dfrac{2}{2n + 1} & \text{if } m = n \end{cases}$$

Proof.   Multiply (3.7) by P_m, obtaining

(3.9) $$(1 - z^2)P''_n P_m - 2zP'_n P_m + n(n + 1)P_n P_m = 0$$

Interchanging the roles of m and n in (3.9) and subtracting, we get

$$(1 - z^2)[P''_n P_m - P''_m P_n] - 2z[P'_n P_m - P'_m P_n] = [m(m + 1) - n(n + 1)]P_n P_m$$

Or,

$$\frac{d}{dz}[(1 - z^2)(P'_n P_m - P'_m P_n)] = [m(m + 1) - n(n + 1)]P_n P_m$$

Integrating from -1 to 1 and noting that $1 - z^2$ equals 0 at $z = \pm 1$ we have

$$0 = [m(m + 1) - n(n + 1)] \int_{-1}^{1} P_n(z)P_m(z)\, dz$$

If $m \neq n$ then we can cancel $m(m + 1) - n(n + 1)$ and obtain

(3.10) $$\int_{-1}^{1} P_m(z)P_n(z)\, dz = 0 \qquad (m \neq n)$$

We now must determine $\|P_n\|_2^2 = \int_{-1}^{1} P_n^2(z)\, dz$. If we replace n by $n - 1$ in

(1.7) and multiplying the resulting equation by $(2n + 1)P_n$ we get

$$n(2n + 1)P_n^2 - (2n - 1)(2n + 1)zP_{n-1}P_n + (n - 1)(2n + 1)P_{n-2}P_n = 0$$

Multiplying (1.7) by $(2n - 1)P_{n-1}$ and subtracting from the equation above we obtain for $n \geq 2$

$$n(2n + 1)P_n^2 + (n - 1)(2n + 1)P_{n-2}P_n$$

$$- (n + 1)(2n - 1)P_{n+1}P_{n-1} - n(2n - 1)P_{n-1}^2 = 0$$

Integrating the equation above over $[-1, 1]$ and using (3.10) we have, for $n \geq 2$,

$$\int_{-1}^{1} P_n^2(z)\, dz = \frac{2n - 1}{2n + 1} \int_{-1}^{1} P_{n-1}^2(z)\, dz$$

Repeatedly applying this formula yields for $n \geq 2$

$$\int_{-1}^{1} P_n^2(z)\, dz = \frac{3}{2n + 1} \int_{-1}^{1} P_1^2(z)\, dz = \frac{2}{2n + 1}$$

Thus for $n \geq 2$

$$\|P_n\|_2^2 = \int_{-1}^{1} P_n^2(z)\, dz = \frac{2}{2n + 1}$$

and direct calculation reveals that the formula above holds for $n = 0, 1$. ∎

Now that we have established the orthogonality of the Legendre polynomials we can form Fourier series with them. Using Definition (1.2), Chapter 2, if f is a piecewise continuous function over $(-1, 1)$ then

(3.11)
$$f \sim \sum_{n=0}^{\infty} c_n P_n(z) \qquad c_n = \frac{2n + 1}{2} \int_{-1}^{1} f(z)P_n(z)\, dz$$

The orthogonal system of Legendre polynomials is known to be complete. We will demonstrate this fact in §8 [or, see (3.18)].

(3.12) Example. Suppose that

$$f(z) = \begin{cases} V & \text{if } 0 < z < 1 \\ 0 & \text{if } -1 < z < 0 \end{cases}$$

where V is a constant. Expand f in a series of Legendre polynomials.

Solution. From (3.11) we have

$$c_n = \frac{(2n + 1)V}{2} \int_0^1 P_n(z)\, dz$$

We see that $c_0 = \frac{1}{2}V$ while for $n \geq 1$ we use Rodrigues' formula

$$c_n = \frac{(2n+1)V}{n!2^{n+1}} \int_0^1 \frac{d^n}{dz^n}(z^2-1)^n \, dz$$

$$= \frac{(2n+1)V}{n!2^{n+1}} \frac{d^{n-1}}{dz^{n-1}}(z^2-1)^n \bigg|_0^1$$

Since $n-1$ derivatives leaves at least one factor of (z^2-1) in every term, the evaluation at $z=1$ vanishes. By the binomial expansion theorem we have

$$c_n = -\frac{(2n+1)V}{n! \, 2^{n+1}} \frac{d^{n-1}}{dz^{n-1}} \left[\sum_{k=0}^{n} (-1)^{n-k}\binom{n}{k} z^{2k} \right]\bigg|_{z=0}$$

$$= -\frac{(2n+1)V}{n! \, 2^{n+1}} \sum_{k=0}^{n} (-1)^{n-k}\binom{n}{k} \frac{d^{n-1}}{dz^{n-1}}(z^{2k})\bigg|_{z=0}$$

Therefore $c_n = 0$ unless $2k = n-1$, that is, unless $n = 2k+1$ is odd. In which case,

$$c_{2k+1} = -\frac{(4k+3)V}{(2k+1)! \, 2^{2k+2}}(-1)^{k+1}\binom{2k+1}{k}(2k)!$$

$$= (-1)^k \frac{(4k+3)V}{4^{k+1}} \frac{(2k)!}{k! \, (k+1)!}$$

Thus

$$f \sim V\left[\frac{1}{2} + \sum_{k=0}^{\infty} \frac{(-1)^k(4k+3)(2k)!}{4^{k+1}(k+1)! \, k!} P_{2k+1}(z) \right]$$

Exercises

(3.13)   Expand the following functions in series of Legendre polynomials.

(a) $f(z) = \begin{cases} 0 & \text{if } 0 < z < 1 \\ 5 & \text{if } -1 < z < 0 \end{cases}$

(b) $f(z) = \begin{cases} 1 & \text{if } 0 < z < 1 \\ -1 & \text{if } -1 < z < 0 \end{cases}$

(c) $f(z) = \begin{cases} V & \text{if } 0 < z < a \quad (a < 1) \\ 0 & \text{otherwise} \end{cases}$ [*Hint*: Use (2.6).]

(d) $f(z) = |z|$

(e) $f(z) = z^2$

(f) $f(z) = \begin{cases} z^2 & \text{if } 0 < z < 1 \\ 0 & \text{if } -1 < z < 0 \end{cases}$

(3.14)   Prove that if p is a polynomial of degree less than n then

$$\int_{-1}^{1} p(z)P_n(z) \, dz = 0$$

(3.15) Show that for every piecewise continuous function f on $(-1, 1)$ we have

$$\sum_{n=0}^{\infty} c_n^2 \frac{2}{2n+1} \le \int_{-1}^{1} f^2(z)\, dz$$

where the c_n's are the Fourier coefficients for f relative to $\{P_n\}$.

(3.16) Assuming the completeness of $\{P_n\}$ prove that

$$\tfrac{1}{2} + \sum_{k=0}^{\infty} \frac{2(4k+3)(2k)!^2}{16^{k+1}(k+1)!^2 k!^2} = 1$$

(3.17) Prove that P_n has n distinct roots, all of which lie in the open interval $(-1, 1)$. [*Hint*: If $n \ge 1$, suppose that less than n roots are in $(-1, 1)$. Let $p(z) = (z - r_1)(z - r_2) \cdots (z - r_m)$ where $r_1, r_2, \ldots, r_m$ are the roots of odd multiplicity of P_n in $(-1, 1)$. Show that $\int_{-1}^{1} p(z) P_n(z)\, dz \ne 0$ contradicting (3.14).]

(3.18) Using (6.1) and (8.17) from Chapter 2, prove the completeness of the Legendre polynomials.

§4. Solution of Dirichlet's Problem by Legendre Polynomials

In this section we will show that, given a special symmetry condition, Dirichlet's problem for a three-dimensional ball has a solution in the form of a series of Legendre polynomials.

Throughout this section we will use the *spherical coordinates* defined by

$$x = r \sin \theta \cos \phi \qquad y = r \sin \theta \sin \phi \qquad z = r \cos \theta$$

Using these coordinates we can transform the Laplacian $\Delta \Psi = \Psi_{xx} + \Psi_{yy} + \Psi_{zz}$ into

(4.1) $\quad \Delta \Psi = \dfrac{1}{r^2} \dfrac{\partial}{\partial r}\left[r^2 \dfrac{\partial \Psi}{\partial r} \right] + \dfrac{1}{r^2 \sin \theta} \dfrac{\partial}{\partial \theta}\left[\sin \theta \dfrac{\partial \Psi}{\partial \theta} \right] + \dfrac{1}{r^2 \sin^2 \theta} \dfrac{\partial^2 \Psi}{\partial \phi^2}$

Formula (4.1) is well known; we shall indicate in the exercises how to derive it, for the reader who wishes to do so.

Using (4.1) we can solve the following problem for the unknown function Ψ

(4.2) $\qquad \Delta \Psi = 0 \qquad (0 \le r < a) \qquad \Psi(a, \theta, \phi) = F(\theta)$

where F is a continuous function. Problem (4.2) is a form of Dirichlet's problem in spherical coordinates for the ball B_a of radius a about the origin. The fact that F depends only θ, and is independent of ϕ, is a special symmetry condition. (See Figure 8.2). In the next section, we will consider Dirichlet's problem without any symmetry assumption. Most of the main ideas, however, will be illustrated as we solve the simpler problem (4.2).

Problem (4.2) can be tackled by separation of variables. From (4.1) we

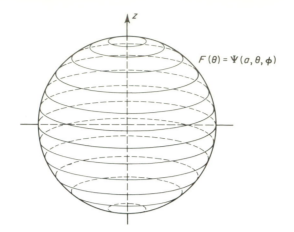

Figure 8.2   The function F is constant for each angle θ.

see that after multiplying by r^2 we need to solve

$$\frac{\partial}{\partial r}\left[r^2\frac{\partial\Psi}{\partial r}\right] + \frac{1}{\sin\theta}\frac{\partial}{\partial\theta}\left[\sin\theta\frac{\partial\Psi}{\partial\theta}\right] + \frac{1}{\sin^2\theta}\frac{\partial^2\Psi}{\partial\phi^2} = 0$$

(4.2′)

$$\Psi(a,\,\theta,\,\phi) = F(\theta)$$

Substituting $\Psi = R(r)T(\theta)H(\phi)$ into the partial differential equation in (4.2′), performing the required differentiations, and dividing out by $R(r)T(\theta)H(\phi)$ we get

(4.3)
$$\frac{r^2R'' + 2rR'}{R} + \frac{1}{T\sin\theta}\frac{d}{d\theta}\left[\sin\theta\frac{dT}{d\theta}\right] + \frac{H''(\phi)}{\sin^2\theta H(\phi)} = 0$$

By substituting any values of θ and ϕ into (4.3) and varying r we see that the first fractional term in (4.3) is a constant, say α. Thus

(4.4)
$$r^2R'' + 2rR' - \alpha R = 0$$

Likewise, substituting any values of r and θ into (4.3) and varying ϕ we see that

(4.5)
$$H''(\phi) = \beta H(\phi)$$

where β is a constant. Using (4.4) and (4.5) in (4.3) we obtain

(4.6)
$$\frac{1}{\sin\theta}\frac{d}{d\theta}\left[\sin\theta\frac{dT}{d\theta}\right] + \left[\alpha + \frac{\beta}{\sin^2\theta}\right]T = 0$$

We now use the symmetry condition $\Psi(a,\,\theta,\,\phi) = F(\theta)$ to simplify the equations above. We assume that since the boundary function F is independent of ϕ our solution Ψ to (4.2) should also be independent of ϕ. Hence, we may as well assume that $\Psi = R(r)T(\theta)$, that is, $H(\phi) = 1$ for all ϕ, in our separation of variables argument. In that case, (4.5) shows that

$\beta = 0$. Hence we must now solve

(4.7) $r^2 R'' + 2rR' - \alpha R = 0$ $\dfrac{1}{\sin \theta} \dfrac{d}{d\theta} \left[\sin \theta \dfrac{dT}{d\theta} \right] + \alpha T = 0$

Since $z = \cos \theta$ we see that

$$-\frac{d}{dz} = \frac{1}{\sin \theta} \frac{d}{d\theta} \qquad (z^2 - 1)\frac{d}{dz} = \sin \theta \frac{d}{d\theta}$$

Using those relations the second equation in (4.7) can be rewritten as

(4.8) $-\dfrac{d}{dz} \left[(z^2 - 1) \dfrac{dP}{dz} \right] + \alpha P(z) = 0$ $[P(z) = T(\cos^{-1} z)]$

Performing the differentiations in (4.8) we get

$$(1 - z^2)\frac{d^2 P}{dz^2} - 2z\frac{dP}{dz} + \alpha P = 0$$

Comparing the equation above with Legendre's equation (3.7) we are led to set $\alpha_n = n(n + 1)$ and take $P_n(z)$ to be the nth Legendre polynomial. Substituting $n(n + 1)$ in place of α in the first equation in (4.7), we now must solve

$$r^2 R_n'' + 2rR_n' - n(n + 1)R_n = 0$$

Trying a solution of the form $R_n(r) = r^p$, where p is some real power, we obtain

$$p(p - 1)r^p + 2pr^p - n(n + 1)r^p = 0$$

Dividing out by r^p and performing some algebra yields $[p - n][p + n + 1] = 0$. Thus, $p = n$ or $-(n + 1)$. Hence $R_n(r) = A_n r^n + B_n r^{-(n+1)}$ where A_n and B_n are arbitrary constants. Thus we can now set

$$\Psi_n(r, \theta, \phi) = [A_n r^n + B_n r^{-(n+1)}]P_n(\cos \theta)$$

Since we want $\Delta\Psi_n = 0$, we must ensure that Ψ_n is continuous in the ball B_a. Therefore, we reject the term $B_n r^{-(n+1)}$, that is, we put $B_n = 0$. Performing superposition of all the separated solutions we obtain

$$\Psi(r, \theta, \phi) = \sum_{n=0}^{\infty} A_n r^n P_n(\cos \theta)$$

Our next step will be to substitute $r = a$; it is therefore convenient to rewrite the series above as

(4.9) $\Psi(r, \theta, \phi) = \displaystyle\sum_{n=0}^{\infty} c_n P_n(\cos \theta)\left(\dfrac{r}{a}\right)^n$

where the coefficients c_n are as yet undetermined.

Substituting $r = a$ into (4.9) we get

$$F(\theta) \sim \sum_{n=0}^{\infty} c_n P_n(\cos \theta)$$

Or, letting $z = \cos \theta$, we desire the expansion

$$F(\cos^{-1} z) \sim \sum_{n=0}^{\infty} c_n P_n(z)$$

Hence, we set

(4.10)

$$c_n = \frac{2n+1}{2} \int_{-1}^{1} F(\cos^{-1} z) P_n(z) \, dz$$

$$= \frac{2n+1}{2} \int_{0}^{\pi} F(\theta) P_n(\cos \theta) \sin \theta \, d\theta$$

Combining (4.9) and (4.10) we propose the following solution to (4.2).

For $0 \le r < a$,

(4.11)
$$\Psi(r, \theta, \phi) = \sum_{n=0}^{\infty} c_n P_n(\cos \theta) \left(\frac{r}{a}\right)^n$$

where

$$c_n = \frac{2n+1}{2} \int_{-1}^{1} F(\theta) P_n(\cos \theta) \sin \theta \, d\theta$$

If $r = a$, then $\Psi(a, \theta, \phi) = F(\theta)$.

For example, suppose that

$$F(\theta) = \begin{cases} V & \text{if } 0 < \theta < \frac{1}{2}\pi \\ 0 & \text{if } \frac{1}{2}\pi < \theta < \pi \end{cases}$$

Then by Example (3.12)

$$F(\cos^{-1} z) \sim V\left[\frac{1}{2} + \sum_{k=0}^{\infty} \frac{(-1)^k (4k+3)(2k)!}{4^{k+1}(k+1)! \, k!} P_{2k+1}(z)\right]$$

hence for $0 \le r < a$

$$\Psi(r, \theta, \phi) = V\left[\frac{1}{2} + \sum_{k=0}^{\infty} \frac{(-1)^k (4k+3)(2k)!}{4^{k+1}(k+1)! k!} P_{2k+1}(\cos \theta) \left(\frac{r}{a}\right)^{2k+1}\right]$$

In later sections we will give proofs that Ψ really does solve $\Delta \Psi = 0$, in Cartesian as well as spherical coordinates. For now, however, we shall merely show that the series in (4.11) does converge. Since

$$|c_n| \le \frac{2n+1}{2} \int_{-1}^{1} |F(\theta)| \, |P_n(\cos \theta)| \, d\theta \le \frac{2n+1}{2} \int_{-1}^{1} |F(\theta)| \, d\theta$$

if we let $M = \int_{-1}^{1} |F(\theta)| \, d\theta$ then

$$\sum_{n=0}^{\infty} \left| c_n P_n(\cos \theta) \left(\frac{r}{a} \right)^n \right| \leq M \sum_{n=0}^{\infty} \frac{2n+1}{2} \left(\frac{r}{a} \right)^n$$

which converges for all $0 \leq r < a$ by the ratio test. Moreover, the convergence is uniform for $0 \leq r \leq a - \epsilon$ when $\epsilon > 0$. Thus Ψ is continuous for $0 \leq r < a$ (letting ϵ tend to 0). Similar arguments can be used to show that Ψ satisfies $\Delta\Psi = 0$ in spherical coordinates, but we leave the details to the reader. The one defect in showing that $\Delta\Psi = 0$ in spherical coordinates is that the Laplacian is not defined along the z axis ($\theta = 0$ or π).

Exercises

(4.12)   Solve problem (4.2) given the following functions for F

(a) $F(\theta) = \begin{cases} \cos \theta & \text{if } 0 < \theta < \frac{1}{2}\pi \\ 0 & \text{if } \frac{1}{2}\pi < \theta < \pi \end{cases}$

(b) $F(\theta) = |\cos \theta|$

(c) $F(\theta) = \begin{cases} \cos^2 \theta & \text{if } 0 < \theta < \frac{1}{2}\pi \\ 0 & \text{if } \frac{1}{2}\pi < \theta < \pi \end{cases}$

(d) $F(\theta) = \cos^2 \theta$

(e) $F(\theta) = \begin{cases} 0 & \text{if } 0 < \theta < \frac{1}{2}\pi \\ \sin^2 \theta & \text{if } \frac{1}{2}\pi < \theta < \pi \end{cases}$

(4.13)   Show that by the chain rule

$$\frac{\partial}{\partial r} = (\sin \theta \cos \phi) \frac{\partial}{\partial x} + (\sin \theta \sin \phi) \frac{\partial}{\partial y} + \cos \theta \frac{\partial}{\partial z}$$

$$\frac{\partial}{\partial \theta} = (r \cos \theta \cos \phi) \frac{\partial}{\partial x} + (r \cos \theta \sin \phi) \frac{\partial}{\partial y} - r \sin \theta \frac{\partial}{\partial z}$$

$$\frac{\partial}{\partial r} = (-r \sin \theta \sin \phi) \frac{\partial}{\partial x} + (r \sin \theta \cos \phi) \frac{\partial}{\partial y}$$

(4.14)   Using (4.13) show that

$$\frac{\partial}{\partial x} = \sin \theta \cos \phi \frac{\partial}{\partial r} + \frac{\cos \theta \cos \phi}{r} \frac{\partial}{\partial \theta} - \frac{\sin \phi}{r \sin \theta} \frac{\partial}{\partial \phi}$$

$$\frac{\partial}{\partial y} = \sin \theta \sin \phi \frac{\partial}{\partial r} + \frac{\cos \theta \sin \phi}{r} \frac{\partial}{\partial \theta} + \frac{\cos \phi}{r \sin \theta} \frac{\partial}{\partial \phi}$$

$$\frac{\partial}{\partial z} = \cos \theta \frac{\partial}{\partial r} - \frac{1}{r} \sin \theta \frac{\partial}{\partial \theta}$$

(4.15)   Using (4.14), verify (4.1).

(4.16)   Prove that the series for Ψ in (4.11) solves $\Delta\Psi = 0$ where $\Delta\Psi$ is expressed in spherical coordinates. [*Hint:* (2.13) and (2.22) will be needed to justify term by term differentiation.]

(4.17) *Harmonic Functions outside a Sphere.* Solve the following problem

$$\Delta\Psi = 0 \quad (a < r < +\infty) \quad \Psi(a, \theta, \phi) = F(\theta)$$

$$\lim_{r \to +\infty} \Psi(r, \theta, \phi) = 0 \quad \text{uniformly for all } \theta \text{ and } \phi$$

(4.18) What are the series solutions to (4.17) for the functions F given in (4.12)?

§5. Dirichlet's Problem inside a Sphere, Laplace's Series

In this section we shall solve the following problem

(5.1) $$\Delta\Psi = 0 \quad (0 \le r < a) \quad \Psi(a, \theta, \phi) = F(\theta, \phi)$$

where F is a continuous function.

If we let $\Psi = R(r)T(\theta)H(\phi)$ then we obtain Eqs. (4.4), (4.5), and (4.6) as we described in §4. A general solution to (4.5) is given by

$$H(\phi) = ae^{\beta^{1/2}\phi} + be^{-\beta^{1/2}\phi}$$

However, since Ψ is defined in the ball B_a we must have H periodic in ϕ with period 2π. The only values of $\beta^{1/2}$ that allow for that are imaginary integer values. Therefore we obtain

(5.2) $$\beta_m = -m^2 \quad H_m(\phi) = e^{im\phi}, e^{-im\phi}$$

for $m = 0, 1, 2, \ldots,$.

Because we desire our solution to (5.1) to reduce to the one found for (4.2) when F depends only on θ, we set $\alpha_n = n(n+1)$ in (4.4) and (4.6). Thus we must solve

(5.3) $$r^2 R_n'' + 2rR_n' - n(n+1)R_n = 0$$

and

(5.4) $$\frac{1}{\sin\theta}\frac{d}{d\theta}\left[\sin\theta\frac{dT}{d\theta}\right] + \left[n(n+1) - \frac{m^2}{\sin^2\theta}\right]T = 0$$

Equation (5.3) was solved in §4 with the result $R_n(r) = r^n,\ r^{-(n+1)}$ and for the same reasons as in §4 we reject $r^{-(n+1)}$.

To solve (5.4) will take several steps. We begin by substituting $z = \cos\theta$, as we did in §4, we obtain

(5.4') $$-\frac{d}{dz}\left[(z^2 - 1)\frac{dP}{dz}\right] + \left[n(n+1) - \frac{m^2}{1-z^2}\right]P = 0 \quad P(z) = T(\cos^{-1}z)$$

Next we make the substitution $P(z) = (1 - z^2)^{m/2}Q(z)$ into (5.4'). After computing the necessary derivatives and canceling a factor of $(1 - z^2)^{m/2}$ Eq. (5.4') becomes

(5.4") $(1 - z^2)Q''(z) - 2(m + 1)zQ'(z) + [n(n + 1) - m(m + 1)]Q(z) = 0$

Finally we observe that if we differentiate Legendre's equation (3.7) once, we obtain

$$(1 - z^2)\frac{d^2}{dz^2}[P_n^{(1)}] - 4z\frac{d}{dz}[P_n^{(1)}] + [n(n + 1) - 2]P_n^{(1)} = 0$$

where $P_n^{(1)} = (d/dz)[P_n]$. But the equation above is just like (5.4") when $m = 1$. More generally, if we differentiate (3.7) m times we get

(5.5)
$$(1 - z^2)\frac{d^2}{dz^2}[P_n^{(m)}] - 2(m + 1)z\frac{d}{dz}[P_n^{(m)}]$$
$$+ [n(n + 1) - m(m + 1)]P_n^{(m)} = 0$$

for $m = 0, 1, \ldots, n$ where $P_n^{(m)} = (d^m/dz^m)[P_n]$.

Comparing (5.5) and (5.4") and backtracking through the discussion above to (5.4), we have solved (5.4) for $n = 0, 1, 2, \ldots,$ and $m = 0, 1, \ldots, n$ by

(5.6) $T_{mn}(\theta) = \sin^m \theta P_n^{(m)}(\cos \theta)$

Combining all of these results we propose that the general solution to $\Delta\Psi = 0$ in the ball B_a should be of the form

$$\Psi(r, \theta, \phi) = \sum_{n=0}^{\infty} \left[\sum_{m=0}^{n} (C_{mn}e^{im\phi} + D_{mn}e^{-im\phi}) \sin^m \theta P_n^{(m)}(\cos \theta) \right]\left(\frac{r}{a}\right)^n$$

Or, equivalently,

(5.7) $$\Psi(r, \theta, \phi) = \sum_{n=0}^{\infty} \left[\sum_{m=-n}^{+n} c_{mn}e^{im\phi} \sin^{|m|} \theta P_n^{(|m|)}(\cos \theta) \right]\left(\frac{r}{a}\right)^n$$

where the constants c_{mn} are as yet undetermined.

Based on (5.7) we define the functions H_{mn} and Y_{mn} as follows, for $n = 0, 1, 2, \ldots,$ and $m = 0, \pm 1, \ldots, \pm n$

(5.8) $H_{mn}(r, \theta, \phi) = Y_{mn}(\theta, \phi)\left(\frac{r}{a}\right)^n$ $Y_{mn}(\theta, \phi) = e^{im\phi} \sin^{|m|} \theta P_n^{(|m|)}(\cos \theta)$

The functions Y_{mn} are called (*surface*) *spherical harmonics*, while the functions H_{mn} are called (*solid*) *spherical harmonics*. In Table 8.2 we list some spherical harmonics. The functions $P_n^{(|m|)}(z)$ are called the *associated Legendre polynomials*.

Spherical harmonics satisfy orthogonality relations that will allow us to determine the constants c_{mn} in (5.7). In the following theorem $\bar{Y}_{mn}$ will denote the function $\bar{Y}_{mn}(\theta, \phi) = e^{-im\phi} \sin^{|m|} \theta P_n^{(|m|)}(\cos \theta)$ which is the complex conjugate of Y_{mn}.

Table 8.2 Some Spherical Harmonics Y_{mn}

$n = 0$	$n = 1$	$n = 2$	$n = 3$
$Y_{00} = 1$	$Y_{\pm 11} = e^{\pm i\phi} \sin \theta$	$Y_{\pm 22} = 3e^{\pm i2\phi} \sin^2 \theta$	$Y_{\pm 33} = 15e^{\pm i3\phi} \sin^3 \theta$
	$Y_{01} = \cos \theta$	$Y_{\pm 12} = 3e^{\pm i\phi} \sin \theta \cos \theta$	$Y_{\pm 23} = 15e^{\pm i2\phi} \sin^2 \theta \cos \theta$
		$Y_{02} = \frac{1}{2}(3 \cos^2 \theta - 1)$	$Y_{\pm 13} = \frac{1}{2}e^{\pm i\phi} \sin \theta (15 \cos^2 \theta - 1)$
			$Y_{03} = \frac{1}{2}(15 \cos^3 \theta - 3 \cos \theta)$

(5.9) Theorem: Orthogonality of Spherical Harmonics. The spherical harmonics $\{Y_{mn}\}$ satisfy

$$\int_0^\pi \int_0^{2\pi} Y_{jk}(\theta, \phi)\bar{Y}_{mn}(\theta, \phi) \sin \theta \, d\phi \, d\theta$$

$$= \begin{cases} 0 & \text{if } (j, k) \neq (m, n) \\ \dfrac{[n + |m|]!}{[n - |m|]!} \dfrac{4\pi}{2n + 1} & \text{if } (j, k) = (m, n) \end{cases}$$

Proof. If $j \neq m$ then the orthogonality of $\{e^{im\phi}\}$ over $[0, 2\pi]$ implies that the integral in question is 0. Therefore, let's assume that $j = m$. Letting $q = |m|$, we have

$$\int_0^\pi \int_0^{2\pi} Y_{mk}(\theta, \phi)\bar{Y}_{mn}(\theta, \phi) \sin \theta \, d\phi \, d\theta$$

(5.10)
$$= 2\pi \int_0^\pi \sin^{2q+1}\theta P_k^{(q)}(\cos \theta)P_n^{(q)}(\cos \theta) \, d\theta$$

$$= 2\pi \int_{-1}^1 (1 - z^2)^q P_k^{(q)}(z)P_n^{(q)}(z) \, dz$$

But $P_n^{(q)}$ satisfies (5.5) with q in place of m, hence upon multiplying that equation by $(1 - z^2)^q P_k^{(q)}$ we obtain

$$(1 - z^2)^{q+1}\frac{d^2}{dz^2}[P_n^{(q)}]P_k^{(q)} - 2(q + 1)z(1 - z^2)^q \frac{d}{dz}[P_n^{(q)}]P_k^{(q)}$$

$$+ [n(n + 1) - q(q + 1)](1 - z^2)^q P_n^{(q)}P_k^{(q)} = 0$$

Interchanging the roles of k and n and subtracting from the equation above, we get,

$$(1 - z^2)^{q+1}\left(\frac{d^2}{dz^2}[P_n^{(q)}]P_k^{(q)} - \frac{d^2}{dz^2}[P_k^{(q)}]P_n^{(q)}\right)$$

$$- 2(q + 1)z(1 - z^2)^q\left(\frac{d}{dz}[P_n^{(q)}]P_k^{(q)} - \frac{d}{dz}[P_k^{(q)}]P_n^{(q)}\right)$$

$$+ [n(n + 1) - k(k + 1)](1 - z^2)^q P_n^{(q)}P_k^{(q)} = 0$$

Rewriting the first two terms above as a single derivative yields

$$\frac{d}{dz}\left\{(1-z^2)^{q+1}\left(\frac{d}{dz}[P_n^{(q)}]P_k^{(q)} - \frac{d}{dz}[P_k^{(q)}]P_n^{(q)}\right)\right\}$$

$$= [k(k+1) - n(n+1)](1-z^2)^q P_n^{(q)}P_k^{(q)}$$

Integrating this last equation from -1 to 1 yields, due to the vanishing of $(1-z^2)^{q+1}$ at $z = \pm 1$,

$$0 = [k(k+1) - n(n+1)]\int_{-1}^{1} (1-z^2)^q P_k^{(q)}(z)P_n^{(q)}(z)\, dz$$

If $k \neq n$ we may cancel the factor $k(k+1) - n(n+1)$ from the equation above, and returning to (5.10), we have proved the orthogonality of $\{Y_{mn}\}$. All that remains is to prove that

(5.11)

$$\|Y_{mn}\|_2^2 = \int_0^\pi \int_0^{2\pi} Y_{mn}(\theta, \phi)\bar{Y}_{mn}(\theta, \phi)\sin\theta\, d\phi\, d\theta$$

$$= \frac{[n+|m|]!}{[n-|m|]!}\frac{4\pi}{2n+1}$$

If $m = 0$ then

$$\|Y_{0n}\|_2^2 = 2\pi \int_{-1}^{1} [P_n(z)]^2\, dz = \frac{4\pi}{2n+1}$$

Using (5.10) for $k = n$ and $q = |m| \geq 1$, we have

(5.12)

$$\int_0^\pi \int_0^{2\pi} Y_{mn}(\theta, \phi)\bar{Y}_{mn}(\theta, \phi)\sin\theta\, d\phi\, d\theta$$

$$= 2\pi \int_{-1}^{1} (1-z^2)^q [P_n^{(q)}(z)]^2\, dz$$

Setting $u = (1-z^2)^q P_n^{(q)}(z)$ and $dv = P_n^{(q)}(z)\, dz$, then

$$du = \left\{(1-z^2)^q \frac{d^2}{dz^2}[P_n^{(q-1)}] - 2qz(1-z^2)^{q-1}\frac{d}{dz}[P_n^{(q-1)}]\right\} dz$$

$$= -(1-z^2)^{q-1}[n(n+1) - q(q+1)]P_n^{(q-1)}(z)\, dz$$

because $P_n^{(q-1)}$ satisfies (5.5) with $q-1$ in place of m. Furthermore, $v = P_n^{(q-1)}$. Using these results we perform an integration by parts on the right side of (5.12) obtaining

$$2\pi \int_{-1}^{1} (1-z^2)^q [P_n^{(q)}(z)]^2\, dz$$

$$= 2\pi[n(n+1) - q(q-1)]\int_{-1}^{1} (1-z^2)^{q-1}[P_n^{(q-1)}(z)]^2\, dz$$

$$= (n+q)(n-q+1)2\pi \int_{-1}^{1} (1-z^2)^{q-1}[P_n^{(q-1)}(z)]^2\, dz$$

Continuing in this way, we obtain

$$2\pi \int_{-1}^{1} (1-z^2)^q [P_n^{(q)}(z)]^2 \, dz$$

$$= (n+q)(n+q-1)\cdots(n-q+2)(n-q+1)2\pi \int_{-1}^{1} [P_n^{(0)}(z)]^2 \, dz$$

$$= \frac{(n+q)!}{(n-q)!} \frac{4\pi}{2n+1}$$

Recalling that $q = |m|$, and combining this last result with (5.12), we see that (5.11) holds. ∎

Using the orthogonality of the spherical harmonics we can define Fourier series in terms of them. This is done in a way that is analogous to Fourier series with complex exponentials. Namely, if F is a function of θ and ϕ then

$$F \sim \sum_{n=0}^{\infty} \left[\sum_{m=-n}^{n} c_{mn} Y_{mn}(\theta, \phi) \right]$$

(5.13)

$$c_{mn} = \frac{[n-|m|]!}{[n+|m|]!} \frac{2n+1}{4\pi} \int_{0}^{\pi} \int_{0}^{2\pi} F(\theta, \phi) \bar{Y}_{mn}(\theta, \phi) \sin\theta \, d\phi \, d\theta$$

Later we shall see that $\{Y_{mn}\}$ *is a complete orthogonal system over the rectangle* $0 \le \theta \le \pi$, $0 \le \phi \le 2\pi$.

Comparing (5.7) with (5.13) for $r = a$ leads us to propose the following solution to problem (5.1).

For $0 \le r < a$, Ψ is given by *Laplace's series*

(5.14) $$\Psi(r, \theta, \phi) = \sum_{n=0}^{\infty} \left[\sum_{m=-n}^{n} c_{mn} Y_{mn}(\theta, \phi) \right] \left(\frac{r}{a}\right)^n$$

where the coefficients c_{mn} are defined in (5.13). For $r = a$ we set $\Psi(a, \theta, \phi) = F(\theta, \phi)$.

In subsequent sections we shall demonstrate that the series in (5.14) converges for $0 \le r < a$. Moreover, we will demonstrate that Ψ satisfies $\Delta\Psi = 0$ both in Cartesian and spherical coordinates, and that Ψ is continuous on the entire closed ball $\bar{B}_a$ centered at the origin whenever F is continuous. In short, (5.14) provides us with a recipe for solving Dirichlet's problem for the ball B_a.

Suppose we want to solve

(5.15) $$\Delta\Psi = 0 \quad (a < r < +\infty) \quad \Psi(a, \theta, \phi) = F(\theta, \phi)$$

which is Dirichlet's problem for the region outside the ball B_a. *If we also require that* $\lim_{r \to +\infty} \Psi(r, \theta, \phi) = 0$ then the analysis above can be modified to yield the following.

For $a < r < +\infty$

(5.16) $\Psi(r, \theta, \phi) = \sum_{n=0}^{\infty} \left[\sum_{m=-n}^{n} c_{mn} Y_{mn}(\theta, \phi) \right] \left(\frac{a}{r} \right)^{n+1}$

where the coefficients c_{mn} are defined in (5.13). For $r = a$ we set
$\Psi(a, \theta, \phi) = F(\theta, \phi)$.

Remark. In the course of proving Theorem (5.9) we found that the associated Legendre polynomials $\{P_n^{(q)}\}_{n=q}^{\infty}$ satisfy the following relations

$$\int_{-1}^{1} P_m^{(q)}(z) P_n^{(q)}(z)(1 - z^2)^q \, dz = \begin{cases} 0 & \text{if } m \neq n \\ \dfrac{(n+q)!}{(n-q)!} \dfrac{2}{2n+1} & \text{if } m = n \end{cases}$$

Because of the relations above we say that $\{P_n^{(q)}\}_{n=q}^{\infty}$ is a *weighted orthogonal system* over $[-1, 1]$ with weight $(1 - z^2)^q$. All of the general results in Chapter 2 for orthogonal systems can be easily adapted to the present case by replacing dx by $(1 - z^2)^q \, dz$. See Exercise (5.23).

In the Miscellaneous Exercises we shall discuss another class of weighted orthogonal functions, the Gegenbauer polynomials.

Exercises

(5.17) Verify (5.5) and derive (5.16).

(5.18) Find the two dimensional analogue of (5.14). [*Hint*: See §2, Chapter 4.]

(5.19) Show that (5.14) may be written in the real form

$$\Psi(r, \theta, \phi) = \sum_{n=0}^{\infty} \left[\sum_{m=0}^{n} A_{mn} u_{mn}(\theta, \phi) + B_{mn} v_{mn}(\theta, \phi) \right] \left(\frac{r}{a} \right)^n$$

Find the appropriate definitions for the real valued functions u_{mn}, v_{mn} and the constants A_{mn}, B_{mn}.

(5.20) Express Y_{m1} for $m = \pm 1$, 0 in terms of the Cartesian coordinates x, y, and z. What about Y_{m2} for $m = \pm 2, \pm 1, 0$?

(5.21) Same problem as (5.20) for H_{m1} and H_{m2}.

(5.22) Show that the associated Legendre polynomial $P_n^{(q)}$ has $n - q$ distinct roots, all of which lie in the interval $(-1, 1)$.

(5.23) Given a piecewise continuous function f on $(-1, 1)$, define what the Fourier series of f must be relative to the weighted orthogonal system $\{P_n^{(q)}\}_{n=q}^{\infty}$. What would Bessel's inequality and the least squares property be in this case?

§6. Spherical Harmonics

In this section we shall study more closely the spherical harmonics Y_{mm} and solid spherical harmonics H_{mn} found in the previous section. Of particular importance are the properties of those functions when Cartesian coordinates are used.

The solid spherical harmonics H_{mn} when expressed in Cartesian coordinates are polynomials of a special kind, called homogeneous harmonic polynomials.

(6.1) Definition. A polynomial $p(x, y, z)$ is called a *homogeneous harmonic polynomial*, of order n, if $p(bx, by, bz) = b^n p(x, y, z)$ for each number b and $\Delta p = p_{xx} + p_{yy} + p_{zz} = 0$. ∎

It is easily seen that a polynomial p satisfies the homogeneity condition $p(bx, by, bz) = b^n p(x, y, z)$ if and only if for each term of p the sum of the powers of x, y, and z is n.

(6.2) Theorem. The solid spherical harmonic H_{mn} is a homogeneous harmonic polynomial of degree n, when expressed in Cartesian coordinates.

Proof. Recall that H_{mn} is defined in (5.8). By Rodrigues' formula

(6.3)
$$P_n^{(q)}(z) = \frac{1}{2^n n!} \frac{d^{n+q}}{dz^{n+q}}(z^2 - 1)^n = \frac{1}{2^n n!} \frac{d^{n+q}}{dz^{n+q}} \sum_{k=0}^{n} \binom{n}{k}(-1)^k z^{2n-2k}$$

$$= \sum_{k=0}^{K} b_k z^{n-2k-q}$$

where K is the largest value of k for which $n - 2k - q$ is nonnegative. The values of the constants b_k are irrelevant to our present discussion.

Furthermore

(6.4)
$$e^{im\phi} = [e^{\pm i\phi}]^{|m|} = [\cos\phi \pm i\sin\phi]^{|m|}$$

$$= \sum_{j=0}^{|m|} d_j \cos^{|m|-j}\phi \sin^j\phi$$

where again the values of the constants d_j (which are easily obtained by the binomial theorem) are not relevant here.

From (6.3) and (6.4) we obtain for some constants C_{jk}

$$H_{mn} = e^{im\phi} \sin^{|m|}\theta P_n^{(|m|)}(\cos\theta)(r/a)^n$$

$$= \sum_{j=0}^{|m|} \sum_{k=0}^{K} C_{jk} r^n \cos^{|m|-j}\phi \sin^j\phi \sin^{|m|}\theta \cos^{n-2k-|m|}\theta$$

$$= \sum_{j=0}^{|m|} \sum_{k=0}^{K} C_{jk} r^{2k}(r\sin\theta\cos\phi)^{|m|-j}(r\sin\theta\sin\phi)^j(r\cos\theta)^{n-2k-|m|}$$

Therefore, using the formulas for x, y, and z in spherical coordinates,

(6.5) $$H_{mn} = \sum_{j=0}^{|m|} \sum_{k=0}^{K} C_{jk} (x^2 + y^2 + z^2)^k x^{|m|-j} y^j z^{n-2k-|m|}$$

From (6.5) we find that $H_{mn}(bx, by, bz) = b^n H_{mn}(x, y, z)$ so H_{mn} satisfies the homogeneity condition of Definition (6.1). To see that H_{mn} is harmonic we observe that $\Delta H_{mn} = 0$ when Δ and H_{mn} are expressed in spherical coordinates (this was discussed in §5). Therefore

(6.6) $$\Delta H_{mn} = (H_{mn})_{xx} + (H_{mn})_{yy} + (H_{mn})_{zz} = 0$$

provided $r \sin \theta \neq 0$. Since H_{mn} is a polynomial it follows that (6.6) holds for all x, y, and z (by continuity, or algebraic identity). ∎

Let $\mathcal{H}_n$ stand for the set of all homogeneous harmonic polynomials of order n. We will now prove that $\{H_{mn}\}_{m=-n}^{n}$ generates $\mathcal{H}_n$.

(6.7) Theorem.   The set $\{H_{mn}\}_{m=-n}^{n}$ is a basis for the vector space $\mathcal{H}_n$ over the complex numbers.

Proof.   The set $\mathcal{H}_n$ is clearly a vector space. Our first step will be to show that the solid spherical harmonics of order n, $\{H_{mn}\}_{m=-n}^{n}$, are linearly independent. Suppose

$$\sum_{m=-n}^{n} c_m H_{mn} = 0$$

for certain complex constants c_m. Replacing each H_{mn} by $Y_{mn}(r/a)^n$ and canceling the common factor of $(r/a)^n$ yields

$$\sum_{m=-n}^{n} c_m Y_{mn}(\theta, \phi) = 0$$

Multiplying the equation above by $\bar{Y}_{kn}(\theta, \phi) \sin \theta$ and integrating over $0 \le \phi \le 2\pi$, $0 \le \theta \le \pi$ we obtain by orthogonality of $\{Y_{mn}\}$ that

$$c_k \frac{(n + |k|)!}{(n - |k|)!} \frac{4\pi}{2n + 1} = 0$$

Therefore each coefficient c_k equals 0 for $k = 0, \pm 1, \pm 2, \ldots, \pm n$. Thus $\{H_{mn}\}_{m=-n}^{n}$ is a linearly independent set and hence the dimension of the vector space $\mathcal{H}_n$ is at least $2n + 1$.

Our second step is to show that the dimension of $\mathcal{H}_n$ is no more than $2n + 1$, which will be sufficient for proving our theorem. Suppose that $p(x, y, z)$ belongs to $\mathcal{H}_n$. Then for some complex constants c_{jkm} we have

(6.8) $$p(x, y, z) = \sum_{j+k+m=n} c_{jkm} x^j y^k z^m$$

Equation (6.8) only expresses the fact that p is homogeneous. We see from (6.8) that the set of all homogeneous polynomials of order n has a basis

Table 8.3   A Basis for the Vector Space of Homogeneous Polynomials of Degree n

x^n	$x^{n-1}y$	$x^{n-2}y^2$	$\cdots$	x^2y^{n-2}	xy^{n-1}	y^n
$x^{n-1}z$	$x^{n-2}yz$	$x^{n-3}y^2z$	$\cdots$	$xy^{n-2}z$	$y^{n-1}z$	
$x^{n-2}z^2$	$x^{n-3}yz^2$	$x^{n-4}y^2z^2$	$\cdots$	$y^{n-2}z^2$		
$\vdots$			$\cdot\cdot$			
x^4z^{n-4}	x^3yz^{n-4}	$x^2y^2z^{n-4}$	$\cdot\cdot$			
x^3z^{n-3}	x^2yz^{n-3}	xy^2z^{n-3}	$\cdot$			
x^2z^{n-2}	xyz^{n-2}	y^2z^{n-2}				
xz^{n-1}	yz^{n-1}					
z^n						

consisting of $\{x^j y^k z^m\}_{j+k+m=n}$ (see Table 8.3). Because p is harmonic we also have

$$0 = \Delta p = \sum_{j'+k'+m'=n-2} d_{j'k'm'} \, x^{j'} y^{k'} z^{m'}$$

$$\textbf{(6.9)} \qquad = \sum_{j'+k'+m'=n-2} [(j'+2)(j'+1)c_{j'+2,\,k',\,m'}$$

$$+ (k'+2)(k'+1)c_{j',\,k'+2,\,m'} + (m'+2)(m'+1)c_{j',\,k',\,m'+2}] x^{j'} y^{k'} z^{m'}$$

where we have taken sums over all indices j', k', m' such that $j' + k' + m' = n - 2$ since the Laplacian Δ lowers two powers of either x, y, or z from every term of p. From (6.9) we infer that

$$\textbf{(6.10)} \qquad c_{j',\,k',\,m'+2} = \frac{-(j'+2)(j'+1)c_{j'+2,\,k',\,m'} - (k'+2)(k'+1)c_{j',\,k'+2,\,m'}}{(m'+2)(m'+1)}$$

Since $j' + k' + m' = n - 2$ we have $j' + k' + (m' + 2) = n$. Hence (6.8) shows that each coefficient c_{jkm} is determined by elements from the row that lies two rows back in the coefficient table shown in Table 8.4. By continually backtracking it follows that the first two rows of coefficients in Table 8.4 determine all the coefficients of p in (6.8). Hence the following linear transformation

$$L: \mathcal{H}_n \to \mathbb{C}^{2n+1}$$

$$p \mapsto \begin{bmatrix} c_{n00} \\ c_{n-1,1,0} \\ \vdots \\ c_{0n0} \\ c_{n-1,0,1} \\ \vdots \\ c_{0,n-1,1} \end{bmatrix}$$

Table 8.4 How the Coefficients of a Homogeneous Harmonic Polynomial Are Generated by the Coefficients of the First Two Rows

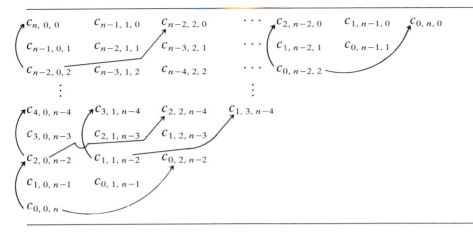

is a one to one transformation (trivial kernel, or null space) of $\mathcal{H}_n$ onto $\mathbb{C}^{2n+1}$. It follows that the dimension of $\mathcal{H}_n$ equals the dimension of $\mathbb{C}^{2n+1}$, which is $2n+1$. Therefore, the linearly independent set $\{H_{mn}\}_{m=-n}^{n}$ of $2n+1$ elements of $\mathcal{H}_n$ is a basis for $\mathcal{H}_n$. ∎

(6.11) Remark. We defined the functions Y_{mn} to be the surface spherical harmonics. We see that if $r = a = 1$ then $H_{mn} = Y_{mn}$, hence the surface spherical harmonics are just the solid spherical harmonics H_{mn} *restricted to the unit sphere ∂B_1. A more general definition of a (surface) spherical harmonic, order n, is a function defined by restricting a homogeneous harmonic polynomial, order n, to the unit sphere ∂B_1.*

Theorem (6.7) will play a key role in relating spherical harmonic expansions to the Poisson integral solution of Dirichlet's problem for balls. Our last theorem in this section will also be needed to describe that relationship, that we will take up in §8.

(6.12) Theorem. Laplace's equation $\Delta\Psi = 0$ is invariant under any orthogonal transformation of $\mathbb{R}^3$.[2]

Proof. If $\mathcal{O}: \mathbb{R}^3 \to \mathbb{R}^3$ is orthogonal,[2] then letting

$$\begin{pmatrix} x' \\ y' \\ z' \end{pmatrix} = \mathcal{O} \begin{pmatrix} x \\ y \\ z \end{pmatrix}$$

where

$$\mathcal{O} = \begin{pmatrix} a_{11} & a_{12} & a_{13} \\ a_{21} & a_{22} & a_{23} \\ a_{31} & a_{32} & a_{33} \end{pmatrix}$$

---

2. The reader may recall from linear algebra that an orthogonal transformation $\mathcal{O}: \mathbb{R}^3 \to \mathbb{R}^3$ is a linear transformation that preserves the dot product: $\mathcal{O}(\mathbf{x}) \cdot \mathcal{O}(\mathbf{x}) = \mathbf{x} \cdot \mathbf{x}$. Or, equivalently, the transpose $\mathcal{O}^T$ of $\mathcal{O}$ equals the inverse $\mathcal{O}^{-1}$ of $\mathcal{O}$.

we have by the chain rule

$$
\begin{pmatrix} \dfrac{\partial}{\partial x} \\[2mm] \dfrac{\partial}{\partial y} \\[2mm] \dfrac{\partial}{\partial z} \end{pmatrix} = \begin{pmatrix} \dfrac{\partial}{\partial x'}a_{11} + \dfrac{\partial}{\partial y'}a_{21} + \dfrac{\partial}{\partial z'}a_{31} \\[2mm] \dfrac{\partial}{\partial x'}a_{12} + \dfrac{\partial}{\partial y'}a_{22} + \dfrac{\partial}{\partial z'}a_{32} \\[2mm] \dfrac{\partial}{\partial x'}a_{13} + \dfrac{\partial}{\partial y'}a_{23} + \dfrac{\partial}{\partial z'}a_{33} \end{pmatrix} = \mathcal{O}^T \begin{pmatrix} \dfrac{\partial}{\partial x'} \\[2mm] \dfrac{\partial}{\partial y'} \\[2mm] \dfrac{\partial}{\partial z'} \end{pmatrix}
$$

Therefore, by a routine check that we leave to the reader [see Exercise (6.14)] we find that $\Psi_{xx} + \Psi_{yy} = \Psi_{zz} = 0$ implies that

(6.13) $$\Psi_{x'x'} + \Psi_{y'y'} + \Psi_{z'z'} = 0$$

due to the orthogonality of $\mathcal{O}$. Equation (6.13) demonstrates the invariance. ∎

Exercises

(6.14) Check (6.13).

(6.15) Express $\{H_{mn}\}$ for $n = 0$ and $n = 1$ in terms of x, y, and z.

(6.16) Find the exact expressions for the coefficients b_k in (6.3) and d_j in (6.4).

(6.17) Using the more general definition of spherical harmonic given in Remark (6.11), prove that for spherical harmonics p and q with different orders

$$\int_{\partial B_1} pq \, dA = 0$$

where the integral above is a surface integral over the unit sphere ∂B_1. [See §6, Chapter 4.] [*Hint:* Express the integral in spherical coordinates.]

(6.18) Let $\mathcal{O} \colon \mathbb{R}^3 \to \mathbb{R}^3$ be an orthogonal transformation. Prove that if h is in $\mathcal{H}_n$ then so is the composition $h \circ \mathcal{O}^{-1}$. Moreover, the mapping $L_{\mathcal{O}} \colon \mathcal{H}_n \to \mathcal{H}_n$ defined by $L_{\mathcal{O}}(h) = h \circ \mathcal{O}^{-1}$ is linear, one to one, onto, and satisfies $L_{\mathcal{O}_1} \circ L_{\mathcal{O}_2} = L_{\mathcal{O}_1 \mathcal{O}_2}$.

(6.19) Consider the following generalizations of results from this section to $\mathbb{R}^m$ for $m > 3$. (a) Show that Laplace's equation $\Delta \Psi = \Psi_{x_1 x_1} + \cdots + \Psi_{x_m x_m} = 0$ is invariant under orthogonal transformations. (b) Find the dimension of $\mathcal{H}_n$ the set of homogeneous harmonic polynomials of order n.

§7. Poisson's Integral

In this section we make a slight detour. We will show that Dirichlet's problem for the ball B_a of radius a centered at the origin can be solved by an integral, called Poisson's integral. In §8 we will show that this solution and the series solution in (5.14) are identical. We will begin by working with the *unit ball B_1* of radius 1 centered at the origin.

To define Poisson's integral we first need to define Poisson's kernel. The use of vector notation makes this easier. Let $\mathbf{x} = (x, y, z)$ denote a point in $\mathbb{R}^3$, for now lying in the unit ball B_1. The vector $\mathbf{u} = (x', y', z')$ will denote a point on the unit sphere ∂B_1 and we let

$$|\mathbf{x}| = (x^2 + y^2 + z^2)^{1/2} \qquad |\mathbf{x} - \mathbf{u}| = [(x - x')^2 + (y - y')^2 + (z - z')^2]^{1/2}$$

Similar notation can be defined for $\mathbb{R}^2$ (or $\mathbb{R}^m$ for $m > 3$).

In Figure 8.3 we illustrate that a reasonable guess[3] for the Poisson kernel p for the unit ball B_1 is

(7.1) $$p(\mathbf{x}; \mathbf{u}) = \frac{1}{4\pi} \frac{1 - |\mathbf{x}|^2}{|\mathbf{x} - \mathbf{u}|^3} = \frac{1}{4\pi} \frac{|\mathbf{u}|^2 - |\mathbf{x}|^2}{|\mathbf{x} - \mathbf{u}|^3}$$

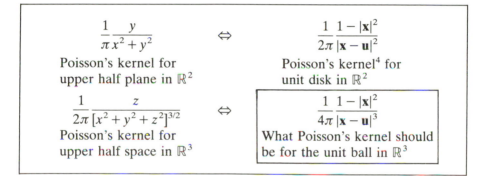

Figure 8.3   Derivation of Poisson's kernel for the unit ball in $\mathbb{R}^3$.

From Example (1.2), Chapter 4, we see that p is a harmonic function of $\mathbf{x}$, that is

(7.2) $$\Delta_\mathbf{x} p = p_{xx} + p_{yy} + p_{zz} = 0$$

for all $\mathbf{x}$ in the unit ball B_1.

Besides being harmonic, Poisson's kernel is also a summation kernel in the sense described in the following Lemma.

(7.3) Lemma.   Poisson's kernel satisfies the following properties

   (A_1) $p(\mathbf{x}; \mathbf{u}) \geq 0$ for all $\mathbf{x}$ in B_1 and $\mathbf{u}$ on ∂B_1

   (A_2) $\displaystyle\int_{\partial B_1} p(\mathbf{x}; \mathbf{u})\, dA_\mathbf{u} = 1$ for all $\mathbf{x}$ in B_1

---

3. Poisson's kernel is usually deduced by considering *Green functions*. See, for example, Bitsadze (1982), Chap. 1 §2.

4. Compare with the kernel P defined in §3 of Chapter 4. We have introduced the constant factor $1/2\pi$ and changed to vector notation above.

(A₃)   For each $\epsilon > 0$, $\delta > 0$, and vector $\mathbf{u}_0$ on ∂B_1 we can have

$$0 \leq p(\mathbf{x}, \mathbf{u}) < \epsilon$$

provided $|\mathbf{u} - \mathbf{u}_0| \geq \delta$ and $\mathbf{x}$ is close enough to $\mathbf{u}_0$.

Proof.   Property (A₁) is obvious considering (7.1). Since $4\pi p$ is harmonic in $\mathbf{x}$, Theorem (6.6), Chapter 4, applied to the ball B_r $(0 < r < 1)$ says that

$$1 = 4\pi p(0; \mathbf{u}) = \frac{1}{r^2} \int_{\partial B_r} p(\mathbf{x}; \mathbf{u}) \, dA_{\mathbf{x}}$$

Setting $\mathbf{x} = r\mathbf{v}$ where $|\mathbf{v}| = 1$ and noting that $dA_{r\mathbf{v}} = r^2 \, dA_{\mathbf{v}}$ we obtain from the equation above

(7.4)
$$1 = \int_{\partial B_1} p(r\mathbf{v}; \mathbf{u}) \, dA_{\mathbf{v}}$$

By an easy calculation we see that $|r\mathbf{v} - \mathbf{u}|^2 = |r\mathbf{u} - \mathbf{v}|^2$. Hence

$$p(r\mathbf{v}; \mathbf{u}) = \frac{1}{4\pi} \frac{1 - r^2}{|r\mathbf{v} - \mathbf{u}|^3} = \frac{1}{4\pi} \frac{1 - r^2}{|r\mathbf{u} - \mathbf{v}|^3} = p(\mathbf{x}'; \mathbf{v})$$

where $\mathbf{x}' = r\mathbf{u}$ is in B_1. Therefore (7.4) becomes

(7.4′)
$$1 = \int_{\partial B_1} p(\mathbf{x}'; \mathbf{v}) \, dA_{\mathbf{v}}$$

Replacing $\mathbf{v}$ by $\mathbf{u}$ and $\mathbf{x}'$ by $\mathbf{x}$ in (7.4′) yields (A₂). We now prove (A₃). If $|\mathbf{u} - \mathbf{u}_0| \geq \delta$ then $|\mathbf{x} - \mathbf{u}| \geq \frac{1}{2}\delta$ if $\mathbf{x}$ is close enough to $\mathbf{u}_0$. Hence

$$0 \leq p(\mathbf{x}, \mathbf{u}) \leq \frac{2}{\pi\delta^3}(1 - |\mathbf{x}|^2) < \epsilon$$

when $\mathbf{x}$ is close enough to $\mathbf{u}_0$ (since $|\mathbf{u}_0| = 1$). ∎

We can now solve Dirichlet's problem; first, for the unit ball B_1 and then for an arbitrary ball $B_a(\mathbf{x}_0)$ of radius a centered at $\mathbf{x}_0$.

(7.5) Theorem.   Given a continuous function f on the unit sphere ∂B_1, the function Ψ defined by

$$\Psi(\mathbf{x}) = \int_{\partial B_1} f(\mathbf{u}) p(\mathbf{x}; \mathbf{u}) \, dA_{\mathbf{u}}$$

for $\mathbf{x}$ in the unit ball B_1 and $\Psi(\mathbf{u}) = f(\mathbf{u})$ for $\mathbf{u}$ on ∂B_1, solves Dirichlet's problem for the ball B_1.

Proof.   By differentiating under the integral sign we see that

$$\Psi(\mathbf{x}) = \int_0^{2\pi} \int_0^{\pi} f(\sin\theta\cos\phi, \sin\theta\sin\phi, \cos\theta)$$

$$\times p(\mathbf{x}; \sin\theta\cos\phi, \sin\theta\sin\phi, \cos\theta) \sin\theta \, d\theta \, d\phi$$

is harmonic in $\mathbf{x}$ in B_1. Since Ψ is continuous in B_1 and is continuous on ∂B_1, it suffices for us to prove that for each $\mathbf{u}_0$ on ∂B_1

(7.6)
$$\lim_{\mathbf{x} \to \mathbf{u}_0} \Psi(\mathbf{x}) = f(\mathbf{u}_0)$$

To prove (7.6) we use Lemma (7.3). We have

$$|\Psi(\mathbf{x}) - f(\mathbf{u}_0)| = \left| \int_{\partial B_1} f(\mathbf{u}) p(\mathbf{x}; \mathbf{u}) \, dA_{\mathbf{u}} - \int_{\partial B_1} f(\mathbf{u}_0) p(\mathbf{x}; \mathbf{u}) \, dA_{\mathbf{u}} \right|$$

(7.7)
$$\leq \int_{\partial B_1} |f(\mathbf{u}) - f(\mathbf{u}_0)| \, p(\mathbf{x}; \mathbf{u}) \, dA_{\mathbf{u}}$$

$$= \int_{|\mathbf{u} - \mathbf{u}_0| \geq \delta} |f(\mathbf{u}) - f(\mathbf{u}_0)| \, p(\mathbf{x}; \mathbf{u}) \, dA_{\mathbf{u}}$$

$$+ \int_{|\mathbf{u} - \mathbf{u}_0| < \delta} |f(\mathbf{u}) - f(\mathbf{u}_0)| \, p(\mathbf{x}; \mathbf{u}) \, dA_{\mathbf{u}}$$

The rest of the proof proceeds in the manner for summation kernels. The integral $\int_{|\mathbf{u} - \mathbf{u}_0| < \delta}$ is made less than ϵ by applying the continuity of f and properties (A_1) and (A_2). While the integral $\int_{|\mathbf{u} - \mathbf{u}_0| \geq \delta}$ is made less than $8\pi M \epsilon$, where

$$M = \underset{\mathbf{u} \text{ on } \partial B_1}{\text{maximum}} |f(\mathbf{u})|$$

by applying property (A_3). ∎

(7.8) Corollary. If f is a continuous function on $\partial B_a(\mathbf{x}_0)$ then the function Ψ defined by

$$\Psi(\mathbf{x}) = \frac{a}{4\pi} \int_{\partial B_1} f(\mathbf{x}_0 + a\mathbf{v}) \frac{a^2 - |\mathbf{x} - \mathbf{x}_0|^2}{|(\mathbf{x} - \mathbf{x}_0) - a\mathbf{v}|^3} \, dA_{\mathbf{v}}$$

for $\mathbf{x}$ in $B_a(\mathbf{x}_0)$ and $\Psi(\mathbf{u}) = f(\mathbf{u})$ for $\mathbf{u}$ on $\partial B_a(\mathbf{x}_0)$, solves Dirichlet's problem for the ball $B_a(\mathbf{x}_0)$.

Proof. See Exercise (7.11). ∎

(7.9) Remark. By the three-dimensional version of the maximum–minimum principle it follows that the function Ψ described in (7.8) is unique for a given continuous function f. We will use the notation P.I.$[f]$, for "Poisson integral of f," to denote the integral used to define $\Psi(\mathbf{x})$ in (7.8). The specific ball over which the harmonic function P.I.$[f]$ is defined will, of course, have to be specified each time we use this notation. In §9, we will examine the general properties of harmonic functions of three variables that are derivable from Poisson's integral.

Exercises

(7.10) Check that $|r\mathbf{v} - \mathbf{u}|^2 = |r\mathbf{u} - \mathbf{v}|^2$ when $\mathbf{u}$, $\mathbf{v}$ lie on ∂B_1.

(7.11) Prove Corollary (7.8). [*Hint*: Make a change of variables in order to apply Theorem (7.5).]

(7.12) Find a solution in integral form to Dirichlet's problem for the outside of a ball of radius a.

*(7.13) Suppose that in spherical coordinates $\mathbf{x} = (r \cos \phi \sin \theta, r \sin \phi \sin \theta, r \cos \theta)$ and $\mathbf{u} = (\cos \phi' \sin \theta', \sin \phi' \sin \theta', \cos \theta')$. Verify that

$$p(\mathbf{x}; \mathbf{u}) = \frac{1}{4\pi} \frac{1 - r^2}{(1 + r^2 - 2r \cos \mu)^{3/2}}$$

where $\cos \mu = \cos \theta \cos \theta' + \sin \theta \sin \theta' \cos(\phi - \phi')$.

*(7.14) Using (7.13), show that P.I.$[f]$ over the unit ball B_1 may be expressed in spherical coordinates as follows

$$\Psi(\mathbf{x}) = \frac{1}{4\pi} \int_0^\pi \int_0^{2\pi} f(\cos \phi' \sin \theta', \sin \phi' \sin \theta', \cos \theta')$$

$$\times \frac{(1 - r^2) \sin \theta'}{(1 + r^2 - 2r \cos \mu)^{3/2}} d\phi' \, d\theta'$$

*(7.15) Generalize (7.14) and (7.15) to show that P.I.$[f]$ over the ball B_a may be expressed in spherical coordinates as

$$\Psi(\mathbf{x}) = \frac{a}{4\pi} \int_0^\pi \int_0^{2\pi} f(a \cos \phi' \sin \theta', a \sin \phi' \sin \theta', a \cos \theta')$$

$$\times \frac{(a^2 - r^2) \sin \theta'}{(a^2 + r^2 - 2ar \cos \mu)^{3/2}} d\phi' \, d\theta'$$

where $\mathbf{x} = (r \cos \phi \sin \theta, r \sin \phi \sin \theta, r \cos \theta)$ for $0 \le r \le a$ and $\cos \mu$ is defined as in (7.13).

§8. Validity of Spherical Harmonic Series Expansions

In this section we will show that the series (5.14), proposed as a solution to Dirichlet's problem for the ball B_a centered at the origin, is identical to the Poisson integral solution described in Corollary (7.8). That identity will prove that the series form is a valid solution.

From Exercise (7.15) we know that P.I.$[f]$ on the ball B_a may be expressed as

$$\textbf{(8.1)} \qquad \Psi(\mathbf{x}) = \frac{a}{4\pi} \int_0^\pi \int_0^{2\pi} F(\theta', \phi') \frac{(a^2 - r^2) \sin \theta'}{(a^2 + r^2 - 2ar \cos \mu)^{3/2}} d\phi' \, d\theta'$$

where

$$F(\theta', \phi') = f(a \cos \phi' \sin \theta', a \sin \phi' \sin \theta', a \cos \theta')$$

$$\mathbf{x} = (r \cos \phi \sin \theta, r \sin \phi \sin \theta, r \cos \theta) \qquad (0 \le r < a)$$

$$\cos \mu = \cos \theta \cos \theta' + \sin \theta \sin \theta' \cos(\phi - \phi')$$

Writing a as a^3/a^2 and factoring a^2 from the numerator and a^3 from the denominator of the fraction in (8.1) we obtain

$$(8.1') \qquad \Psi(\mathbf{x}) = \frac{1}{4\pi} \int_0^\pi \int_0^{2\pi} F(\theta', \phi') \frac{(1 - \rho^2) \sin \theta'}{(1 + \rho^2 - 2\rho \cos \mu)^{3/2}} d\phi' d\theta'$$

where $\rho = r/a$. Note that $|\rho| < 1$.

Combining formulas (2.14) and (2.16), and using ρ in place of r, we obtain

$$\frac{1 - \rho^2}{(1 + \rho^2 - 2\rho z)^{3/2}} = (1 - \rho^2) \sum_{n=0}^\infty P'_{n+1}(z)\rho^n$$

$$= P_0(z) + 3P_1(z)\rho + \sum_{n=2}^\infty [P'_{n+1}(z) - P'_{n-1}(z)]\rho^n$$

Applying the recursion formula (2.6) we get

$$(8.2) \qquad \frac{1 - \rho^2}{(1 + \rho^2 - 2\rho z)^{3/2}} = \sum_{n=0}^\infty (2n + 1)P_n(z)\rho^n$$

Putting $z = \cos \mu$ in (8.2) and returning to (8.1') we have

$$(8.3) \qquad \begin{aligned} \Psi(\mathbf{x}) &= \frac{1}{4\pi} \int_0^\pi \int_0^{2\pi} F(\theta', \phi') \left[\sum_{n=0}^\infty (2n + 1)P_n(\cos \mu)\rho^n \right] \sin \theta' \, d\phi' \, d\theta' \\ &= \sum_{n=0}^\infty \left[\frac{2n + 1}{4\pi} \int_0^\pi \int_0^{2\pi} F(\theta', \phi')P_n(\cos \mu) \sin \theta' \, d\phi' \, d\theta' \right] \rho^n \end{aligned}$$

The interchange of summation and integration in (8.3) is justified by Weierstrass' M-test, using (2.12).

We will have accomplished our conversion of Poisson's integral into a series of spherical harmonics when we apply the following result to (8.3).

(8.4) Theorem: Addition Formula for Spherical Harmonics. If $\cos \mu = \cos \theta \cos \theta' + \sin \theta \sin \theta' \cos(\phi - \phi')$ then

$$P_n(\cos \mu) = \sum_{m=-n}^n \frac{(n - |m|)!}{(n + |m|)!} \bar{Y}_{mn}(\theta', \phi')Y_{mn}(\theta, \phi)$$

Proof. Fix momentarily the vector $\mathbf{v}$ on ∂B_1. Suppose that $\mathbf{v}$ and $\mathbf{u}$ are described in spherical coordinates by

$$\mathbf{v} = (\cos \phi \sin \theta, \sin \phi \sin \theta, \cos \theta)$$

$$\mathbf{u} = (\cos \phi' \sin \theta', \sin \phi' \sin \theta', \cos \theta')$$

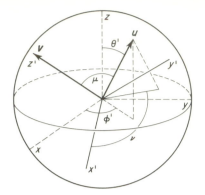

Figure 8.4   Spherical coordinates with polar axis **v**.

and that **u** varies over the sphere ∂B_1. It is easy to see that $\cos \mu$ equals the cosine of the angle that **u** makes with the axis determined by **v**. Introducing two other axes perpendicular to **v** we can form a Cartesian coordinate system x', y', z'. (See Figure 8.4). Letting v be the angle in the $x'-y'$ plane, we have the following equations for spherical coordinates relative to this new system

(8.5) $x' = \rho \cos v \sin \mu$ $y' = \rho \sin v \sin \mu$ $z' = \rho \cos \mu$

Hence, in this new system of coordinates, $h = \rho^n P_n(\cos \mu)$ is a spherical harmonic of order n. By Theorem (6.12)

$$h_{xx} + h_{yy} + h_{zz} = h_{x'x'} + h_{y'y'} + h_{z'z'} = 0$$

which implies that h is a spherical harmonic in the coordinates x, y, z, the standard Cartesian coordinate system. Therefore, using the solid spherical harmonics $\{H_{mn}\}$ as a basis for expanding h, we have upon changing to spherical coordinates

(8.6) $$P_n(\cos \mu)\rho^n = \sum_{m=-n}^{n} a_{mn} Y_{mn}(\theta', \phi')\rho^n$$

By the orthogonality of the spherical harmonics $\{Y_{mn}\}$ we have

(8.7) $$a_{mn} = \frac{(n - |m|)!}{(n + |m|)!} \frac{2n + 1}{4\pi} \int_0^\pi \int_0^{2\pi} P_n(\cos \mu) \bar{Y}_{mn}(\theta', \phi') \sin \theta' \, d\phi' \, d\theta'$$

On the other hand, letting **v** vary, we see that the homogeneous harmonic polynomial $\rho^n \bar{Y}_{mn}(\theta, \phi) = (r/a)^n e^{-im\phi} \sin^{|m|} \theta P_n^{(|m|)}(\cos \theta)$ solves Dirichlet's problem for the ball B_a where $f = \bar{H}_{mn} = \bar{Y}_{mn}$ on ∂B_a. Therefore, by formula (8.3) and the uniqueness of solutions to Dirichlet's problem, we must have

(8.8)
$$\rho^n \bar{Y}_{mn}(\theta, \phi) = \sum_{k=0}^{\infty} \left[\frac{2k + 1}{4\pi} \int_0^\pi \int_0^{2\pi} \bar{Y}_{mn}(\theta', \phi') \right.$$
$$\left. \times P_k(\cos \mu) \sin \theta' \, d\phi' \, d\theta' \right] \rho^k$$

Comparing powers of ρ in (8.8), we obtain in view of (8.7)

$$a_{mn} = \frac{(n - |m|)!}{(n + |m|)!} \bar{Y}_{mn}(\theta, \phi)$$

Substituting this last formula into (8.6) and cancelling ρ^n yields the addition formula (after applying complex conjugation and noting that $P_n(\cos \mu)$ is *real valued*). ∎

Because of the addition formula for spherical harmonics, formula (8.3) becomes

$$\Psi(\mathbf{x}) = \sum_{n=0}^{\infty} \left[\sum_{m=-n}^{n} c_{mn} Y_{mn}(\theta, \phi) \right] (r/a)^n$$

where

$$c_{mn} = \frac{(n - |m|)!}{(n + |m|)!} \frac{2n + 1}{4\pi} \int_0^{\pi} \int_0^{2\pi} F(\theta', \phi') \bar{Y}_{mn}(\theta', \phi') \sin \theta' \, d\phi' \, d\theta'$$

Thus, in view of the three equations following (8.1), we have proved that for a given continuous function f on B_a the solution to Dirichlet's problem is given by (8.1) in integral form, or by (5.14) in series form. In particular, (5.14) is now seen to be a mathematically justified form of solving problem (5.1).

One consequence of these results is that the Legendre polynomials $\{P_n\}_{n=0}^{\infty}$ are a complete orthogonal system over the interval $[-1, 1]$.

(8.9) Theorem. The Legendre polynomials are a complete orthogonal system over the interval $[-1, 1]$.

Proof. We will give a proof that can be generalized to prove that $\{Y_{mn}\}$ is complete [see Remark (8.13)]. Suppose g is a given continuous function on $[-1, 1]$. Let f be the continuous function on ∂B_1 defined for $\mathbf{u} = (x, y, z)$ by $f(\mathbf{u}) = g(z)$. The solution Ψ, to Dirichlet's problem on B_1 with boundary function f, is continuous on the compact ball $\bar{B}_1$. Therefore, Ψ is uniformly continuous, hence for $\mathbf{x} = r\mathbf{u}$

(8.10) $$\lim_{r \to 1} \Psi(r\mathbf{u}) = f(\mathbf{u}) = g(z)$$

holds uniformly for all $\mathbf{u}$ on ∂B_1. However, since $f(\mathbf{u})$ depends only on the angle θ in spherical coordinates, formula (5.14) reduces to

(8.10′) $$\Psi(r\mathbf{u}) = \sum_{n=0}^{\infty} c_n P_n(\cos \theta) r^n \qquad (0 \le r < 1)$$

where

$$c_n = c_{0n} = \frac{2n + 1}{2} \int_0^{\pi} g(\cos \theta) P_n(\cos \theta) \sin \theta \, d\theta$$

Substituting $z = \cos \theta$ into (8.10′) we have from (8.10)

(8.11) $$\left| g(z) - \sum_{n=0}^{\infty} c_n P_n(z) r^n \right| < \epsilon$$

if r is taken sufficiently close to 1. In (8.11) the coefficients c_n are equal to $[(2n+1)/2] \int_{-1}^{1} g(z) P_n(z) \, dz$, so they are the coefficients in the Legendre polynomial series for g.

Choose $r = r_0$ so that (8.11) holds. Since $|P_n(z)| \le 1$ and

$$|c_n| \le \frac{2n+1}{2} \int_{-1}^{1} |g(z)| \, dz \le (2n+1)M$$

where

$$M = \underset{-1 \le z \le 1}{\text{maximum}} |g(z)|$$

it follows that

(8.11′)
$$\left| \sum_{n=N+1}^{\infty} c_n P_n(z) r_0^n \right| \le \sum_{n=N+1}^{\infty} M(2n+1)r_0^n < \epsilon$$

provided N is taken sufficiently large. From (8.11) and (8.11′) we conclude that

$$\left\| g - \sum_{n=0}^{N} c_n r_0^n P_n \right\|_2 \le \left\| g - \sum_{n=0}^{\infty} c_n P_n r_0^n \right\|_2 + \left\| \sum_{n=N+1}^{\infty} c_n P_n r_0^n \right\|_2$$

$$< \left[\int_{-1}^{1} \epsilon^2 \, dz \right]^{1/2} + \left[\int_{-1}^{1} \epsilon^2 \, dz \right]^{1/2} = 2\sqrt{2}\epsilon$$

By the least squares property $\|g - \sum_{n=0}^{N} c_n P_n\|_2 < 2\sqrt{2}\epsilon$ provided N is sufficiently large. Thus

$$\lim_{N \to \infty} \left\| g - \sum_{n=0}^{N} c_n P_n \right\|_2 = 0$$

which proves the completeness of $\{P_n\}$. ∎

(8.12) Corollary: Weierstrass Approximation Theorem. Let g be a continuous function on the closed interval $[a, b]$. Given $\delta > 0$ we can find a polynomial p_δ such that $|g(z) - p_\delta(z)| < \delta$ for all z in $[a, b]$.

Proof. Suppose first that $[a, b] = [-1, 1]$. By combining (8.11) and (8.11′) we have $|g(z) - \sum_{n=0}^{N} c_n r_0^n P_n(z)| < 2\epsilon$ which is clearly sufficient for proving the theorem for the interval $[-1, 1]$, since $\sum_{n=0}^{N} c_n r_0^n P_n(z)$ is a polynomial. For other intervals we use a simple change of variables to reduce the problem to the interval $[-1, 1]$. [See Exercise (8.16).] ∎

(8.13) Remark. The completeness relation

(a) $\quad \lim_{N \to \infty} \int_{0}^{\pi} \int_{0}^{2\pi} \left| F(\theta, \phi) - \sum_{n=0}^{N} \left[\sum_{m=-n}^{n} c_{mn} Y_{mn}(\theta, \phi) \right] \right|^2 \sin \theta \, d\phi \, d\theta = 0$

for every continuous F on the rectangle $0 \le \phi \le 2\pi$, $0 \le \theta \le \pi$ can be proved in a manner similar to the proof of Theorem (8.9). We proved Theorem (8.9) because, although it is less general than (a), all of the essential ideas involved in proving (a) are used in proving that simpler theorem. The proof of (a) is outlined in Exercises (8.18) through (8.20).

Exercises

(8.14) Prove that $\cos \mu = \cos \theta \cos \theta' + \sin \theta \sin \theta' \cos(\phi - \phi')$ equals the cosine of the angle which $\mathbf{v} = (\cos \phi \sin \theta, \sin \phi \sin \theta, \cos \theta)$ makes with $\mathbf{u} = (\cos \phi' \sin \theta', \sin \phi' \sin \theta', \cos \theta')$. (*Hint*: Use the dot product definition of cosine.]

(8.15) Without comparing powers of ρ, show that the series on the right side of (8.8) reduces to

$$\frac{2n+1}{4\pi} \int_0^\pi \int_0^{2\pi} \bar{Y}_{mn}(\theta', \phi') P_n(\cos \mu) \sin \theta' \, d\phi' \, d\theta' \, \rho^n$$

(8.16) Complete the proof of (8.12). [*Hint*: Transform g on $[a, b]$ into h on $[-1, 1]$ by a linear substitution.]

(8.17) Suppose that a function Ψ is harmonic on the ball $B_R(\mathbf{x}_0)$ and is identically zero on the smaller ball $B_{R'}(\mathbf{x}_0)$ where $R' < R$. Prove that Ψ is identically zero on the whole disk $B_R(\mathbf{x}_0)$.

(8.18) Express the spherical harmonic addition formula in the real form

$$P_n(\cos \mu) = u_{m0}(\theta', \phi') u_{m0}(\theta, \phi) + 2 \sum_{m=1}^n \frac{(n-m)!}{(n+m)!} [u_{mn}(\theta', \phi') u_{mn}(\theta, \phi)$$
$$+ v_{mn}(\theta', \phi') v_{mn}(\theta, \phi)]$$

for appropriate real valued functions u_{mn}, v_{mn}. [See Exercise (5.19).]

(8.19) Prove that the least squares property holds for series of the type

$$\sum_{n=0}^\infty \left[\sum_{m=0}^n a_{mn} u_{mn}(\theta, \phi) + b_{mn} v_{mn}(\theta, \phi) \right]$$

where a_{mn}, b_{mn} are arbitrary real constants, in comparison with the Fourier series

$$f \sim \sum_{n=0}^\infty \left[\sum_{m=0}^n A_{mn} u_{mn}(\theta, \phi) + B_{mn} v_{mn}(\theta, \phi) \right]$$

where A_{mn}, B_{mn} are defined as in the solution to Exercise (5.19).

(8.20) Prove the completeness relation (8.13a). [*Hint*: First, let $F(\theta, \phi) = f(\cos \phi \sin \theta, \sin \phi \sin \theta, \cos \theta)$ for f continuous and real valued on ∂B_1 and examine formula (8.3) to get inequalities analogous to those in the proof of (8.9). The general case is treated by approximating an arbitrary continuous (real valued) F by continuous functions such as the first type.]

§9. Properties of Harmonic Functions

The properties of harmonic functions of two variables, which we discussed in §5 of Chapter 4, generalize nicely to harmonic functions of three variables. In this section we will briefly discuss these properties.

The key idea is that every harmonic function Ψ is locally a Poisson integral.

(9.1) Theorem.   Suppose that Ψ is a harmonic function on an open set U in $\mathbb{R}^3$. For each closed ball $\bar{B}_R(\mathbf{x}_0)$ contained in U, the Poisson integral P.I.$[\Psi]$ equals Ψ on $B_R(\mathbf{x}_0)$.

Proof.   The proof is analogous to the proof of Theorem (4.5), Chapter 4. The uniqueness of solutions to Dirichlet's problem for $B_R(\mathbf{x}_0)$ guarantees that the function $\Psi^{\#}$ defined by

$$\Psi^{\#}(\mathbf{x}) = \begin{cases} \text{P.I.}[\Psi](\mathbf{x}) & \text{if } \mathbf{x} \text{ is in } B_R(\mathbf{x}_0) \\ \Psi(\mathbf{x}) & \text{if } \mathbf{x} \text{ is on } \partial B_R(\mathbf{x}_0) \end{cases}$$

equals Ψ on $\bar{B}_R(\mathbf{x}_0)$. In particular, in the ball $B_R(\mathbf{x}_0)$, $\Psi = \text{P.I.}[\Psi]$.   ∎

In the same way that all of the properties of two-dimensional harmonic functions followed from Theorem (4.5), Chapter 4, we find that analogous properties for three-dimensional harmonic functions follow from Theorem (9.1).

We already know that Gauss's mean value theorem holds [see Theorem (6.6), Chapter 4]. Recall that that theorem says that

$$\Psi(\mathbf{x}_0) = \frac{1}{4\pi R^2} \int_{\partial B_R(\mathbf{x}_0)} \Psi(\mathbf{u}) \, dA_{\mathbf{u}}$$

whenever the closed ball $\bar{B}_R(\mathbf{x}_0)$ is contained in an open set U over which Ψ is harmonic. Or, writing $\mathbf{u} = x_0 + R\mathbf{v}$ where $\mathbf{v}$ is on ∂B_1 we also have

$$\Psi(\mathbf{x}_0) = \frac{1}{4\pi} \int_{\partial B_1} \Psi(\mathbf{x}_0 + R\mathbf{v}) \, dA_{\mathbf{v}}$$

since $dA_{\mathbf{u}} = R^2 \, dA_{\mathbf{v}}$. This latter form of Gauss's mean value theorem will be useful in proving Harnack's inequality.

(9.2) Theorem: Harnack's Inequality.   Let Ψ be a harmonic function on an open set U in $\mathbb{R}^3$. Suppose that Ψ has only nonnegative values. If the closed disk $\bar{B}_R(\mathbf{x}_0)$ is contained in U, then for a point $\mathbf{x}$ in the open disk $B_R(\mathbf{x}_0)$, which lies at a distance r from $\mathbf{x}_0$, we have

$$R\frac{R-r}{(R+r)^2} \Psi(\mathbf{x}_0) \leq \Psi(\mathbf{x}) \leq R\frac{R+r}{(R-r)^2} \Psi(\mathbf{x}_0)$$

Proof.   We have for $\mathbf{v}$ on ∂B_1

$$\frac{R}{4\pi} \frac{R^2 - |\mathbf{x} - \mathbf{x}_0|^2}{|(\mathbf{x} - \mathbf{x}_0) - R\mathbf{v}|^3} = \frac{R}{4\pi} \frac{R^2 - r^2}{[R^2 + r^2 - 2rR\cos\mu]^{3/2}}$$

where $\cos \mu$ is the cosine of the angle between $\mathbf{x} - \mathbf{x}_0$ and $R\mathbf{v}$. Since $-1 \leq \cos \mu \leq 1$ we have

$$\frac{R}{4\pi} \frac{R-r}{(R+r)^2} \leq \frac{R}{4\pi} \frac{R^2 - |\mathbf{x} - \mathbf{x}_0|^2}{|\mathbf{x} - \mathbf{x}_0 - R\mathbf{v}|^3} \leq \frac{R}{4\pi} \frac{R+r}{(R-r)^2}$$

Multiplying the inequalities above by $\Psi(\mathbf{x}_0 + R\mathbf{v})$ and integrating over ∂B_1 we obtain Harnack's inequality when we apply Gauss's mean value theorem on the ends of the inequalities and the definition of Poisson's integral in the middle. ∎

(9.3) Remark.   Since

$$R\frac{R-r}{(R+r)^2} \leq R\frac{R-r'}{(R+r')^2}$$

and

$$R\frac{R+r'}{(R-r')^2} \leq R\frac{R+r}{(R-r)^2}$$

for $r' \leq r$, it follows that Harnack's inequality holds for all points $\mathbf{x}$ in the ball $B_R(\mathbf{x}_0)$ that are at a distance *less than or equal to* r from $\mathbf{x}_0$.

From Harnack's inequality we immediately obtain Picard's theorem.

(9.4) Theorem: Picard.   Suppose that Ψ is harmonic on all of $\mathbb{R}^3$. If Ψ is bounded below (above) by a constant, then Ψ is a constant function.

Proof.   See Exercise (9.9). ∎

Another corollary of Harnack's inequality is Harnack's first convergence theorem.

(9.5) Theorem: Harnack's First Convergence Theorem.   Let $\{\Psi_n\}$ be a sequence of harmonic functions on an open set U in $\mathbb{R}^3$. If Ψ_n converges uniformly to the function g on every compact subset of U, then g is harmonic on U.

Proof.   We generalize the proof of Theorem (5.5), Chapter 4 by simply using $\mathbf{x}_0$ in place of (c, d). The key inequality is then [for $|\mathbf{x} - \mathbf{x}_0| = r < R$]

$$|\text{P.I.}[g](\mathbf{x}) - \text{P.I.}[\Psi_m](\mathbf{x})| \leq \frac{\epsilon R}{4\pi} \int_{\partial B_1} \frac{R^2 - r^2}{|\mathbf{x} - \mathbf{x}_0 - R\mathbf{v}|^3} \, dA_\mathbf{v}$$

$$= \epsilon \int_{\partial B_1} p(R^{-1}(\mathbf{x} - \mathbf{x}_0); \mathbf{v}) \, dA_\mathbf{v}$$

$$= \epsilon$$

because of Lemma (7.3), property (A_2). Hence

$$g(\mathbf{x}) = \lim_{m \to \infty} \Psi_m(\mathbf{x}) = \lim_{m \to \infty} \text{P.I.}[\Psi_m](\mathbf{x})$$

$$= \text{P.I.}[g](\mathbf{x})$$

Since $g = \text{P.I.}[g]$ over every ball in U it follows that g is harmonic on U. ∎

All the other theorems in §§5 and 8 of Chapter 4 generalize in a similar fashion to harmonic functions of three variables. By a subharmonic function g we mean that g is continuous on an open set U in $\mathbb{R}^3$ and for each $\mathbf{x}_0$ in U there exists a radius R such that for all radii $r < R$

$$g(\mathbf{x}_0) \leq \frac{1}{4\pi} \int_{\partial B_1} g(\mathbf{x}_0 + r\mathbf{v}) \, dA_{\mathbf{v}} = \frac{1}{4\pi r^2} \int_{\partial B_r(\mathbf{x}_0)} g(\mathbf{u}) \, dA_{\mathbf{u}}$$

The details of these generalizations are left to the reader to fill in. Some of them will be given as exercises below.

Exercises

(9.6) Consider formula (4.12), Chapter 5. Prove that if f_1 is absolutely integrable over the rectangular region $[0, \pi] \times [0, \pi]$, then Ψ is harmonic inside the cube $(0, \pi) \times (0, \pi) \times (0, \pi)$.

(9.7) Consider formula (4.12), Chapter 5, again. Prove that if f_1 is C^2 on $[0, \pi] \times [0, \pi]$, then Ψ rigorously solves problem (4.9), Chapter 5.

(9.8) Discuss Dirichlet's problem for the cube $C = (0, \pi) \times (0, \pi) \times (0, \pi)$.

(9.9) Prove Theorem (9.4) and generalize Corollary (5.8), Chapter 4, to three dimensions.

(9.10) Prove the following three-dimensional version of Schwarz's Reflection Principle.

Theorem. Suppose that Ψ is harmonic in the upper half space $\mathbb{R}^3_+ = \{(x, y, z): z > 0\}$ and continuous on the closure $\overline{\mathbb{R}}^3_+ = \{(x, y, z): z \geq 0\}$, and $\Psi = 0$ on the x–y plane $\mathbb{R}^3_+$. Then the function $\Psi^{\#}$ defined by $\Psi^{\#}(x, y, z) = \Psi(x, y, z)$ if $z \geq 0$ and $\Psi^{\#}(x, y, z) = -\Psi(x, y, -z)$ if $z < 0$ is harmonic on $\mathbb{R}^3$.

(9.11) Conclude from (9.10) that if Ψ_1 and Ψ_2 are harmonic and bounded on the upper half space, are continuous on its closure, and agree on the x–y plane, then $\Psi_1 = \Psi_2$ on the upper half space.

(9.12) We say that a continuous function g defined on an open set U in $\mathbb{R}^3$ has the *Mean Value Property* if for each closed ball $\bar{B}_R(\mathbf{x}_0)$ contained in U

$$g(\mathbf{x}_0) = \frac{1}{4\pi R^2} \int_{\partial B_R(\mathbf{x}_0)} g(\mathbf{u}) \, dA_{\mathbf{u}} = \frac{1}{4\pi} \int_{\partial B_1} g(\mathbf{x}_0 + R\mathbf{v}) \, dA_{\mathbf{v}}$$

Generalize Theorem (5.12), Chapter 4, to three dimensions.

(9.13) Generalize Theorems (8.2) and (8.3), Chapter 4, to three dimensions.

(9.14) Suppose that Ψ is a non-negative harmonic function on a connected open set U in $\mathbb{R}^3$. Prove that if Ψ is not constantly zero, then Ψ is always positive in value.

(9.15) Suppose that Ψ_1 and Ψ_2 are harmonic on a connected open set in $\mathbb{R}^3$. Prove that if $\Psi \geq \Psi_2$ on U and $\Psi_1 \neq \Psi_2$, then $\Psi_1 > \Psi_2$.

(9.16) Give some examples of three-dimensional subharmonic functions that are not harmonic.

(9.17) Generalize Theorems (8.13) and (8.14), Chapter 4, to three dimensions.

(9.18) Give examples of open, connected sets in $\mathbb{R}^3$ that have barriers at every point of their boundaries. Give some examples of open, connected sets in $\mathbb{R}^3$ that do not possess that property.

§10. Miscellaneous Exercises

Gegenbauer Polynomials

These polynomials, also known as *ultraspherical polynomials*, play a role analogous to the Legendre polynomials for dimensions greater than 3. In the exercises below the reader is asked to derive a few of their properties. For more details, see Erdélyi et al. (1953, 1954).

(10.1) Define $G_n^\lambda(z)$ by

$$[1 + r^2 - 2rz]^{-\lambda} = \sum_{n=0}^{\infty} G_n^\lambda(z)r^n \qquad (\lambda > 0)$$

Show that $G_0^\lambda(z) = 1$, $G_1^\lambda(z) = 2\lambda z$, and $G_2^\lambda(z) = 2\lambda(\lambda + 1)z^2 - \lambda$.

(10.2) Derive the recursion relation

$$(n + 1)G_{n+1}^\lambda(z) - 2(n + \lambda)zG_n^\lambda(z) + (n + 2\lambda - 1)G_{n-1}^\lambda(z) = 0$$

(10.3) Derive the differential equation

$$(1 - z^2)\frac{d^2}{dz^2}G_n^\lambda(z) - (2\lambda + 1)z\frac{d}{dz}G_n^\lambda(z) + n(n + 2\lambda)G_n^\lambda(z) = 0$$

(10.4) Prove that if $m \neq n$ then

$$\int_{-1}^1 G_m^\lambda(z)G_n^\lambda(z)[1 - z^2]^{\lambda - 1/2} \, dz = 0$$

Remark. Exercise (10.4) shows that $\{G_n^\lambda(z)\}_{n=0}^{\infty}$ is a weighted orthogonal set of functions over $[-1, 1]$. [See the Remark following (5.16).] Using (10.4) we can expand a piecewise continuous function f over $(-1, 1)$ in a generalized Fourier series

$$f \sim \sum_{n=0}^{\infty} c_n G_n^\lambda(z)$$

where

$$c_n = \left[\int_{-1}^{1} f(z) G_n^\lambda(z)(1 - z^2)^{\lambda - 1/2} \, dz \right] \bigg/ \left[\int_{-1}^{1} (G_n^\lambda(z))^2 (1 - z^2)^{\lambda - 1/2} \, dz \right]$$

The rationale for this definition of c_n consists in formally integrating the series above, after multiplying by $G_n(z)(1 - z^2)^{\lambda - 1/2}$ and applying (10.4).

(10.5)   Prove that $G_n^\lambda(z)$ has n distinct roots all of which lie in $(-1, 1)$ and that G_n^λ is odd (even) if n is odd (even).

(10.6)   Derive the relation $(d/dz)G_n^\lambda(z) = 2\lambda G_{n-1}^{\lambda+1}(z)$. What does that relation imply concerning Gegenbauer polynomials and the associated Legendre polynomials?

References

For more information on spherical harmonics, see Hobson (1931), MacRobert (1967), Lebedev (1965), Dym and McKean (1972), Stein and Weiss (1971), and Erdélyi et al. (1953, 1954). For further applications, see Jackson (1962) or Landau and Lifshitz (1977). Legendre, associated Legendre, and Gegenbauer polynomials are treated as examples in the theory of orthogonal polynomials in Davis (1975), Natanson (1964, 1965), Wilf (1962) and Szegö (1939).

9

Some Other Transforms

There are many other useful transforms besides the Fourier transform. In this chapter we will discuss two of them. First, we shall briefly discuss the *Laplace transform*. Since this transform and its applications are extensively treated elsewhere, we shall settle for a brief introduction. One important point that some readers may not have seen before is the proof of the inversion theorem for Laplace transforms that we discuss in §2. Second, we shall treat in more detail the *Radon transform*. This transform has been the object of much recent study due primarily to its applications in medicine. Its principal use in medical technology is in the construction of CAT scanners. The reader who prefers to study the Radon transform first may turn immediately to §4.

Part A. The Laplace Transform

§1. Definition and Examples

The Laplace transform is particularly useful in initial value problems where a variable, x say, is assumed greater than or equal to 0. Suppose f is a given piecewise continuous function. The Laplace transform of f will be denoted by $f^>$ and is defined as a function of z by

(1.1)
$$f^>(z) = \int_0^{+\infty} e^{-zx} f(x)\, dx$$

Besides $f^>$ we might write $\mathcal{L}[f]$. We shall assume that x is real and $z = u + iv$ is complex with real part $\operatorname{Re} z = u$ and imaginary part $\operatorname{Im} z = v$.

The integral in (1.1) is shorthand for the improper integral

(1.1′)
$$f^>(z) = \lim_{\substack{b \to +\infty \\ a \to 0+}} \int_a^b e^{-zx} f(x)\, dx$$

Furthermore, f may be complex valued.

(1.2) Examples

(a)   Suppose $f(x) = 1$ for $x > 0$. Then

$$\int_0^{+\infty} e^{-zx}\, dx = \lim_{\substack{b \to +\infty \\ a \to 0+}} \frac{1}{z}(e^{-az} - e^{-bz})$$

If $\operatorname{Re} z > 0$ then e^{-bz} will tend to 0 as b tends to $+\infty$. Hence

$$f^>(z) = \frac{1}{z} \qquad (\operatorname{Re} z > 0)$$

(b)   Suppose $f(x) = e^{cx}$ where c is a complex constant. Then

$$\int_0^{+\infty} e^{-zx} e^{cx}\, dx = \lim_{\substack{b \to +\infty \\ a \to 0+}} \left[\frac{e^{-(z-c)a}}{z - c} - \frac{e^{-(z-c)b}}{z - c} \right]$$

If $\operatorname{Re}(z - c) > 0$ then $e^{-(z-c)b}$ will tend to 0 as b tends to $+\infty$. Thus

$$f^>(z) = \mathscr{L}[e^{cx}](z) = \frac{1}{z - c} \qquad (\operatorname{Re} z > \operatorname{Re} c)$$

(c)   Suppose $f(x) = x$. Then, using Kronecker's rule,

$$\int_0^{+\infty} x e^{-zx}\, dx = \lim_{\substack{b \to +\infty \\ a \to 0+}} \left[\frac{ae^{-az}}{z} - \frac{be^{-bz}}{z} + \frac{e^{-az}}{z^2} + \frac{e^{-bz}}{z^2} \right]$$

When $\operatorname{Re} z > 0$ we will have be^{-bz} and e^{-bz} tending to 0 as b tends to $+\infty$. Therefore

$$f^>(z) = \frac{1}{z^2} \qquad (\operatorname{Re} z > 0)$$

(d)   Suppose $f(x) = \sin bx$ for b a real constant. Writing $\sin bx$ as $(i/2)e^{-ibx} - (i/2)e^{ibx}$ and applying example (b) we obtain

$$f^>(z) = \frac{b}{z^2 + b^2} \qquad (\operatorname{Re} z > 0)$$

Sometimes we shall use the notation $f(x) \underset{\mathscr{L}}{\supset} f^>(z)$ to denote the Laplace transform process. The examples above can then be summarized as follows.

(1.3) $1 \underset{\mathscr{L}}{\supset} \dfrac{1}{z}$ $e^{cx} \underset{\mathscr{L}}{\supset} \dfrac{1}{z - c}$ $x \underset{\mathscr{L}}{\supset} \dfrac{1}{z^2}$ $\sin bx \underset{\mathscr{L}}{\supset} \dfrac{b}{z^2 + b^2}$

(1.4) Remark. In most of the examples above the function f was *not* absolutely integrable. Yet the Laplace transform $f^>$ was defined. This enlargement of the class of transformable functions beyond those that are absolutely integrable allows the Laplace transform to be applied in some situations where the Fourier transform cannot be used.

We close this section with a theorem that supplies us with a sufficient condition for when the Laplace transform integral in (1.1) is defined.

(1.5) Theorem. If there exists a positive constant B and a real constant c such that the piecewise continuous function f satisfies

$$|f(x)| \leq Be^{cx} \qquad (x > 0)$$

then $f^>$ is defined for $\operatorname{Re} z > c$.

Proof. The theorem follows from the fact that $|f(x)e^{-zx}| \leq Be^{-(\operatorname{Re} z - c)x}$ and the latter function has a finite integral over $[0, +\infty)$ when $\operatorname{Re} z > c$. ∎

Remark. Theorem (1.5) shows that the Laplace transform $f^>$ is always defined for some half-plane region ($\operatorname{Re} z > c$) in the complex plane. The condition on f in Theorem (1.5) is usually abbreviated as $f = 0(e^{cx})$.

Exercises

(1.6) Find the Laplace transforms of the following functions.

(a) x^2
(b) $\sinh bx$
(c) $\cosh bx$
(d) $\cos bx$
(e) xe^{cx} (c complex)

(1.7) Find the Laplace transforms of the following functions.

(a) $f(x) = \begin{cases} 1 & \text{if } 0 \leq x \leq 4 \\ 0 & \text{if } 4 < x < +\infty \end{cases}$

(b) $f(x) = \begin{cases} x & \text{if } 0 \leq x \leq 2 \\ 0 & \text{if } 2 < x < +\infty \end{cases}$

(1.8) Let f satisfy the conditions of Theorem (1.5). Show that if $f^> = F + iG$ where $z = u + iv$ then the following equations are continuously satisfied for $u > c$

(a) $\dfrac{\partial F}{\partial u} = \dfrac{\partial G}{\partial v} \qquad \dfrac{\partial F}{\partial v} = -\dfrac{\partial G}{\partial u}$

Remark. The equations in (1.8a) are the *Cauchy–Riemann equations*. Their satisfaction (in a continuous fashion) guarantees that $f^>$ is an analytic function of z for $\operatorname{Re} z > c$. [See Knopp (1945), or LePage (1980), or Ahlfors (1966).]

§2. Properties of the Laplace Transform

In this section we shall briefly describe some of the more elementary properties of the Laplace transform. The simplest properties, such as linearity and change of scale, we leave as exercises and begin with the relation between the Laplace transform and differentiation.

(2.1) Theorem. Suppose that f is continuous and f' is piecewise continuous over $[0, +\infty)$[1] and $f(x) = 0(e^{cx})$ then

$$f'(x) \underset{\mathscr{L}}{\supset} zf^>(z) - f(0+) \qquad (\operatorname{Re} z > c)$$

Proof. Computing $\mathscr{L}[f']$ we have

$$\int_0^{+\infty} e^{-zx}f'(x)\,dx = \lim_{\substack{b \to +\infty \\ a \to 0+}} \int_a^b e^{-zx}f'(x)\,dx$$

$$= \lim_{\substack{b \to + \\ a \to 0+}} \left[e^{-bz}f(b) - e^{-az}f(a) + z\int_a^b e^{-zx}f(x)\,dx \right]$$

where an integration by parts was performed to obtain the last line. Since $f(x) = 0(e^{cx})$ we have for some positive constant B

$$|e^{-bz}f(b)| \le Be^{-b(\operatorname{Re} z - c)}$$

Hence for $\operatorname{Re} z > c$ we have $\lim_{b \to +\infty} e^{-bz}f(b) = 0$ and $f^>(z)$ is defined. Therefore

$$\int_0^{+\infty} e^{-zx}f'(x)\,dx = zf^>(z) - f(0+) \qquad \blacksquare$$

For example, since $(d/dx)(x^2) = 2x$ and $2x \underset{\mathscr{L}}{\supset} (2/z^2)$ we must have

(2.2) $$x^2 \underset{\mathscr{L}}{\supset} \frac{2}{z^3} \qquad (\operatorname{Re} z > 0)$$

(2.3) Corollary. If $f^{(j)}$ for $j = 0, 1, \ldots, n-1$ are all continuous and $0(e^{cx})$ over $[0, +\infty)$ and $f^{(n)}$ is piecewise continuous over that interval, then

$$f^{(n)}(x) \underset{\mathscr{L}}{\supset} z^n f^>(z) - z^{n-1}f(0+) - z^{n-2}f'(0+) - \cdots - f^{(n-1)}(0+)$$

Proof. The proof is straightforward and we leave it to the reader. ∎

Just as with Fourier transforms we have a theorem that shows how multiplication by x transforms into a differentiation operation.

(2.4) Theorem. If f is piecewise continuous and $0(e^{cx})$ over $[0, +\infty)$, then

$$xf(x) \underset{\mathscr{L}}{\supset} -\frac{df^>}{dz} \qquad (\operatorname{Re} z > c)$$

---

1. That is, f is continuous and f' is piecewise continuous over $[0, A]$ for all finite numbers A.

Proof. Differentiate $\int_0^{+\infty} e^{-zx} f(x)\, dx$ under the integral sign. ∎

(2.5) Theorem. If f is piecewise continuous and $0(e^{cx})$ over $[0, +\infty)$ then

$$x^n f(x) \underset{\mathscr{L}}{\supset} (-1)^n \frac{d^n f^>}{dz^n} \qquad (\mathrm{Re}\, z > c) \qquad ∎$$

For example, since $e^{cx} \underset{\mathscr{L}}{\supset} (z - c)^{-1}$ it follows that $xe^{cx} \underset{\mathscr{L}}{\supset} (z - c)^{-2}$ (for $\mathrm{Re}\, z > \mathrm{Re}\, c$ when c is a complex constant).

Like the Fourier transform, the Laplace transform turns convolution into multiplication. For the Laplace transform, however, we need to define the convolution $f * g$ of two functions f and g as follows

$$f * g(x) = \int_0^x f(s) g(x - s)\, ds$$

The reader may show [see Exercise (2.16)] that $f * g = g * f$. Moreover, we have the following theorem.

(2.6) Theorem: Convolution Theorem. If f and g are $0(e^{cx})$ then

$$f * g(x) \underset{\mathscr{L}}{\supset} f^>(z) g^>(z)$$

Proof. We have

$$f^>(z) g^>(z) = \int_0^{+\infty} e^{-zu} f(u)\, du \int_0^{+\infty} e^{-zv} g(v)\, dv$$

$$= \int_0^{+\infty} \left[\int_0^{+\infty} f(u) e^{-z(u+v)}\, du \right] g(v)\, dv$$

Letting $w = u + v$ in the inner integral above yields

$$f^>(z) g^>(z) = \int_0^{+\infty} \left[\int_v^{+\infty} f(w - v) e^{-zw}\, dw \right] g(v)\, dv$$

Reversing the order of integration we obtain

$$f^>(z) g^>(z) = \int_0^{+\infty} e^{-zw} \left[\int_0^w f(w - v) g(v)\, dv \right] dw$$

$$= \int_0^{+\infty} e^{-zw} f * g(w)\, dw \qquad ∎$$

In addition to enjoying similar properties, the Laplace and Fourier transforms are closely linked as shown by the following theorem.

(2.7) Theorem. Let f be piecewise continuous and $0(e^{cx})$ then for $z = u + iv$

$$f^>(z) = \mathscr{F}[g(x; u)] \left(\frac{v}{2\pi} \right) \qquad (\mathrm{Re}\, z > c)$$

where $g(x; u) = e^{-ux} f(x)$ if $x \geq 0$ and $g(x; u) = 0$ if $x < 0$.

Proof.   For $z = u + iv$ we have

$$f^>(z) = \int_0^{+\infty} e^{-zx} f(x)\, dx = \int_0^{+\infty} e^{-ivx} e^{-ux} f(x)\, dx$$

$$= \int_{-\infty}^{+\infty} e^{-i2\pi(v/2\pi)x} g(x; u)\, dx = \mathscr{F}[g(x; u)]\left(\frac{v}{2\pi}\right) \quad \blacksquare$$

Theorem (2.7) provides us with the means for inverting Laplace transforms using Fourier inversion.

(2.8) Theorem: Inversion of Laplace Transforms.   Let f be piecewise smooth on $(0, +\infty)$ and $0(e^{cx})$. Then for every positive x value and any fixed u value greater than c,

$$\tfrac{1}{2}[f(x+) + f(x-)] = \lim_{R \to +\infty} \frac{1}{2\pi i} \int_{u-iR}^{u+iR} e^{zx} f^>(z)\, dz$$

If $x = 0$ then inversion yields $\tfrac{1}{2} f(0+)$.

Proof.   Treating $v/2\pi$ as a variable w in Theorem (2.7) we obtain by Fourier inversion for each positive x value (fixing some u value)

$$\tfrac{1}{2}[f(x+) + f(x-)]e^{-ux} = \mathrm{P.V.} \int_{-\infty}^{+\infty} e^{i2\pi wx} f^>(z)\, dw$$

$$= \lim_{R \to +\infty} \int_{-R}^{R} e^{ivx} f^>(z)\, \frac{dv}{2\pi}$$

where $z = u + iv$. [If $x = 0$ then the left side above is just $\tfrac{1}{2} f(0+)$.] Multiplying both sides above by e^{ux} we obtain

$$\tfrac{1}{2}[f(x+) + f(x-)] = \lim_{R \to +\infty} \frac{1}{2\pi} \int_{-R}^{R} e^{(u+iv)x} f^>(z)\, dv$$

For $z = u + iv$ we have $dz = i\,dv$ because u is fixed, hence

$$\tfrac{1}{2}[f(x+) + f(x-)] = \lim_{R \to +\infty} \frac{1}{2\pi i} \int_{u-iR}^{u+iR} e^{zx} f^>(z)\, dz$$

And, since the left side will be $\tfrac{1}{2} f(0+)$ if $x = 0$, our theorem is proved. $\blacksquare$

(2.9) Theorem.   If f and g are two piecewise continuous functions both of which are $0(e^{cx})$ and $f^> = g^>$, then $f = g$ except for a finite number of points in every closed interval $[0, +\infty)$. $\blacksquare$

The inversion integral

$$\lim_{R \to +\infty} \frac{1}{2\pi i} \int_{u-iR}^{u+iR} e^{zx} f^>(z)\, dz$$

is amenable to treatment by methods of analytic function theory. The method of residues is particularly important. See Scott (1955), Chapter 2, §§9–11.

Exercises

(2.10) Find the Laplace transforms of the following functions.

(a) $x \cos bx$
(b) $e^{ax} \cos bx$
(c) x^3
(d) $x \sin bx$
(e) x^n

(2.11) *Linearity.* Prove that if $f(x) \underset{\mathscr{L}}{\supset} f^>(z)$ and $g(x) \underset{\mathscr{L}}{\supset} g^>(z)$ then for all complex constants a and b we have $af(x) + bg(x) \underset{\mathscr{L}}{\supset} af^>(z) + bg^>(z)$.

(2.12) *Change of Scales.* Suppose that $f(x) \underset{\mathscr{L}}{\supset} f^>(z)$ and c is a positive constant. Prove that $f(cx) \underset{\mathscr{L}}{\supset} (1/c)f^>(z/c)$.

(2.13) *Modulation.* Suppose that $f(x) \underset{\mathscr{L}}{\supset} f^>(b)$ and b is a real constant. Prove that

$$f(x)e^{ibx} \underset{\mathscr{L}}{\supset} f^>(z - ib)$$

$$f(x) \cos bx \underset{\mathscr{L}}{\supset} \tfrac{1}{2}f^>(z - ib) + \tfrac{1}{2}f^>(z + ib)$$

$$f(x) \sin bx \underset{\mathscr{L}}{\supset} (i/2)f^>(z + ib) - (i/2)f^>(z - ib)$$

(2.14) *Shifting.* Suppose that $f(x) \underset{\mathscr{L}}{\supset} f^>(z)$ and c is a positive constant. Prove that

$$f(x - c) \underset{\mathscr{L}}{\supset} e^{-cz}f^>(z)$$

where $f(x - c) = 0$ for $0 \le x < c$.

(2.15) Using the properties from (2.11) through (2.14), as well as the other properties described in this section, find the Laplace transforms of the following functions.

(a) $x + \tfrac{1}{2}x^2 e^{ibx}$
(b) $f(x) = \begin{cases} x & \text{if } x > 2 \\ 0 & \text{if } 0 \le x < 2 \end{cases}$
(c) $H_a(x) = \begin{cases} 1 & \text{if } x > a \\ 0 & \text{if } 0 \le x < a \end{cases}$
(d) $H_a(x)e^{cx}$

(2.16) Show that $f * g = g * f$.

§3. Some Remarks on Applications

As we said in the introduction to this chapter, there are many fine books on the applications of Laplace transforms. In this section we will present one simple application and then discuss some of the references that the reader might consult.

Suppose we want to solve the following initial value problem in ordinary differential equations

(3.1) $y'' + 3y' + 2y = f(x)$ $y(0) = 0$ $y'(0) = 0$

Solving (3.1) will allow us to solve

(3.1') $y'' + 3y' + 2y = f(x)$ $y(0) = a_0$ $y'(0) = a_1$

by adding a general solution $y_H(x) = c_0 e^{-x} + c_1 e^{-2x}$ of the homogeneous equation

(3.2) $$y'' + 3y' + 2y = 0$$

to the solution for (3.1) and using the constants c_0 and c_1 to match with the given initial values a_0 and a_1. As for (3.1), if we let

$$y^>(z) = \int_0^{+\infty} e^{-zx} y(x)\, dx$$

and use Corollary (2.3), we obtain from (3.1)

$$(z^2 + 3z + 2)y^>(z) = f^>(z)$$

Hence

(3.3) $$y^>(z) = \frac{1}{z^2 + 3z + 2} f^>(z)$$

Expanding in partial fractions reveals that

(3.4) $$\frac{1}{z^2 + 3z + 2} = \frac{1}{z+1} - \frac{1}{z+2}$$

Using the fact that

$$e^{-x} \underset{\mathscr{L}}{\supset} \frac{1}{z+1} \qquad e^{-2x} \underset{\mathscr{L}}{\supset} \frac{1}{z+2}$$

and the convolution theorem, we obtain from (3.3) that

(3.5) $$y(x) = \int_0^x f(s)\Phi(x-s)\, ds \qquad \Phi(x) = e^{-x} - e^{-2x}$$

The function Φ is called the *fundamental solution* of (3.1). Notice that it is a linear combination of solutions to the homogeneous equation (3.2).

Using the generalized Leibniz rule we can verify that (3.5) does solve (3.1). For

$$y' = \int_0^x f(s)\Phi'(x-s)\,ds \qquad y'' = f(x)\Phi'(0) + \int_0^x f(s)\Phi''(x-s)\,ds$$

$$= f(x) + \int_0^x f(s)\Phi''(x-s)\,ds$$

and substituting these relations into (3.1) yields

$$y'' + 3y' + 2y = f(x) + \int_0^x f(s)[\Phi''(x-s) + 3\Phi'(x-s) + 2\Phi(x-s)]\,ds$$

$$= f(x) + \int_0^x f(s)\,0\,ds = f(x)$$

$$y(0) = \int_0^0 f(s)\phi(x-s)\,ds = 0$$

$$y'(0) = \int_0^0 f(s)\Phi'(x-s)\,ds = 0$$

Thus (3.5) describes the solution to (3.1), provided the differentiations under the integral sign above are justified, which is the case if f is continuous. Moreover, the reader can easily verify that

$$y = (2a_0 + a_1)e^{-x} - (a_0 + a_1)e^{-2x} + \int_0^x f(s)\Phi(x-s)\,ds$$

solves (3.1').

The ideas above can certainly be seen to apply to any linear ordinary differential equation with constant coefficients and given initial values

(3.6)
$$y^{(n)} + b_{n-1}y^{(n-1)} + \cdots + b_1 y^{(1)} + b_0 = f(x)$$
$$y^{(n-1)}(0) = a_{n-1}, \ldots, y^{(1)}(0) = a_1 \qquad y(0) = a_0$$

This is especially true when the characteristic equation

(3.7)
$$z^n + b_{n-1}z^{n-1} + \cdots + b_1 z + b_0 = 0$$

has n distinct roots. If (3.7) does not have n distinct roots, then the partial fraction expansion of $(z^n + b_{n-1}z^{n-1} + \cdots + b_1 z + b_0)^{-1}$ is more difficult and inverting the Laplace transform is more involved, but these problems can be solved and we are still able to solve (3.6). This more general case is discussed in Scott (1955), or Braun (1983), or Hochstadt (1964).

We could not hope to do justice in a brief section like this to the many applications of Laplace transforms. Many of those applications involve interesting questions in residue theory of analytic functions of complex variables. For a brief, but lucid, treatment of basic applications in ordinary differential equations, see Boyce and DiPrima (1977), Chapter 6, or

Hochstadt (1964), Chapter 2, §6. More extensive applications, including a discussion of residue theory, can be found in Scott (1955). Chapter VI in that book, which discusses *difference equations,* is especially interesting. A complete theoretical treatment of the Laplace transform can be found in Widder (1946).

Exercises

(3.8)   Solve $y'' + y' - 6y = f(x)$ $y(0) = 1$ $y'(0) = 2$.

(3.9)   Solve $y^{(3)} - 3y^{(2)} + y^{(1)} - 3y = f(x)$ $y(0) = y'(0) = 0$ $y''(0) = 1$.

(3.10)   Solve $y'' - 4y' + 4y = f(x)$ $y(0) = 0$ $y'(0) = 0$.

(3.11)   Solve $y'' + 3y' + 2y = \delta(x)$ $y(0) = 0$ $y'(0) = 0$ and interpret (3.5) in the light of your results. [$\delta(x)$ is the Dirac δ-function.]

(3.12)   Solve $y'' - 4y' + 4y = \delta(x)$ $y(0) = 0$ $y'(0) = 0$ and interpret the solution to (3.10) in the light of your results. Generalize.

Part B. The Radon Transform

§4. Introduction

Unlike the Fourier and Laplace transforms, the Radon transform has its origins in the twentieth century. Its mathematical roots lie in the work of Funk and Radon in the first quarter of this century. A major application of Radon transforms in medical technology, the method of CAT scanning,[2] has emerged only in the last 20 years. Much research is presently being done in improving the performance of CAT scanners; the work is related to the problem of finding practical means of inverting Radon transforms.

   We will now derive the definition of the two-dimensional Radon transform. Suppose that a narrow beam (small cross-sectional area) of X-rays passes through a small length Δs of homogeneous material. (See Figure 9.1). The beam intensity I is observed to decrease according to the formula

$$I = I_0 e^{-a\rho \Delta s}$$

where I_0 is the intensity when the X-ray enters the material, I is the observed intensity of the X-ray leaving the material, ρ is the (linear) density of the material, and a is a positive constant dependent upon other features.[3] If the X-ray passes through two distinct materials—where the distance Δs_1 through medium 1 is characterized by a_1 and ρ_1 and the distance Δs_2

---

2. CAT stands for computerized axial tomography.

3. Such as the nuclear composition of the material.

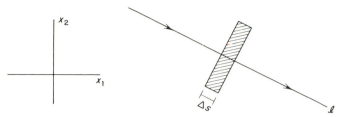

Figure 9.1   X-Rays traveling along a line ℓ through some material.

through medium 2 is characterized by a_2 and ρ_2—then

$$I = I_0 e^{-[a_1\rho_1 \Delta s_1 + a_2 \rho_2 \Delta s_2]}$$

Hence several media yield

$$I = I_0 e^{-\sum_{j=1}^{n} a_j \rho_j \Delta s_j}$$

which leads, upon passage to the limit, to

(4.1) $$I = I_0 e^{-\int_\ell a(\mathbf{x})\rho(\mathbf{x})\, ds_\mathbf{x}}$$

Here $\mathbf{x} = (x_1, x_2)$ is in $\mathbb{R}^2$ and $\int_\ell a(\mathbf{x})\rho(\mathbf{x})\, ds_\mathbf{x}$ is a *line integral* of the function $a\rho$ along the line ℓ that describes the X-ray path.

Letting $f = a\rho$ we can rewrite formula (4.1) as

(4.2) $$-\ln(I/I_0) = \int_\ell f(\mathbf{x})\, ds_\mathbf{x}$$

The line integral on the right side of (4.2) is called a *Radon transform* of f. There are several kinds of Radon transforms.

(a) If f is a function over $\mathbb{R}^2$ then the *two-dimensional Radon transform* will be denoted by $\mathscr{R}_2[f]$. We have

$$\mathscr{R}_2[f](\ell) = \int_\ell f(\mathbf{x})\, ds_\mathbf{x}$$

is a function of *lines*, which for each line ℓ in $\mathbb{R}^2$ takes as its value the number given by the line integral of f over ℓ.

(b) If f is a function over $\mathbb{R}^3$ then its *three-dimensional Radon transform* will be denoted by $\mathscr{R}_3[f]$. We have

$$\mathscr{R}_3[f](\mathscr{P}) = \int_\mathscr{P} f(\mathbf{x})\, dA_\mathbf{x} \qquad (\mathbf{x} \text{ in } \mathbb{R}^3)$$

is a function of *planes*, which for each plane $\mathscr{P}$ in $\mathbb{R}^2$ takes as its value the number given by the plane integral of f over $\mathscr{P}$.

(c) Generalizing the two cases above, if f is a function over $\mathbb{R}^n$ then its *n-dimensional Radon transform* will be denoted by $\mathscr{R}_n[f]$. For $n > 3$, we have

$$\mathscr{R}_n[f](\mathscr{H}) = \int_\mathscr{H} f(\mathbf{x})\, dV_\mathbf{x}$$

is a function of $n-1$ dimensional hyperplanes that for each hyperplane in $\mathbb{R}^n$ takes as its value the number given by the surface (hyperplane) integral of f over $\mathcal{H}$. This last case may be viewed by the reader as either an actual generalization or as simply a convenient way of discussing the first two cases above simultaneously.

In the three examples above we see that Radon transforms, unlike most functions the reader may have encountered previously, are functions of lines, or planes, rather than points. That property makes them especially interesting mathematically. In the next section we shall, however, describe a way of viewing Radon transforms in terms of points (i.e., in a coordinatized form).

To ensure that the transforms described in (a) through (c) are defined we will generally assume that the functions f to be transformed belong to the set $C_c(\mathbb{R}^n)$, $n = 2, 3, \ldots$, which consists of all continuous functions over $\mathbb{R}^n$ that are zero outside of some compact set. Sometimes we will require that f be k times continuously differentiable as well, in which case we shall say that f belongs to $C_c^k(\mathbb{R}^n)$. If f belongs to $C_c^k(\mathbb{R}^n)$ for every k then we say that f belongs to $C_c^\infty(\mathbb{R}^n)$. Although all of our theorems will be stated for $C_c(\mathbb{R}^n)$ or $C_c^k(\mathbb{R}^n)$ these theorems will always remain true if we consider the set of functions $\mathscr{S}(\mathbb{R}^n)$.[4] By $\mathscr{S}(R^n)$ we mean the set of all functions f on $\mathbb{R}^n$ that are infinitely continuously differentiable and for which

$$\left| x_1^{j_1} \cdots x_n^{j_n} \frac{\partial^{m_1 + \cdots + m_n} f}{\partial_{x_1}^{m_1} \cdots \partial_{x_n}^{m_n}} (x_1, \ldots, x_n) \right|$$

is bounded for all $\mathbf{x} = (x_1, \ldots, x_n)$ in $\mathbb{R}^n$ and all nonnegative integers $j_1, \ldots, j_n$ and $m_1, \ldots, m_n$. The set (vector space) $\mathscr{S}(\mathbb{R}^n)$ has certain theoretical advantages in connection with the Radon transform that we shall mention later. The sets $C_c(\mathbb{R}^n)$ or $C_c^k(\mathbb{R}^n)$ are somewhat closer to practical needs, since densities of objects such as the human brain are definitely zero outside a certain region. Some specific functions, such as $f(\mathbf{x}) = e^{-\pi|\mathbf{x}|^2}$, however, can be seen to belong to $\mathscr{S}(\mathbb{R}^n)$ and have easily calculated transforms.

Exercises

(4.3) Show that $e^{-\pi(x_1^2 + x_2^2)}$ is in $\mathscr{S}(\mathbb{R}^2)$ and $e^{-\pi|\mathbf{x}|^2}$ is in $\mathscr{S}(\mathbb{R}^n)$ for $n \geq 3$. [Here $|\mathbf{x}|^2 = x_1^2 + \cdots + x_n^2.$]

(4.4) Show that for every polynomial p that $p(x_1, \ldots, x_n)e^{-\pi|\mathbf{x}|^2}$ is in $\mathscr{S}(\mathbb{R}^n)$.

(4.5) Give an example of a function f which is in $C_c(\mathbb{R}^2)$ but is not in $C_c^1(\mathbb{R}^2)$.

---

4. $\mathscr{S}(R^n)$ is called *Schwartz's space* after Laurent Schwartz who showed its importance in Fourier analysis [see Schwartz (1959) or Gel'fand and Shilov (1964)].

(4.6) Suppose f and g belong to $C_c(\mathbb{R}^2)$, or $C_c^k(\mathbb{R}^2)$, then the convolution $f * g$ belongs to $C_c(\mathbb{R}^2)$, or $C_c^{2k}(\mathbb{R}^2)$. [$f * g$ as in Chapter 6 §6].

(4.7) Generalize (4.6) to $C_c(\mathbb{R}^n)$, $C_c^k(\mathbb{R}^n)$, and $\mathscr{S}(\mathbb{R}^n)$.

§5. Coordinatized Form for the Radon Transform

In order to work with Radon transforms it is helpful to have a coordinatized form for them. We will now describe how this can be done.

Let's begin with the two-dimensional Radon transform $\mathscr{R}_2$. Each line in $\mathbb{R}^2$ can be parameterized by a real number r and a unit vector $\mathbf{u}$ on the unit circle ∂B_1 ($|\mathbf{u}| = 1$). We require the following

(5.1) $$\mathbf{x} \text{ lies on } \ell \Leftrightarrow \mathbf{x} \cdot \mathbf{u} = r$$

Here $\cdot$ stands for the usual dot product in $\mathbb{R}^2$. Figure 9.2 illustrates formula (5.1). This correspondence between $(r, \mathbf{u})$ and ℓ is 2 to 1 because (5.1) implies that

(5.2) $$\ell \Leftrightarrow (r, \mathbf{u}) \qquad \text{and} \qquad \ell \Leftrightarrow (-r, -\mathbf{u})$$

The second pairing in (5.2) follows from $\mathbf{x} \cdot (-\mathbf{u}) = -\mathbf{x} \cdot \mathbf{u} = -r$. Nevertheless, we can express the Radon transform $\mathscr{R}_2[f]$ as

$$\mathscr{R}_2[f](r, \mathbf{u}) = \int_{\mathbf{x} \cdot \mathbf{u} = r} f(\mathbf{x}) \, ds_{\mathbf{x}}$$

if we consider $\mathscr{R}_2[f]$ as an *even function*. That is

$$\mathscr{R}_2[f](-r, -\mathbf{u}) = \mathscr{R}_2[f](r, \mathbf{u})$$

Sometimes we shall use the symbol $f^\#$ in place of $\mathscr{R}_2[f]$. Thus

$$f^\#(r, \mathbf{u}) = \int_{\mathbf{x} \cdot \mathbf{u} = r} f(\mathbf{x}) \, ds_{\mathbf{x}}$$

and we shall also write $f(\mathbf{x}) \underset{\mathscr{R}_2}{\supset} f^\#(r, \mathbf{u})$ to denote the operation of taking a two-dimensional Radon transform.

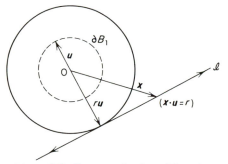

Figure 9.2 Parameterization of lines in $\mathbb{R}^2$. The vector $\mathbf{u}$ is a *unit normal* for the line ℓ.

Radon transforms over $\mathbb{R}^3$ ($\mathbb{R}^n$) can be handled similarly. Each plane $\mathscr{P}$ in $\mathbb{R}^3$ (hyperplane $\mathscr{H}$ in $\mathbb{R}^n$) can be parameterized by a real number r and a unit vector $\mathbf{u}$ on the unit sphere ∂B_1. We require that

(5.3) $\mathbf{x}$ lies on $\mathscr{P} \Leftrightarrow \mathbf{x} \cdot \mathbf{u} = r$ ($\mathbf{x}$ in $\mathscr{H} \Leftrightarrow \mathbf{x} \cdot \mathbf{u} = r$)

Then

$$\mathscr{R}_3[f](r, \mathbf{u}) = f^{\#}(r, \mathbf{u}) = \int_{\mathbf{x} \cdot \mathbf{u} = r} f(\mathbf{x}) \, dA_{\mathbf{x}}$$

or for $\mathbb{R}^n$

$$\mathscr{R}_n[f](r, \mathbf{u}) = f^{\#}(r, \mathbf{u}) = \int_{\mathbf{x} \cdot \mathbf{u} = r} f(\mathbf{x}) \, dV_{\mathbf{x}}$$

Again, we consider $f^{\#}$ to be even in the sense that $f^{\#}(-r, -\mathbf{u}) = f^{\#}(r, \mathbf{u})$.

We close this section with an example of a Radon transform. More examples will be treated in the next section in conjunction with some of the basic properties of Radon transforms.

(5.2) Example.   Suppose f is the function on $\mathbb{R}^2$ defined by

$$f(\mathbf{x}) = \begin{cases} 1 & \text{if } |\mathbf{x}| < 1 \\ 0 & \text{if } |\mathbf{x}| > 1 \end{cases}$$

Find the Radon transform $\mathscr{R}_2[f]$.

Solution.   It is not hard to see that for a line ℓ parameterized by $(r, \mathbf{u})$ that the line integral $\int_{\ell} f(\mathbf{x}) \, ds_{\mathbf{x}}$ is just the length of the chord that ℓ describes by crossing the unit disk B_1 in $\mathbb{R}^2$. Thus $\mathscr{R}_2[f]$ is given by

$$f^{\#}(r, \mathbf{u}) = \begin{cases} 0 & \text{if } |r| \geq 1 \\ 2(1 - r^2)^{1/2} & \text{if } 0 \leq r < 1 \end{cases}$$

This example easily generalizes to $\mathbb{R}^3$. If f is the function on $\mathbb{R}^3$ defined by $f(\mathbf{x}) = 1$ if $|\mathbf{x}| < 1$ and $f(\mathbf{x}) = 0$ if $|\mathbf{x}| > 1$ then $\mathscr{R}_3[f]$ is given by

(5.5) $$f^{\#}(r, \mathbf{u}) = \begin{cases} 0 & \text{if } |r| \geq 1 \\ \pi(1 - r^2) & \text{if } |r| < 1 \end{cases}$$

Exercises

(5.6)   Check (5.5).

(5.7)   Show that the two-dimensional Radon transform $f^{\#}$ may be expressed in polar coordinates as

$$f^{\#}(r, \theta) = \int_{-\infty}^{+\infty} f(r \cos \theta - t \sin \theta, \, r \sin \theta + t \cos \theta) \, dt$$

and interpret the integral above geometrically.

(5.8) Express the three-dimensional Radon transform in spherical coordinates.

(5.9) Prove that if f is a radial function on $\mathbb{R}^2$ then so is $\mathcal{R}_2[f]$, that is, $f^\#(r, \mathbf{u}) = g(r)$ for some function g. Show that this result generalizes to $\mathbb{R}^3$ ($\mathbb{R}^n$).

(5.10) Show that if f is in $C_c(\mathbb{R}^n)$ then $f^\#$ is a continuous function of $(r, \mathbf{u})$. [*Hint*: For $n = 2$ or 3 use (5.7) or (5.8).]

§6. Some Properties of Radon Transforms

In this section we shall describe some of the basic properties of the Radon transform and apply them to more examples of specific transforms.

The most important aspect of Radon transforms is their connection to the Fourier transform. If f is a function on $\mathbb{R}^n$ then for $\mathbf{v}$ and $\mathbf{x}$ in $\mathbb{R}^n$

$$
\textbf{(6.1)} \qquad \hat{f}(\mathbf{v}) = \int_{\mathbb{R}^n} f(\mathbf{x}) e^{-i2\pi \mathbf{v} \cdot \mathbf{x}} \, d\mathbf{x}
$$

is the n-dimensional Fourier transform. [Here $\int_{\mathbb{R}^n} d\mathbf{x}$ is short for $\int_{-\infty}^{+\infty} \int_{-\infty}^{+\infty} \cdots \int_{-\infty}^{+\infty} dx_1 \, dx_2 \cdots dx_n$.] Sometimes we write $\mathscr{F}_n[f]$ instead of $\hat{f}$. Note that for the double and triple Fourier transforms formula (6.1) matches the definitions given in §7 of Chapter 6.

(6.2) Theorem: Central Slice Theorem. Suppose that $f \underset{\mathscr{R}_n}{\supset} f^\#$ and $\mathbf{v}$ in $\mathbb{R}^n$ is written as $s\mathbf{u}$ for $\mathbf{u}$ on the unit sphere ∂B_1. Then

$$
\mathscr{F}_n[f](s\mathbf{u}) = \mathscr{F}_1[f^\#(r, \mathbf{u})](s)
$$

The Fourier transform $\mathscr{F}_1$ on the right is taken with respect to r.

Proof. By (6.1) we have

$$
\mathscr{F}_n[f](s\mathbf{u}) = \int_{\mathbb{R}^n} f(\mathbf{x}) e^{-i2\pi \mathbf{x} \cdot (s\mathbf{u})} \, d\mathbf{x}
$$

$$
= \int_{-\infty}^{+\infty} \left[\int_{\mathbf{x} \cdot \mathbf{u} = r} f(\mathbf{x}) e^{-i2\pi \mathbf{x} \cdot (s\mathbf{u})} \, dV_\mathbf{x} \right] dr
$$

where the last integration was obtained by performing integration over $\mathbb{R}^n$ along the hyperplanes that lie orthogonal to $\mathbf{u}$ (i.e., $\mathbf{x} \cdot \mathbf{u} = r$ for all r).

Hence, writing $e^{-i2\pi \mathbf{x} \cdot (s\mathbf{u})}$ as $e^{-i2\pi sr}$ for $\mathbf{x} \cdot \mathbf{u} = r$, we obtain

$$
\mathscr{F}_n[f](s\mathbf{u}) = \int_{-\infty}^{+\infty} \left[\int_{\mathbf{x} \cdot \mathbf{u} = r} f(\mathbf{x}) \, dV_\mathbf{x} \right] e^{-i2\pi sr} \, dr
$$

$$
= \mathscr{F}_1[f^\#(r, \mathbf{u})](s) \qquad \blacksquare
$$

(6.3) Example. Compute the two-dimensional Radon transform of $e^{-\pi(x_1^2 + x_2^2)}$.

Solution. We have $\mathcal{F}_2[e^{-\pi(x_1^2+x_2^2)}] = e^{-\pi(v_1^2+v_2^2)} = e^{-\pi s^2}$ for $\mathbf{v} = s\mathbf{u}$. But $\mathcal{F}_1(e^{-\pi r^2}) = e^{-\pi s^2}$. Therefore, by Theorem (6.2) we must have

$$e^{-\pi(x_1^2+x_2^2)} \underset{\mathscr{R}_2}{\supset} e^{-\pi r^2}$$

The reader may easily see this last result extends without change to n dimensions

(6.4) $$e^{-\pi|\mathbf{x}|^2} \underset{\mathscr{R}_n}{\supset} e^{-\pi r^2}$$

Like the Fourier and Laplace transforms, the Radon transform possesses simple properties of linearity, change of scale, and shifting.

(6.5) Theorem. The Radon transform $\mathscr{R}_n$ enjoys the following properties.

(a) *Linearity.* If $f \underset{\mathscr{R}_n}{\supset} f^{\#}$ and $g \underset{\mathscr{R}_n}{\supset} g^{\#}$ then for all complex numbers a and b

$$af + bg \underset{\mathscr{R}_n}{\supset} af^{\#} + bg^{\#}$$

(b) *Change of Scale.* If $f \underset{\mathscr{R}_n}{\supset} f^{\#}$ and c is a positive constant, then

$$f(c\mathbf{x}) \underset{\mathscr{R}_n}{\supset} \frac{1}{c^{n-1}} f^{\#}(cr, \mathbf{u})$$

(c) *Shifting.* If $f \underset{\mathscr{R}_n}{\supset} f^{\#}$ and $\mathbf{a}$ is a constant element of $\mathbb{R}^n$ then

$$f(\mathbf{x} - \mathbf{a}) \underset{\mathscr{R}_n}{\supset} f^{\#}(r - \mathbf{a} \cdot \mathbf{u}, \mathbf{u})$$

Proof. The proof of (a) is straightforward and we leave it to the reader. As for (b) we have

$$f(c\mathbf{x}) \underset{\mathscr{R}_n}{\supset} \int_{\mathbf{x} \cdot \mathbf{u} = r} f(c\mathbf{x}) \, dV_{\mathbf{x}}$$

Letting $\mathbf{y} = c\mathbf{x}$ we have $d\mathbf{y} = c^n \, d\mathbf{x}$ hence $dV_{\mathbf{y}} = c^{n-1} \, dV_{\mathbf{x}}$. Thus

$$f(c\mathbf{x}) \underset{\mathscr{R}_n}{\supset} \int_{c^{-1}\mathbf{y} \cdot \mathbf{u} = r} f(\mathbf{y}) \frac{1}{c^{n-1}} \, dV_{\mathbf{y}}$$

$$= \frac{1}{c^{n-1}} \int_{\mathbf{y} \cdot \mathbf{u} = cr} f(\mathbf{y}) \, dV_{\mathbf{y}} = \frac{1}{c^{n-1}} f^{\#}(cr, \mathbf{u})$$

To prove (c) we make a substitution of $\mathbf{y} = \mathbf{x} - \mathbf{a}$, obtaining

$$f(\mathbf{x} - \mathbf{a}) \underset{\mathscr{R}_n}{\supset} \int_{(\mathbf{y}+\mathbf{a}) \cdot \mathbf{u} = r} f(\mathbf{y}) \, dV_{\mathbf{y}}$$

$$= \int_{\mathbf{y} \cdot \mathbf{u} = r - \mathbf{a} \cdot \mathbf{u}} f(\mathbf{y}) \, dV_{\mathbf{y}} = f^{\#}(r - \mathbf{a} \cdot \mathbf{u}, \mathbf{u}) \qquad \blacksquare$$

(6.6) Examples.   Using the theorem above we have

$$e^{-\pi[(x_1-2)^2+(x_2+3)^2]} \underset{\mathcal{R}_2}{\supset} e^{-\pi(r-2u_1+3u_2)^2}$$

Or, in polar coordinates the Radon transform on the right is $e^{-\pi(r-2\cos\theta+3\sin\theta)^2}$.

Furthermore, if $f(\mathbf{x}) = 0$ for $|\mathbf{x}| > c$ and $f(\mathbf{x}) = 1$ for $|\mathbf{x}| < c$ where $\mathbf{x}$ is in $\mathbb{R}^3$ then

$$f(\mathbf{x}) \underset{\mathcal{R}_3}{\supset} f^{\#}(r, \mathbf{u}) = \begin{cases} 0 & \text{if } |r| \ge c \\ \pi(c^2 - r^2) & \text{if } |r| < c \end{cases}$$

Like Fourier and Laplace transforms, Radon transforms also interact nicely with derivatives.

(6.7) Theorem.   Suppose f is in $C_c^1(\mathbb{R}^n)$ then for $k = 1, 2, \ldots, n$

$$\frac{\partial f}{\partial x_k} \underset{\mathcal{R}_n}{\supset} u_k \frac{\partial f^{\#}}{\partial r}$$

Proof.   If $\mathbf{e}_1 = (1, 0, \ldots, 0)$, $\mathbf{e}_2 = (0, 1, 0, \ldots, 0), \ldots, \mathbf{e}_n = (0, \ldots, 0, 1)$ then

$$\frac{\partial f}{\partial x_k} = \lim_{h\to 0} \frac{f(\mathbf{x} + h\mathbf{e}_k) - f(\mathbf{x})}{h}$$

Hence,

(6.8) $$\mathcal{R}_n\left[\frac{\partial f}{\partial x_k}\right] = \int_{\mathbf{x}\cdot\mathbf{u}=r} \lim_{h\to 0} \frac{f(\mathbf{x}+h\mathbf{e}_k) - f(\mathbf{x})}{h} \, dV_{\mathbf{x}}$$

Since f belongs to $C_c^1(\mathbb{R}^n)$ the integral $\int_{\mathbf{x}\cdot\mathbf{u}=r}$ is only over a compact set K (not the whole hyperplane described by $\mathbf{x} \cdot \mathbf{u} = r$). Therefore, since $\partial f/\partial x_k$ is continuous, interchanging the limit and integral in (6.8) is justified. Hence using linearity and shifting, we obtain

$$\mathcal{R}_n\left[\frac{\partial f}{\partial x_k}\right] = \lim_{h\to 0} \int_{\mathbf{x}\cdot\mathbf{u}=r} h^{-1}f(\mathbf{x}+h\mathbf{e}_k) - h^{-1}f(\mathbf{x}) \, dV_{\mathbf{x}}$$

$$= \lim_{h\to 0} \frac{f^{\#}(r + hu_k, \mathbf{u}) - f^{\#}(r, \mathbf{u})}{h}$$

$$= \lim_{h\to 0} u_k \frac{f^{\#}(r + hu_k, u) - f^{\#}(r, u)}{hu_k} = u_k \frac{\partial f^{\#}}{\partial r}(r, \mathbf{u}) \qquad \blacksquare$$

Here is a simple instance of Theorem (6.7); more examples will be given in the exercises. If $f(\mathbf{x}) = x_1 e^{-\pi|\mathbf{x}|^2}$ for $\mathbf{x} = (x_1, x_2)$ in $\mathbb{R}^2$ then

$$f^{\#}(r, \mathbf{u}) = \frac{-1}{2\pi}\left[u_1 \frac{\partial}{\partial r}(e^{-\pi r^2})\right] = u_1 r e^{-\pi r^2}$$

For this example the reader may have noticed that f did not belong to

$C_c^1(\mathbb{R}^2)$. However, f does belong to $\mathscr{S}(\mathbb{R}^2)$ and Theorem (6.7) remains true for $\mathscr{S}(\mathbb{R}^n)$ with only slight modifications of the proof.[5]

An important concept in Radon transform theory is the relation between transforms and convolutions.

(6.9) Theorem: Convolution Theorem.   Suppose f and g belong to $C_c(\mathbb{R}^n)$ then

$$f * g \underset{\mathcal{R}_n}{\supset} f^{\#} *_r g^{\#}$$

where $*_r$ means convolution with respect to the variable r only.

Proof.   An interchange of integrals[6] along with shifting shows us that

$$f * g \underset{\mathcal{R}_n}{\supset} \int_{\mathbf{x}\cdot\mathbf{u}=r} \left[\iint_{\mathbb{R}^n} f(\mathbf{x}-\mathbf{y})g(\mathbf{y})\,d\mathbf{y}\right] dV_{\mathbf{x}}$$

$$= \int_{\mathbb{R}^n} \left[\iint_{\mathbf{x}\cdot\mathbf{u}=r} f(\mathbf{x}-\mathbf{y})\,dV_{\mathbf{x}}\right] g(\mathbf{y})\,d\mathbf{y}$$

$$= \int_{\mathbb{R}^n} f^{\#}(r-\mathbf{y}\cdot\mathbf{u},\mathbf{u})g(\mathbf{y})\,d\mathbf{y}$$

Using a central slice argument [see the proof of Theorem (6.2)]

$$f * g \underset{\mathcal{R}_n}{\supset} \int_{-\infty}^{+\infty} \left[\iint_{\mathbf{y}\cdot\mathbf{u}=s} f^{\#}(r-\mathbf{y}\cdot\mathbf{u},\mathbf{u})g(\mathbf{y})\,dV_{\mathbf{y}}\right] ds$$

$$= \int_{-\infty}^{+\infty} \left[\iint_{\mathbf{y}\cdot\mathbf{u}=s} g(\mathbf{y})\,dV_{\mathbf{y}}\right] f^{\#}(r-s,\mathbf{u})\,ds$$

$$= \int_{-\infty}^{+\infty} f^{\#}(r-s,\mathbf{u})g^{\#}(s,\mathbf{u})\,ds = f^{\#} *_r g^{\#}(r,\mathbf{u}) \qquad \blacksquare$$

Exercises

(6.10)   Prove that $\Delta_{\mathbf{x}} f \underset{\mathcal{R}_n}{\supset} (\partial^2 f^{\#}/\partial r^2)$ if f is in $C_c^2(\mathbb{R}^n)$ [where $\Delta_{\mathbf{x}} = \sum_{j=1}^{n}(\partial^2/\partial x_j^2)$].

(6.11)   Find the Radon transforms of the following functions.

(a) $f(\mathbf{x}) = x_2 e^{-\pi|\mathbf{x}|^2}$ ($\mathbf{x}$ in $\mathbb{R}^3$)
(b) $f(\mathbf{x}) = [|\mathbf{x}|^2 - 5]e^{-\pi|\mathbf{x}|^2}$ ($\mathbf{x}$ in $\mathbb{R}^n$)
(c) $f(\mathbf{x}) = \begin{cases} 2+3e^{-|\mathbf{x}|^2} & \text{if } |\mathbf{x}| < 1 \\ 3e^{-|\mathbf{x}|^2} & \text{if } |\mathbf{x}| > 1 \end{cases}$ ($\mathbf{x}$ in $\mathbb{R}^2$ or $\mathbb{R}^3$)
(d) $f(\mathbf{x}) = \begin{cases} 1 & \text{if } |\mathbf{x}-\mathbf{a}| < 4 \\ 0 & \text{if } |\mathbf{x}-\mathbf{a}| > 4 \end{cases}$ ($\mathbf{x}$ in $\mathbb{R}^2$ or $\mathbb{R}^3$)

(6.12)   Prove the following uniqueness theorem for Radon transforms.

---

5. Outside of a large enough compact set K the integrals $\int_K dV_{\mathbf{x}}$ of all functions involved will be negligible, hence the interchange of limit and integral in (6.8) will remain valid.
6. The interchange is justified because $f * g$ belongs to $C_c(\mathbb{R}^n)$ hence the two integrals involved are integrals over compact sets.

Theorem.   Suppose that f and g belong to $C_c(\mathbb{R}^n)$ [or $\mathscr{S}(\mathbb{R}^n)$] and $f^\# = g^\#$ identically in r and $\mathbf{u}$, then $f = g$ identically.

§7. Poisson's Equation

Our discussion of Radon inversion in the next section requires the classic solution of Poisson's equation that we shall now describe. Let the function w_n be defined by[7]

$$w_n(\mathbf{y}) = \begin{cases} -(4\pi\,|\mathbf{y}|)^{-1} & \text{if } n = 3 \text{ and } \mathbf{y} \text{ is in } \mathbb{R}^3 \\ \dfrac{1}{2\pi}\ln|\mathbf{y}| & \text{if } n = 2 \text{ and } \mathbf{y} \text{ is in } \mathbb{R}^2 \end{cases}$$

The function w_n is called the *fundamental solution* to Poisson's equation $\Delta_{\mathbf{x}}\Psi(\mathbf{x}) = f(\mathbf{x})$ for the following reason.

(7.1) Theorem.   Given f in $C_c^1(\mathbb{R}^n)$ for $n = 2$, or 3, the function

$$\Psi(\mathbf{x}) = \int_{\mathbb{R}^n} f(\mathbf{y})w_n(\mathbf{y} - \mathbf{x})\,d\mathbf{y} = f * w_n(\mathbf{x})$$

solves Poisson's equation.

   Proof.   We will prove the theorem for $n = 3$, leaving the case of $n = 2$ to the reader as Exercise (7.6). If we differentiate Ψ under the integral sign we obtain

(7.2) $$\frac{\partial\Psi}{\partial x_j} = \frac{-1}{4\pi}\int_{\mathbb{R}^3} f(\mathbf{y})(y_j - x_j)\,|\mathbf{y} - \mathbf{x}|^{-3}\,d\mathbf{y} \qquad (j = 1, 2, 3)$$

We note that since $|(y_j - x_j)\,|\mathbf{y} - \mathbf{x}|^{-3}| \le |\mathbf{y} - \mathbf{x}|^{-2}$ is integrable over the compact support of f, the differentiation under the integral sign in (7.2) is justified. From (7.2) we obtain by changing variables and then differentiating under the integral sign [which is justified since f belongs to $C_c^1(\mathbb{R}^n)$]

$$\Delta_{\mathbf{x}}\Psi = \frac{-1}{4\pi}\sum_{j=1}^{3}\frac{\partial}{\partial x_j}\int_{\mathbb{R}^3} f(\mathbf{y} + \mathbf{x})y_j\,|\mathbf{y}|^{-3}\,d\mathbf{y}$$

$$= \frac{-1}{4\pi}\sum_{j=1}^{3}\int_{\mathbb{R}^3} f_{x_j}(\mathbf{y} + \mathbf{x})y_j\,|\mathbf{y}|^{-3}\,d\mathbf{y}$$

$$= \frac{-1}{4\pi}\sum_{j=1}^{3}\int_{\mathbb{R}^n} f_{y_j}(\mathbf{y} + \mathbf{x})y_j\,|\mathbf{y}|^{-3}\,d\mathbf{y}$$

The last equality holding because as functions $f_{x_j}(\mathbf{y} + \mathbf{x}) = f_{y_j}(\mathbf{y} + \mathbf{x})$. But

---

7. For $n > 3$, see John (1955), p. 10.

then

$$(7.3) \qquad \Delta_{\mathbf{x}}\Psi = \frac{-1}{4\pi} \lim_{r\to 0+} \sum_{j=1}^{3} \int_{|\mathbf{y}|>r} f_{y_j}(\mathbf{y}+\mathbf{x})y_j\, |\mathbf{y}|^{-3}\, d\mathbf{y}$$

Since $f_{y_j}=0$ for $|\mathbf{y}|\ge K$, K sufficiently large, we can apply Green's identity (6.4a), Chapter 4 to (7.3). Using $v(\mathbf{y})=f(\mathbf{y}+\mathbf{x})$ and $u(\mathbf{y})=-|\mathbf{y}|^{-1}$ we get

$$\sum_{j=1}^{3} \int_{|\mathbf{y}|>r} f_{y_j}(\mathbf{y}+\mathbf{x})y_j\,|\mathbf{y}|^{-3}\, d\mathbf{y} = \int_{|\mathbf{y}|=r} f(\mathbf{y}+\mathbf{x})\frac{\partial(-|\mathbf{y}|^{-1})}{\partial\eta}\, dA_{\mathbf{y}}$$

$$-\int_{|\mathbf{y}|>r} f(\mathbf{y}+\mathbf{x})\Delta_{\mathbf{y}}(-|\mathbf{y}|^{-1})\, d\mathbf{y}$$

Since $\Delta_{\mathbf{y}}(-|\mathbf{y}|^{-1})=0$ and

$$\frac{\partial}{\partial\eta} = \sum_{j=1}^{3} -\frac{y_j}{r}\frac{\partial}{\partial y_j}$$

on $\{\mathbf{y}: |\mathbf{y}|=r\}$ we obtain

$$\sum_{j=1}^{3} \int_{|\mathbf{y}|>r} f_{y_j}(\mathbf{y}+\mathbf{x})y_j\,|\mathbf{y}|^{-3}\, d\mathbf{y} = \int_{|\mathbf{y}|=r} f(\mathbf{y}+\mathbf{x})(-|\mathbf{y}|^{-1}/r)\, dA_{\mathbf{y}}$$

$$= \frac{-1}{r^2}\int_{|\mathbf{y}|=r} f(\mathbf{y}+\mathbf{x})\, dA_{\mathbf{y}}$$

Returning from this last result to (7.3), we have

$$\Delta_{\mathbf{x}}\Psi = \lim_{r\to 0+} \frac{1}{4\pi r^2}\int_{|\mathbf{y}|=r} f(\mathbf{y}+\mathbf{x})\, dA_{\mathbf{y}} = f(\mathbf{x})$$

because of the continuity of f. ∎

The solution $\Psi=f*w_3=w_3*f$ to Poisson's equation is unique given the extra assumption that $\lim_{|\mathbf{x}|\to\infty}\Psi(\mathbf{x})=0$. [See Exercise (7.7)]. It is also known that $\Psi=f*w_n=w_n*f$ solves Poisson's equation under the milder assumption that f is in $C_c(\mathbb{R}^n)$ and f is Lipschitz, order α, with respect to each variable.[8] Furthermore, by modifying slightly the proof above we can show that $\Psi=f*w_n=w_n*f$ solves Poisson's equation if f is in $\mathscr{S}(\mathbb{R}^n)$.

Our discussion of Radon inversion will also require the following two lemmas.

(7.4) Lemma.   Let $\mathbf{y}$ be in $\mathbb{R}^3$ and $\mathbf{u}$ be on the unit sphere ∂B_1 in $\mathbb{R}^3$. Then if we integrate the function $|\mathbf{y}\cdot\mathbf{u}|$ over the unit sphere ∂B_1 we obtain

$$\int_{\partial B_1} |\mathbf{y}\cdot\mathbf{u}|\, dA_{\mathbf{u}} = 2\pi\, |\mathbf{y}|$$

Proof.   The integral $I(\mathbf{y})=\int_{\partial B_1}|\mathbf{y}\cdot\mathbf{u}|\, dA_{\mathbf{u}}$ is invariant under any orthogonal transformation $\mathcal{O}: \mathbb{R}^3\to\mathbb{R}^3$. That is $I(\mathcal{O}\mathbf{y})=I(\mathbf{y})$. Hence we may as

---

8. This is also called a *Hölder condition* on f, see Kellogg, p. 156.

well assume that **y** points along the z axis. In which case $|\mathbf{y} \cdot \mathbf{u}| = r|\cos \theta|$ if $\mathbf{y} = (0, 0, r)$ and $\mathbf{u} = (\sin \theta \cos \phi, \sin \theta \sin \phi, \cos \theta)$ in spherical coordinates. Converting $I(\mathbf{y})$ to spherical coordinates yields

$$I(\mathbf{y}) = \int_0^{2\pi} \int_0^{\pi} r|\cos \theta| \sin \theta \, d\theta \, d\phi = 2\pi r \int_{-1}^{1} |z| \, dz = 2\pi r$$

Since $2\pi r = 2\pi |\mathbf{y}|$ we are done. ∎

(7.5) Lemma.   Let **y** be in $\mathbb{R}^2$ and **u** be on the unit circle ∂B_1 in $\mathbb{R}^2$. Then if we integrate the function $\log |\mathbf{y} \cdot \mathbf{u}|$ over ∂B_1 we obtain

$$\int_{\partial B_1} \log |\mathbf{y} \cdot \mathbf{u}| \, ds_{\mathbf{u}} = 2\pi \log |\mathbf{y}| + A$$

where A is a constant.

Proof.   See Exercise (7.6). ∎

Exercises

(7.6)   Prove Theorem (7.1) for $n = 2$. And prove Lemma (7.5).

(7.7)   Show that $\Psi = f * w_3$ is the unique solution to

$$\Delta_{\mathbf{x}} \Psi = f(\mathbf{x}) \qquad \lim_{|\mathbf{x}| \to +\infty} \Psi(\mathbf{x}) = 0$$

for a given function f in $C_c^1(\mathbb{R}^3)$.

*(7.8)   Extend Theorem (7.1) to f in $\mathscr{S}(\mathbb{R}^n)$.

(7.9)   Interpret Theorem (7.1) in terms of the Dirac δ-function.

(7.10)   Verify $I(\mathcal{O}\mathbf{y}) = I(\mathbf{y})$ in the proof of (7.4). [*Hint*: $dA_{\mathcal{O}\mathbf{u}} = dA_{\mathbf{u}}$.]

(7.11)   Let $n = 3$ and write $\mathbf{x} = (r \sin \theta \cos \phi, r \sin \theta \sin \phi, r \cos \theta)$ and $\mathbf{y} = (\rho \sin \theta' \cos \phi', \rho \sin \theta' \sin \phi', \rho \cos \theta')$. Expand $w_3(\mathbf{y} - \mathbf{x})$ into a spherical harmonic series (see Chapter 8).

§8. Inversion of Radon Transforms

In this section we shall consider the inversion of two- and three-dimensional Radon transforms. Inverting Radon transforms is of great practical importance. To see this, recall our derivation of the Radon transform $\mathscr{R}_2$ in §4 [See especially Figure 9.1 and (4.2).] If we pass X-rays through planar sections of some body, say a human skull, then using formula (4.2) we can determine $\mathscr{R}_2[f]$ where f corresponds to the density of material (brain matter and skull bone). The problem of inversion then consists in recovering f, the density of the skull and its interior brain matter, from the Radon transform $\mathscr{R}_2[f]$. (See Figure 9.3.)

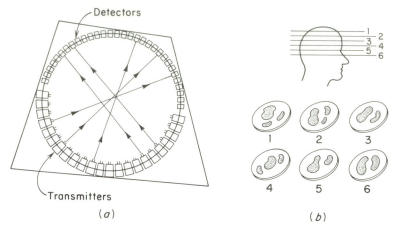

Figure 9.3 (*a*) A schematic diagram of the X-ray scanning along plane sections of a skull. (*b*) A schematic diagram of six sections and inversions of the transforms along these sections. For better figures, see Smith, Solmon, and Wagner (1977), p. 1231, or see Shepp and Kruskal (1978).

Before we can prove these inversion theorems we need two preliminary results.

(8.1) Theorem.   Suppose f is in $C_c^1(\mathbb{R}^3)$. Then

$$f(\mathbf{x}) = -\frac{(\Delta_\mathbf{x})^2}{16\pi^2} \int_{\mathbb{R}^3} \left[\int_{\partial B_1} f(\mathbf{y}) \, |(\mathbf{y} - \mathbf{x}) \cdot \mathbf{u}| \, dA_\mathbf{u} \right] d\mathbf{y}$$

where $(\Delta_\mathbf{x})^2$ symbolizes two applications of the Laplacian.

Proof.   First, we observe that

(8.2) $$\Delta_\mathbf{x} |\mathbf{y} - \mathbf{x}| = 2 \, |\mathbf{y} - \mathbf{x}|^{-1}$$

Hence, using Lemma (7.4) and (8.2) we have

$$(\Delta_\mathbf{x})^2 \int_{\mathbb{R}^3} \left[\int_{\partial B_1} f(\mathbf{y}) \, |(\mathbf{y} - \mathbf{x}) \cdot \mathbf{u}| \, dA_\mathbf{u} \right] d\mathbf{y} = (\Delta_\mathbf{x})^2 \int_{\mathbb{R}^3} f(\mathbf{y}) \cdot 2\pi \, |\mathbf{y} - \mathbf{x}| \, d\mathbf{y}$$

$$= -16\pi^2 \Delta_\mathbf{x} \int_{\mathbb{R}^3} f(\mathbf{y}) \frac{|\mathbf{y} - \mathbf{x}|^{-1}}{-4\pi} \, d\mathbf{y}$$

where the differentiation under the integral sign in the second equality is justified because the integrand is absolutely integrable with compact support. Applying our solution to Poisson's equation from Theorem (7.1) we have

$$-\frac{(\Delta_\mathbf{x})^2}{16\pi^2} \int_{\mathbb{R}^3} \left[\int_{\partial B_1} f(\mathbf{y}) \, |(\mathbf{y} - \mathbf{x}) \cdot \mathbf{u}| \, dA_\mathbf{u} \right] = f(\mathbf{x}) \quad \blacksquare$$

A similar argument, using Lemma (7.5) and Theorem (7.1), yields the following result. The details of its proof we leave to the reader as Exercise (8.18).

(8.3) Theorem.   Suppose f is in $C_c^1(\mathbb{R}^2)$. Then

$$f(\mathbf{x}) = \frac{\Delta_\mathbf{x}}{4\pi^2} \int_{\mathbb{R}^2} \left[\int_{\partial B_1} f(\mathbf{y}) \log |(\mathbf{y} - \mathbf{x}) \cdot \mathbf{u}| \, ds_\mathbf{u} \right] dy$$

We are now in a position where we can prove our inversion theorems.

(8.4) Theorem: Three-Dimensional Radon Inversion.   If f is in $C_c^1(\mathbb{R}^3)$ then

$$f(\mathbf{x}) = \frac{-1}{8\pi^2} \Delta_\mathbf{x} \int_{\partial B_1} f^{\#}(\mathbf{x} \cdot \mathbf{u}, \mathbf{u}) \, dA_\mathbf{u}$$

Proof.   Since f has compact support and is continuous we may interchange the two integrals that yield f in Theorem (8.1). Thus

$$f(\mathbf{x}) = -\frac{(\Delta_\mathbf{x})^2}{16\pi^2} \int_{\partial B_1} \left[\int_{\mathbb{R}^3} f(\mathbf{y}) \, |(\mathbf{y} - \mathbf{x}) \cdot \mathbf{u}| \, dy \right] dA_\mathbf{u}$$

$$= -\frac{(\Delta_\mathbf{x})^2}{16\pi^2} \int_{\partial B_1} \left[\int_{\mathbb{R}^3} f(\mathbf{y} + \mathbf{x}) \, |\mathbf{y} \cdot \mathbf{u}| \, dy \right] dA_\mathbf{u}$$

Expressing the integral over $\mathbb{R}^3$ as an integration over all planes of the form $\mathbf{y} \cdot \mathbf{u} = r$ for $-\infty < r < +\infty$ we have

$$f(\mathbf{x}) = -\frac{(\Delta_\mathbf{x})^2}{16\pi^2} \int_{\partial B_1} \left[\int_{-\infty}^{+\infty} \left(\int_{\mathbf{y} \cdot \mathbf{u} = r} f(\mathbf{y} + \mathbf{x}) \, |\mathbf{y} \cdot \mathbf{u}| \, dA_\mathbf{y} \right) dr \right] dA_\mathbf{u}$$

(8.5)

$$= -\frac{(\Delta_\mathbf{x})^2}{16\pi^2} \int_{\partial B_1} \left[\int_{-\infty}^{+\infty} \left(\int_{\mathbf{y} \cdot \mathbf{u} = r} f(\mathbf{y} + \mathbf{x}) \, dA_\mathbf{y} \right) |r| \, dr \right] dA_\mathbf{u}$$

Recognizing that the innermost integral above is the Radon transform of $f(\mathbf{y} + \mathbf{x})$, we apply the shift property and get one kind of inversion

(8.6) $$f(\mathbf{x}) = -\frac{(\Delta_\mathbf{x})^2}{16\pi^2} \int_{\partial B_1} \left(\int_{-\infty}^{+\infty} |r| f^{\#}(r + \mathbf{x} \cdot \mathbf{u}, \mathbf{u}) \, dr \right) dA_\mathbf{u}$$

Our next move is to prove that

(8.7) $$\Delta_\mathbf{x} \int_{-\infty}^{+\infty} |r| f^{\#}(r + \mathbf{x} \cdot \mathbf{u}, \mathbf{u}) \, dr = 2f^{\#}(\mathbf{x} \cdot \mathbf{u}, \mathbf{u})$$

Since $f^{\#}$ is continuous, we can bring one Laplacian inside the integral sign in (8.6). Using (8.7) would then yield

$$f(\mathbf{x}) = -\frac{\Delta_\mathbf{x}}{8\pi^2} \int_{\partial B_1} f^{\#}(\mathbf{x} \cdot \mathbf{u}, \mathbf{u}) \, dA_\mathbf{u}$$

Thus, we are done as soon as we verify (8.7). We have

$$\Delta_\mathbf{x} \int_{-\infty}^{+\infty} |r| f^{\#}(r + \mathbf{x} \cdot \mathbf{u}, \mathbf{u}) \, dr = \Delta_\mathbf{x} \int_{-\infty}^{+\infty} |r - \mathbf{x} \cdot \mathbf{u}| \, f^{\#}(r, \mathbf{u}) \, dr$$

$$= \Delta_\mathbf{x} \left[\int_{\mathbf{x} \cdot \mathbf{u}}^{+\infty} (r - \mathbf{x} \cdot \mathbf{u}) f^{\#}(r, \mathbf{u}) \, dr \right.$$

$$\left. - \int_{-\infty}^{\mathbf{x} \cdot \mathbf{u}} (r - \mathbf{x} \cdot \mathbf{u}) f^{\#}(r, \mathbf{u}) \, dr \right]$$

If we fix any vector $\mathbf{u}$ on ∂B_1, then by the generalized Leibniz rule

$$\Delta_{\mathbf{x}}\left[\int_{\mathbf{x}\cdot\mathbf{u}}^{+\infty}(r-\mathbf{x}\cdot\mathbf{u})f^{\#}(r,\mathbf{u})\,dr-\int_{-\infty}^{\mathbf{x}\cdot\mathbf{u}}(r-\mathbf{x}\cdot\mathbf{u})f^{\#}(r,\mathbf{u})\,dr\right]$$

$$=\sum_{j=1}^{3}\frac{\partial}{\partial x_j}\left[\int_{\mathbf{x}\cdot\mathbf{u}}^{+\infty}(-u_j)f^{\#}(r,\mathbf{u})\,dr+\int_{-\infty}^{\mathbf{x}\cdot\mathbf{u}}u_jf^{\#}(r,\mathbf{u})\,dr\right]$$

$$=2\sum_{j=1}^{3}u_j^2 f^{\#}(\mathbf{x}\cdot\mathbf{u},\mathbf{u})=2f^{\#}(\mathbf{x}\cdot\mathbf{u},\mathbf{u})$$

which proves (8.7). ∎

In the course of proving Theorem (8.4) we obtained the following result.

(8.8) Corollary.   If f is in $C_c^1(\mathbb{R}^3)$ then Radon inversion can be performed by

$$f(\mathbf{x})=-\frac{(\Delta_{\mathbf{x}})^2}{16\pi^2}\int_{\partial B_1}\left(\int_{-\infty}^{+\infty}|r|f^{\#}(r+\mathbf{x}\cdot\mathbf{u},\mathbf{u})\,dr\right)dA_{\mathbf{u}}$$

The following result we leave for the reader to prove as an exercise.

(8.9) Theorem.   If f is in $C_c^2(\mathbb{R}^3)$ then Radon inversion can be performed by

$$f(\mathbf{x})=-\frac{1}{8\pi^2}\int_{\partial B_1}\frac{\partial^2 f^{\#}}{\partial r^2}(\mathbf{x}\cdot\mathbf{u},\mathbf{u})\,dA_{\mathbf{u}}$$

We will now prove a two-dimensional Radon inversion theorem.

(8.10) Theorem: Two-Dimensional Radon Inversion.   If f is in $C_c^2(\mathbb{R}^2)$ then

$$f(\mathbf{x})=\frac{-1}{4\pi^2}\int_{\partial B_1}\left(\fint_{-\infty}^{+\infty}\frac{(\partial f^{\#}/\partial r)(r,\mathbf{u})}{r-\mathbf{x}\cdot\mathbf{u}}\,dr\right)ds_{\mathbf{u}}$$

where $\fint_{-\infty}^{+\infty}$ denotes a principal value integral.[9]

Proof.   We begin with the result from Theorem (8.3) and by interchanging integrals, changing variables to obtain $f(\mathbf{y}+\mathbf{x})$, and performing a central slice argument as in the proof above of (8.4) we get the following form of inversion

$$f(\mathbf{x})=\frac{\Delta_{\mathbf{x}}}{4\pi^2}\int_{\partial B_1}\left[\int_{-\infty}^{+\infty}f^{\#}(r+\mathbf{x}\cdot\mathbf{u},\mathbf{u})\log|r|\,dr\right]ds_{\mathbf{u}}$$

$$=\frac{1}{4\pi^2}\int_{\partial B_1}\left[\Delta_{\mathbf{x}}\int_{-\infty}^{+\infty}f^{\#}(r+\mathbf{x}\cdot\mathbf{u},\mathbf{u})\log|r|\,dr\right]ds_{\mathbf{u}}$$

By Exercise (6.10) and the chain rule (*note*: $|\mathbf{u}|^2=1$)

$$\Delta_{\mathbf{x}}f^{\#}(r+\mathbf{x}\cdot\mathbf{u},\mathbf{u})=\frac{\partial^2 f^{\#}}{\partial r^2}(r+\mathbf{x}\cdot\mathbf{u},\mathbf{u})$$

---

9. $\fint_{-\infty}^{+\infty}dr=\lim_{\epsilon\to 0+}[\int_{\mathbf{x}\cdot\mathbf{u}+\epsilon}^{+\infty}dr+\int_{-\infty}^{\mathbf{x}\cdot\mathbf{u}-\epsilon}dr]$.

hence

(8.11) $$f(\mathbf{x}) = \frac{1}{4\pi^2} \int_{\partial B_1} \left[\int_{-\infty}^{+\infty} \frac{\partial^2 f^\#}{\partial r^2} (r + \mathbf{x} \cdot \mathbf{u}, \mathbf{u}) \log |r| \, dr \right] ds_{\mathbf{u}}$$

If we perform a simple change of variables, we have

$$\int_{-\infty}^{+\infty} \frac{\partial^2 f^\#}{\partial r^2} (r + \mathbf{x} \cdot \mathbf{u}, \mathbf{u}) \log |r| \, dr = \int_{-\infty}^{+\infty} \frac{\partial^2 f^\#}{\partial r^2} (r, \mathbf{u}) \log |r - \mathbf{x} \cdot \mathbf{u}| \, dr$$

$$= \lim_{\epsilon \to 0+} \left[\int_{\mathbf{x} \cdot \mathbf{u} + \epsilon}^{+\infty} \frac{\partial^2 f^\#}{\partial r^2} \log |r - \mathbf{x} \cdot \mathbf{u}| \, dr \right.$$

$$\left. + \int_{-\infty}^{\mathbf{x} \cdot \mathbf{u} - \epsilon} \frac{\partial^2 f^\#}{\partial r^2} \log |r - \mathbf{x} \cdot \mathbf{u}| \, dr \right]$$

Performing an integration by parts, and noting that $f = 0$ outside a large disk implies that $f^\#$ and hence $\partial^2 f^\# / \partial r^2$ are 0 for sufficiently large r, we obtain

(8.12)
$$\int_{-\infty}^{+\infty} \frac{\partial^2 f^\#}{\partial r^2} (r + \mathbf{x} \cdot \mathbf{u}, \mathbf{u}) \log |r| \, dr = \lim_{\epsilon \to 0+} \left[\log |r - \mathbf{x} \cdot \mathbf{u}| \frac{\partial f^\#}{\partial r} \Big|_{r = \mathbf{x} \cdot \mathbf{u} + \epsilon}^{r = \mathbf{x} \cdot \mathbf{u} - \epsilon} \right]$$

$$- \lim_{\epsilon \to 0+} \left[\int_{\mathbf{x} \cdot \mathbf{u} + \epsilon}^{+\infty} \frac{(\partial f^\# / \partial r)}{r - \mathbf{x} \cdot \mathbf{u}} \, dr + \int_{-\infty}^{\mathbf{x} \cdot \mathbf{u} - \epsilon} \frac{(\partial f^\# / \partial r)}{r - \mathbf{x} \cdot \mathbf{u}} \, dr \right]$$

We leave it to the reader to verify that

(8.13) $$\lim_{\epsilon \to 0+} \log |r - \mathbf{x} \cdot \mathbf{u}| \frac{\partial f^\#}{\partial r} (r, \mathbf{u}) \Big|_{r = \mathbf{x} \cdot \mathbf{u} + \epsilon}^{r = \mathbf{x} \cdot \mathbf{u} - \epsilon} = 0$$

[See Exercise (8.19).] From (8.13) and (8.12), in view of (8.11), we have that

$$f(\mathbf{x}) = -\frac{1}{4\pi^2} \int_{\partial B_1} \left[\lim_{\epsilon \to 0+} \left(\int_{\mathbf{x} \cdot \mathbf{u} + \epsilon}^{+\infty} \frac{(\partial f^\# / \partial r)}{r - \mathbf{x} \cdot \mathbf{u}} \, dr + \int_{-\infty}^{\mathbf{x} \cdot \mathbf{u} - \epsilon} \frac{(\partial f^\# / \partial r)}{r - \mathbf{x} \cdot \mathbf{u}} \, dr \right) \right]$$

which is the required result. ∎

In the course of the proof above we obtained the following result.

(8.14) Corollary.   If f is in $C_c^2(\mathbb{R}^2)$ then Radon inversion can be performed by

$$f(\mathbf{x}) = \frac{\Delta_{\mathbf{x}}}{4\pi^2} \int_{\partial B_1} \left[\int_{-\infty}^{+\infty} f^\#(r + \mathbf{x} \cdot \mathbf{u}, \mathbf{u}) \log |r| \, dr \right] ds_{\mathbf{u}}$$

and

$$f(\mathbf{x}) = \frac{1}{4\pi^2} \int_{\partial B_1} \left[\int_{-\infty}^{+\infty} \frac{\partial^2 f^\#}{\partial r^2} (r + \mathbf{x} \cdot \mathbf{u}, \mathbf{u}) \log |r| \, dr \right] ds_{\mathbf{u}}$$

Theorem (8.10) is a more commonly stated in terms of the *Hilbert transform.* The Hilbert transform $\mathcal{H}[g]$ of a function g over $\mathbb{R}$ is defined by

$$\mathcal{H}[g](x) = \frac{1}{\pi} \int_{-\infty}^{+\infty} \frac{g(s)}{s - x} \, ds$$

Hence the result of Theorem (8.10) can be expressed as

(8.15) $f(\mathbf{x}) = -\frac{1}{4\pi} \int_{\partial B_1} \mathcal{H}\left[\frac{\partial f^{\#}}{\partial r}(r, \mathbf{u})\right](\mathbf{x} \cdot \mathbf{u}) \, dA_{\mathbf{u}}$

*(8.16) Remark. For the readers familiar with Riemann–Stieltjes integrals,[10] we can write (8.10) in the form

(8.17) $f(\mathbf{x}) = -\frac{1}{4\pi^2} \int_{\partial B_1} \left[\fint_{-\infty}^{+\infty} \frac{d_r f^{\#}(r, \mathbf{u})}{r - \mathbf{x} \cdot \mathbf{u}}\right] dA_{\mathbf{u}}$

where

$$\fint_{-\infty}^{+\infty} \frac{d_r f^{\#}(r, \mathbf{u})}{r - \mathbf{x} \cdot \mathbf{u}} = \lim_{\epsilon \to 0+} \left[\int_{\mathbf{x} \cdot \mathbf{u} + \epsilon}^{+\infty} \frac{d_r f^{\#}(r, \mathbf{u})}{r - \mathbf{x} \cdot \mathbf{u}} + \int_{-\infty}^{\mathbf{x} \cdot \mathbf{u} - \epsilon} \frac{d_r f^{\#}(r, \mathbf{u})}{r - \mathbf{x} \cdot \mathbf{u}}\right]$$

is a principal value for the Riemann–Stieltjes integral of $(r - \mathbf{x} \cdot \mathbf{u})^{-1}$ with respect to $f^{\#}(\mathbf{r}, \mathbf{u})$ as a function of r. The inversion formula (8.17) has an advantage theoretically over (8.10) in that it depends only upon the data $f^{\#}$ and not on $\partial f^{\#}/\partial r$.

Finally, we note that all of the inversion theorems stated above still hold if f is in $\mathscr{S}(\mathbb{R}^n)$ for $n = 2$ or 3 (with no essential changes in the proofs).

Exercises

(8.18) Verify (8.2) and prove Theorems (8.3) and (8.9).

(8.19) Verify (8.13). [*Hint*: Use the mean value theorem to show that

$$\left|\frac{\partial f^{\#}}{\partial r}(\mathbf{x} \cdot \mathbf{u} - \epsilon, \mathbf{u}) - \frac{\partial f^{\#}}{\partial r}(\mathbf{x} \cdot \mathbf{u} + \epsilon, \mathbf{u})\right| \leq A\epsilon$$

for some constant A, then use $\lim_{\epsilon \to 0+} \epsilon \log \epsilon = 0$.]

(8.20) Explain why, except under exceptional circumstances, it is *not* possible to replace the principal value integral in (8.10) with an ordinary integral.

*(8.21) Prove that if f is in $C_c(\mathbb{R}^2)$ then the Riemann–Stieltjes integrals

$$\int_{\mathbf{x} \cdot \mathbf{u} + \epsilon}^{+\infty} \frac{d_r f^{\#}(r, \mathbf{u})}{r - \mathbf{x} \cdot \mathbf{u}} \qquad \int_{-\infty}^{\mathbf{x} \cdot \mathbf{u} - \epsilon} \frac{d_r f^{\#}(r, \mathbf{u})}{r - \mathbf{x} \cdot \mathbf{u}}$$

both exist (for each $\epsilon > 0$).

---

10. See Bartle (1964), Chap. VI.

§9. Further Aspects of Inversion

In this section we shall describe two further aspects of inversion theory for Radon transforms: (1) the concept of approximate inversion, and (2) the modern operational form of the inversion theory described in the previous section.

The theory of approximate inversion is a response to a rather obvious defect of the inversion theorems described so far. What if f does not belong to $C_c^1(\mathbb{R}^3)$ or $C_c^2(\mathbb{R}^2)$, then how do we apply the inversion theorems from §8? One way is to use approximate inversion involving an appropriate (summation) kernel. We will sketch how this works without dwelling on rigorous technicalities, which the reader might find instructive to work out. In this section when we write $\mathcal{R}_n$ or $\mathbb{R}^n$ the reader should think of $\mathcal{R}_2$ or $\mathcal{R}_3$ or $\mathbb{R}^2$ or $\mathbb{R}^3$.

Let $_cW(\mathbf{s})$ stand for the Gauss–Weierstrass kernel $c^{-n}e^{-\pi|\mathbf{s}|^2/c^2}$ where $\mathbf{s}$ is in $\mathbb{R}^n$ and $c > 0$. Then

$$_cW * f \underset{\mathcal{R}_n}{\supset} {}_cW^{\#} *_r f^{\#}$$

where $_cW^{\#}(r, \mathbf{u}) = (1/c)e^{-\pi c^2 r^2}$. If f is in $C_c(\mathbb{R}^n)$, then $_cW * f$ is easily seen to be in $\mathscr{S}(\mathbb{R}^n)$. Hence we can apply Radon inversion to $_cW^{\#} *_r f^{\#}$. Symbolizing this inversion by $\mathcal{R}_n^{-1}$ we have

(9.1) $$\mathcal{R}_n^{-1}({}_cW^{\#} *_r f^{\#}) = {}_cW * f \doteq f$$

By $_cW * f \doteq f$ we mean that $\lim_{c \to 0+} {}_cW * f(\mathbf{x}) = f(\mathbf{x})$ uniformly on $\mathbb{R}^n$.

One defect of this last example is that $_cW * f$ belongs to $\mathscr{S}(\mathbb{R}^n)$, instead of $C_c^k(\mathbb{R}^n)$ for instance. To ensure that f belongs to $C_c^k(\mathbb{R}^n)$ we need to use a summation kernel that belongs to $C_c^k(\mathbb{R}^n)$. For instance, if g is some function in $C_c^k(\mathbb{R}^n)$ then so is the kernel $_\tau K(\mathbf{s}) = \tau^{-n}g(\tau^{-1}\mathbf{s})$ for each $\tau > 0$. Moreover, the support of $_\tau K(\mathbf{s})$ shrinks as τ tends to 0. In fact, given any $\epsilon > 0$, if τ is sufficiently small the support of $_\tau K$ will be contained in a ball B_ϵ of radius ϵ about the origin. We then have that $_\tau K * f$ will have support nearly identical to the support of f when τ is taken small enough. Performing Radon inversion on $_\tau K^{\#} *_r f^{\#}$ yields (provided $\int_{\mathbb{R}^n} g(\mathbf{x})\, d\mathbf{x} = 1$)

(9.2) $$\mathcal{R}_n^{-1}({}_\tau K^{\#} *_r f^{\#}) = {}_\tau K * f \doteq f$$

In formulas (9.1) and (9.2) we write $\mathcal{R}_n^{-1}$ as shorthand for the processes of inversion described in the previous section. Those theorems all involved complicated differentiations. The modern formulation of Radon inversion involves expressing $\mathcal{R}_n^{-1}$ in a more operational form. We will concentrate on $\mathcal{R}_3^{-1}$. Similar results apply for $\mathcal{R}_2^{-1}$ but we will refer the reader to the literature for these [see Remark (9.6)].

Suppose G is a function of planes in $\mathbb{R}^3$, like the Radon transform $\mathcal{R}_3[f]$. Then we may write $G(\mathcal{P}) = g(r, \mathbf{u})$ where the plane $\mathcal{P}$ is defined by $\mathbf{x} \cdot \mathbf{u} = r$. For example $\mathcal{R}_3[f](\mathcal{P}) = f^{\#}(r, \mathbf{u})$. The *dual transform* $\mathcal{R}_3^*$ of G is defined by

$$\mathcal{R}_3^*[G](\mathbf{x}) = \int_{\partial B_1} g(\mathbf{x} \cdot \mathbf{u}, \mathbf{u})\, dA_{\mathbf{u}}$$

For example, $\mathscr{R}_3^*[f^\#](\mathbf{x}) = \int_{\partial B_1} f^\#(\mathbf{x} \cdot \mathbf{u}, \mathbf{u}) \, dA_{\mathbf{u}}$. We can now write the operation $\mathscr{R}_3^{-1}$ as

(9.3)
$$\mathscr{R}_3^{-1}[f^\#](\mathbf{x}) = -\frac{\Delta_{\mathbf{x}}}{8\pi^2} \mathscr{R}_3^*[f^\#](\mathbf{x})$$

We can also put the Laplacian in (9.3) into an operator form. First, note that if f is in $C_c^2(\mathbb{R}^3)$ then

$$\Delta_{\mathbf{x}} f \underset{\mathscr{F}_3}{\supset} -4\pi^2 |\mathbf{y}|^2 \hat{f}(\mathbf{y})$$

where $\mathscr{F}_3$ and $\hat{f}$ denote the three-dimensional Fourier transform operation. The operator $\mathscr{Y}^2$ is defined by

(9.4)
$$\mathscr{Y}^2 f = \int_{\mathbb{R}^3} |\mathbf{y}|^2 \hat{f}(\mathbf{y}) e^{i2\pi \mathbf{y} \cdot \mathbf{x}} \, d\mathbf{y}$$

hence, using Fourier inversion, we can write (9.3) as

(9.5)
$$\mathscr{R}_3^{-1}[f^\#] = \tfrac{1}{2} \mathscr{Y}^2 \circ \mathscr{R}_3^*[f^\#]$$

For some applications the operator form (9.5) of Radon inversion is more convenient than the more classical Radon inversion theorem (8.4).

(9.6) Remark. For further details, the reader should consult the excellent article by K. T. Smith, "Reconstruction Formulas in Computed Tomography," which appears on pp. 7–23 of Shepp (1983). In that article, Smith discusses the X-ray transform and divergent beam X-ray transforms, using the same methods we have used in this section for the Radon transform. We should also note that there are methods for performing Radon inversion by applying Fourier inversion to (6.2). Here again, approximate methods are necessary. See the article by Smith, or Shepp and Kruskal (1978), or Iizuka (1985), pp. 302–311.

Exercises

(9.7) Verify the statements made between formulas (9.1) and (9.2).

(9.8) Describe how approximate Fourier inversion might be performed using summation kernels.

§10. The Wave Equation, Solution by Plane Waves

Radon transforms are an important tool in the field of partial differential equations. In this section we shall discuss the solution to the three-dimensional wave equation using Radon transforms. For many more applications, see John (1955).

Let's consider the following problem involving the three-dimensional wave equation over $\mathbb{R}^3$ with initial position and velocity conditions

(10.1)
$$\Psi_{xx} + \Psi_{yy} + \Psi_{zz} = \frac{1}{c^2}\Psi_{tt}$$

$$\Psi(\mathbf{x}, 0+) = f(\mathbf{x}) \qquad \Psi_t(\mathbf{x}, 0+) = g(\mathbf{x})$$

Let $\Psi^{\#}$ be defined by

$$\Psi^{\#}(r, \mathbf{u}, t) = \int_{\mathbf{x}\cdot\mathbf{u}=r} \Psi(\mathbf{x}, t)\, dA_{\mathbf{x}}$$

Then (10.1) transforms into

(10.2)
$$\Psi_{rr}^{\#} = \frac{1}{c^2}\Psi_{tt}^{\#} \qquad \Psi^{\#}(r, \mathbf{u}, 0+) = f^{\#}(r, \mathbf{u})$$

$$\Psi_t^{\#}(r, \mathbf{u}, 0+) = g^{\#}(r, \mathbf{u})$$

But (10.2) is a one-dimensional wave equation problem for the unknown function $\Psi^{\#}$. Using D'Alembert's form of solution [see (1.28), Chapter 7] we have

(10.3)
$$\Psi^{\#}(r, \mathbf{u}, t) = \tfrac{1}{2}[f^{\#}(r + ct, \mathbf{u}) + f^{\#}(r - ct, \mathbf{u})] + (2c)^{-1}\int_{r-ct}^{r+ct} g^{\#}(s, \mathbf{u})\, ds$$

$$= \tfrac{1}{2}[f^{\#}(r + ct, \mathbf{u}) + f^{\#}(r - ct, \mathbf{u})] + (2c)^{-1}\int_{-ct}^{ct} g^{\#}(s - r, \mathbf{u})\, ds$$

which solves (10.2) provided $f^{\#}$ is C^2 and $g^{\#}$ is C^1 in r.

If we can then apply Radon inversion to (10.3) we will obtain the desired solution to (10.1). To simplify matters, let's assume that f and g belong to $C_c^2(\mathbb{R}^3)$ or $\mathscr{S}(\mathbb{R}^3)$. Then we can apply Theorem (8.9) to obtain

$$\Psi(\mathbf{x}, t) = \frac{-1}{16\pi^2}\left[\int_{\partial B_1} \frac{\partial^2 f^{\#}}{\partial r^2}(r + ct, \mathbf{u})_{|r=\mathbf{x}\cdot\mathbf{u}} + \frac{\partial^2 f^{\#}}{\partial r^2}(r - ct, \mathbf{u})_{|r=\mathbf{x}\cdot\mathbf{u}}\, dA_{\mathbf{u}} \right.$$

$$\left. + \frac{1}{c}\int_{\partial B_1}\int_{-ct}^{ct} \frac{\partial^2 g^{\#}}{\partial r^2}(s - r, \mathbf{u})_{|r=\mathbf{x}\cdot\mathbf{u}}\, ds\, dA_{\mathbf{u}} \right]$$

$$= \frac{-1}{16\pi^2}\int_{\partial B_1}\left[\frac{\partial^2 f^{\#}}{\partial r^2}(\mathbf{x}\cdot\mathbf{u} + ct, u) + \frac{\partial^2 f^{\#}}{\partial r^2}(\mathbf{x}\cdot\mathbf{u} - ct, \mathbf{u}) \right.$$

$$\left. + \frac{1}{c}\int_{-ct}^{ct} \frac{\partial^2 g^{\#}}{\partial r^2}(s - \mathbf{x}\cdot\mathbf{u}, \mathbf{u})\, ds\right] dA_{\mathbf{u}}$$

If we let

$$f^{\times}(r, \mathbf{u}) = \frac{-1}{16\pi^2}\frac{\partial^2 f^{\#}}{\partial r^2}(r, \mathbf{u})$$

and define $g^{\times}$ similarly, then we have that

(10.4)
$$\Psi(\mathbf{x}, t) = \int_{\partial B_1} f^{\times}(\mathbf{x}\cdot\mathbf{u} + ct, \mathbf{u})\, dA_{\mathbf{u}} + \int_{\partial B_1} f^{\times}(\mathbf{x}\cdot\mathbf{u} - ct, \mathbf{u})\, dA_{\mathbf{u}}$$

$$\frac{1}{c}\int_{\partial B_1}\int_{-ct}^{ct} g^{\times}(s - \mathbf{x}\cdot\mathbf{u}, \mathbf{u})\, ds\, dA_{\mathbf{u}}$$

solves (10.1) when f and g belong to $C_c^2(\mathbb{R}^3)$ or $\mathscr{S}(\mathbb{R}^3)$. Formula (10.4) shows that Ψ is a sum of the dual transforms of a family of plane functions (also called *plane waves*) moving at the speed c in opposite directions along the axis determined by **u**.

Exercises

(10.5)   Solve the two-dimensional form of (10.1).

(10.6)   Check that (10.4) does solve (10.1).

References

For more information on the Laplace transform, see the references cited in §3, and Krabbe (1975). For more information on the Radon transform, see Shepp (1983), Shepp and Kruskal (1978), Deans (1983), Helgason (1980), and John (1955).

10

A Brief Introduction to Bessel Functions

Our final chapter is concerned with some elementary properties of the Bessel functions. These functions generate orthogonal series expansions in a manner very similar to the spherical harmonics. For instance, they satisfy a differential equation, called Bessel's equation, and they satisfy recursion relations and have an interesting integral form. The orthogonal series, called Fourier–Bessel series, enable us to solve some elementary problems in mathematical physics. We will confine our discussion to Bessel functions of integral order of the first kind. Other kinds of Bessel functions are treated in the References for this chapter.

§1. Bessel's Equation

For a real constant p, *Bessel's equation of order p* is the following differential equation

$$(1.1) \qquad x^2 y'' + xy' + (x^2 - p^2)y = 0$$

To solve (1.1) we use the power series method of Frobenius. Letting

$$x^p z = y \qquad (z = x^{-p} y)$$

we have

$$y' = px^{p-1}z + x^p z' \qquad y'' = p(p-1)x^{p-2}z + 2px^{p-1}z' + x^p z''$$

Hence (1.1) can be written as

$$(1.2) \qquad z'' + \frac{2p+1}{x} z' + z = 0$$

We now express z in a power series form

$$z = a_0 + a_1 x + a_2 x^2 + \cdots + a_m x^m + \cdots$$

Substituting this series for z into (1.2) and differentiating term by term, we are led to the equation

$$\frac{2p+1}{x}a_1 + [2a_2 + (2p+1)2a_2 + a_0] + \cdots$$

$$+ [(m+1)(m+2)a_{m+2} + (2p+1)(m+2)a_{m+2} + a_m]x^m + \cdots = 0$$

From this last equation we infer that $a_1 = 0$ and, in general,

$$(m+1+2p+1)(m+2)a_{m+2} + a_m = 0$$

Therefore, we have the following *recurrence equation*

(1.3) $$a_{m+2} = \frac{-a_m}{(m+2)(2p+m+2)} \qquad (m = 0, 1, 2, \ldots)$$

Since a_1 equals 0 it follows from (1.3) that a_m equals 0 whenever m is odd. For even m we have

$$a_2 = \frac{-a_0}{2(2p+2)}$$

$$a_4 = \frac{a_0}{2 \cdot 4(2p+2)(2p+4)}$$

$$\vdots$$

$$a_{2k} = \frac{(-1)^k a_0}{2 \cdot 4 \cdots 2k(2p+2)(2p+4) \cdots (2p+2k)} = \frac{(-1)^k a_0}{2^{2k} k! \, (p+1) \cdots (p+k)}$$

Thus

$$z = a_0 \left[\sum_{k=0}^{\infty} \frac{(-1)^k x^{2k}}{2^{2k} k! \, (p+1) \cdots (p+k)} \right] \qquad [\textit{Note:} \text{ let } (p+1) \cdots (p+0) = 1]$$

Because $y = x^p z$ we have

(1.4)

$$y = a_0 \left[\sum_{k=0}^{\infty} \frac{(-1)^k x^{2k+p}}{2^{2k} k! \, (p+1) \cdots (p+k)} \right] \qquad [\textit{Note:} \text{ let } (p+1) \cdots (p+0) = 1]$$

Using the ratio test, it is easily checked that the series in (1.4) and all its derivatives converge uniformly and absolutely on every finite interval (differentiation being performed term by term). It follows that y is infinitely continuously differentiable for all x values and satisfies (1.1).

We will concern ourselves from now on with *nonnegative integer values* of p and shall write n instead of p. For an integer $n \geq 0$ it is customary to assign the value $1/(2^n n!)$ to a_0. In that case, the function y in (1.4) is denoted by J_n and

(1.5) $$J_n(x) = \sum_{k=0}^{\infty} \frac{(-1)^k (x/2)^{2k+n}}{k! (n+k)!} \qquad (n = 0, 1, 2, \ldots)$$

The function J_n is called the *nth-order Bessel function of the first kind*. We have found that J_n satisfies

(1.6) $$x^2 J_n''(x) + x J_n'(x) + (x^2 - n^2) J_n(x) = 0$$

for all x values. From (1.5) we see immediately that J_n is odd (even) if n is odd (even) and

$$J_0(0) = 1 \qquad J_n(0) = 0 \qquad (n = 1, 2, 3, \ldots)$$

Exercises

★(1.7)   Let λ be a real number. Show that if y equals $J_n(\lambda x)$ then y solves

$$x^2 y'' + x y' + (\lambda^2 x^2 - n^2) y = 0$$

[*Hint*: First let $y = J_n(t)$ where $t = \lambda x$ and use Bessel's equation, order n.]

(1.8)   Show that $(d/dx)[x J_1(x)] = x J_0(x)$.

(1.9)   Suppose that $p = \frac{1}{2}$. Show that if $a_0 = (2/\pi)^{1/2}$ then (1.4) yields $y = \sqrt{2}(\pi x)^{-1/2} \sin x$. This function is usually denoted by $J_{1/2}(x)$.

(1.10)   Suppose $p = -\frac{1}{2}$. Show that if $a_0 = (2/\pi)^{1/2}$ then (1.4) yields $y = \sqrt{2}(\pi x)^{-1/2} \cos x$. This function is usually denoted by $J_{-1/2}(x)$.

(1.11)   Prove that $J_n(x)/x^n$ tends to $1/(2^n n!)$ when x tends to 0.

§2. Recursion Relations, Integral Form

The Bessel functions satisfy some interesting recursion relations. The most basic are the following two

(2.1a) $$\frac{d}{dx}[x^{-n} J_n(x)] = -x^{-n} J_{n+1}(x) \qquad (n = 0, 1, 2, \ldots)$$

(2.1b) $$\frac{d}{dx}[x^n J_n(x)] = x^n J_{n-1}(x) \qquad (n = 1, 2, 3, \ldots).$$

We will prove (2.1a) and leave (2.1b) as an exercise.

By differentiating (1.5) term by term, and making a change of index, we obtain

$$\frac{d}{dx}[x^{-n} J_n(x)] = \frac{d}{dx}\left[\sum_{k=0}^{\infty} \frac{(-1)^k x^{2k}}{2^{2k+n} k!(n+k)!}\right]$$

$$= -x^{-n} \sum_{k=1}^{\infty} \frac{(-1)^{k-1} x^{2k+n-1}}{2^{2k+n-1}(k-1)!(n+k)!}$$

$$= -x^{-n} \sum_{j=0}^{\infty} \frac{(-1)^j x^{2j+n+1}}{2^{2j+n+1} j!(j+n+1)!}$$

$$= -x^{-n} J_{n+1}(x)$$

Thus (2.1a) is proved.

By applying (2.1a) we can prove that J_n has the following integral form.

(2.2) Theorem: Integral Form for Bessel Functions. The Bessel function J_n has the following integral form

$$J_n(x) = \frac{1}{2\pi} \int_{-\pi}^{\pi} e^{ix \sin \phi} e^{-in\phi} \, d\phi \qquad (n = 0, 1, 2, \ldots)$$

Proof. Let

$$H_n(x) = \frac{1}{2\pi} \int_{-\pi}^{\pi} e^{ix \sin \phi} e^{-in\phi} \, d\phi$$

We will prove that $H_0 = J_0$ and that

(2.3) $$\frac{d}{dx}[x^{-n}H_n(x)] = -x^{-n}H_{n+1}(x) \qquad (n = 0, 1, 2, \ldots)$$

from which $H_n = J_n$ follows by mathematical induction [by comparison of (2.3) with (2.1a)].

Using the Maclaurin expansion of e^z for z equal to $ix \sin \phi$ we have

$$H_0(x) = \frac{1}{2\pi} \int_{-\pi}^{\pi} e^{ix \sin \phi} \, d\phi = \frac{1}{2\pi} \int_{-\pi}^{\pi} \sum_{m=0}^{\infty} \frac{i^m x^m \sin^m \phi}{m!} \, d\phi$$

Since $|\sin^m \phi| \le 1$ it follows that the series inside the integral converges uniformly with respect to ϕ for each fixed x value. Therefore

$$H_0(x) = \sum_{m=0}^{\infty} \frac{i^m x^m}{m!} \frac{1}{2\pi} \int_{-\pi}^{\pi} \sin^m \phi \, d\phi$$

$$= \sum_{k=0}^{\infty} \frac{(-1)^k x^{2k}}{(2k)!} \frac{1}{2\pi} \int_{-\pi}^{\pi} \sin^{2k} \phi \, d\phi$$

since $\sin^m \phi$ is odd if m is odd. Letting b_{2k} stand for

$$\frac{(-1)^k}{(2k)!} \frac{1}{2\pi} \int_{-\pi}^{\pi} \sin^{2k} \phi \, d\phi$$

we see that $b_0 = 1$. Moreover, performing an integration by parts (with $u = \sin^{2k+1} \phi$ and $dv = \sin \phi \, d\phi$) yields

$$b_{2k+2} = \frac{(-1)^{k+1}}{(2k+2)!} \frac{1}{2\pi} \int_{-\pi}^{\pi} \sin^{2k+2} \phi \, d\phi$$

$$= \frac{(-1)^{k+1}(2k+1)}{(2k+2)!} \frac{1}{2\pi} \int_{-\pi}^{\pi} \sin^{2k} \phi \cos^2 \phi \, d\phi$$

Since $\cos^2 \phi = 1 - \sin^2 \phi$ we obtain

$$b_{2k+2} = \frac{(-1)^{k+1}(2k+1)}{(2k+2)!} \frac{1}{2\pi} \int_{-\pi}^{\pi} \sin^{2k} \phi - \sin^{2k+2} \phi \, d\phi$$

$$= \frac{-1}{2k+2} b_{2k} - (2k+1)b_{2k+2}$$

Therefore,

$$b_0 = 1 \qquad b_{2k+2} = \frac{-b_{2k}}{(2k+2)^2}$$

which proves, by comparison with (1.3) for $m = 2k$ and $p = 0$, that H_0 and J_0 have identical power series. Thus $H_0 = J_0$. Moreover,

$$\frac{d}{dx}[x^{-n}H_n(x)] = -x^{-n}\left[\frac{n}{x}H_n(x) - H_n'(x)\right]$$

$$= -x^{-n}\left[\frac{n}{2\pi x}\int_{-\pi}^{\pi}e^{ix\sin\phi}e^{-in\phi}\,d\phi\right.$$

$$\left. - \frac{1}{2\pi}\int_{-\pi}^{\pi}i\sin\phi e^{ix\sin\phi}e^{-in\phi}\,d\phi\right]$$

$$= \frac{-x^{-n}}{2\pi}\int_{-\pi}^{\pi}\left[i\frac{d}{d\phi}\left(\frac{e^{ix\sin\phi}e^{-in\phi}}{x}\right) + \cos\phi\, e^{ix\sin\phi}e^{-in\phi}\right.$$

$$\left. - i\sin\phi\, e^{ix\sin\phi}e^{-in\phi}\right]d\phi$$

Since $\cos\phi - i\sin\phi = e^{-i\phi}$, and the first term in the last integral vanishes after evaluation at $\pm\pi$, we have

$$\frac{d}{dx}[x^{-n}H_n(x)] = \frac{-x^{-n}}{2\pi}\int_{-\pi}^{\pi}e^{ix\sin\phi}e^{-i(n+1)\phi}\,d\phi = -x^{-n}H_{n+1}(x) \qquad \blacksquare$$

Theorem (2.2) has several interesting consequences. For example, since

$$|J_n(x)| \le \frac{1}{2\pi}\int_{-\pi}^{\pi}|e^{ix\sin\phi}e^{-in\phi}|\,d\phi$$

$$= \frac{1}{2\pi}\int_{-\pi}^{\pi}1\,d\phi = 1$$

we have for all x values

$$|J_n(x)| \le 1 \qquad (n = 0, 1, 2, 3, \ldots)$$

Similarly, for all x values and all nonnegative integers k

(2.4) $$\left|\frac{d^k}{dx^k}J_n(x)\right| \le 1 \qquad (n = 0, 1, 2, 3, \ldots)$$

Finally, for each fixed x value, if we set

$$f(\phi) = \begin{cases} e^{ix\sin\phi} & \text{for } |\phi| \le \pi \\ 0 & \text{for } |\phi| > \pi \end{cases}$$

then the Riemann–Lebesgue lemma [see (5.21), Chapter 6] implies that

(2.5) $$\lim_{n\to\infty}J_n(x) = 0$$

Besides providing useful results about Bessel functions, Theorem (2.2) also provides a natural generalization of these functions. For instance, for a negative integer $-n$ we can define

$$J_{-n}(x) = \frac{1}{2\pi} \int_{-\pi}^{\pi} e^{ix \sin \phi} e^{in\phi} \, d\phi$$

Exercises

*(2.6)   Prove that

(a) $xJ_n'(x) + nJ_n(x) = xJ_{n-1}(x)$
(b) $xJ_n'(x) - nJ_n(x) = -xJ_{n+1}(x)$
(c) $J_{n-1}(x) + J_{n+1}(x) = (2n/x)J_n(x)$
(d) $J_{n-1}(x) - J_{n+1}(x) = 2J_n'(x)$

*(2.7)   Using (2.1a), (2.1b), and Rolle's theorem, prove the following statements:

(a)   Between every two positive roots of J_n there lies a root of J_{n+1}.
(b)   Between every two nonnegative roots of J_{n+1} there lies a root of J_n.

(2.8)   Prove that $J_{-n}(x) = (-1)^n J_n(x)$.

(2.9)   Prove (2.4), (2.1a), (2.1b), and (2.6) for J_{-n} where $-n$ is a negative integer.

(2.10)   Prove that

$$J_n(x) = \frac{1}{\pi} \int_0^{\pi} \cos(n\phi - x \sin \phi) \, d\phi \qquad \text{for } n = 0, \pm 1, \pm 2, \ldots$$

This is *Bessel's Integral Form* for J_n.

(2.11)   *Poisson's Representation of J_n.*   Prove that

$$J_n(x) = \frac{(2x)^n n!}{(2n)!} \frac{2}{\pi} \int_{-1}^{1} e^{ixs} (1 - s^2)^{n-1/2} \, ds \qquad (n = 0, 1, 2, \ldots)$$

[*Hint*: For $n = 0$, substitute $s = \sin \phi$ into the integral on the right, then check for recursion as was done in proving Theorem (2.2).]

(2.12)   Prove that $\lim_{x \to \infty} J_0(x) = 0$ [*Hint*: Use (2.11) and Riemann–Lebesgue Lemma for $\int_{-c}^{c}$ when c is close to 1.]

(2.13)   Prove that $\lim_{x \to \infty} J_n(x) = 0$ [*Hint*: First, integrate by parts in the integral in (2.11), then proceed as in (2.12).]

*(2.14)   Prove that $\lim_{n \to \infty} P_n[\cos(x/n)] = J_0(x)$ where P_n is the nth Legendre polynomial. [*Hint*: Use (2.2) and Laplace's integral for P_n.]

(2.15)   Prove that

$$e^{ix \sin \phi} = \sum_{n=-\infty}^{+\infty} J_n(x) e^{in\phi} \qquad e^{ix \cos \phi} = J_0(x) + 2 \sum_{n=1}^{\infty} i^n J_n(x) \cos n\phi$$

(2.16) Prove that

$$\cos(x \sin \phi) = J_0(x) + 2 \sum_{m=1}^{\infty} J_{2m}(x) \cos 2m\phi$$

$$\sin(x \sin \phi) = 2 \sum_{m=0}^{\infty} J_{2m+1}(x) \sin(2m + 1)\phi$$

***(2.17)** Addition Formula.*   Prove the identities

$$J_0[(x^2 + y^2 - 2xy \cos \theta)^{1/2}] = J_0(x)J_0(y) + 2 \sum_{n=1}^{\infty} J_n(x)J_n(y) \cos n\theta$$

$$J_0(x + y) = J_0(x)J_0(y) + 2 \sum_{n=1}^{\infty} (-1)^n J_n(x)J_n(y)$$

$$J_0(x - y) = J_0(x)J_0(y) + 2 \sum_{n=1}^{\infty} J_n(x)J_n(y)$$

[*Hint*: Consider the integral $\int_0^{2\pi} e^{ix \cos \phi} e^{-iy \cos(\phi - \theta)} \, d\phi$ in two different ways, and use the second series expansion in (2.15).]

§3. Orthogonality, Fourier–Bessel Series

Let λ and μ be two distinct positive numbers. Using the chain rule, we see from Exercise (1.7) that

$$\lambda^2 x^2 J_n''(\lambda x) + \lambda x J_n'(\lambda x) + (\lambda^2 x^2 - n^2)J_n(\lambda x) = 0$$
$$\mu^2 x^2 J_n''(\mu x) + \mu x J_n'(\mu x) + (\mu^2 x^2 - n^2)J_n(\mu x) = 0$$

If we multiply the first equation by $x^{-1}J_n(\mu x)$ and the second equation by $x^{-1}J_n(\lambda x)$ and subtract we get

$$x[\lambda^2 J_n''(\lambda x)J_n(\mu x) - \mu^2 J_n''(\mu x)J_n(\lambda x)] + [\lambda J_n'(\lambda x)J_n(\mu x) - \mu J_n(\lambda x)J_n'(\mu x)]$$
$$= (\mu^2 - \lambda^2)x J_n(\lambda x)J_n(\mu x)$$

Or

$$\frac{d}{dx}[\lambda x J_n'(\lambda x)J_n(\mu x) - \mu x J_n(\lambda x)J_n'(\mu x)] = (\mu^2 - \lambda^2)x J_n(\lambda x)J_n(\mu x)$$

Hence, for λ and μ distinct positive numbers

(3.1) $$\int_0^1 J_n(\lambda x)J_n(\mu x)x \, dx = \frac{\lambda J_n'(\lambda)J_n(\mu) - \mu J_n(\lambda)J_n'(\mu)}{\mu^2 - \lambda^2}$$

From this last result we are led to the following theorem.

(3.2) Theorem.   Suppose that λ and μ are positive numbers. Then

$$\int_0^1 J_n(\lambda x)J_n(\mu x)x \, dx = 0$$

provided that one of the following three alternatives holds

(a) λ and μ are distinct roots of J_n
(b) λ and μ are distinct roots of J_n'
(c) λ and μ are distinct roots of $xJ_n'(x) - HJ_n(x)$ where H is a constant.

Proof.   Cases (a) and (b) clearly give the desired result because of (3.1). For case (c) we have

$$\lambda J_n'(\lambda) - HJ_n(\lambda) = 0 \qquad \mu J_n'(\mu) - HJ_n(\mu) = 0$$

Multiplying the left equation by $J_n(\mu)$ and the right equation by $J_n(\lambda)$ and subtracting yields

$$\lambda J_n'(\lambda)J_n(\mu) - \mu J_n'(\mu)J_n(\lambda) = 0$$

which gives the desired result because of (3.1). ∎

From now on, we will refer to the three alternatives in Theorem (3.2) as cases (a), (b), and (c). Theorem (3.2) is the basis for forming various orthogonal series expansions known as Fourier–Bessel series. To define Fourier–Bessel series we need an asymptotic formula for Bessel functions; we will now derive this well known formula.

Substitute $y = x^{-1/2}z$ into (1.1) obtaining

(3.3)
$$z'' + \left[1 - \frac{p^2 - \frac{1}{4}}{x^2}\right]z = 0$$

For large x values (3.3) becomes essentially

$$z'' + z = 0$$

Hence for large x values we expect z to approximately equal $a\cos(x + \theta)$. Letting $p = n$ and $y = J_n$ we obtain for large x values

(3.4)
$$J_n(x) \doteq a_n x^{-1/2}\cos(x + \theta_n)$$

A more precise relation is

(3.5)
$$J_n(x) = \left(\frac{2}{\pi x}\right)^{1/2}\cos\left(x - \frac{n\pi}{2} - \frac{\pi}{4}\right) + 0(x^{-3/2})$$

Here $0(x^{-3/2})$ means that

(3.5′)
$$\left|J_n(x) - \left(\frac{2}{\pi x}\right)^{1/2}\cos\left(x - \frac{n\pi}{2} - \frac{\pi}{4}\right)\right| \le \frac{C}{x^{3/2}}$$

for some constant C.

We will omit the proof of (3.5). The interested reader will find a proof in Lemma (3.11) on p. 158 of Stein and Weiss (1971). A careful analysis of the proof in Stein and Weiss (1971) shows that it suffices to let C equal 8 in (3.5′).

Because [see Exercise (2.6a)]

$$J_n'(x) = J_{n-1}(x) - (n/x)J_n(x)$$

we have

(3.6) $$J_n'(x) = \left(\frac{2\pi}{x}\right)^{1/2} \cos\left(x - \frac{n\pi}{2} + \frac{\pi}{4}\right) + 0(x^{-3/2})$$

By combining (3.5) and (3.6) we have

(3.7) $$xJ_n'(x) - HJ_n(x) = (2\pi x)^{1/2} \cos\left(x - \frac{n\pi}{2} + \frac{\pi}{4}\right) + 0(x^{-1/2})$$

From (3.5) it follows that J_n has infinitely many positive roots that can be listed as follows[1]

(3.8) (a) $\lambda_1 < \lambda_2 < \cdots < \lambda_m < \lambda_{m+1} < \cdots$

(b) $\displaystyle\lim_{m\to\infty} (\lambda_{m+1} - \lambda_m) = \pi$

Similarly, because of (3.6) and (3.7), the positive roots of J_n' and the function $xJ_n'(x) - HJ_n(x)$ can be arranged in a fashion like (3.8).

Based on these last results concerning the roots of J_n, J_n', and $xJ_n'(x) - HJ_n(x)$, we can define Fourier–Bessel series for each of the cases (a), (b), and (c) in Theorem (3.2).

(3.9) Theorem: Fourier–Bessel Series. Let (3.8) describe the positive roots of either (a) J_n, (b) J_n' or (c) $xJ_n'(x) - HJ_n(x)$. In each case, we have that $\{J_n(\lambda_m x)\}_{m=1}^{\infty}$ is an orthogonal set with weight x over the interval $(0, 1)$ and

$$f \sim \sum_{m=1}^{\infty} c_m J_n(\lambda_m x) \qquad c_m = \frac{\displaystyle\int_0^1 f(x)J_n(\lambda_m x)x\,dx}{\displaystyle\int_0^1 J_n^2(\lambda_m x)x\,dx}$$

for each piecewise continuous function f over $(0, 1)$. Moreover, for case (a)

$$c_m = \frac{2}{J_{n+1}^2(\lambda_m)} \int_0^1 f(x)J_n(\lambda_m x)x\,dx$$

for case (b)

$$c_m = \frac{2\lambda_m^2}{(\lambda_m^2 - n^2)J_n^2(\lambda_m)} \int_0^1 f(x)J_n(\lambda_m x)x\,dx$$

and for case (c)

$$c_m = \frac{2\lambda_m^2}{[\lambda_m^2 J_n'^2(\lambda_m) + (\lambda_m^2 - n^2)J_n^2(\lambda_m)]} \int_0^1 f(x)J_n(\lambda_m x)x\,dx$$

Proof. The weighted orthogonality of $\{J_n(\lambda_m x)\}_{m=1}^{\infty}$, in each case, follows by combining (3.8a) with Theorem (3.2). Another way of putting

---

1. See Exercise (3.16) for more discussion of (3.8).

this is that $\{x^{1/2}J_n(\lambda_m x)\}_{m=1}^{\infty}$ is an orthogonal set of functions over the interval $(0, 1)$, then the definition of the generalized Fourier series

$$x^{1/2}f(x) \sim \sum_{m=1}^{\infty} c_m x^{1/2}J_n(\lambda_m x) \qquad c_m = \frac{\displaystyle\int_0^1 f(x)J_n(\lambda_m x)x\, dx}{\displaystyle\int_0^1 J_n^2(\lambda_m x)x\, dx}$$

is precisely the type discussed in Chapter 2. Dropping the factor $x^{1/2}$ in the series correspondence above yields the series correspondence required by the Theorem. The values of c_m for each of the three cases (a), (b), and (c) depend upon evaluating $\int_0^1 J_n^2(\lambda_m x)\, dx$ for each case. To do that, let $\mu \to \lambda$ in (3.1). By l'Hôpital's rule we have

(3.10)
$$\int_0^1 J_n^2(\lambda x)x\, dx = \frac{\lambda J_n'^2(\lambda) - \lambda J_n(\lambda)J_n''(\lambda) - J_n(\lambda)J_n'(\lambda)}{2\lambda}$$

$$= \tfrac{1}{2}[J_n'^2(\lambda) - J_n(\lambda)J_n''(\lambda) - \lambda^{-1}J_n(\lambda)J_n'(\lambda)]$$

Regarding J_n as a function of λ, it satisfies Bessel's equation of order n

$$\lambda^2 J_n''(\lambda) + \lambda J_n'(\lambda) + (\lambda^2 - n^2)J_n(\lambda) = 0$$

Therefore

$$-J_n(\lambda)J_n''(\lambda) - \lambda^{-1}J_n(\lambda)J_n'(\lambda) = [1 - (n^2/\lambda^2)]J_n^2(\lambda)$$

Thus, (3.10) becomes

(3.11)
$$\int_0^1 J_n^2(\lambda x)x\, dx = \tfrac{1}{2}\{J_n'^2(\lambda) + [1 - (n^2/\lambda^2)]J_n^2(\lambda)\}$$

If λ_m is a positive root of J_n, then

$$\int_0^1 J_n^2(\lambda_m x)x\, dx = \tfrac{1}{2}J_n'^2(\lambda_m)$$

But, substituting λ_m for x in (2.6b), yields $J_n'(\lambda_m) = -J_{n+1}(\lambda_m)$. Hence, for case (a), we have

(3.12a)
$$\int_0^1 J_n^2(\lambda_m x)x\, dx = \tfrac{1}{2}J_{n+1}^2(\lambda_m)$$

For cases (b) and (c), (3.11) leads immediately to

(3.12b)
$$\int_0^1 J_n^2(\lambda_m x)x\, dx = \frac{\lambda_m^2 - n^2}{2\lambda_m^2}J_n^2(\lambda_m)$$

(3.12c)
$$\int_0^1 J_n^2(\lambda_m x)x\, dx = \frac{\lambda_m^2 J_n'^2(\lambda_m) + (\lambda_m^2 - n^2)J_n^2(\lambda_m)}{2\lambda_m^2}$$

respectively. The values listed for c_m in cases (a), (b), and (c) follow easily from (3.12a), (3.12b), and (3.12c), respectively. ∎

(3.13) Corollary. Let f be piecewise continuous over the interval $(0, L)$. Then f can be expanded in a Fourier–Bessel series

$$f \sim \sum_{m=1}^{\infty} c_m J_n(\lambda_m x/L)$$

where

(a) $$c_m = \frac{2}{L^2 J_{n+1}^2(\lambda_m)} \int_0^L f(x) J_n(\lambda_m x/L) x\, dx$$

(b) $$c_m = \frac{2\lambda_m^2}{L^2(\lambda_m^2 - n^2) J_n^2(\lambda_m)} \int_0^L f(x) J_n(\lambda_m x/L) x\, dx$$

(c) $$c_m = \frac{2\lambda_m^2}{L^2[\lambda_m^2 J_n'^2(\lambda_m) + (\lambda_m^2 - n^2) J_n^2(\lambda_m)]} \int_0^L f(x) J_n(\lambda_m x/L) x\, dx$$

for each of the three cases (a), (b), and (c) for $\{\lambda_m\}_{m=1}^{\infty}$.

Proof.   Define the function g by $g(t) = f(Lt)$. Expand g using Theorem (3.9) and then substitute $t = x/L$. ■

The following theorem shows that Fourier–Bessel series satisfy properties similar to those of ordinary (trigonometric) Fourier series.

(3.14) Theorem.   In either case (a), (b), or (c) [(c) when $n > H$] the orthogonal system $\{x^{1/2} J_n(\lambda_m x/L)\}_{m=1}^{\infty}$ is complete over the interval $(0, L)$ [$\{J_n(\lambda_m x/L)\}_{m=1}^{\infty}$ is complete as a weighted orthogonal system over $(0, L)$]. Moreover, if f is piecewise smooth over $(0, L)$ then the Fourier–Bessel series for f converges to $\frac{1}{2}[f(x+) + f(x-)]$ for all x values in the open interval $(0, L)$.

The proof of Theorem (3.14) is quite difficult and is beyond the scope of this text. The reader who desires a proof might consult Vladimorov (1971), Watson (1944), or Titchmarsh (1946). It is also true that if f is piecewise smooth on $(0, L)$ and f is continuous on a closed interval $[a, b]$ contained in $(0, L)$, then the Fourier–Bessel series for f converges uniformly to f on every closed interval $[a + \delta, b - \delta]$, $\delta > 0$. That result is proved in Watson (1944).

We conclude this section with an example of a Fourier–Bessel expansion.

(3.15) Example.   Expand $f(x) = x^3$ in a Fourier–Bessel series of type (a) over the interval $(0, 2)$ using $\{J_3(\lambda_m x/2)\}_{m=1}^{\infty}$.

Solution.   Using (3.13a) we get

$$x^3 \sim \sum_{m=1}^{\infty} c_m J_3(\lambda_m x/2) \qquad c_m = \frac{\int_0^2 x^3 J_3(\lambda_m x/2) x\, dx}{2 J_4^2(\lambda_m)}$$

Substituting $t = x/2$ we have

$$c_m = \frac{16}{J_4^2(\lambda_m)} \int_0^1 t^4 J_3(\lambda_m t)\, dt$$

By substituting $x = \lambda_m t$ and then using (2.1b), with 4 in place of n, we get

$$c_m = \frac{16}{\lambda_m^5 J_4^2(\lambda_m)} \int_0^{\lambda_m} x^4 J_3(x)\, dx$$

$$= \frac{16}{\lambda_m^5 J_4^2(\lambda_m)} \int_0^{\lambda_m} \frac{d}{dx}[x^4 J_4(x)]\, dx$$

$$= \frac{16}{\lambda_m J_4(\lambda_m)}$$

Thus, in view of Theorem (3.14),

$$x^3 = 16 \sum_{m=1}^{\infty} \frac{J_3(\lambda_m x/2)}{\lambda_m J_4(\lambda_m)} \qquad (0 < x < 2)$$

Exercises

(3.16) *Roots of J_n.* Use (3.5) to prove (3.8b). [*Note*: (3.5) implies that for x sufficiently large, $J_n(x) = 0$ for $x \doteq [(n/2) + k + \frac{1}{2}]\pi + \frac{1}{4}\pi$ for some integer k, that these are the only such roots for large x follows from (3.6).]

Remark. Because of (3.16) and the fact that J_n is an analytic function (convergent power series for all x), (3.8a) must hold. This is because of the well known result[2] that an analytic function has only a finite number of roots in any bounded set.

(3.17) Prove the following Theorem.

Theorem: Interlacing of Roots. Between every two consecutive positive roots of J_n there lies one and only one root of J_{n+1}.

(3.18) Prove a similar interlacing of roots for the Legendre polynomials on $[-1, 1]$.

(3.19) Explain why (3.8) should hold for the roots of J_n' and $xJ_n'(x) - HJ_n(x)$.

(3.20) Use (3.5) to find approximate values for the first four roots of $[2J_1(x)/x]^2$ [*Hint*: Ignore $0(x^{-3/2})$ in (3.5); the actual roots are approximately 3.83, 7.02, 10.18, and 13.33.]

(3.21) Use (3.5) and (1.5) to sketch the graph of $[2J_1(x)/x]^2$. (Compare with Figure 7.13(*a*), Chapter 7.)

(3.22) Expand $f(x) = x^n$ in a Fourier–Bessel series of type (a) over $(0, L)$ using $\{J_n(\lambda_m x/L)\}_{m=1}^{\infty}$. [*Answer*: $x^n = 2L^n \sum_{m=1}^{\infty} [\lambda_m J_{n+1}(\lambda_m)]^{-1} J_n(\lambda_m x/L)$ for $0 < x < L$.]

(3.23) Expand $f(x) = x^n$ in a Fourier–Bessel series of type (b) over $(0, L)$

---

2. See, e.g., Knopp (1945), Ahlfors (1966), or Saks and Zygmund (1971).

using $\{J_n(\lambda_m x/L)\}_{m=1}^{\infty}$. [*Answer*:

$$x^n = 2L^n \sum_{m=1}^{\infty} \frac{\lambda_m J_{n+1}(\lambda_m)}{(\lambda_m^2 - n^2)J_n^2(\lambda_m)} J_n(\lambda_m x/L)$$

for $0 < x < L$.]

(3.24)   Multiply the result of (3.22) by x^{n+1}, integrate from 0 to x, and use (2.1b), to obtain

$$x^{n+1} = 4(n+1)L^n \sum_{m=1}^{\infty} \frac{J_{n+1}(\lambda_m x/L)}{\lambda_m^2 J_{n+1}(\lambda_m)} \qquad (0 < x < L)$$

(3.25)   Multiply the result of (3.24) by x^{n+2} integrate from 0 to x, to obtain

$$x^{n+2} = 8(n+1)(n+2)L^{n+1} \sum_{m=1}^{\infty} \frac{J_{n+2}(\lambda_m x/L)}{\lambda_m^3 J_{n+1}(\lambda_m)} \qquad (0 < x < L)$$

(3.26)   Prove that the series

$$\sum_{m=1}^{\infty} \frac{J_n(\lambda_m x)}{m^2}$$

converges uniformly and absolutely for all x values [here $\{\lambda_m\}$ corresponds to either case (a), (b), or (c)].

(3.27)   Prove that the series in (3.24) and (3.25) converge uniformly for all x values. [*Hint*: Use (3.5) to estimate $J_{n+1}(\lambda_m)$ for large m.]

§4. Applications

In this last section we shall examine applications of Fourier–Bessel series to some of the principal equations in mathematical physics.

Heat Flow

We will begin with a problem in heat conduction. Suppose that a circular cylinder has radius L, has temperature 0 on its lateral surface, and is infinitely long.[3] Suppose that the initial temperature f in the cylinder is a function of only the radial distance r from the central axis of the cylinder. Since f depends only upon r we may assume that the subsequent temperature u inside the cylinder depends only on r and the time t. Assuming that the central axis of the cylinder lies along the z axis, we can express the Laplacian $\Delta u = u_{xx} + u_{yy} + u_{zz}$ in terms of cylindrical coordinates

$$x = r \cos \phi \qquad y = r \sin \phi \qquad z = z$$

---

3. For the case of a cylinder of finite length, see Exercise (4.11).

obtaining

$$\Delta u = u_{rr} + (1/r)u_r + (1/r^2)u_{\phi\phi} + u_{zz}$$

Thus, since u depends only on r and t and not on ϕ and z, we are led to solving the following problem for u

(4.1)
$$\begin{cases} u_t = a^2[u_{rr} + (1/r)u_r] & \text{(heat equation, } 0 < r < L, \ 0 < t) \\ u(L, t) = 0 & (t \geq 0) \\ u(r, 0+) = f(r) & \text{(initial temperature, } 0 \leq r \leq L) \end{cases}$$

Problem (4.1) is solved by separation of variables. Letting u equal $R(r)T(t)$ leads us to

(4.2)   (a) $\begin{cases} rR''(r) + R'(r) + \lambda^2 rR(r) = 0 \\ R(L) = 0 \end{cases}$

  (b) $T'(t) = -\lambda^2 a^2 T(t)$

where λ^2 is a real constant (at this point allowed to be either positive or negative; we have written λ^2 as a convenience for what comes next).

Letting $n = 0$ and $x = r$ in the equation in (1.7) we have, after dividing by r,

$$r\frac{d^2}{dr^2}[J_0(\lambda r)] + \frac{d}{dr}[J_0(\lambda r)] + \lambda^2 rJ_0(\lambda r) = 0$$

It follows that (4.2a) is solved by $R(r) = J_0(\lambda r)$ when $J_0(\lambda L) = 0$. Therefore, we take λL to be one of the positive roots $\{\lambda_m\}_{m=1}^{\infty}$ of the function J_0. Then (4.2a) is solved by $R_m(r) = J_0(\lambda_m r/L)$ for $m = 1, 2, 3, \ldots$, and (4.2b) becomes

$$T'(t) = -(\lambda_m a/L)^2 T(t)$$

which is solved by $T_m(t) = e^{-(\lambda_m a/L)^2 t}$. Hence superposition yields

(4.3)
$$u(r, t) = \sum_{m=1}^{\infty} c_m J_0(\lambda_m r/L)e^{-(\lambda_m a/L)^2 t}$$

Letting t tend to 0 in (4.3) and using the initial temperature condition in (4.1) yields (formally)

(4.4)
$$f(r) = \sum_{m=1}^{\infty} c_m J_0(\lambda_m r/L) \qquad (0 \leq r \leq L)$$

Because of (4.3) and (4.4), viewed in the light of Corollary (3.13), we state the following as a solution to problem (4.1).

Problem (4.1) is solved by

$$u(r, t) = \sum_{m=1}^{\infty} c_m J_0(\lambda_m r/L) e^{-(\lambda_m a/L)^2 t} \qquad (0 \leq r \leq L \qquad 0 < t)$$

(4.5)

$$c_m = \frac{2}{L^2 J_1^2(\lambda_m)} \int_0^L f(r) J_0(\lambda_m r/L) r \, dr$$

$[\{\lambda_m\}$ are positive roots of $J_0]$

We shall now briefly discuss the verification of this solution. Due to the asymptotic formula (3.5) for $n = 1$, it follows that $\{1/J_1^2(\lambda_m)\}_{m=1}^{\infty}$ is bounded. Hence, because f and J_0 are bounded, it follows that $\{c_m\}$ is bounded. Therefore, for some positive constant A

$$|c_m J_0(\lambda_m r/L) e^{-(\lambda_m a/L)^2 t}| \leq A e^{-(\lambda_m a/L)^2 t}$$

Due to (3.8b) we have $(\lambda_{m+1} a/L) - (\lambda_m a/L)$ tending to $\pi a/L$ as m tends to ∞. It follows that $\sum_{m=1}^{\infty} e^{-(\lambda_m a/L)^2 t}$ converges uniformly for $t \geq \delta > 0$. Weierstrass' M-test implies that $u(r, t)$ is continuous for $0 \leq r \leq L$ and $t > 0$ (letting $\delta \to 0+$).

Similarly, because the derivatives of J_0 are bounded [see (2.4)] and for each $k = 0, 1, 2, 3, \ldots$

$$\sum_{m=1}^{\infty} A(\lambda_m a/L)^k e^{-(\lambda_m a/L)^2 t}$$

converges uniformly whenever $t \geq \delta > 0$, it follows that u may be differentiated term by term any number of times with respect to r or t. In particular, the heat equation in (4.1) is satisfied.

When f is piecewise smooth and continuous on $[0, L]$, then combining the remark following Theorem (3.14) about uniform convergence with an argument like the one used for the heat conduction problem in Chapter 4 §4, we obtain

(4.6)
$$\lim_{t \to 0+} u(r, t) = f(r) \qquad (0 \leq r \leq L)$$

Thus, with the assumption above about f, (4.1) is solved by (4.5).

Some further examples of heat conduction problems are given in the exercises.

Vibrations of a Circular Membrane

Suppose that an elastic membrane is stretched with uniform tension over a fixed circular frame of radius L, as in a circular drum, for example. Let the membrane be initially displaced to a position described by $z = f(r, \phi)$,

$0 \le r \le L$, $-\pi \le \phi \le \pi$. It can be shown that the subsequent displacement of the membrane satisfies a two-dimensional wave equation $z_{tt} = c^2 \Delta z$ for $t > 0$, where the Laplacian Δ is described in polar coordinates by $\Delta z = z_{rr} + (1/r)z_r + (1/r^2)z_{\phi\phi}$.[4] Therefore, to find the subsequent displacement of the membrane, *given that it has zero initial velocity*, we must solve the following problem for z.

(4.7)
$$z_{tt} = c^2(z_{rr} + (1/r)z_r + (1/r^2)z_{\phi\phi}) \qquad \text{(wave equation)}$$
$$z(L, \phi, t) = 0 \qquad \text{(fixed edge)}$$
$$z(r, \phi, 0+) = f(r, \phi) \qquad z_t(r, \phi, 0+) = 0 \qquad \text{(initial conditions)}$$

Problem (4.7) can be solved by separation of variables. Letting $z = R(r)T(t)F(\phi)$ the wave equation becomes

(4.8)
$$\frac{T''(t)}{c^2 T(t)} = \frac{R''(r) + r^{-1}R'(r)}{R(r)} + \frac{1}{r^2}\frac{F''(\phi)}{F(\phi)}$$

Setting $T''(t)/c^2 T(t) = -\lambda^2$ a constant, the right side of (4.8) can be separated in its variables (by multiplying by r^2). Hence we set $F''(\phi)/F(\phi)$ equal to a constant $-\mu^2$ and we are led to

$$r^2 R''(r) + rR'(r) + (\lambda^2 r^2 - \mu^2)R(r) = 0$$
$$R(L) = 0$$

(4.9)
$$F''(\phi) = -\mu^2 F(\phi) \qquad T''(t) = -\lambda^2 c^2 T(t)$$
$$F \text{ has period } 2\pi$$

Since F must have period 2π, the equation for F above is solvable only when $-\mu^2 = -n^2$ ($n = 0, 1, 2, 3, \ldots$) and then

$$F_n(\phi) = A \cos n\phi + B \sin n\phi$$

where A and B are arbitrary constants. For $x = r$ and $-\mu^2 = -n^2$, Exercise (1.7) shows that

$$r^2 \frac{d^2}{dr^2}[J_n(\lambda r)] + r \frac{d}{dr}[J_n(\lambda r)] + (\lambda^2 r^2 - n^2)J_n(\lambda r) = 0$$

Hence, if λL is one of the positive roots $\{\lambda_{mn}\}_{m=1}^{\infty}$ of the function J_n, then the problem in (4.9) for R is solved by defining

$$R_{mn}(r) = J_n(\lambda_{mn} r/L)$$

Finally, the problem in (4.9) for T when $-(\lambda_{mn}/L)^2$ replaces $-\lambda^2$ is solved by either $\cos(\lambda_{mn} ct/L)$ or $\sin(\lambda_{mn} ct/L)$. Using only $\cos(\lambda_{mn} ct/L)$ is suitable for satisfying the initial conditions in (4.7). Thus, we obtain by superposition of all the separated solutions that a solution to (4.7) should be given by the following formula.

---

4. See Morse (1948), pp. 173–174.

Problem (4.7) is solved by

(4.10) $z(r, \phi, t) = \sum_{n=0}^{\infty} \sum_{m=1}^{\infty} \{[A_{mn} \cos n\phi$

$$+ B_{mn} \sin n\phi]J_n(\lambda_{mn}r/L) \cos(\lambda_{mn}ct/L)\}$$

where

$$B_{mn} = \frac{2}{\pi L^2 J_{n+1}^2(\lambda_{mn})} \int_0^L \int_{-\pi}^{\pi} f(r, \phi)rJ_n(\lambda_{mn}r/L) \sin n\phi \, d\phi \, dr$$

$$(m, n = 1, 2, \ldots)$$

$$A_{mn} = \frac{2}{\pi L^2 J_{n+1}^2(\lambda_{mn})} \int_0^L \int_{-\pi}^{\pi} f(r, \phi)rJ_n(\lambda_{mn}r/L) \cos n\phi \, d\phi \, dr$$

$$A_{m0} = \frac{1}{\pi L^2 J_1^2(\lambda_{m0})} \int_0^L \int_{-\pi}^{\pi} f(r, \phi)rJ_0(\lambda_{m0}r/L) \, d\phi \, dr \qquad (m = 1, 2, \ldots)$$

That (4.10) actually solves (4.7) rigorously is somewhat difficult to verify. However, if f is continuously differentiable a sufficient number of times, then it can be shown that convergence and term by term differentiation of z is justified and, hence, that z solves (4.7).[5]

We close this section by noting that for the circular membrane we have found harmonics (eigenfunctions) that have the form

$$J_n(\lambda_{mn}r/L) \cos(\lambda_{mn}ct/L) \cos n\phi \qquad J_n(\lambda_{mn}r/L) \cos(\lambda_{mn}ct/L) \sin n\phi$$

with time frequencies $\lambda_{mn}c/L$ $(m = 1, 2, 3, \ldots; n = 0, 1, 2, \ldots)$. These harmonics have *nodal lines* (lines where the height of the membrane is stationary at $z = 0$) at the roots of $J_n(\lambda_{mn}r/L)$, $\cos n\phi$, and $\sin n\phi$. (See Figure 10.1.)

Unlike the vibrating string (see §2, Chapter 3), the time frequencies $\{\lambda_{mn}c/L\}$ are *not* integral multiplies of one another. This is why drums do not hold a tune; they are only used to provide rhythm in music, not melody.

Exercises

(4.11) Suppose a circular cylinder has a radius L, has temperature 0 on its lateral surface, and length A along its central axis (which we assume lies along the z axis from $z = 0$ to $z = A$). If the initial temperature inside the cylinder is described by $u(r, \phi, z, 0) = f(r, z)$ and the top $(z = A)$ and

---

5. It is sufficient to assume that f is twice continuously differentiable in ϕ and four times continuously differentiable in r.

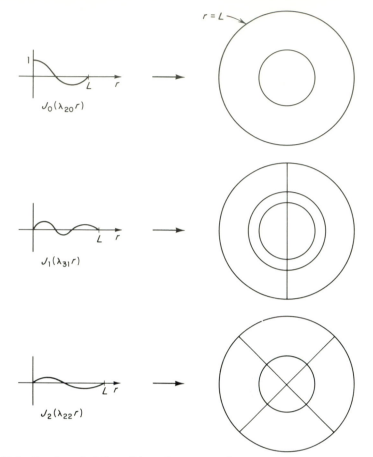

Figure 10.1 Graphs of $J_n(\lambda_{mn}r/L)$ and corresponding nodal lines for the harmonics $J_n(\lambda_{mn}r/L) \cos n\phi \cos(\lambda_{mn}ct/L)$.

bottom $(z = 0)$ are held at temperature 0, find the subsequent temperature $u(r, z, t)$ for $t > 0$.

$$\left[Answer: u(r, \phi, t) = \sum_{m,n=1}^{\infty} c_{mn}J_0(\lambda_m r/L) \sin \frac{n\pi z}{A} e^{-[(\lambda_m/L)^2+(n\pi/A)^2]a^2 t} \right.$$

$$\left. c_m = \frac{4}{L^2 A J_1^2(\lambda_m)} \int_0^A \int_0^L f(r, z)J_0(\lambda_m r/L)r \sin \frac{n\pi z}{A} \, dr \, dz \right]$$

(4.12) Suppose an infinitely long circular cylinder of radius L with central axis along the z axis has its lateral surface free to exchange heat with the surrounding medium. If the initial temperature in the cylinder is described by $u(r, \phi, z, 0) = f(r)$, then the subsequent temperature u satisfies

$$u_t = a^2[u_{rr} + (1/r)u_r] \qquad (t > 0)$$

$$u(r, 0+) = f(r)$$

$$u_r(L, t) + hu(L, t) = 0 \qquad (h \text{ a positive constant})$$

Solve the problem above for u.

[*Answer*:

$$u(r, t) = \sum_{m=1}^{\infty} c_m J_0(\lambda_m r/L) e^{-(a\lambda_m/L)^2 t}$$

where $\{\lambda_m\}_{m=1}^{\infty}$ are the positive roots of $xJ_0'(x) + hLJ_0(x)$ and

$$c_m = \frac{2}{L^2[J_0'^2(\lambda_m) + J_0^2(\lambda_m)]} \int_0^L rf(r)J_0(\lambda_m r/L) \, dr]$$

(4.13) Suppose that Ψ is harmonic inside a cylinder of radius L and length A along its central axis (which we assume to be the z axis). Let $\Psi = 0$ at $z = 0$ (bottom) and at $r = L$ (lateral surface), and suppose that $\Psi(r, A) = f(r)$ for $0 < r < L$. Show that for $0 \le r \le L$ and $0 < z < A$

$$\Psi(r, z) = \sum_{m=1}^{\infty} c_m J_0(\lambda_m r/L) \frac{\sinh(\lambda_m z/L)}{\sinh(\lambda_m A/L)}$$

where $\{\lambda_m\}_{m=1}^{\infty}$ are the positive roots of J_0 and

$$c_m = \frac{2}{L^2 J_1^2(\lambda_m)} \int_0^L f(r)J_0(\lambda_m r/L)r \, dr$$

(4.14) Prove that the series for Ψ in (4.13) defines Ψ as a harmonic function inside the cylinder.

(4.15) Discuss Dirichlet's problem for the inside of a circular cylinder of finite length. Show that the function Ψ found in (4.13) satisfies

$$\lim_{z \to A^-} \Psi(r, z) = f(z) \qquad (0 \le r < L)$$

provided that f is continuous and $f(L) = 0$.

(4.16) Suppose that Ψ is harmonic inside a half cylinder described by $0 \le r \le L$, $0 \le z \le A$, and $0 \le \phi \le \pi$. Suppose that $\Psi = 0$ when $z = 0$, or $\phi = 0$, or $\phi = \pi$, or $r = L$. Let $\Psi(r, \phi, A) = f(r, \phi)$. Show that Ψ is described inside the cylinder by

$$\Psi(r, \phi, z) = \sum_{m=1}^{\infty} \sum_{n=1}^{\infty} c_{mn} J_n(\lambda_{mn} r/L) \sin n\phi \frac{\sinh(\lambda_{mn} z/L)}{\sinh(\lambda_{mn} A/L)}$$

where $\{\lambda_{mn}\}_{m=1}^{\infty}$ are the positive roots of J_n for $n = 1, 2, 3, \ldots$, and

$$c_{mn} = \frac{4}{L^2 \pi J_{n+1}^2(\lambda_{mn})} \int_0^{\pi} \int_0^L f(r, \phi) \sin n\phi J_n(\lambda_{mn} r/L)r \, dr \, d\phi$$

(4.17) Prove that the function Ψ found in (4.16) is harmonic in the given cylinder and that $\lim_{z \to A^-} \Psi(r, \phi, z) = f(r, z)$ provided that f is continuous and $f = 0$ when $\phi = 0$, $\phi = \pi$, or $r = L$.

(4.18) Solve the vibration problem (4.7) if the initial velocity is described by $z_t(r, \phi, 0+) = g(r, \phi)$ (instead of $z_t = 0$).

References

For further study of Bessel functions and their applications, see Sommerfeld (1949), Lebedev, Skalskaya, and Uflyand (1979), Morse (1948), Lebedev (1965), and Watson (1944). Bessel functions also appear in the Fourier series expansions of elliptical orbits in astronomy, see Pollard (1976), Chap. 1, §12, and Wintner (1947), pp. 204–222.

Appendix A:
Divergence of Fourier Series

The purpose of this appendix is to prove that there are continuous functions that have divergent Fourier series. *Before studying this appendix the reader should first completely read Chapter 2.*

(A.1) Theorem. There exist continuous functions of period 2π whose Fourier series diverge at $x = 0$.

Proof. The example we describe is due to Fejér. For all positive integers $N > n$ consider the following function of x

(A.2)
$$Q(x, N, n) = 2 \sin Nx \sum_{k=1}^{n} \frac{\sin kx}{k}$$

By an elementary trigonometric identity

(A.3)
$$Q = \frac{\cos(N - n)x}{n} + \frac{\cos(N - n + 1)x}{n - 1} + \cdots + \frac{\cos(N - 1)x}{1}$$
$$- \frac{\cos(N + 1)x}{1} - \cdots - \frac{\cos(N + n)x}{n}$$

Since the partial sums of the series $\sum_{k=1}^{\infty} (\sin kx)/k$ are uniformly bounded by $\frac{1}{2}\pi + 1$ [Exercise (7.26), Chapter 2] it follows from (A.2) that Q is uniformly bounded for all x, n, and N. In particular,

$$|Q(x, N, n)| \le \pi + 2 \qquad \text{(all } x,\ N,\ \text{and } n\text{)}$$

But, at $x = 0$ the sum of the last n terms of Q satisfies

$$|-1 - \tfrac{1}{2} - \cdots - 1/n| = 1 + \tfrac{1}{2} + \cdots + 1/n > \ln n$$

which tends to $+\infty$ as n tends to $+\infty$.

Let $\{N_k\}$ and $\{n_k\}$ be two sequences of positive integers such that $n_k < N_k$ for each k, and let $\sum_{k=1}^{\infty} a_k$ be a convergent series of positive numbers. Then

421

by Weierstrass' M-test, the series $\sum_{k=1}^{\infty} a_k Q(x, N_k, n_k)$ converges uniformly to a continuous function, with period 2π, which we shall denote by f.

Let's now require that $N_k + n_k < N_{k+1} - n_{k+1}$ for every k, then it is clear that for different values of k the functions $Q(x, N_k, n_k)$ *have no like terms.* In particular, for some constants b_m (which are the Fourier coefficients of f)

$$f(x) = \sum_{m=1}^{\infty} b_m \cos mx$$

Letting S_M^f stand for the Mth partial sum of the Fourier series for f, we have

$$S_{N_m+n_m}^f(x) = \sum_{k=1}^{m} a_k Q(x, N_k, n_k)$$

$$S_{N_m}^f(x) = \sum_{k=1}^{m-1} a_k Q(x, N_k, n_k) + a_m \left[\frac{\cos(N_m - n_m)x}{n_m} + \cdots + \frac{\cos(N_m - 1)x}{1} \right]$$

We know that $\lim_{m\to\infty} S_{N_m+n_m}^f(x) = f(x)$ for all x values, in particular, for $x = 0$. But,

$$|S_{N_m+n_m}^f(0) - S_{N_m}^f(0)| = \left| a_m \left[\frac{-1}{1} - \cdots - \frac{1}{n_m} \right] \right| > a_m \ln(n_m)$$

Therefore, $\lim_{k\to\infty} S_k^f(0)$ *will not exist* if $a_m \ln(n_m)$ does not tend to zero as m tends to $+\infty$.

For example, if

$$a_k = k^{-2} \qquad N_k = 2 \cdot 2^{k^3} \qquad n_k = 2^{k^3}$$

then $S_{N_m}^f(0) \to +\infty$ as $m \to +\infty$. Hence the Fourier series for

$$f(x) = \sum_{k=1}^{\infty} k^{-2} Q(x, 2 \cdot 2^{k^3}, 2^{k^3})$$

diverges at $x = 0$. (In an infinitely oscillatory manner, because $S_{N_m+n_m}^f(0) \to f(0)$ which is finite.) ∎

(A.4) Remark. If $a_k = k^{-2}$, $N_k = 2 \cdot 2^{k^2}$, $n_k = 2^{k^2}$ then it can be shown that $|S_{N_m+n_m}^f(0) - S_{N_m}^f(0)|$ tends to a positive limit as m tends to $+\infty$. From that it follows that the Fourier series for

$$f(x) = \sum_{k=1}^{\infty} k^{-2} Q(x, 2 \cdot 2^{k^2}, 2^{k^2})$$

diverges at $x = 0$ in a finite oscillatory manner (i.e., the partial sums are bounded but do not converge at $x = 0$).

The point of using unspecified constants a_k, N_k, n_k in the proof above is that we exhibit infinitely many distinct examples of continuous functions whose Fourier series diverge at $x = 0$.

By modifying the proof of Theorem (A.1) we obtain the following stronger result, which was obtained by Du Bois Reymond in 1875.

(A.5) Theorem. There are continuous functions whose Fourier series diverge at all rational multiples of π.

Remark. The theorem is amazing because every interval $[a, b]$, no matter how small, contains infinitely many rational multiples of π.

Proof. Define P_k by $P_k(x) = Q(k!\, x, N_k, n_k)$ where, to be specific, $N_k = 2 \cdot 2^{k^3}$ and $n_k = 2^{k^3}$. Then let $a_k = k^{-2}$ and define g by

$$g(x) = \sum_{k=1}^{\infty} a_k P_k(x)$$

The function g is continuous for all x values, for the same reasons that f was in the proof of Theorem (A.1). But, we have

$$|S^g_{(N_m + n_m)m!}(0) - S^g_{N_m m!}(0)| > m^{-2} \ln(2^{m^3}) = m \ln 2$$

which tends to $+\infty$ as m tends to $+\infty$. It follows that the Fourier series for $g = \sum_{k=1}^{\infty} a_k P_k$ diverges at $x = 0$. However, the same thing is true for each tail of the series $g_M = \sum_{k=M}^{\infty} a_k P_k$. Each term P_k has a period of $2\pi/k!$, consequently each tail g_M has a period $2\pi/M!$. Therefore, g_M has a divergent Fourier series at every x equal to an integral multiple of $2\pi/M!$. It follows that g has a divergent Fourier series at every integral multiple of each x equal to $2\pi/M!$ for $M = 1, 2, 3, \ldots$. Thus, the Fourier series for g diverges at all rational multiples of π. ∎

(A.6) Remark. By varying the constants a_k, N_k, and n_k in the proof of (A.5) we generate an infinite number of distinct continuous functions whose Fourier series diverge at all rational multiples of π.

The reader who desires further discussion might begin by consulting Rudin (1974), Chapter 5, §11. Rudin gives an elegant discussion showing that divergent Fourier series arise from the fact that the Dirichlet kernel satisfies

$$\int_{-\pi}^{\pi} |D_n(x)|\, dx \to +\infty \text{ as } n \to +\infty$$

See also Katznelson (1968), Chapter II, where it is shown that there exist functions whose Fourier series diverge at *all* values of x.

Appendix B:
Brief Tables of Fourier Series and Fourier Transforms

More extensive tables can be found in Abramowitz and Stegun (1972), Campbell and Foster (1948), Erdélyi et al. (1953, 1954), and Ryshik and Gradstein (1957).

Fourier Series

1. $$\sum_{n=1}^{\infty} \frac{\sin nx}{n} = \frac{\pi - x}{2} \qquad (0 < x < 2\pi)$$

2. $$\sum_{n=1}^{\infty} \frac{\cos nx}{n} = -\ln\left(2 \sin \frac{x}{2}\right) \qquad (0 < x < 2\pi)$$

3. $$\sum_{n=1}^{\infty} (-1)^{n+1} \frac{\sin nx}{n} = \frac{x}{2} \qquad (-\pi < x < \pi)$$

4. $$\sum_{n=1}^{\infty} (-1)^{n+1} \frac{\cos nx}{n} = \ln\left(2 \cos \frac{x}{2}\right) \qquad (-\pi < x < \pi)$$

5. $$\sum_{n=0}^{\infty} \frac{\sin(2n + 1)x}{2n + 1} = \frac{\pi}{4} \qquad (0 < x < \pi)$$

6. $$\sum_{n=0}^{\infty} \frac{\cos(2n + 1)x}{2n + 1} = -\frac{1}{2} \ln\left(\tan \frac{x}{2}\right) \qquad (0 < x < \pi)$$

7. $$\sum_{n=1}^{\infty} \frac{\cos nx}{n^2} = \frac{3x^2 - 6\pi x + 2\pi^2}{12} \qquad (0 \le x \le 2\pi)$$

8. $$\sum_{n=1}^{\infty} \frac{\sin nx}{n^2} = -\int_0^x \ln\left(2 \sin \frac{s}{2}\right) ds \qquad (0 \le x \le 2\pi)$$

9. $$\sum_{n=1}^{\infty} (-1)^{n+1} \frac{\cos nx}{n^2} = \frac{\pi^2 - 3x^2}{\pi} \qquad (-\pi \le x \le \pi)$$

10. $$\sum_{n=1}^{\infty} (-1)^{n+1} \frac{\sin nx}{n^2} = \int_0^x \ln\left(2 \cos \frac{s}{2}\right) ds \qquad (-\pi \le x \le \pi)$$

424

11. $\displaystyle\sum_{n=0}^{\infty} \frac{\cos(2n+1)x}{(2n+1)^2} = \frac{\pi^2 - 2\pi x}{8}$ $\qquad (0 \leq x \leq \pi)$

12. $\displaystyle ax^2 + bx + c = \frac{4a\pi^2}{3} + b\pi + c + 4a \sum_{n=1}^{\infty} \frac{\cos nx}{n^2}$

$$- (4\pi a + 2b) \sum_{n=1}^{\infty} \frac{\sin nx}{n} \qquad (0 < x < 2\pi)$$

13. $\displaystyle ax^2 + bx + c = \frac{a\pi^2}{3} + c + 4a \sum_{n=1}^{\infty} (-1)^n \frac{\cos nx}{n^2}$

$$- 2b \sum_{n=1}^{\infty} (-1)^n \frac{\sin nx}{n} \qquad (-\pi < x < \pi)$$

Fourier Transforms

1. $\Pi(x) = \begin{cases} 1 & \text{if } |x| < \frac{1}{2} \\ 0 & \text{if } |x| > \frac{1}{2} \end{cases} \supset \operatorname{sinc} u = \dfrac{\sin \pi u}{\pi u}$

2. $e^{-\pi x^2} \supset e^{-\pi u^2}$

3. $\Lambda(x) = \begin{cases} 1 - |x| & \text{if } |x| \leq 1 \\ 0 & \text{if } |x| > 1 \end{cases} \supset \operatorname{sinc}^2 u = \dfrac{\sin^2 \pi u}{(\pi u)^2}$

4. $e^{-2\pi|x|} \supset \dfrac{1}{\pi}\dfrac{1}{1+u^2}$

5. $\operatorname{sinc}^2 x \supset \Lambda(u)$

6. $e^{-|x|}\dfrac{\sin x}{x} \supset \tan^{-1}\left(\dfrac{1}{2\pi^2 u^2}\right)$

7. $e^{-|x|} \cdot \begin{cases} 1 & \text{if } x > 0 \\ 0 & \text{if } x < 0 \end{cases} \supset \dfrac{1 - i2\pi u}{1 + (2\pi u)^2}$

8. $\operatorname{sech} \pi x \supset \operatorname{sech} \pi u$

9. $\operatorname{sech}^2 \pi x \supset 2u \operatorname{cosech} \pi u$

10. $\delta(x) \supset 1$

Bibliography

Abramowitz, M. and I. Stegun. *Handbook of Mathematical Functions with Formulas, Graphs, and Mathematical Tables*. National Bureau of Standards, Wiley, New York, 1972.

Ahlfors, L. V., *Complex Analysis, 2nd ed*. McGraw-Hill, New York, 1966.

Allan, W. B., *Fibre Optics, Theory and Practice*. Plenum, London, 1973.

Ash, J. M. (Ed.). *Studies in Harmonic Analysis, Vol. 13, M.A.A. Studies in Mathematics*. Mathematical Association of America, Washington, D.C., 1976.

Bachman, G. *Elements of Abstract Harmonic Analysis*. Academic Press, New York, 1964.

Bartle, R. W., *The Elements of Real Analysis*. Wiley, New York, 1964.

Bary, N. *A Treatise on Trigonometric Series, Vols. 1 and 2*. Pergamon, New York, 1964.

Bell, R. J., *Introductory Fourier Transform Spectroscopy*. Academic Press, New York, 1972.

Bellman, R. *A Brief Introduction to Theta Functions*. Holt, Rhinehart & Winston, New York, 1961.

Berg, P. W., and J. L. McGregor. *Elementary Partial Differential Equations*. Holden-Day, San Francisco, 1966.

Bitsadze, A. V., *Equations of Mathematical Physics*. Mir, Moscow, 1980.

Blackman, R. B., and J. W. Tukey. *The Measurement of Power Spectra*. Dover, New York, 1959.

Bochner, S., *Lectures on Fourier Integrals*. Princeton University Press, Princeton, N.J., 1959.

————. *Harmonic Analysis and the Theory of Probability*. University of California Press, Berkeley, 1960.

Bochner, S., and K. Chandrasekharan. *Fourier Transforms*. Princeton University Press, Princeton, N.J., 1949.

Born, M., and E. Wolf. *Principles of Optics, 3rd ed*. Pergamon, Oxford, 1965.

Boyce, W. E., and R. C. DiPrima. *Elementary Differential Equations and Boundary Value Problems, 3rd ed*. Wiley, New York, 1977.

Bracewell, R. N., *The Fourier Transform and Its Applications, 2nd ed*. McGraw-Hill, New York, 1978.

Braun, M. *Differential Equations and Their Applications, 3rd ed*. Springer-Verlag, New York, 1983.

Brigham, E. O., *The Fast Fourier Transform*. Prentice-Hall, Englewood Cliffs, N.J., 1974.

Buck, R. C. *Advanced Calculus, 3rd ed*. McGraw-Hill, New York, 1978.

Campbell, G. A., and R. M. Foster. *Fourier Integrals for Practical Applications*. Van Nostrand, Princeton, N.J., 1948.

Carslaw, H. S., *An Introduction to the Theory of Fourier's Series and Integrals*. Dover, New York, 1950.

Carslaw, H. S., and J. C. Jaeger. *Conduction of Heat in Solids,* 2nd ed. Oxford University Press, London, 1959.

Churchill, R. V., and J. W. Brown. *Fourier Series and Boundary Value Problems,* 3rd ed. McGraw-Hill, New York, 1978.

————. *Complex Variables and Applications,* 4th ed. McGraw-Hill, New York, 1984.

Conway, J. B. *Functions of One Complex Variable,* 2nd ed. Springer-Verlag, New York, 1978.

Copson, E. T. *An Introduction to the Theory of Functions of a Complex Variable.* Oxford University Press, Oxford, 1978.

————. *Partial Differential Equations.* Cambridge University Press, Cambridge, 1975.

Courant, R., and D. Hilbert. *Methods of Mathematical Physics, Vol. I.* Wiley-Interscience, New York, 1953.

Cramér, H. *Mathematical Methods of Statistics.* Princeton University Press, Princeton, N.J., 1974.

Crank, J., *The Mathematics of Diffusion.* Oxford University Press, Oxford, 1956.

Davis, H. F., *Fourier Series and Orthogonal Functions.* Allyn and Bacon, Boston, 1963.

Davis, P. J., *Interpolation and Approximation.* Dover, New York, 1975.

Davis, P. J., and R. Hersh. *The Mathematical Experience.* Houghton Mifflin, Boston, 1982.

Deans, S. R., *The Radon Transform and Some of Its Applications.* Wiley, New York, 1983.

Doob, J. L., *Classical Potential Theory and Its Probabilistic Counterpart.* Springer-Verlag, New York, 1984.

Dym, H., and H. P. McKean. *Fourier Series and Integrals.* Academic Press, New York, 1972.

Elliot, D. F., and K. R. Rao. *Fast Fourier Transforms: Algorithms, Analyses, and Applications.* Academic Press, New York, 1982.

Erdélyi, A., W. Magnus, F. Oberhettinger, and F. G. Tricomi. *Higher Transcendental Functions* and *Tables of Integral Transforms* (in five volumes), based, in part, on notes left by Harry Bateman. McGraw-Hill, New York, 1953 and 1954.

Eves, H. *An Introduction to the History of Mathematics,* 5th ed. Saunders, New York, 1983.

Feynman, R. P., and A. R. Hibbs. *Path Integrals and Quantum Mechanics.* McGraw-Hill, New York, 1965.

Fourier, J. *The Analytical Theory of Heat,* translated by A. Freeman. Dover, New York, 1955.

Garabedian, P. R., *Partial Differential Equations.* Wiley, New York, 1964.

Gel'fand, I. M., and G. E. Shilov. *Generalized Functions, Vol. 1.* Academic Press, New York, 1964.

Gnedenko, B. V. *The Theory of Probability.* Mir, Moscow, 1978.

Goodman, J. W. *Introduction to Fourier Optics.* McGraw-Hill, San Francisco, 1968.

Hamming, R. W. *Coding and Information Theory.* Prentice-Hall, Englewood Cliffs, N.J., 1980.

Harburn, G., C. A. Taylor, and T. R. Welberry. *Atlas of Optical Transforms.* Cornell University Press, Ithaca, N.Y. 1975.

Hecht, E., and A. Zajac, *Optics.* Addison-Wesley, Reading, Mass., 1974.

Helgason, S., *The Radon Transform.* Birkhäuser, Boston, 1980.

Helmholtz, H. *On the Sensations of Tone.* Dover, New York, 1954.

Helms, L. L. *Introduction to Potential Theory.* Wiley-Interscience, New York, 1969.

Higgins, J. R. Five Short Stories about the Cardinal Series. *Bull. (New Ser.) Am. Math. Soc.* **12:** 45 (1985).

Hjellming, R. M., and R. C. Bignell. Radio Astronomy with the Very Large Array. *Science* **216:** (1982).

Hobson, E. W. *The Theory of Spherical and Ellipsoidal Harmonics.* Cambridge University Press, London, 1931.

Hochstadt, H. *Differential Equations, A Modern Approach.* Holt, Rhinehart & Winston, New York, 1964 (reprinted by Dover, New York, 1975).

Iizuka, K., *Engineering Optics.* Springer-Verlag, New York, 1985.

Igari, S. "Lectures on Fourier Series of Several Variables," University of Wisconsin Lecture Notes, Madison, Wisc., 1968.

Jackson, J. D. *Classical Electrodynamics.* Wiley, New York, 1962.

John, F. *Partial Differential Equations*, 4th ed. Springer-Verlag, New York, 1982.

————. *Plane Waves and Spherical Means Applied to Partial Differential Equations*. Wiley-Interscience, New York, 1955.

Kapany, N. S. *Fiber Optics*. Academic Press, New York, 1967.

Kaplan, W. *Advanced Calculus*, 3rd ed. Addison-Wesley, Boston, 1984.

Katznelson, Y., *An Introduction to Harmonic Analysis*. Wiley, New York, 1968 (reprinted by Dover, New York, 1976).

Kellogg, O. D., *Foundations of Potential Theory*. Springer, Berlin, 1928 (reprinted by Dover, New York, 1953).

Kemble, E. C. *The Fundamental Principles of Quantum Mechanics*. McGraw-Hill, New York, 1937.

Kline, M. *Mathematical Thought from Ancient to Modern Times*. Oxford University Press, New York, 1972.

Knopp, K. *Theory of Functions, Parts I and II*. Dover, New York, 1945.

Krabbe, G. *Operational Calculus*. Plenum, New York, 1975.

Kramer, E. E. *The Nature and Growth of Modern Mathematics*. Princeton University Press, Princeton, N.J., 1981.

Landau, L. D. and E. M. Lifshitz. *Quantum Mechanics* (*Nonrelativistic Theory*). Pergamon, Oxford, 1977.

Langer, R. E. *Fourier Series*. Slaught Memorial Paper. *Am. Math. Monthly,* Supplement to **54:** 1 (1947).

Lebedev, N. N., *Special Functions and Their Applications*. Prentice-Hall, Englewood cliffs, N.J., 1965 (reprinted by Dover, New York, 1972).

Lebedev, N. N., I. P. Skalskaya and Y. S. Uflyand. *Worked Problems in Applied Mathematics*. Dover, New York, 1979.

Leigh Silver, A. L., Musimatics or the Nun's Fiddle. *Am. Math. Monthly* **78:** 351 (1971).

LePage, W. R. *Complex Variables and the Laplace Transform for Engineers*. McGraw-Hill, New York, 1961 (reprinted by Dover, New York, 1980).

Lipson, H. (Ed.), *Optical Transforms*. Academic Press, London, 1972.

Lipson, H., and W. Cochran. *The Determination of Crystal Structures*. Cornell University Press, Ithaca, N.Y., 1966.

Lipson, S. G., and H. Lipson. *Optical Physics, 2nd ed.* Cambridge University Press, Cambridge, 1981.

Mackey, G. W. Harmonic Analysis as the Exploitation of Symmetry—A Historical Survey. *Bull.* (*New Ser.*) *Am. Math. Soc.* **3:** 543 (1980).

MacRobert, T. M. *Spherical Harmonics, 3rd ed.* (with I. N. Sneddon). Pergamon, Oxford, 1967.

Marks, R. J. Multi-dimensional Signal Sample Dependency at Nyquist Densities. *J. Opt. Soc. Am.* **3**(5): 268 (1986).

Mathews, M. V., and J. R. Pierce. The Computer as a Musical Instrument. *Sci. Am.* **256**(2): 126 (1987).

Messiah, A. *Quantum Mechanics, Vols. I and II*. Wiley, New York, 1958.

Meyer, R. E. *Introduction to Mathematical Fluid Dynamics*. Wiley-Interscience, New York, 1971 (reprinted by Dover, New York, 1982).

Monforte, J. The Digital Reproduction of Sound. *Sci. Am.* **251**(6):78 (1984).

Morse, P. M. *Vibration and Sound, 2nd ed.* McGraw-Hill, New York, 1948 (reprinted by Acoustical Society of America, 1983).

Natanson, I. P. *Constructive Function Theory, Vols. I and II*. Ungar, New York, 1964 (Vol. I) and 1965 (Vol. II).

Nevanlinna, R., and V. Paatero. *Complex Analysis*. Chelsea, New York, 1982.

Nussbaumer, H. J., *Fast Fourier Transform and Convolution Algorithms*. Springer-Verlag, New York, 1982.

Olson, H. F. *Music, Physics and Engineering, 2nd ed.* Dover, New York, 1967.

Papoulis, A. *The Fourier Integral and Its Applications*. McGraw-Hill, New York, 1962.

————. *Systems and Transforms with Applications in Optics*. McGraw-Hill, New York, 1968.

Pauli, *Optics and the Theory of Electrons*. MIT Press, Cambridge, Mass., 1973.

Pollard, H. *Celestial Mechanics*. Carus Mathematical Monographs, no. 18. Mathematical Association of America, Washington, D.C., 1976.

Press, W. H., B. P. Flannery, S. A. Teukolsky, and W. T. Vetterling. *Numerical Recipes*. Cambridge University Press, Cambridge, 1986.

Raisbeck, G. *Information Theory*. M.I.T. Press, Cambridge, Mass., 1963.

Rayleigh, J. W. S. *The Theory of Sound, Vols. I and II*. Dover, New York, 1945.

Reitz, J. R., F. J. Milford, and R. W. Christy. *Foundations of Electromagnetic Theory, 3rd ed.* Addison-Wesley, Reading, Mass., 1979.

Royden, H. L. *Real Analysis, 2nd ed.* Macmillan, New York, 1968.

Rudin, W. *Principles of Mathematical Analysis, 2nd ed.* McGraw-Hill, New York, 1964.

———. *Real and Complex Analysis, 2nd ed.* McGraw-Hill, New York, 1974.

Ryshik, I. M., and I. S. Gradstein. *Tables of Series, Products, and Integrals*. VEB, Deutscher Verlag der Wissenschaften, Berlin, 1957.

Saks, S., and A. Zygmund. *Analytic Functions, 3rd ed.* Elsevier, Amsterdam, 1971.

Schiff, L. I. *Quantum Mechanics, 3rd ed.* McGraw-Hill, New York, 1968.

Schwartz, L. *Theorie des distributions*. Hermann, Paris, 1959.

Scott, E. J. *Transform Calculus*. Harper & Row, New York, 1955.

Shannon, C. E. Communication in the Presence of Noise. *Proc. IRE* **37**: 10 (1948).

Shepp, L. A. (Ed.). *Computed Tomography*. Proc. of Symp. in Appl. Math., Vol. 27. American Mathematics Society, Providence, R.I., 1983.

Shepp, L. A., and J. B. Kruskal. Computerized Tomography: The New Medical X-Ray Technology. *Am. Math. Monthly* **85**: 420 (1978).

Shilov, G. E. *Generalized Functions and Partial Differential Equations*. Gordon and Breach, New York, 1968.

Slater, J. C, and N. H. Frank. *Electromagnetism*. McGraw-Hill, New York, 1947 (reprinted by Dover, New York, 1969).

Smith, K. T., D. C. Solmon, and S. L. Wagner. Practical and Mathematical Aspects of the Problem of Reconstructing Objects from Radiographs. *Bull. Am. Math. Soc.* **83**: 1227 (1977).

Sneddon, I. N. *Fourier Transforms*. McGraw-Hill, New York, 1951.

Sommerfeld, A. *Partial Differential Equations in Physics*. Academic Press, New York, 1949.

———. *Optics*. Academic Press, New York, 1949.

Stein, E. M. *Singular Integrals and Differentiability Properties of Functions*. Princeton University Press, Princeton, N.J., 1970.

Stein, E. M., and G. Weiss. *Fourier Analysis on Euclidean Spaces*. Princeton University Press, Princeton, N.J., 1971.

Struik, D. J. *A Concise History of Mathematics, 3rd rev. ed.* Dover, New York, 1967.

Szegö, G. *Orthonormal Polynomials*. Amer. Math. Soc. Colloq. Public., Vol. 23. American Mathematics Society, Providence, R.I., 1939.

Terras, A. *Harmonic Analysis of Symmetric Spaces and Applications, Vol. I*. Springer-Verlag, New York, 1985.

Thomas, G. B., Jr., and R. L. Finney. *Calculus and Analytic Geometry, 6th ed.* Addison-Wesley, Reading, Mass., 1984.

Titchmarsh, E. C. *Theory of Functions, 2nd ed.* Oxford University Press, Oxford, 1939.

———. *Eigenfunction Expansions Associated with Second Order Differential Equations, Vol. 1*. Oxford University Press, Oxford, 1946.

———. *Introduction to the Theory of Fourier Integrals*. Oxford University Press, Oxford, 1937.

Tolstov, G. P. *Fourier Series*. Prentice-Hall, Englewood Cliffs, N.J., 1962 (reprinted by Dover, New York, 1976).

Unwin, N., and R. Henderson. The Structure of Proteins in Biological Membranes. *Sci. Am.* **250**(2): 78 (1984).

Vladimirov, V. S. *Equations of Mathematical Physics*. Dekker, New York, 1971.

Watson, G. N. *Theory of Bessel Functions, 2nd ed.* Cambridge University Press, Cambridge, 1944.

Weinberger, H. F. *A First Course in Partial Differential Equations*. Wiley, New York, 1965.

Weiner, N. *The Fourier Integral and Certain of Its Applications*. Cambridge University Press, Cambridge, 1933 (reprinted by Dover, New York, 1958).

Wheatley, P. J. *The Determination of Molecular Structure, 2nd ed.* Oxford University Press, Oxford, 1968 (reprinted by Dover, New York, 1981).

Wheeden, R. L., and A. Zygmund. *Measure and Integral*. Dekker, New York, 1977.

White, R. L. *Basic Quantum Mechanics*. McGraw-Hill, New York, 1968.

Widder, D. V. *The Heat Equation*. Academic Press, New York, 1975.

———. *The Laplace Transform*. Princeton University Press, Princeton, N.J., 1946.

Wilf, H. S. *Mathematics for the Physical Sciences*. Wiley, New York, 1962 (reprinted by Dover, New York, 1978).

Wintner, A. *The Analytical Foundations of Celestial Mechanics*. Princeton University Press, Princeton, N.J. 1947.

Woolfson, M. M. *X-Ray Crystallography*. Cambridge University Press, Cambridge, 1970.

Zygmund, A. *Trigonometric Series, Vols. I and II*. Cambridge University Press, Cambridge, 1968.

Index